Practical Foundation of Applied Systems Analysis

Practical Foundation of Applied Systems Analysis
From Basic Statistics to Systems Modeling

First edition published by Seoul National University Press in 2024

Seoul National University Press
1 Gwanak-ro, Gwanak-gu, Seoul 08826, Republic of Korea
Tel: +82-2-880-5252
Fax: +82-2-889-0785
Homepage: http://www.snupress.com
Facebook: @snupress1947
Instagram: @snupress
E-mail: snubook@snu.ac.kr

ISBN 978-89-521-3098-3 (93560)

IIASA

PRACTICAL FOUNDATION OF APPLIED SYSTEMS ANALYSIS

From Basic Statistics to Systems Modeling

Chulwook Park

SNUPRESS

Contents

Acknowledgements

This publication was made possible through the support of the Basic Science Research Program, funded by the Ministry of Education and facilitated by the National Research Foundation of Korea (grant number 2020R1I1A1A01056967). It also received support under the International Cooperation Program, managed by the National Research Foundation of Korea and funded by the Ministry of Science (grant number 2016K2A9A1A02952017). The development of the ideas and discussions in this book occurred during my periods partially undertaken as a postdoctoral researcher at IIASA. The publication is my initiative, not commissioned by the institute above. Permission for the use of affiliation and logo has been kindly granted. This work is dedicated to my family. Without their presence and support, the ordeals during the process would not have been overcome. That includes my lovely daughter Jiyoon and my son Geonwoo.

Foreword

Welcome to *Applied Systems Analysis*. This book offers an introductory three-part course in research methodologies, emphasizing programming within the realm of systems analysis. The first section delves into fundamental descriptive statistics and the principles of inferential reasoning. The second section explores applied time series analysis, extending from probability theories to stochastic diffusion processes. The third section is dedicated to systems analysis, encompassing critical concepts such as vectors, matrices, agent-based models, and beyond.

This book caters to those intending to apply these competencies and insights in their research endeavors. Each section is enriched with diverse examples and datasets, providing ample opportunities for practical application. You will acquire hands-on techniques applicable across various fields, laying a solid groundwork for advanced proficiency in systems analysis, statistics, and data science. Applied systems analysis transcends mere programming and statistical knowledge. It involves a deep engagement with practical research analysis, necessitating an understanding of problem domains and the accurate interpretation of domain-specific methodologies. The examples in this book, predominantly sourced from real datasets, will equip you to employ systems analysis and predictive techniques in real-world situations.

While individuals with a background in data analysis or statistics may find these concepts more accessible, this book is specifically tailored for those new to statistics or data science. Regular application of programming and visualization techniques to novel problems is a hallmark of this book. No previous experience in analysis is required to benefit from its content, yet many of the statistical and analytical methods presented here are not typically found in standard statistics curricula. These methods extensively utilize sampling and simulation, demanding computational execution. For optimal practice, it is strongly advised to use a Jupyter Notebook integrated with a Python workflow.

Each part of the book includes reading and practice sections. While the practice exercises are not graded, they are integral to the learning process. Commitment to the course material is essential. I encourage you to approach every task with integrity, avoiding replication or duplication of efforts. To maximize learning, it is advisable to engage with the practice assignments progressively, rather than postponing them to the end.

Features of the Book

Technical details: There are many books on systems analysis, but most are written for a popular audience. They usually skip the technical details, which is frustrating for more advanced readers. This book presents the mathematics, the code, and any other material you will need to understand the work.

Data programming: Each part provides a Jupyter Notebook (or Python) that includes the code from the section, additional examples (especially animation), and suggestions for experiments you can run with minor changes to the code.

Exercises and solutions: At the end of each part, I suggest activities you may want to work on and include explanations.

Contributor List

If you have a suggestion or correction, please e-mail me at pcw8531@snu.ac.kr (or pcw8531gmail.com). If I make a change based on your feedback, I will add you to the contributor list (unless you ask to remain anonymous).

If you include at least part of the sentence in which the error appears, it will be easier for me to find it. Page and section numbers are OK too. Thanks!

Preface

The concept of a paradigm shift, as defined by Kuhn in 1962, occurs when foundational assumptions in a field are altered or when one theoretical framework supersedes another. Classic examples include the transition from the phlogiston theory to the oxygen-based model of combustion, and the advent of relativity theory. The evolution of systems analysis, as I see it, does not represent a direct replacement of an antiquated methodology, but rather a gradual shift in evaluative criteria and an expanded acceptance of various modeling approaches.

Traditional models are often law-based, articulated through equations, and solved using mathematical methods. In contrast, models within the scope of systems analysis typically rely on rule-based structures, articulated through computational descriptions, and are computed rather than traditionally analyzed. However, these models do not always meet universal approval. For instance, in 'Sync' by Abrams and Strogatz (2003), the authors explore spontaneous synchronization in certain firefly species, showcasing a simulation that illustrates this phenomenon.

This book is tailored not only for researchers and scientists, who may naturally gravitate towards computational models, but also for a broader audience, including laypersons seeking to expand their understanding of systems thinking. It presents diverse perspectives, extending beyond mere statistical analysis, to pose novel questions and utilize varied tools for in-depth exploration or validation of concepts. These perspectives are rooted in value judgments, and as such, unanimity is not a given expectation.

Nonetheless, the scientific community has generally coalesced around certain norms regarding what constitutes legitimate analytical tools and what falls into the categories of fringe science, pseudoscience, or non-science (Kolstø, 2001). The central thesis of this discussion is that the criteria underpinning this consensus are subject to gradual evolution. The emergence and integration of research methodologies founded on these evolving principles are indicative of this progressive shift.

Lecture 1. Probability

Probability is a branch of mathematics in which you start by making assumptions about how the world works (Smith, 1968). Well, not the whole world, but a particular part of the world. Given this exact specification or this set of assumptions about how a random process evolves, you make predictions about the outcomes.

Let us learn the basic rules of probability. Probability describes the chance that some event or outcome in a random process will occur (Olkin et al., 1980). The lowest probability is 0, which means that the event or outcome is impossible. The highest probability is 1 (often expressed as 100%), which means that the event or outcome is certain. When you flip a coin, the probability that it will come up as both heads and tails is 0. The chance that it will come up heads or tails is 1. If an event has a probability of 70%, then the probability that it will not happen is one minus 70% or 100% minus 70% (30%). In general, either an event occurs or it does not, so the chance that it will happen and the chance that it will not happen have to add up to 1, which is why we get this '1 minus' formula. When there are equally likely outcomes, and they all have the same probability of occurring, then the chance of an event A is the ratio of the number of results possible: A divided by the total number of outcomes.

Probability serves as a fundamental branch of mathematics that facilitates predictions regarding the outcomes of random processes, based on specific assumptions. With a probability scale ranging from 0 (impossibility) to 1 (certainty), this concept provides a comprehensive understanding of the likelihood of events, employing the '1 minus' formula to account for the complementary probabilities of events occurring or not occurring.

Lecture 2. Statistics

Now we move on to the field of statistics, but the approach we take focuses on sampling.

Probability is a branch of mathematics that enables you to compute what will happen when you run an experiment based on assumptions about how the world works. Statistics goes in the opposite direction. Statisticians try to reason about what the world is like based on the outcome of an investigation. Statistics is a method of understanding from observation (Wilcox, 2009). Sampling ties the two together. When we sample, we use the rules of probability to understand what might happen; in this book, we use sampling to reason about how the outcome of an experiment might be different based on our observations.

That allows us to reason about how the world works or to deduce certain facts about a population. Statistical inference involves making conclusions based on data and random samples (Lindsey, 1996). For example, we might want to use data to guess the value of an unknown number. This is just a fact about the population that we sampled from, but it is a fact we do not know yet. When we guess the value of an unknown number, we are estimating an unknown quantity. The unknown number is a particular value, but the calculation we make might depend on the random sample we happened to get. If we ran a different experiment trying to figure out this same unknown number, we might get a different estimate, even using the same procedure, because we drew a different random sample. Before we go any further, let us define some critical terminology. A *setting* is a number associated with a population, whereas a *statistic* is a number calculated from a sample from that population. Usually all we have is a sample, and we are trying to guess what is going on in the population, so we can compute the statistic directly, but it might be different if we have a different sample. The *parameter* is what we would like to know, and we use our example to estimate what that parameter might be.

Thus, we can use a statistic to estimate a parameter, and by doing so we are performing statistical inference. Here we create an estimate—a view of a particular event. Let us say we roll four dice. What is the chance of our getting at least one 6? The population is all of the different outcomes that could result from rolling four dice. Each possible set of four outcomes has a probability, and only some of the sets contain a 6. This number that we compute through the rules of chance is a parameter of the population of all four rolls of the dice. We can calculate this probability by running a simulation, but the estimate could come out differently. I would call this success divided by trials—estimation by simulation—and here we can choose the number of tests. What if we make many such estimates? One thousand times, I am going to create a new assessment by taking my old list of ratings and appending on an original forecast by simulation using 10,000 trials.

This computation is going to take a moment, because 1,000 times we are computing 10,000 trials of rolling four dice, so we have to run 40 million dice to finish all of the estimates. What we are looking at is called the *empirical distribution* of a statistic. The empirical distribution is based on simulated values of the statistic, which we might estimate multiple times to get a sense of how much they vary. It consists of all observed values of the statistic and the proportion of times each value appears. According to the law of averages, we would expect that as long as we perform this simulation many times, the empirical distribution of the statistic will resemble the probability distribution of that same statistic (Araújo & New, 2007). This means that we do not need to compute the probability distribution; we can just perform the simulation and see that most values are right around the parameter. Understanding how statistics vary as you draw multiple samples is critical: When you learn about the world through observation, you need to understand how much variability might deceive you in the sampling procedure itself. You need to be able to express this variability when you perform statistical inference.

Lecture 3. Data Science

Data science is more than just an industry buzzword. It involves taking large data sets and trying to make them useful or informative, especially for understanding the world or making informed decisions (Provost & Fawcett, 2013). To do this, we need to use ideas from computing, ideas from statistics, and domain knowledge that informs what the data represent.

What do I mean by *domain knowledge*? You cannot analyze the legal domain without understanding something about the law. You have to know what is really going on behind the numbers in a data set as well as what these numbers represent about the world.

What is data science? What do you get when you combine computing, statistics, and domain knowledge? You get a science that is about drawing useful conclusions from data using computation as the primary tool. Data science, as a practice, has three core activities, all of which we study in this book:

- Exploration: Identifying patterns in information (uses visualization)
- Inference: Quantifying whether patterns are reliable (uses randomization)
- Prediction: Making informed guesses (uses simulation)

Exploration involves figuring out what patterns exist in data (Zuur et al., 2010). When you have many observations about a phenomenon, what can you conclude about the event itself? Instead of just looking at large tables of numbers, we frequently visualize the data, because it is much easier to interpret a lot of information at once when it is presented visually.

Once we have found a pattern, we need to perform statistical inference. This is because some patterns are there just by chance, and some are there because they reflect an underlying process about the world that is interesting. The goal of statistical inference is to quantify whether the patterns we observe during the exploration phase are reliable. If we collected more data, would we see this pattern again or not? The primary tool we have in this stage is randomization, because by simulating random processes, we can see what kinds of patterns appear just by chance. If the pattern we observe is not due to chance, then we can conclude that it is due to a robust or reliable model in the underlying phenomenon we want to study.

Finally, we make a prediction. In this stage we have incomplete information about something we want to know, and we want to guess about the things we do not know yet. We can make informed, quantitative guesses using a discipline called *machine learning*. Commonly when we write programs, we focus on the particular logic of what the computer should do. Machine learning is about not programming every detail but instead involves using the data to make decisions or choices in a program (Quinlan,

2014). For instance, when we write a program to recognize speech or automatically translate languages or control a car or a robot, we do not write down all of the details of what to do but instead use examples from the world to help the computer automatically learn how to behave.

In this book, we first talk about how to identify patterns; then we talk about quantifying whether those patterns are reliable. Finally, based on the trends we discover, we can use the reliable ones to make informed guesses about the information we wish we knew. In the process of doing all of these things, we must also learn how to program a computer, because computing underlies each step of the way. Once we are able to do all that, we will be well on our way to conducting systems analysis.

Lecture 4. Systems Analysis

A system comprises a collection of interrelated components functioning together to generate specific patterns over time. System dynamics, with its extensive applicability, has found relevance in diverse disciplines, including control systems, communication systems, tropical systems, and various system modes. While the term 'system' is widely used, the crucial questions of its meaning and the factors that render a system dynamic are often posed. This book does not adhere to a strict, inflexible definition; instead, it aims to convey the concept intuitively through engaging discussions.

A system is an abstract or vague entity. It rejects a strict codification (Davis & Yen, 2019). However, we may be able to understand it more fully by defining a nonsystem. A nonsystem is a set of separate parts that do not interact with one another or a collection of related parts whose relations have no implications for their properties and behaviors. Acknowledging the inherent ambiguity in defining certain concepts using words, we shall attempt to clarify the understanding of these concepts within the context of dynamic systems. We can consider a system as a collection of individual elements, as it is uncommon to discuss a system consisting of only a single component. Most likely, these elements exhibit nontrivial interactions as having coherent behavior (this is not always positively synergetic) to move or work together [$\chi \rightarrow \vec{f} \rightarrow f(x_i)$]. Here χ denotes an element composed of many individuals; f is a function toward a particular direction ($\rightarrow$), such as a goal; and $f(x_i)$ denotes coherence, which has specific individuals (x_n) as opposed to other individuals underlying a very well-defined goal boundary (f). Another essential assumption of a system is that such well-defined individuals must interact not just within themselves but also with everything outside, which we can call the *environment* [$x_t \rightarrow f(x_i) \rightarrow y_t$]. In this equation, x_i can have two kinds of interactions with the outside: The outside exerts some impact (x) on x_i and is influenced (y) by x_i at a specific time (t) (Ashby, 1947). Thus, intuitively, a system can be defined as many interactive individual elements embedded within particular coherent behavior. This appears to be a mechanism by which an organism can change its behavior dynamically, allowing the real, material system to be examined objectively with variables that can be specified numerically.

A system's dynamics involve interrelated elements whose interactions generate patterns. The connections between a system's components reveal how they function collectively to establish structure or adapt to environmental changes. Nevertheless, examining the dynamics often necessitates managing a considerable number of elements and multiple components. This book proposes specific syntheses and principles derived from the study of diverse phenomena, focusing on fundamental mechanisms that can be illustrated on a particular screen, thereby extracting simplicity from complex dynamics.

Remarks

Up to this point, we have discussed systems analysis within the realm of science, but it is important to recognize that dynamics are both a driving force and a consequence of changes in engineering and the organization of systems:

- Centralized systems may be conceptually straightforward and easier to analyze, but decentralized systems can offer enhanced robustness. (Decentralization)

- Traditional engineering approaches tackle the complexity of large systems by compartmentalizing components and minimizing interactions. (Interaction)

- In numerous communication systems, broadcast services are being supplemented or even replaced by services that empower users to interact, create, share, and modify content. (Many-to-many interaction)

- Social, political, and economic systems increasingly witness activities typically reserved for central organization now emerging as grassroots movements. Even military forces, which exemplify hierarchical structure, are gravitating towards devolved command and control. (Bottom-up activity)

- In traditional engineering, our analytical capacity constrains the realm of viable designs. However, contemporary computer-aided design and analysis tools enable the creation of virtually anything we can envision. (Computation)

- Engineering is often characterized as a quest for solutions within a landscape of potential designs, and increasingly, this search process can be automated. (Search)

While these trends may not be all-encompassing or absolute, they signify a shift in prevailing viewpoints. They serve as both catalysts and outcomes of systems analysis. For instance, highly abstract models are more widely accepted now, as there is less insistence on a unique, correct model for every system.

This section has provided a glimpse of the themes we will explore throughout this book. Some concepts may not fully resonate until you delve deeper. Once you reach the end, revisiting this section could prove insightful.

Good luck, and enjoy the journey!

Part 1.

Applied Statistics Analysis

Contents

Part 1. Applied Statistics Analysis

[1.1]

Fundamentals of Statistics

Overview of Statistics, Terminology, Data

Statistics, as a scientific discipline, is intricately involved in the design and study of methodologies for collecting, analyzing, interpreting, and disseminating empirical data. Its interdisciplinary nature allows its research findings to permeate almost all scientific domains, further fueling the evolution of statistical methodologies and theories. The systems analysis perspective embeds this discipline into a broader context, integrating it with other fields to better comprehend and address complex issues (Box, 1976).

1. Overview of Statistics

Statisticians draw on a variety of mathematical and computational tools to develop methods and study theory that involves summarizing a large set of information (data) and estimating parameters from a sample.

+ **Characteristic of Statistics:** A *statistic* is a characteristic of a sample. Statistics enables you to observe and make an educated guess about a population parameter (a parameter is a characteristic of a population) based on a statistic computed from a sample randomly drawn from that population (Agresti & Finlay, 2009). It uses evidence to draw a conclusion, for example:

- The number of people who participate in regular exercise is increasing
- The experimental group showed better performance → there is a treatment effect

Two fundamental ideas in the field of statistics are uncertainty and variation (Moore, 1990). In many situations in science (or in life more generally), the outcome is uncertain. In some cases, there is uncertainty because the outcome in question has not yet been determined (e.g., we may not know whether it will rain tomorrow). In other cases, although the outcome has already been determined, we are not aware of what it is (e.g., we may not know whether we passed an exam).

+ **Main Domains:** It may be separated into three primary categories: probability, inferential statistics, and descriptive statistics. The characteristics of a collection of data are outlined and described using descriptive statistics. Mathematics' study of random events and quantification of an event's possibility is the subject matter of probability.

On the other hand, inferential statistics uses the data gathered to draw conclusions or generalizations about a larger population from a smaller sample of the population.

- Probability is a mathematical language we can use to discuss uncertain events, and it plays a crucial role in statistics (see section 2.1 for more detail). Any measurement or data collection effort is subject to many sources of variation. That is, if the same analysis were repeated, the answer would likely change. Thus, statisticians attempt to understand and control whenever possible the sources of variation in any situation.
- Descriptive statistics are used to organize or summarize a particular set of measurements (Wilson, 2004). In other words, a descriptive statistic describes information that is compiled to explain a population at a given point in time. What descriptive statistics have in common is that they organize, summarize, and represent a set of measurements.
- Inferential statistics use data gathered from a sample to make inferences about the larger population from which the sample was drawn (Wilson, 2004). For example, say you want to know the mean income of subscribers to a particular magazine—a parameter of a population. You bring together a random sample of 100 subscribers and determine that their mean income (a statistic) is $27,500. You conclude that the population mean income is likely to be close to $27,500 as well. This example is one of statistical inference.

Together, these three domains are used to interpret the connections and trends shown in data and to make data-driven decisions based on the following symbolic words.

2. Terminology

A variable represents a quantity (a symbolic term) associated with a value and whose associated value may change. Different symbols are used to denote statistical variables and parameters.

+ **Sample and Population:** There is nothing fancy to these terms. Assume we want to know what the average salary is. We go and ask every individual (i.e., approximately 50 million people) what they make, then calculate their average salary.

Population: In this case, we have observed the whole population with not even a single person left unobserved. Thus, we can say that the population includes all the possible elements that form a set of data.

- A measurable characteristic (mean or standard deviation), is called a *parameter*.

Sample: Having defined the population, we should see why observing the whole population is impossible under practical circumstances. Hence, the best way we can get an idca about a population is to take a bunch of random people from that population.

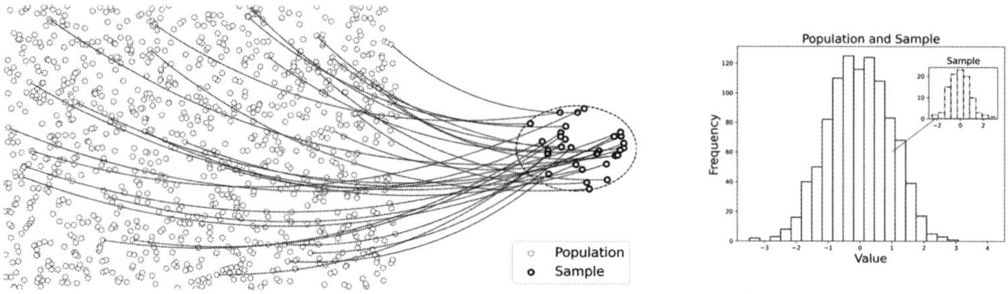

Figure 1.1.1. Schematic representation of a population and sample. Set of the left side shows their characteristics relation, while the set of the right denotes their statistical relation

A random group from a population is known as a sample (Figure 1.1.1). Every member of a population has the same probability of being sampled → The sample represents the population. We are focusing on the word *random* because it is crucial that all n objects of the sample are equally likely to be chosen. Imagine that the sample consists of a bunch of software engineers working at Google, Microsoft, or Facebook.

- Sampling error: the difference between population and sample values (as N increases, sampling error decreases)
- Sampling bias: When random sampling is not performed, the sample does not represent the population.

This will not be a precise representation of the whole population. In such a case, we will have a skewed sample. This is undesirable; thus, we can apply an appropriate way of sampling (from the population) as below (Figure 1.1.2).

	random_id	values
0	1	10.248
1	2	9.931
2	3	10.324
3	4	10.762
4	5	9.883
5	6	9.883
6	7	10.790
7	8	10.384
8	9	9.765
9	10	10.271

Random sampling

	random_id	values
1	2	9.931
4	5	9.883
5	6	9.883
8	9	9.765

Systematic sampling

	random_id	values
0	1	10.248
3	4	10.762
6	7	10.790
9	10	10.271

Cluster sampling

	product_id	measure	cluster_id
2	3	10.324	2
3	4	10.762	2
6	7	10.790	4
7	8	10.384	4

Figure 1.1.2. Different ways of sampling (right) from a population (left)

- Random sampling: Randomly assigning subjects to groups (in experimental research) → What about generalizations when subjects are not randomly sampled from the population?
- Cluster sampling: Sampling a cluster or clusters is useful when a specific group (cluster) can represent the population. A random sampling of clusters is possible.
- Systematic sampling: Randomly select the first sample and choose every *k*th one on the list. If there is no specific pattern in population members by order, this is equivalent to random sampling.

A measurable characteristic of a sample, such as a mean or standard deviation, is also called as a statistic. We denote the sample mean as $\bar{X}$ and the sample standard deviation as S (Table 1.1.1).

[Table 1.1.1] Comparison of Sample and Population

	Sample statistic	Population parameter
Mean	$\bar{X}$	μ
Standard deviation	s	σ
Variance	s^2	σ^2

+ **Common Biases in Sampling:** A sample is a subset of data drawn from a large data set, or population. The quality of data often matters more than the quantity of data. Acquiring quality data is critical to drawing valid conclusions, as many errors can lurk in statistical models and experiments, leading to invalid results.

Statistical bias: Statistical bias refers to measurement or sampling errors that are systematic in nature and produced by the measurement or sampling process. An important distinction should be made between errors due to random chance and errors due to bias. Random chance bias is a result of measurement error, is not skewed in a specific direction, and is usually independent for each sample. However, error due to bias is somehow correlated and has a centric tendency around a fixed value. For example, all samples have the added error of a constant. This type of bias can indicate that a statistical or machine-learning model has been mis specified or an important variable has been left out (Bruce et al., 2020).

Many sampling methods have been proposed to combat this type of bias, but at the heart of them all lies random sampling. *Random sampling* is the process in which each available member of a population has the same chance to be selected at each draw. Sampling can be done through replacement, in which each selected sample is included again in the population for each draw, or it can be done without replacement, in which once a member of the population is selected, it is not available for the next draw.

Sometimes, even in random sampling, the sample might not reflect the true population. Imagine that a population consists of two types of members: type I (99%) and type II (1%). If we conduct random sampling, the chance that a type II member is included in our sample is negligible. To preserve the proportionality in the sample, we can use stratified sampling. In *stratified sampling*, the population is divided into strata, and random samples are taken from each stratum. Thus, each member of a type has the same chance as in the population data.

Selection bias: Selection biases are statistical biases in which each member of a population—consciously or unconsciously—is chosen selectively, which leads to a misleading conclusion.

- One selection bias is data snooping. *Data snooping* happens when you mix the real pattern with what is supposed to be noise.

This frequently happens when you examine the data and try to discern patterns without checking if they are repeatable. If a pattern you discover is not provable in many cases or in repeatable experiments, it is considered a random phenomenon, which is usually called *noise*. This is common when a data set is small (Kahneman, 2011).

- Another form of selection bias is of particular interest in data analysis.

It is what John Elder (founder of Elder Research, a respected data mining consultancy) calls the *vast research effect* (Giele & Elder, 1998). If you repeatedly run different models and ask different questions with a large data set, you are bound to find something interesting. But is what you find truly something interesting, or is it a chance outlier? You can minimize this kind of bias by using the holdout set to test the predictability of your model. Elder also suggests using what he calls *target shuffling*, which in essence is the permutation test.

- Other selection biases can be the result of cherry-picking data, selecting time intervals that accentuate a particular statistical effect, or stopping an experiment when the results look interesting.

+ **Constants and Variables:** In order to effectively express particular details regarding statistical analysis in general, there are a few fundamental terms that are crucial.

Constants: A constant, denoted as C, represents an attribute or condition consistent across all subjects within a research. For example, if examining the influence of two distinct methodologies on novice-level participants, the proficiency level would be a constant, C. Here, every participant is characterized by novice status.

Variables: Denoted by V, variables signify attributes that manifest variably among subjects. They might differ in magnitude or context relative to the individual to which they pertain. When evaluating participants' performance via two discrete methodologies, it is improbable for all to garner identical ratings. Hence, the derived score becomes a variable, V.

- Dependent Variable (Y): This is a variable whose values are contingent upon another.

For instance, the participants' performance score is a dependent variable, Y, as it is influenced by other factors.

- Independent Variable (X): Often serving as categorical parameters, these are variables that exert influence on the dependent variables.

In the aforementioned context, the teaching methodology represents an independent variable, X, as it potentially alters the performance score. In mathematical relations, $Y = f(X)$, where Y is the dependent variable influenced by the independent variable X. Dependent variables might be influenced by extraneous factors, e.g., the gender of the teacher: male or female. Conversely, independent variables serve as determinants for dependent variables, e.g., the efficacy score among novice participants.

+ **Scales of Measurement:** In statistics, the term *measurement* is used broadly and is more appropriately called a *scale of measurement* (Labovitz, 1967). Scales of measurement describe how variables and values are defined and categorized. Each scale of measurement has specific properties that in turn determine its appropriateness for use in accurate statistical analyses. The four scales of measurement are nominal, ordinal, interval, and ratio:

- Nominal scale: measures without order, simply indicates values (e.g., grade, uniform number)
- Ordinal scale: measures with order, is different and can be ranked (e.g., rank, order)
- Interval scale: measures with order and numerically equal distances (e.g., temperature, index)
- Ratio scale: contains an absolute or true zero point (e.g., weight, length)

The values in the 'Mathematical Operators,' 'Advanced Operations,' and 'Central Tendency' columns of the table below (Table 1.1.2.) are only the ones this level of measurement introduces.

[Table 1.1.2] Scales of Measurement

Scale	Measurement property	Mathematical operators	Advanced operations	Central tendency
Nominal	Classification, membership	$=, \neq$	Grouping	Mode
Ordinal	Comparison, level	$>, <$	Sorting	Median
Interval	Difference, affinity	$+, -$	Yardstick	Mean, deviation
Ratio	Magnitude, amount	$\times, /$	Ratio	Geometric mean, coefficient of variation

The complete list includes the values of previous levels. This is inverted for the 'Measurement Property' column.

3. Data

According to common knowledge (i.e., in Wikipedia), a data set is a collection of values that convey information, other basic units of meaning, or simply sequences of symbols that may be further interpreted (a datum is an individual value in a set of data).

+ **Data Types:** As we noticed above, a variable denotes information about a sample (i.e., individuals, elements) that is acquired through measurements (scales), such as length, time, diameter, strength, weight, temperature, density, thickness, pressure, or height. With variables, we can define trends in samples (Table 1.1.3), mainly recognizing two types of data (Phillips & Stawarski, 2008).

[Table 1.1.3] Types of Data

Sample (individual)	Variables		
	Categorical	Numerical	
	Gender	Age	Score
A	Male	15	50
B	Male	14	55
C	Female	16	53
D	Male	13	40
E	Female	15	49
F	Female	12	35

- Numerical data: these data are expressed with digits. They can be measured. These data are further divided into two subcategories: discrete and continuous (height, speed, age, weight, sales, cost, etc.).
- Categorical data: these data are collections of information that is divided into groups. Qualitative data can be classified into specific categories (gender, age group, product category, education level, etc.).

When we find patterns, or trends, in the data, then we are able to discuss more basics of the fundamentals of statistics that are crucial in data analysis.

+ **Data Mining:** Several tools and techniques can be used for row data mining, but we will work with a couple of famous programing libraries (i.e., Pandas, NumPy) that are used to manage data structures.

- Library: a collection of pre-written code in a particular programming language
- Data structure: a specialized format for organizing, storing, and managing data enabling efficient access and modification of the data

These libraries provide some useful operations for data preprocessing and cleaning.

- Array: ordered collection of elements of the same data type, often used for storing and manipulating numerical data

First, we need to import the library (i.e., pandas) to the workspace that can be used to perform specific tasks. The series denotes a data structure in the form of an array, with incremental labels starting at 0 as follows:

```
[In]
import pandas as pd
from pandas import Series, DataFrame
from pandas import Series
obj = Series([1, 2, 3, 4, 5, 6])
obj
```

→

```
[Out]
0    1
1    2
2    3
3    4
4    5
5    6
```

Here, we imported the series object from the library and create a series object with a list of values. We can also create our own labels.

```
[In]
from pandas import DataFrame, Series
obj = Series([4, 7, -5, 3], index=['a',
'b', 'c', 'd'])
obj
```

→

```
[Out]
a    3
b    4
c    5
d    6
```

Using some more useful features, we can convert dictionaries to series.

```
[In]
s = Series({'age':23, 'name':
'anuradha'})
s
```

→

```
[Out]
age       20
name      pandas
```

We can use direct index assignment.

```
[In]
obj.index = ['attr 1', 'attr 2', 'attr
3', 'attr 4']
obj
```

→

```
[Out]
attr 1    3
attr 2    4
attr 3    5
attr 4    6
```

And the indices can have an attribute name.

```
[In]
obj.name = 'Sample'
obj
```

→

```
[Out]
attr 1    3
attr 2    4
attr 3    5
attr 4    6
Name: Sample
```

Now let us make the data frame (structure) for the data set. This is a (two)dimensional, size-mutable, and heterogeneous tabular data structure with labeled axes (rows and columns), commonly used for data manipulation and analysis (here in Python using the pandas library) which is quite similar to a spreadsheet and can store large amounts of data.

```
[In]
data = {'city': ['Seoul', 'NewYork', 'Paris', 'Tokyo', 'Beijing', 'London'],
        'year': [2021, 2021, 2021, 2021, 2021, 2021],
        'pop': [0.99, 0.82, 1.1, 3.7, 2.0, 0.94]}
Data
```

With the following data frame:

```
[In]
frame = DataFrame(data)
frame
```

→

[Out]

	city	year	pop
0	Seoul	2021	0.99
1	NewYork	2021	0.82
2	Paris	2021	1.10
3	Tokyo	2021	3.70
4	Beijing	2021	2.00
5	London	2021	0.94

There are many more techniques we should know for the data structures, but let us get into the next step, filtering, following a piece of code:

```
[In]
from pandas import DataFrame, Series
import numpy as np
data =
DataFrame(np.arange(36).reshape(6,6),
index=['Seoul', 'NewYork', 'Paris',
'Tokyo', 'Beijing', 'London'],
columns=['one', 'two', 'three',
'four', 'five', 'six'])
data
```

→

[Out]

	one	two	three	four	five	six
Seoul	0	1	2	3	4	5
NewYork	6	7	8	9	10	11
Paris	12	13	14	15	16	17
Tokyo	18	19	20	21	22	23
Beijing	24	25	26	27	28	29
London	30	31	32	33	34	35

We can use np.arange(36) to create an array of elements from 0 to 35 and use reshape(6,6) to make it a 6 × 6 matrix. Just a set of dummy data.

```
[In]
#Boolean comparisons
data > 5
```

→

[Out]

	one	two	three	four	five	six
Seoul	False	False	False	False	False	False
NewYork	True	True	True	True	True	True
Paris	True	True	True	True	True	True
Tokyo	True	True	True	True	True	True
Beijing	True	True	True	True	True	True
London	True	True	True	True	True	True

We can use the query function data < 5 and replacement operation = −1 to replace values less than 5 with −1.

[In]
```
data[data < 5] = -1
data
```
→ [Out]

	one	two	three	four	five	six
Seoul	-1	-1	-1	-1	-1	5
NewYork	6	7	8	9	10	11
Paris	12	13	14	15	16	17
Tokyo	18	19	20	21	22	23
Beijing	24	25	26	27	28	29
London	30	31	32	33	34	35

A direct application of the library is able to handle missing values. Most of the time the data we get in the real world are not clean. They contain missing or inconsistent values. Thus, we need to either filter them or substitute another value in their place. To filter out missing values, we need to create a sample array with missing values (NaN values or Not a Number values):

[In]
```
a = np.array([1,2,3,np.nan,4])
a
```
→ [Out]
```
array([ 1.,  2.,  3., nan,  4.])
```

See the array as a data frame. Once we have the data frame with missing values? NaNs, we can fill them in.

[In]
```
df = DataFrame(a)
df

df.fillna(-1)
```
→ [Out]

	0
0	1.0
1	2.0
2	3.0
3	NaN
4	4.0

→ [Out]

	0
0	1.0
1	2.0
2	3.0
3	-1.0
4	4.0

This is a simple application of just one common platform, the ‘pandas’ library. Please feel free to discover more before you move on to the next steps.

Remarks: Statistics is a fundamental and important part of systems analysis. It offers the methods and instruments required to analyze, characterize, and forecast data. For every sector that requires dealing with data, including business, health, and social sciences, it is important to comprehend fundamental statistical concepts like descriptive statistics, probability, and inferential statistics. Statistics’ significance has grown as a result of the development of data mining techniques since they enable the extraction of useful patterns and insights from massive and complicated data sets. A thorough framework for making data-based choices is offered by the integration of descriptive, inferential, and data mining approaches.

[1.2]

Descriptive Statistics

A Guide to the Concepts of Descriptive Statistics, Distribution, Central Tendency, Variability

Descriptive statistics provide information about the data you want to analyze and describe them in some manner. More specifically, they collect from the target data set (population) certain detailed measurements (parameters).

1. A Guide to Concepts of Descriptive Statistics

Descriptive statistics differ from inferential statistics, which aims to make assumptions and inferences about a population based on a smaller representation (sample) of it (Marshall & Jonker, 1998). Often the size of a population is far too large to be manipulated, and you must instead rely on a smaller representation of it. This smaller representation is called a *sample*.

[Table 1.2.1] Examples of Descriptive Statistics (artificial data: $N = 100$)

Variable	*Mean*	*SD*	*V*
1-mile run (min)	9.0	1.4	0.16
Stress (0 ~ 100)	15.3	21.6	1.41
Age (years)	17.8	0.6	0.03
Sex (male = 1, female = 2)	1.6	0.5	0.31

From your sample you can compute sample measurements (sample mean, sample variance, etc.,) that will be estimates (or statistics) of the real parameters.

- What can you conclude about age?
- What can you conclude about stress?
- How many females are there?
- On which attribute are the subjects most alike?

The table and questions (Table 1.2.1.) show that descriptive statistics summarize, describe, and present data effectively.

Now let us get into the following well-known concepts as follow.

+ **Distribution:** A distribution is a particular value or range of values for a variable (Figure 1.2.1).

Range	Frequency
0–5	100
5–10	70
10–15	30
15–20	20
20–25	5
25–30	1

→

Figure 1.2.1. Data (left) and its distribution (right)

Here we created a histogram as a prototype of the distribution, using 'hist' function with 'Frequency' as the data and 'bin_edges' as the bins.

```
[In]
import matplotlib.pyplot as plt
import numpy as np
%matplotlib inline

Frequency = [27, 20, 21, 22, 23, 24, … ,  1, 2, 3, 4, 5]
bin_edges = [0,5,10,15,20,25,30]

plt.hist(Frequency, bins=bin_edges, density=False)
```

The 'density=False' argument specifies that the histogram should show the raw frequency of the data in each bin, rather than the relative proportion of the data. The intricacies and functions of this parameter will be thoroughly discussed in various sections of this book, particularly in sections 1.3 and 2.1, as we develop solutions for a wide range of scenarios.

2. Central Tendency

A central tendency is a summary statistic that represents the typical value of a data set (Mertler & Reinhart, 2016). These measures indicate the center of a data set, where most values fall.

- Summarizing many values as one value to represent the whole group of values

The three most common measures of central tendency are the mean, median, and mode. Each of these measures uses a different method to calculate the location of the central point.

+ **Mode**: The mode is the value that appears most frequently in a data set. If a value appears repeatedly throughout the data, it will influence the average toward the value; hence, the mode is an important measure of the central tendency of data.

$$x = [1\,6\,4\,0\;\;13\;\;4\,1\,4]$$
$$x = [0\,1\,1\;\;\underbrace{4\,4\,4}_{Mode=4}\;\;6\,13]$$

Example of the mode: two ways

```
[In]
import numpy as np
x=1,6,4,0,13,4,1,4
mode = max(x, key = x.count)
print(mode)

[Out] 4
```

```
[In]
import statistics
x=1,6,4,0,13,4,1,4
mode = statistics.mode(x)
print(mode)

[Out] 4
```

The left side code uses the 'numpy' library and the 'max()' function to find the most common value (i.e., mode) of a dataset stored in the x list. The code of the right side is similar to the left side but it uses the 'statistics' module from the standard library to calculate the mode of the x list.

+ **Median**: Median is the value located at the midpoint when all values are ordered by magnitude. In other words, it is the value that lies in the middle of a list of ordered numbers (the numbers may be in ascending or descending order).

$$x = [1\,6\,4\,0\,13\,]$$
$$x = [0\,1\;\;\;\underbrace{4}_{Median}\;\;\;6\,13]$$

Example of the mode: two ways

```
[In]
import numpy as np
x = 1,6,4,0,13
median = sorted(x) [len(x) // 2]
print(median)

[Out] 4
```

```
[In]
import statistics
x = 1,6,4,0,13
median = statistics.median(x)
print(median)

[Out] 4
```

The code of the left side uses the 'numpy' library and the 'sorted()' function to calculate the middle value (i.e., median) of the x list. Instead, the right side code uses the 'statistics' module from the library to calculate the median of the x list.

If the number of values is even, the value in the middle position is obtained as the average of the two central values.

$$x = [0\ 1\quad \underbrace{4\ 5}_{(4+5)\neq 4.5=Median}\quad 6\ 13]$$

+ **Mean**: the mean is an arithmetic average (most commonly used measure of central tendency), or the sum of all values divided by the total number of values. This (or average) is the most popular and well-known measure of central tendency. It can be used with both discrete and continuous data, although it is most often used with continuous data.

The mean is equal to the sum of all values in the data set divided by the number of values in the data set. If we have n values in a data set, and they have values $x_1 + x_2 + \cdots + x_n$, the sample mean, usually denoted $\bar{x}$ (pronounced 'x bar'), is

$$\bar{x} = \frac{(x_1 + x_2 + \cdots + x_n)}{n}$$

Example of the mode: two ways

[In] import numpy as np x=1,6,4,0,13,4,1,4 mean = sum(x)/len(x) print(mean) [Out] 4.125	[In] import statistics x=1,6,4,0,13,4,1,4 mean = statistics.mean(x) print(mean) [Out] 4.125

The code of the left side also uses the 'numpy' library to calculate the average (i.e., mean) of the x list, but the code of the right side uses the 'statistics' module from the library to calculate the mean of the x list. This formula is usually written in a slightly different manner using the Greek capital letter sigma, Σ, which means "sum of":

$$\bar{x} = \frac{\sum x}{n}$$

This formula is used to find the sample mean. Why is it called a *sample mean*? It is because in statistics, samples and populations are very different, and these differences are significant, even if, in the case of the mean, they are calculated in the same way. To compute the population mean and not the sample mean, we use the Greek lowercase letter mu, μ:

$$\mu = \frac{\sum x}{n}$$

The mean is essentially a model of a data set. It is the most common value. You will notice, however, that the mean is not often one of the actual values in the data set. However, one of its essential properties is that it minimizes error in the prediction of any one value in the data set. That is, it is the value among all values in the data set that produces the lowest amount of error.

3. Variability

In describing a distribution, it is necessary to know something about its dispersion, or the spread of the values.

In general assumptions, the gap between the greatest and minimum values in a collection of data is represented by the term 'range,' but the term 'variability' describes how widely these values are spread out.

- Variance, which indicates the spread of the data, is the average of the squared deviations from the mean,
- whereas deviation measures the distance between a single data point and the mean.

A measure of the variability in the data set is provided by the standard deviation, which is the positive square root of the variance. In order to provide a relative measure of variability, the coefficient of variation (V) quantifies the standard deviation as a percentage of the mean.

To be specific, variability refers to how spread out a group of values is (Stokes, 1995). For example, consider the graphs below (Figure 1.2.2.). These graphs represent the scores of two samples of students on a quiz. The mean score for each sample is 7.0.

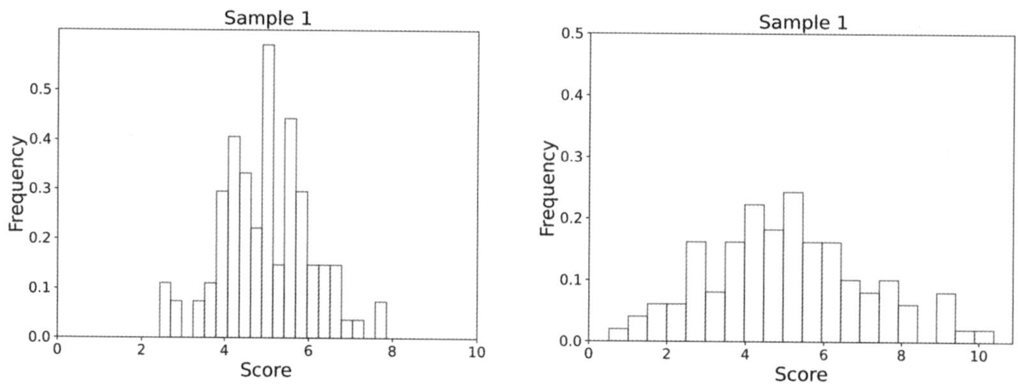

Figure 1.2.2. Scores of two samples of students on a quiz

Despite the equality of means, you can see that the distributions are quite different. Specifically, the scores for sample 1 are more densely packed, and those for sample 2 are more spread out. The differences among students were much greater for sample 2 than sample 1.

The terms variability, spread, and dispersion are synonyms and refer to how spread-out data are. Let us look at several measures of variability in more detail.

+ **Range:** The range is the simplest measure of variability to calculate and one you have probably encountered many times in your life experiences (Murphy et al., 2014). The range is simply the highest value minus the lowest value. Let us consider the following group of numbers:

$$10, 2, 5, 6, 7, 3, 4$$

What is the range? The highest number is 10, and the lowest number is 2, and 10 - 2 = 8. The range is 8. Let us take another example. Here is a data set with 10 numbers:

$$99, 45, 23, 67, 45, 91, 82, 78, 62, 51$$

What is the range? The highest number is 99, and the lowest number is 23, and 99 - 23 = 76. The range is 76. Now consider the two samples. In sample 1, the lowest score is 5 and the highest score is 9. The range is 4. The range for sample 2 is larger: The lowest score is 4 and the highest score is 10. Therefore, the range is 6.

Example of the range: two ways

```
[In]
x = range(5, 9)
for n in x:
    print(n)

[Out] 5, 6, 7, 8
```

```
[In]
for i in range(4, 10):
    print(i, end =" ")
print()

[Out] 4, 5, 6, 7, 8, 9
```

The code of the left side uses a loop to iterate over the elements in the x list and print each element to the console. The code of the right side uses the 'range()' function and a loop to generate and print a sequence of integers.

+ **Interquartile range (IQR):** The interquartile range is the range of the middle 50% of the values in a distribution. It is computed as follows:

$$\text{IQR} = 75^{th} percentail - \ 25^{th} percentail$$

For sample 1, the 75th percentile is 8 and the 25th percentile is 6. The interquartile range is therefore 2. For sample 2, which has a greater spread, the 75th percentile is 9 and the 25th percentile is 5. The interquartile range is therefore 4. In a representation,

- the 75th percentile is called the *upper hinge*, and
- the 25th percentile is called the *lower hinge*.

Using this terminology, we can refer to the interquartile range as the H-spread. A related measure of variability is the semi-interquartile range, which is the interquartile range divided by 2. If a distribution is symmetrical, the median plus or minus the semi-interquartile range contains half of the values in the distribution.

Let us see how to create a code example of IQR detection. Import library and arbitrarily define array of data

```
[In]
import numpy as np
data = np.array([14, 19, 20, 22, 24, 26, 27, 30, 30, 31, 36, 38, 44, 47])
```

Calculate interquartile range

```
[In]
q3, q1 = np.percentile(data, [75 ,25])
iqr = q3 - q1
```

Display the IQR range

```
[In]                                        [Out]
print(iqr)                          →       12.25
```

We used the 'numpy' library to calculate the interquartile range (IQR) of a dataset stored in the 'data' array, and calculated by first calling the 'percentile()' function which returns the values of the specified percentiles (i.e., 25th and 75th) of the 'data' array. These values are assigned to the variables 'q1' and 'q3' respectively and the calculated IQR is printed to the console using the 'print()' function.

+ **Variance:** Variability can also be defined in terms of how close values in the distribution are to the middle of the distribution (Sawyer, 2009).

Using the mean as the measure of the middle of the distribution, we can define the variance as the average squared difference of the scores from the mean. The formula for the variance is

$$\sigma^2 = \frac{\Sigma(x-\mu)^2}{N}$$

where σ^2 is the variance, μ is the mean, and N is the number of values. For sample 1, $\mu = 7$ and $N = 20$. If the variance in a sample is used to estimate the variance in a population, the following formula should be used to avoid underestimating the variance:

$$s^2 = \frac{\Sigma(X-M)^2}{N-1}$$

where s^2 is the estimate of the variance and M is the sample mean. Note that M is the sample mean taken from a population with a mean of μ. Because the variance is usually computed in a sample, this formula is most often used. The simulation 'estimating variance' illustrates the bias in the formula with N in the denominator.

Let us take a concrete example. Assume the values 1, 2, 4, and 5 were sampled from a larger population.

To estimate the variance in the population you would compute s^2 as follows:

$$M = \frac{1+2+4+5}{4} = \frac{12}{4} = 3$$

$$s^2 = [(1-3)^2 + (2-3)^2 + (4-3)^2 + (5-3)^2]/(4-1)$$

$$= \frac{4+1+1+4}{3}$$

$$= \frac{10}{3} = 3.333$$

There are alternative formulas that are easier to use if you are doing your calculations with a hand calculator. You should note that these formulas are subject to rounding error if your values are very large and/or you have an extremely large number of observations.

$$\sigma^2 = \frac{\sum X^2 - \frac{(\sum X)^2}{N}}{N}, \qquad s^2 = \frac{\sum X^2 - \frac{(\sum X)^2}{N}}{N-1}$$

For this example,

$$\sum X^2 = 1^2 + 2^2 + 4^2 + 5^2 = 46$$

$$\rightarrow \frac{(\sum X)^2}{N} = \frac{(1+2+4+5)^2}{4} = \frac{144}{4} = 36$$

$$\rightarrow \sigma^2 = \frac{(46-36)}{4} = 2.5$$

$$\rightarrow s^2 = \frac{(46-36)}{3} = 3.333$$

+ **Standard Deviation:** The standard deviation is simply the square root of the variance (Streiner, 1996). The standard deviation is an especially useful measure of variability when the distribution is normal (or approximately normal) because the proportion of the distribution within a given number of standard deviations from the mean can be calculated.

For example, 68% of the distribution is within 1 standard deviation of the mean, and approximately 95% of the distribution is within 2 standard deviation of the mean. Therefore, if you had a normal distribution with a mean of 50 and a standard deviation of 10, then 68% of the distribution would be between 50 – 10 = 40 and 50 + 10 = 60. Similarly, about 95% of the distribution would be between 50 – 2 × 10 = 30 and 50 + 2 × 10 = 70.

The symbol for the population standard deviation is σ; the symbol for an estimate computed in a sample is s.

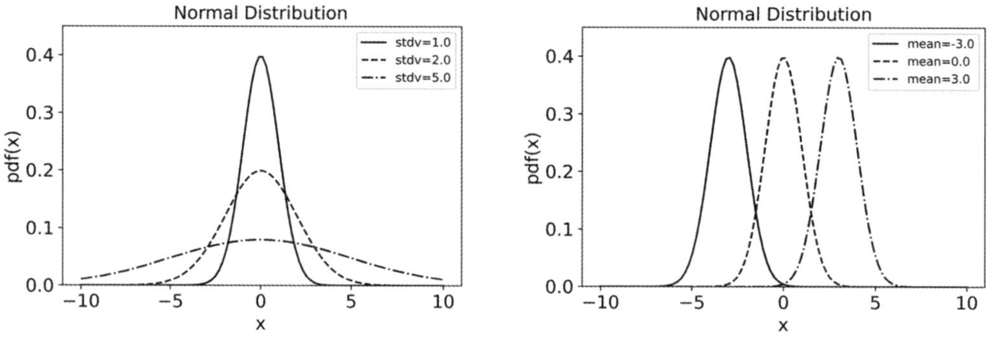

Figure 1.2.3. Normal distributions with different means (left) and standard deviations (right)

The figure above (Figure 1.2.3.) shows two sets of normal distributions. The left distributions have means of −.3, 0, and 3 and a standard deviation of 1; the right distributions have means of 0 and standard deviations of 1, 2, and 5 (pdf = probability density function).

+ **Standard Scores:** The standard score (more commonly referred to as a *Z* score or a *T* score) is a handy statistic because it (a) allows us to calculate the probability of a score occurring within a normal distribution and (b) enables us to compare two scores from different distributions (Johnson, 1948). It does this by standardizing scores in a normal distribution to *Z* scores in what becomes a standard normal distribution (Figure 1.2.4). To show this in simple terms, let us use an example.

- *Z* score: this is obtained by using the standard deviation as the unit of measure and then indicating the relative position of a single score in an entire distribution of scores in terms of the mean and standard deviation:

$$Z = (x - mean)/SD$$

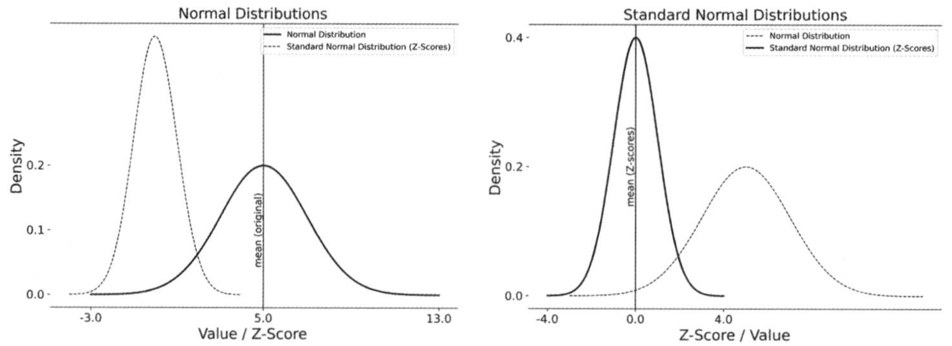

Figure 1.2.4. Distribution depicted by a normal curve (left), for the $(x - \mu)$ sample, when $(x - \mu)$ is oriented by the Z score (right)

- T score: This is a standard Z score shifted and scaled to have a mean of 50 and a standard deviation of 10 (Figure 1.2.5).

$$T = 10 \times z + 50$$

	x	$(x - \text{mean})$	Z
S1	3	–1	–0.5
S2	7	3	1.5
S3	4	0	0
S4	5	1	0.5
S5	1	–3	–1.5
Mean	4	0	0
SD	2	2	1

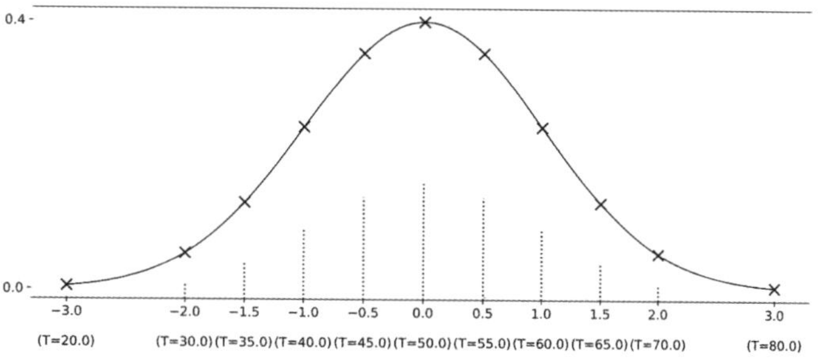

Figure 1.2.5. Schematic representation of the values (left) and distribution depicted by a normal curve (right; *SD* and (*T* score))

+ **Coefficient of Variation**: This is the standard deviation relative to the mean of the data set. An even better unitless measure of dispersion is the coefficient of variation.

The coefficient of variation is obtained by dividing the standard deviation by the mean of the sample.

$$Coefficient\ of\ variation = \text{standard deviation}/mean$$

Consider the following example of five people's heights (in inches): 66, 72, 68, 70, 69
The mean height will be determined first:

$$mean(66 + 72 + 68 + 70 + 69)\ /\ 5 = 68$$

The standard deviation will then be determined:

$$Sqrt([(66 - 68)2 + (72 - 68)2 + (68 - 68)2 + (70 - 68)2 + (69 - 68)2]\ /\ 5) = 2.236$$

The coefficient of variation will then be determined.

$$Coefficient\ of\ variation = 2.236/68$$

Here, the Standard deviation divided by mean equals 2.236/68 (coefficient of variance), or 0.0329 or 3.29 percent.

The coefficient of variation tells us how variable a data collection is in relation to its mean. This is the ratio of the standard deviation to the mean expressed as a percentage. Imagine you need to compare the dispersion of data sets of different units.

The coefficient of variation is useful in such a situation. A series with a greater coefficient of variation is more variable (i.e., less uniform, less stable, or less consistent). A series with a smaller coefficient of variation is less variable (i.e., more uniform, more stable, or more consistent).

+ **What Is a *Z* Score, and How Do You Use It:** A *Z* score is a statistical measure that tells you how far a data point is from the rest of a data set.

In more technical terms, the *Z* score tells you how many standard deviations away a given observation is from the mean (Pontius et al., 1996).

For example, a *Z* score of 2.5 means that the data point is 2.5 standard deviations away from the mean. Because it is far from the center, it is flagged as an outlier or anomaly.

How does it work:

- The *Z* score is a parametric measure that involves two parameters (the mean and standard deviation).
- Once you calculate these two parameters, finding the *Z* score of a data point is easy.

$$Z - score = (x - mean)/standard\ deviation$$

Note that the mean and standard deviation are calculated for the whole data set, whereas x represents every single data point. Thus, every data point will have its own *Z* score, whereas the mean and standard deviation will remain the same everywhere.

Below is a code (in Python) implementation of the *Z* score with a few sample data points.

First, we import module, and create random data points to calculate the *Z* score

```
[In]
import numpy as np
data = [5, 5, 5, -99, 5, 5, 5, 5, 5, 5, 88, 5, 5, 5]
```

This code creates a one-dimensional array by 'data' containing a series of values [] which can be used for various data manipulation and analysis tasks.

So, we calculate mean

```
[In]
mean = np.mean(data) # calculate standard deviation
sd = np.std(data) # determine a threshold
threshold = 2
```

The 'np.mean()' function is called calculate the mean of the 'data' array, and the 'np.std()' function is then called to calculate the standard deviation of the 'data' array.

In addition, we can create empty list to store outliers

```
[In]
outliers = []
```

detect outlier by using the *Z* score calculation

```
[In]
for i in data:
    z = (i-mean)/sd              # calculate z-score
    if abs(z) > threshold:       # identify outliers
        outliers.append(i)       # add to the empty list
```

We used a 'for' loop to iterate over the values in the 'data' array and identify any outliers using a *Z*-score method (see the section 1.3' Outlier section for more detail). For each iteration, the *Z* is calculated by subtracting the mean of the 'data' array from the current value and dividing the result, then it compared to the 'threshold' value using the 'abs()' function to ensure that both positive and negative outliers are detected.

Let us print outliers

```
[In]
print("The detected outliers are: ", outliers)
→
[Out]
The detected outliers are: [-99, 88]
```

If you play with these data, you will notice a few things. There are 14 data points, and the *Z* score correctly detects two outliers (−99 and 88). However, if you remove five data points from the list, it detects only one outlier (−99). This means you need to have a certain number of data points for the *Z* score to work. In large production data sets, the *Z* score works best if the data are normally distributed (Gaussian distribution). We used an arbitrary threshold of 2, beyond which all data points are flagged as outliers. The rule of thumb is to use 2, 2.5, 3, or 3.5 as a threshold. Finally, the *Z* score is sensitive to extreme values because the mean itself is sensitive to extreme values.

Remarks: With the use of descriptive statistics, we may analyze and summarize data in order to learn more about the characteristics of a dataset. We can comprehend the distribution of the data, including measures of central tendency and variability, with the use of descriptive statistics. The distribution of the data is crucial because it enables us to see any trends (or outliers) that could be connected to our research issue. Measures of variability, such as variance and standard deviation, reflect the spread or dispersion of the data, whereas measures of central tendency, such as the mean, median, and mode, give information about the typical or representative value in the dataset. Overall, descriptive statistics is a strong tool for summarizing and interpreting data, allowing us to make inferences and decisions that are based on accurate information.

[1.3]

Bell Curve

A Guide to the Concepts of Bell Curve, Empirical Measure of Central Tendency, Empirical Measure of Dispersion, Central Limit Theorem, Outlier

The bell curve is a statistical distribution (also known as the Gauss or Normal distribution) that naturally occurs in various situations (Herrnstein & Murray, 2010). This pattern is commonly observed in standardized tests such as the Scholastic Assessment Test. The majority of students tend to achieve average scores C, while fewer score B or D. An even smaller percentage of students attain either an A or an F. The resulting distribution takes on a shape reminiscent of a bell. The term 'bell curve' was coined due to this distinctive shape, and the concept was first introduced by mathematician Carl Friedrich Gauss.

1. A Guide to Concepts of Bell Curve

The bell curve, also known as the normal distribution, is symmetrical in shape, with its peak representing the mean value. Half of the data points fall to the left of the mean, while the other half fall to the right, creating a balanced distribution. This symmetry indicates that the average and median values are equal, making it an ideal model for various statistical analyses.

A normal distribution is not a single distribution with a specific mean and standard deviation; instead, it is smooth and symmetrical and follows the general shape of the distribution (Micceri, 1989). The specific normal distribution depends on characteristics such as its variability. Features of the normal distribution include the following:

- Theoretical distribution
- Continuous distribution
- Probability distribution
- Symmetrical distribution
- Mean, median, and mode coincide

Here we look at normal distribution features with respect to the Bell Curve (in descriptive statistics). Let us use an available data set (Iris.txt: see the Supplement data set) with the statistical packages in Python.

To achieve this, we open a file called 'iris.txt' from a specified file path ('C:/ ~), reads the contents of the file,

```
[In]
iris = open('C:/ …  /iris.txt').read().split()
print(iris)
```

and splits the contents into a list of individual words but use only one feature ('sepal_length').

```
[In]
iris = sns.load_dataset('iris')
X = iris.drop('species',axis=1)
x=X['sepal_length']
```

The 'iris' dataset is loaded by using the 'sns.load_dataset()' function from the 'seaborn' library. The 'drop()' method of the 'iris' is then called to create a new data called x that does not contain the 'species' column. and select the 'sepal_length' column from the x using the indexing operator []. Now, we are going to focus on two descriptive statistics first—measures of central tendency and measures of dispersion—separately reviewing the basic concepts using the imported actual data set.

2. Empirical Measures of Central Tendency

The bell curve of data also suggests something about the shape of the mean, median, and mode, all of which describe the behavior of data around their middle areas. Let us think of these again (see the 1.2.2 Central Tendency section) as representing the distribution of data to cluster around a middle value.

+ **Mean:** The mean, or the average value of data, is easy to compute. Just take the sum of your values, divide it by the total number of values, and you have your mean. We can compute it manually with the code below.

```
[In]
mean=x.sum()/len(x)
print('Mean: {}'.format(mean))
```

→

```
[Out]
Mean: 5.843333333333333
```

or use the built-in function in the statistics module.

```
[In]
import statistics
print('Mean:
{}'.format(x.mean()))
```

→

```
[Out]
Mean: 5.843333333333333
```

+ **Median:** The median is the value that lies in the middle of a list of ordered numbers (the numbers may be in ascending or descending order). Let us see what is the median value for our data.

```
[In]
print('Median: {}'.format(x.median()))
→
[Out]
Median: 5.8
```

+ **Mode:** The mode is the value that appears most frequently in a data set. If a value appears repeatedly throughout the data, it will influence the average toward the modal value; hence, the mode is an important measure of the central tendency of data. Let us compute it for our data.

```
[In]
mode=x.mode()
print('Mode: {}'.format(mode))
→
[Out]
Mode: 5.0
```

A normal distribution is symmetrical; hence, mean = median = mode. However, if mode > median > mean or mean > median > mode, we have an asymmetrical distribution.

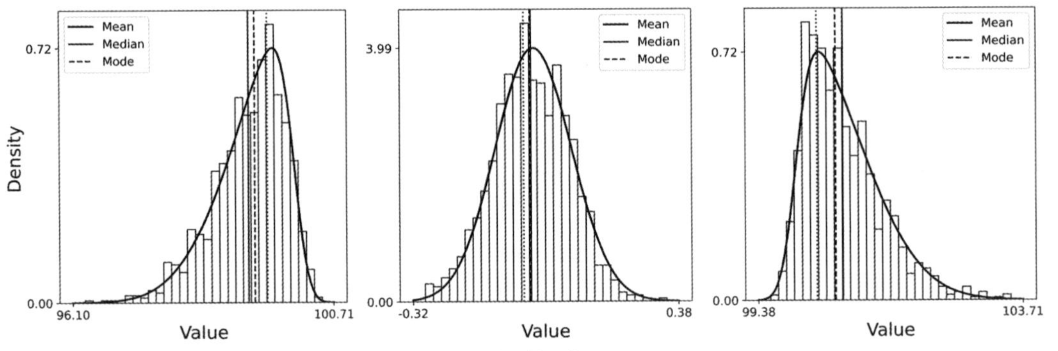

Figure 1.3.1. Symmetrical and asymmetrical distributions

The graph (Figure 1.3.1.) on the left (mean > median > mode) shows a positive asymmetrical distribution, whereas the one on the right (mode > median > mean) shows negative asymmetry. A way to check the symmetry of a distribution is by measuring skewness.

$$\gamma 1 = E\left[\left(\frac{X-\mu}{\sigma}\right)^3\right] = \frac{\mu^3}{\sigma^3} = \frac{E[(X-\mu)^3]}{(E[(X-\mu)^2])^{3/2}} = \frac{k^3}{k_2^{3/2}}$$

The left-hand side of the equation represents the skewness of the variable $X(\gamma 1)$, while the right-hand side is its numerical value in terms on the mean μ, standard deviation σ, and the third and second moments of the k. It is also common that is used the notation $E(X)$ for mean and $Var(X)$ for variance. Positive skewness indicates positive asymmetry, whereas negative skewness indicates negative asymmetry. Recall the parameters of our data set:

$$Mean: 5.843, \quad Median: 5.8, \quad Mode: 5.0$$

We can see that mean > median > mode; hence, the distribution should exhibit positive skewness.

Let us check it out.

[In]

```
f, (ax_box, ax_hist) =
plt.subplots(2, sharex=True,
gridspec_kw= {'height_ratios':
(0.5, 1)})
sns.boxplot(x, ax=ax_box)
sns.distplot(x, ax=ax_hist)

plt.show()
```

→ [Out]

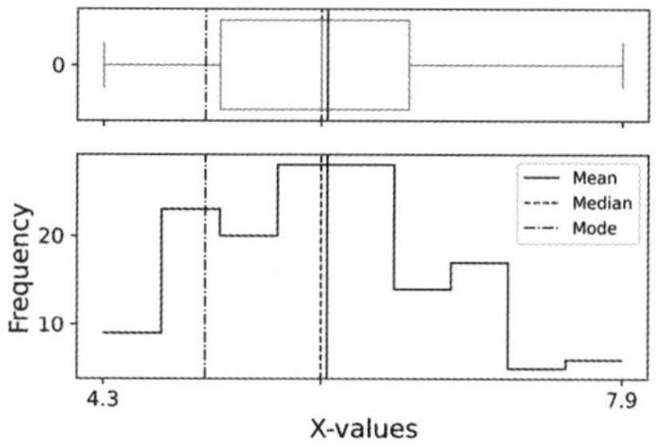

Figure 1.3.2. Skewness (mean, median, and mode)

We created a figure (Figure 1.3.2) with two subplots, 'ax_box' and 'ax_hist'. Each subplot is a plot of the data contained in the variable. Then, we added the skewness computation as below.

[In]

```
from scipy.stats import skew
print('Skewness:
{}'.format(skew(x)))
```

→ [Out]

```
Skewness: 0.3117530585022963
```

As you can see, the skewness is greater than 0; hence, it indicates positive asymmetry, which confirms the fact that mean > median > mode.

3. Empirical Measures of Dispersion

Measures of central tendency, which reveal the average characteristics of data, differ from measures of dispersion that concentrate on the variability within the dataset.

As we briefly noticed in previous section (1.2), specifically, the central tendency encompasses the mode, median, and mean, serving to represent the central or typical value of a dataset. In contrast, dispersion is an independent concept that pertains to the spread or fluctuation of data. Prominent measures of dispersion consist of range, variance, and standard deviation.

+ **Range:** The range is the most intuitive measure of dispersion. It is computed as the difference between the maximum and minimum values of a data set, and it suggests how spread out the data are.

[In]

```
r=x.max()-x.min()
print('Range: ',r)
```

→ [Out]

```
Range: 3.6000000000000005
```

The code calculates the range of the data in the variable x. The range is defined as the difference between the maximum and minimum values of the data. We can also visualize dispersion with the aid of a box plot (Figure 1.3.3).

```
[In]
fig1, ax1 = plt.subplots()
ax1.set_title('BoxPlot')
ax1.boxplot(x)
```

→

[Out]

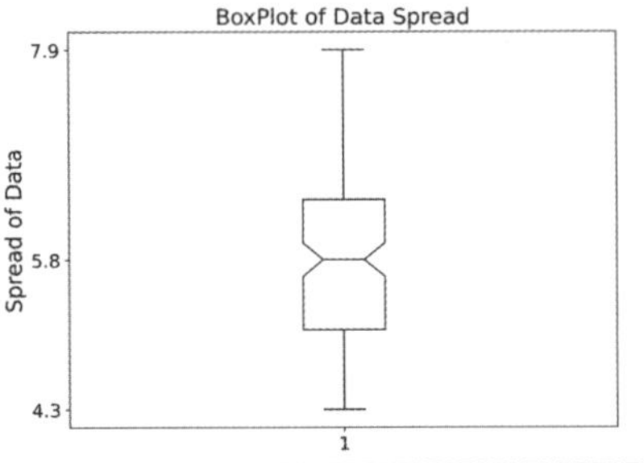

Figure 1.3.3. Representation of dispersion with a box plot

Created figure is a box plot using the 'boxplot()' function which data contained in the variable x.

+ **Interquartile Range:** Quantiles are cut points that divide the range of a probability distribution into continuous intervals with equal probabilities. More specifically, the x quantile (also called the x *percentile*) leaves at its left x% of the probability. Furthermore, the 0, 0.25, 0.5, 0.75, and 1 quartile are known, respectively, as the zero, first, second, third, and fourth quartiles. Note that the second quartile (Q2) corresponds to the median value, as it leaves 50% of the probability to its left and 50% of the probability to its right.

```
[In]
Q1=np.percentile(x, 25)
Q2=np.percentile(x, 50)
Q3=np.percentile(x, 75)

print('Q1: {}'.format(Q1))
print('Q2: {}'.format(Q2))
print('Q3: {}'.format(Q3))
```

→

```
[Out]
Q1: 5.1
Q2: 5.8
Q3: 6.4
```

Let us visualize this on our distribution graph.

```
[In]
f, (ax_box, ax_hist) =
plt.subplots(2, sharex=True,
gridspec_kw= {'height_ratios':
(0.5, 1)})

sns.boxplot(x, ax=ax_box)

plt.show()
```

→

[Out]

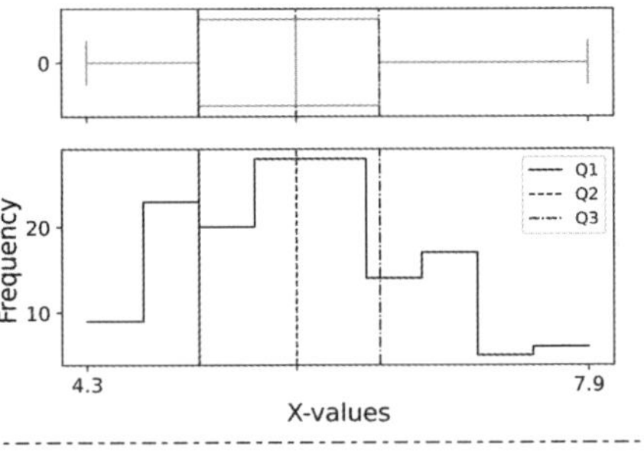

Figure 1.3.4. Representation of quartiles

An interquartile range is the difference between two quantiles. In the box plot above (Figure 1.3.4), you can see the interquartile range, computed as the difference between Q3 and Q1.

```
[In]                                        [Out]
IQR=Q3-Q1                            →      Interquartile Range:  1.300000000000
print('Interquartile Range: ',IQR)          0007
```

The result [out] represent the upper and lower quartiles of the data. The upper quartile, 'Q3' is the value that separates the data into the top 75% and the bottom 25%. The operator of 'IQR' is the difference between 'Q3' and 'Q1', printed to the console using the 'print()' function.

+ **Variance:** The variance summarizes how much the data differ from the mean. It is computed as follows:

$$\sigma^2 = \sum \frac{(X-\mu)^2}{N}$$

where μ is the mean of the population. However, because it often happens that the square of a unit of measurement is meaningless, a more manageable measure is used: the standard deviation, which is nothing but the square root of the variance.

+ **Standard Deviation:** The standard deviation is a measure of how spread out the data points are from the mean (μ) of the data set.

$$\sigma = \sqrt{\sum \frac{(X-\mu)^2}{N}}$$

The standard deviation gives us an idea of how much variation there is in the data set. A small standard deviation indicates that the data points are close to the mean, while a large standard deviation shows that the data points are more spread out.

Let us compute the standard deviation of our data.

```
[In]                                        [Out]
sigma=np.std(x)                      →      Standard Deviation: 0.8253012917
print('Standard Deviation: ',sigma)
```

We used the 'np.std()' function from the 'numpy' library which data contained in x as an input and returns the standard deviation as a float. As you can see, descriptive statistics with respect to the normal distribution are used to describe the basic features of data.

They provide simple summaries of the data, and together with simple graphics analysis, they form the basis of statistical analysis. Hence, it is always good practice to start with these simple computations before diving into the building of complex models.

How Do Variance and Standard Deviation Differ?

In statistics, variance and standard deviation help us measure the variability in data, or how values fluctuate around the mean. Variance is the average of squared differences from the mean. Standard deviation is the square root of the variance (Bland & Altman, 1996). Usually, standard deviation is more intuitive than variance.

To understand what this means, let us consider an example of 12 products:

```
[In]
products = [150, 150, 200, 200, 200, 250, 250, 250, 300, 300, 350, 2000]
products_df = pd.DataFrame(products, columns=["products"])
```

We created a DataFrame called 'products_df' which list is a one-dimensional array containing a series of values with the 'columns' argument to specify the name of the column in the data.

We draw the plots as follows to visualize our artificial data set:

```
[In]
fig, ax = plt.subplots(nrows=2, ncols=1, figsize=(4,7))
…
plt.show()
```

[Out]

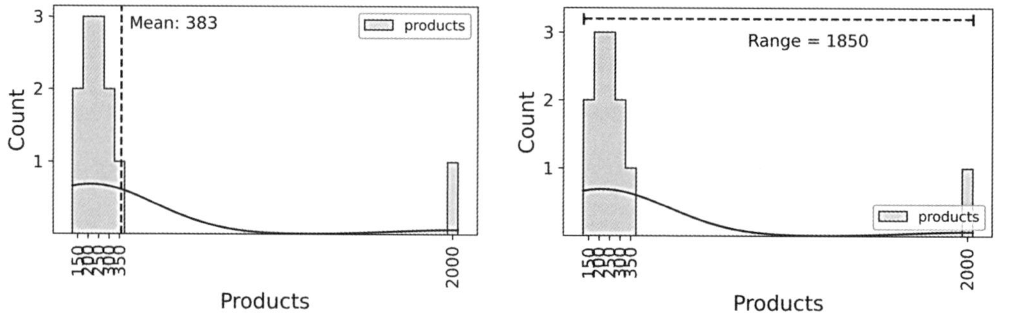

Figure 1.3.5. Products range and domain

Notice that most values are concentrated around 150 and 350, but there is an extreme value (an outlier) of 2,000 that pushes the mean up to 383 and dilates the range to 1,850.

We can write the following code to obtain the mean of a (Pandas) data frame

```
[In]                                  [Out]
products_df.mean()          →         383.333333
```

We can use this simple line of code to obtain the variance.

```
[In]                                  [Out]
products_df.var(ddof=0)     →         240972.222222
```

After we identify the variance, finding the standard deviation is straightforward.

It is the square root of the variance, and we can calculate it as follows:

```
[In]                                    [Out]
products_df.std(ddof=0)          →      490.889216
```

Notice that 490 is more connected to the values in our list, but what does it mean? It means that the products in our list are on average 490 units away from the mean. The closer the values are to the mean, the smaller the standard deviation. In this case, the standard deviation was stretched out because we had an outlier.

4. Central Limit Theorem

Statistics is must-have knowledge for systems analysis, and the central limit theorem is the cornerstone of it (Rice, 2006). The central limit theorem posits that the distribution of sample means approximates a normal distribution as the sample size increases, regardless of the population's distribution. Thus, a key aspect of the central limit theorem is that the average of the sample means and standard deviations will equal the population mean and standard deviation.

According to the central limit theorem, in some situations, when independent random variables are added, their properly normalized sum tends toward a normal distribution even if the original variables themselves are not normally distributed.

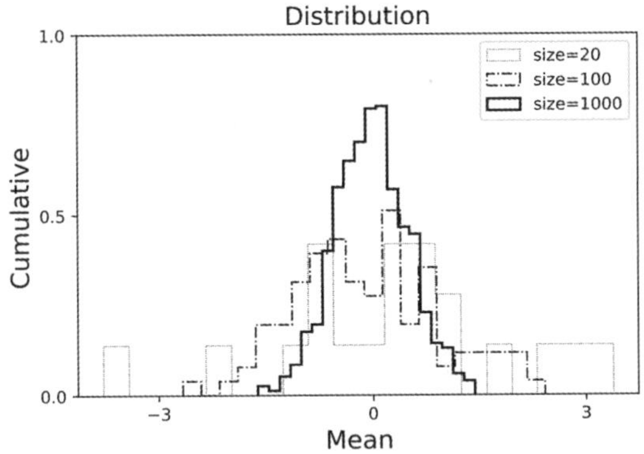

Figure 1.3.6. A normal distribution with different sample sizes for the critical concept in data analysis

Figure 1.3.6 suggests that probabilistic and statistical techniques applicable to bell curve (or normal distributions) can be employed for various issues involving other distribution types. Consider the average life expectancy as an example. Assuming the global average life expectancy (across 186 countries, to be exact) is 73 years, this figure becomes more informative when placed within its broader context. This context consists of the statistical concepts we discussed earlier.

Thus, let us briefly review them and explore their practical implications.

+ **Population and Sample:** The population contains all data, elements, individuals, and measurements from an experimental space. In this example, our experimental space is the world, and our population is life expectancy data from all countries. N denotes the population size. The sample is a randomly selected subset of a population. n denotes the sample size.

+ **Distribution:** The distribution describes the range of data/population/sample and how data are spread out over that range. The distribution of life expectancy in the world as of 2018 is as follows (Figure 1.3.7's left side):

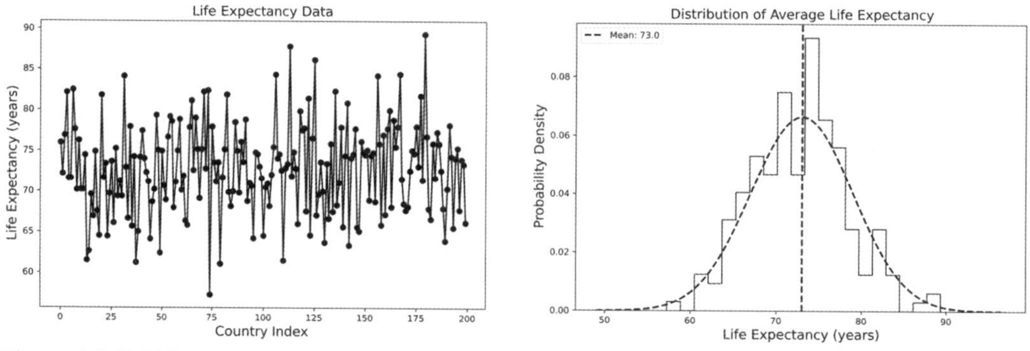

Figure 1.3.7. Life expectancy (plot of the left side) and the distribution of average life expectancy (plot of the right side). Recreated data from https://www.who.life-expectancy-at-birth-(years)

+ **Mean and Standard Deviation:** The mean is the average value of all data from a population or sample. In this example, the population mean is 73 (range 50 – 84). This is denoted μ (mu) for populations and $\bar{x}$ (x bar) for samples. Not all countries' average life expectancy is 73 (range 50 – 84). Values for life expectancy are distributed in a range, as we see above. The standard deviation is a measure of how spread out a population is. It is denoted σ (sigma).

+ **Normal Distribution (Bell-Shaped Curve or Gaussian Distribution):** When the distribution of a population is perfectly symmetrical, with σ standard deviation around the mean, the result is a bell-shaped curve (Figure 1.3.7's right side). What kind of distribution does the average life expectancy in the world as of 2018 have? Is it a normal distribution or some other type? It is not a perfectly normal distribution, as it seems to be pulled to the left side, known as skewed (the graph above, on the left).

Now let us take some samples from the population. In other words, we take 1,000 subsets of size 150 from the population and plot the distribution of each sample mean ($\bar{x}$; the graph above, on the right). This is called the *sampling distribution*: It is a perfectly normal distribution. Any sampling distribution from any population has a normal distribution because the central limit theorem states that regardless of the initial shape of the population distribution, the sampling distribution will approximate a normal distribution. As the sample size increases, the sampling distribution will get narrower and more normal.

The central limit theorem allows us to calculate the standard deviation—known as the standard error—of the sampling distribution with the following formula:

$$SE = \frac{\sigma}{\sqrt{n}}, \qquad \sigma = standar\ deviation, \qquad n = number\ of\ samples$$

We can see whether this calculation holds for our population and sampling distribution.

```
[In]
Standard_deviation = round(population_2018['life_expectancy_years'].std(), 2)
n = 150

standard_error = (standard_deviation / np.sqrt(n))
print('Calculated standard error of the sampling distribution is:',
round(standard_error, 2))

print('Standard error of the sampling distribution is:',
round(sampling_distributionstd(), 2))
```

↓

```
[Out]
Calculated standard error of the sampling distribution is 0.57.
                Standard error of the sampling distribution is 0.58.
```

In this results [out], the standard deviation of the 'life_expectancy_years' data is calculated using the 'std()' method of the 'life_expectancy_years' column, which is accessed using dot notation. The n variable represents the sample size which is the number of data points in the sample. The intuitive explanation of this formula is this: As the sample size increases, the sampling distribution will become a more accurate representation of the population parameters (mean and standard deviation).

In any data analysis, the given data set represents a sample of the whole population. Using this sample, we try to identify the main patterns in the data. Then we try to generalize the trends in the sample to the population while making predictions. The central limit theorem helps us make inferences about the sample and population parameters and construct better machine-learning models using them. Moreover, the theorem can tell us whether a sample possibly belongs to a population by looking at the sampling distribution. The central limit theorem is at the heart of statistical significance and hypothesis testing.

+ **Practical Applications:** Assume we start with a somewhat messy and unstructured the bell-curve. This could represent a lot of different things: the outcomes of a dice throw or the distribution of heights within an inhomogeneous population. The only requirement is that the distribution have a well-defined mean and variance (as you can see in this example, this does not mean the distribution has to follow a bell curve):

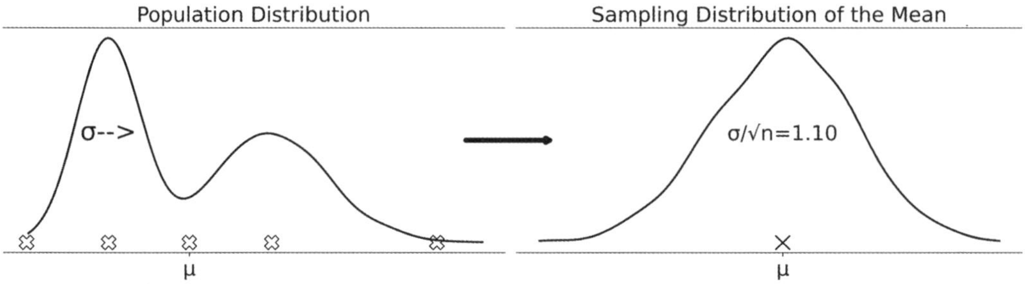

Figure 1.3.8. A distribution with a bell curve. Left = population distribution. Right = sampling distribution. μ = *mean* σ = *standard deviation*, n = *sample size*

The curve above (Figure 1.3.8, left plot) could signify the average height in a mixed population of men and women, in which a larger proportion of the population is male. The distribution has two peaks, with the second one being at the larger average height. Independent of the details of the distribution, let us assume we can repeatedly sample from this distribution N times (say by randomly walking up to N people from the population and measuring their height). For this sample of N people, it is then straightforward to compute the sample mean by adding up all samples and dividing by N.

The important point here is that in practice we do not need to know the underlying population distribution before we can sample from it. The central limit theorem simply does not care: It states that if you draw N samples from this population again and again, the mean of the sample means, divided by the standard errors, will converge to the normal distribution (Figure 1.3.8, right plot).

In more formal terms, the central limit theorem states that if we have a set of xi, with i from 1 … N of identically distributed random variables with finite positive variance and finite expectations. Then the error of the sample means, weighted by the square root of the sample size versus the covariance of the population distribution, converges to the standard normal distribution:

$$\sqrt{n}(\bar{X}_n - \mu) \to N(0, \sigma^2)$$

This is quite remarkable and surprising, considering we did not make any assumptions about the random variables that go into the sum, and it is hard for us to fully come to terms with it (why does such neat order arise even out of the most chaotic and skewed distributions).

5. Outliers

An *outlier* is an observation that lies an abnormal distance from other values in a random sample from a population. As you work more with statistics, you will notice that your data sets will have outliers. These outliers pose a real threat to systems analysis (Billor et al., 2000). They are mostly due to two main factors:

- Mistakes and errors: These are obvious outliers. If you spot such errors or encoding mistakes, you must improve your process to prevent them from occurring again.
- Exceptional demand: Although some demand observations are real, this does not mean that they are not exceptional and should not be cleaned or smoothed.

If you think back to the idea to analyze, we still have a minor issue. The statistics you compute will be based on the data set, including any outliers. Outliers drive the error variation upward, so the acceptable descriptive statistics is biased and overestimated. To correct this, you can shrink the outlier, not to the statistics calculated based on the original demand data set, but rather to a limit calculated on a data set without this specific outlier. Here is how. Populate a first forecast against the historical demand.

- Compute the error, the error mean, and the error standard deviation.
- Compute the lower and upper acceptable thresholds (based on the error mean and standard deviation).
- Identify outliers.
- Recompute the error mean and standard deviation excluding the outliers.
- Update the lower and upper acceptable thresholds based on these new values.
- Update the outlier values based on the new threshold.

+ **Technique with the Standard Deviation:** One approach is to look at the variation around the historical average and exclude values that are exceptionally far from this average. Let us define the demand standard deviation as

$$\sigma = \sqrt{\frac{\Sigma(x-\mu)^2}{n}}$$

where n is the number of observations. If we assume that the data are normally distributed around the historical mean, we can compute the probability of x being between two thresholds. These two thresholds will be centered on the average (μ) with a spread of x times the standard deviation (σ) in both directions. The more chaotic the demand (i.e., when σ is big), the wider the thresholds.

Min threshold	< Probability <	Max threshold
$\mu - 1.23\sigma$ <	95%	< $\mu + 1.23\sigma$
$\mu - 2.33\sigma$ <	98%	< $\mu + 2.33\sigma$
$\mu - 2.58\sigma$ <	99%	< $\mu + 2.58\sigma$

→

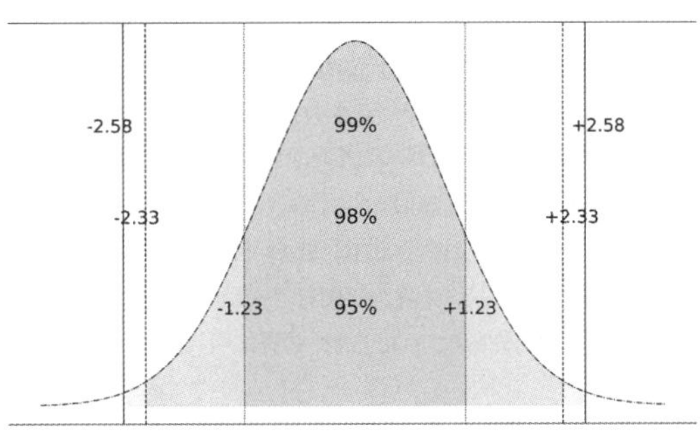

Figure 1.3.9. Schematic representation of the average and standard deviation

For example, we have a 98% probability of being in the range: demand average ± 2.33 times the standard deviation (as in the figure above; Figure 1.3.9). Some basic methods can be used to detect outliers from the distribution plot. One is to examine the distribution of the underlying data set through a distribution plot. This can be done using the following code:

Read in the necessary libraries for the data analysis

```
[In]
import pandas as pd
import numpy as np
import statsmodels.api as sm
import matplotlib.pyplot as plt
import seaborn as sns
```

and import data set (see the Supplement data)

```
[In]
pokemon = pd.read_csv("data…/pokemon.csv")
```

For this, using the pokemon data set above, the focus here will be on the arbitrary designated value of x. One of the first steps is to plot the distribution of the variable to see whether it is potentially skewed and thus whether any anomalies can be identified. We can do this using the Seaborn library as follows: Create the plot

```
[In]
sns.displot(pokemon["x"],
kde = True)

plt.show()
```

→

[Out]

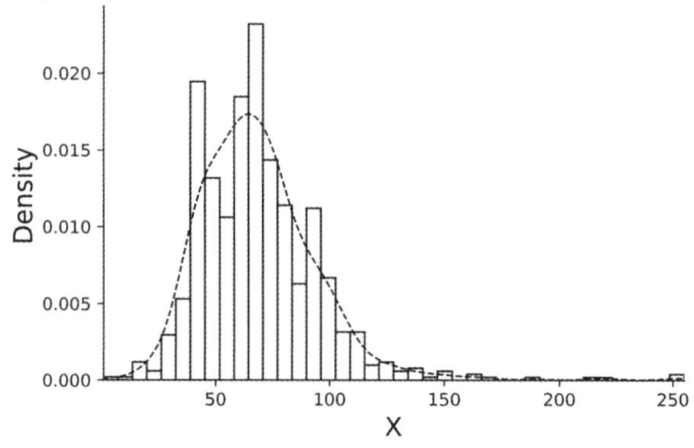

Figure 1.3.10. Outlier detection with data

Here, the 'sns' is a visualization library for plotting (Figure 1.3.10). The 'displot()' function create a distribution plot of a single variable. The x parameter is passed to the x-axis of the plot, and the data is the values present in the column x of the 'pokemon' data. The 'kde' parameter is set to 'True' which means that the plot will show a kernel density estimate for the data.

Using the distribution function, we could suggest that anything greater than 150 HP be suggested to lie outside the existing distribution and thus be classified as an outlier. An alternative method of detecting outliers is through a box plot, which shows the mean, lower quartile, and upper quartile and extends to points that lie within 1.5 interquartile ranges of the lower and upper quartiles. Observations that fall outside this range are displayed independently and are suggested to be potential outliers in the data: Create the boxplot

[Out]

```
[In]
ax = sns.boxplot(x =
pokemon["x"])

plt.show()
```

→

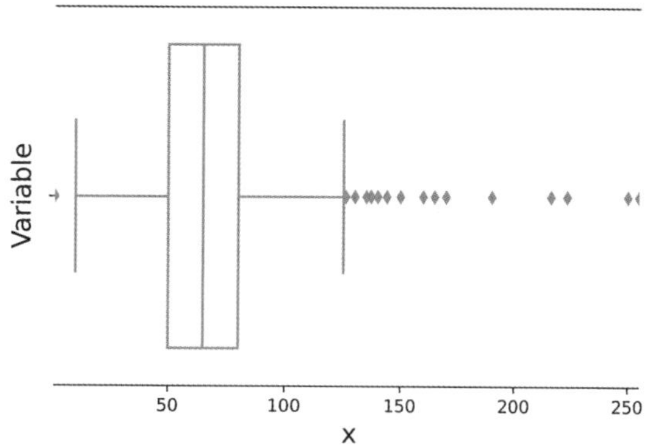

Figure 1.3.11. Outlier detection with box plot

Here, the 'boxplot()' function returns an axes object (Figure 1.3.11), which is assigned to variable 'ax'. The box is created from the first quartile (Q1) to the third quartile (Q3) of the data, which contains the majority of the observations. The box represents the median value of the data, the 'whisker's extend out to the minimum and maximum values of the data, and the horizontal line inside the outliers are plotted as individual points outside the whiskers.

This more clearly suggests that there are several outliers in the data, including ones at both the upper and lower ends of the scale, whereas examining the original distribution suggested outliers at only the upper end of the scale.

We can extract these points directly from the table or by calculating the interquartile range and then extracting the points above a certain value. Another method follows the logic of the standard deviation. We can examine the standard deviation of the data from the mean, with extreme values being those with an absolute Z score greater than 3, as follows. This score is calculated as

$$Z = \frac{x - \mu}{\sigma}$$

where outliers greater than 3 can be extracted as follows:

```
[In]
pokemon_x_outliers = pokemon[abs(pokemon["x_z"]) >= 3]
```

This can be implemented as follows (calculate the Z score): The code contained the 'pokemon' data frame and creates a distribution plot of the standardized data using the 'seaborn' library. The 'kde' parameter specifies whether a kernel density estimate plot should be drawn which is useful for visualizing the distribution of the data (Figure 1.3.12).

[In]

```
pokemon["x_z"] = (pokemon["x"]
-
pokemon["x"].mean())/pokemon["
x"].std()

#show the distribution plot
sns.displot(pokemon["x_z"],
            kde = True)


plt.show()
```

→

[Out]

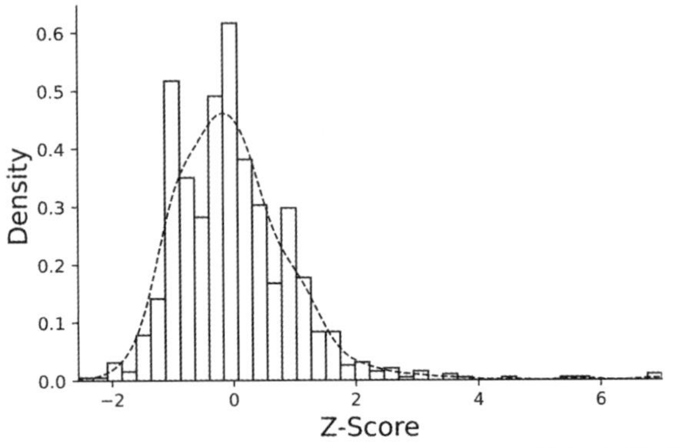

Figure 1.3.12. Outlier detection with distribution

If you can spot outliers and smooth them out in your descriptive statistics (and any other data analyses), you will have better outcomes. There are numerous examples of analysis error being reduced by a couple of percent just by outlier cleaning. The bigger the data set, the more critical it is to automate this detection and cleaning.

Normalization and Standardization with the Outliers

A very common question in regard to systems analysis, especially for those who have just started their research journey, might be, what is normalization and standardization? We have learned these concepts previously, but we still do not know for sure: When should we use them? Quite often, we feel the need to transform data from an original range to another scaling like [0, 1], [-1, 1]. If not, then we may find bias in our data. To control this issue, we use normalization and standardization, the most popular feature-scaling techniques.

[Table 1.3.1] Summary of Normalization and Standardization

	Normalization	Standardization
Transformation	Brings the data into a common range [0, 1], [−1, 1] etc.	Brings the data to mean 0 and unit variance
Pros	Useful when we do not know about the distribution and there are no outliers	Useful when the data set is normally distributed
Cons	Sensitive to outliers	Sensitive to outliers

First, let us consider a popular normalization method. We scale our data in the range [0,1]. Each value is subtracted with the min value and divided by the difference between max and min.

$$x_i' = x_i - x_{min}/x_{max} - x_{min}$$

When we do not want to miss any of our data points, it must be the best way to transform, keeping the original distribution of the data. However, this method is sensitive to outliers because it uses the minimum and maximum values. Next, let us move on to standardization. This is a popular technique in data scaling because we should take care of the normal as well as near-normal distribution of the data. Simply, we can calculate the mean and standard deviation to be used during the transformation of the data set.

$$x_i' = x_i - x_{mean}/x_{std}$$

If the features are normally distributed, then this method will be the best for your data set (only when it satisfies a normal distribution). However, outliers influence this transformation as well because the mean is sensitive to outliers. Thus, if there are outliers in the data, then you can remove those outliers and use either scaling technique. As an alternative, instead of using the minimum and maximum values, we can use the interquartile range (IQR) method, which is robust to outliers. The calculation formula is shown below.

$$x_i' = x_i - \frac{x_{med}}{IQR}, \qquad IQR = x_{75} - x_{25}$$

As we can notice, this contains outliers as it is less sensitive to outliers than the previous normalization.

Remarks: Following up on the descriptive portion before, bell-shaped curve is a fundamental idea in statistics and are frequently used to simulate a variety of real-world occurrences. The metrics of central tendency for the bell-curves (or normal distribution) are mean, median, and mode, with mean being the most widely applied. We may learn about the spread of the data via measures of dispersion, such as standard deviation, which can be useful for drawing conclusions and making predictions. One crucial characteristic of the bell curves that makes them so helpful in statistical inference is the central limit theorem, which asserts that the sample mean of a large number of independent and identically distributed random variables will tend to be normally distributed. Finally, outliers, or data points that depart greatly from the rest of the data, can drastically affect the parameters of the bell curve and should be carefully scrutinized to see if they are real data points or mistakes. In general, a thorough grasp of distributions' properties and features is necessary for statistical analysis and inference in a wide range of disciplines.

Inferential Statistics

A Guide to the Concepts of Inferential Statistics, Sampling for Inference, Hypothesis Testing

Descriptive statistics describe data (e.g., a chart or graph), whereas inferential statistics allow you to make predictions (inferences) from those data (Agresti & Finlay, 2009). Inferential statistics can be defined as a field of statistics that uses analytical tools for drawing conclusions about a population by examining random samples. Thus, we use data from samples to make generalizations about a population. For example, you might stand in a mall and ask a sample of 100 people whether they like shopping at department stores.

1. A Guide to Concepts of Inferential Statistics

Inferential statistics makes use of various methods to draw inferences about population data from sample data. It may involve the t test, analysis of variance (ANOVA), the chi-square (χ^2) test, regression, factor analysis, and so on. The key terms related to this concept follow:

- Population: the totality of all elements, subjects, or members
- Sample: a subset of the population under study
- Parameter: a characteristic or measure of a population

What is the point of inference? Inference involves making conclusions about a larger population based on a sample of observed data.

- Parameter estimation: using a statistic from your sample data (e.g., the sample mean) to say something about a population parameter (e.g., the population mean).
- Hypothesis testing: using sample data to answer research questions. For example, you might be interested in knowing whether a new cancer drug is effective or whether eating breakfast helps children perform better in school.

You could make a bar chart of 'yes' and 'no' answers (descriptive statistics), or you could use your research to reason that around 75 – 80% of the population (all shoppers in all malls) like shopping at department stores (inferential statistics). For example,

- According to a survey, 51% of people favor the Seoul party. The standard error for this statistic is ± 3% with 95% confidence (48 – 54%).

We may want to know the country's opinion of the president. Obviously, you cannot ask every single citizen what he or she thinks; instead, you must make an inference about the larger population using data from a sample of it.

+ **Preliminary Checkup for Inferential Statistics:** Let us estimate mean and variance (standard deviation). The sample mean $\bar{x}$ is the best estimate of the population mean μ. Sample variance (s^2) that uses the sample size (n) as the denominator underestimates the population variance (σ^2). When $n - 1$ is used instead, the sample variance is unbiased.

$$\sigma^2 = \frac{\Sigma(X-\mu)^2}{n}, \qquad s^2 = \frac{\Sigma(X-\bar{X})^2}{n-1}$$

The sample mean is different from the population mean. This difference is called the *sampling error of the mean* ($\mu \neq \bar{X}$). Given any distribution of a variable X with mean μ and standard deviation σ, the sample mean X based on random samples of size n is distributed $N(\mu, \sigma^2/n)$ as n increases.

$$\sigma_{\bar{X}} = \frac{\sigma}{\sqrt{n}}, \qquad s_{\bar{X}} = \frac{s}{\sqrt{n}}$$

The sampling distribution of the mean has a mean equal to the population mean, which reflects its shape, location, and variability. The central limit theorem describes the sampling distribution of the means (see 1.3 for more detail).

+ **Confidence Interval of the Mean:** The confidence interval is a range of values that is likely to include a population value with a certain degree of confidence. It is often expressed as a percentage whereby a population mean lies between an upper and lower interval.

Commonly used confidence levels in research are 95% and 99%. If a 95% confidence interval is used, the researcher wants an interval for the scale of measurement that includes 95% of the area of the sampling distribution. A 95% confidence interval means that 95% of the sample mean lies within $\pm 1.96\sigma_{\bar{X}}$ units of the population mean.

$$(\mu - 1.96\sigma_{\bar{X}}) \leq \bar{X} \leq (\mu + 1.96\sigma_{\bar{X}})$$

The equation states that, with a certain level of confidence, the true population mean μ will be in the interval $[(\mu - 1.96\sigma_{\bar{X}}), (\mu + 1.96\sigma_{\bar{X}})]$, where $\bar{X}$ is the sample mean, $\sigma_{\bar{X}}$ is the sample standard deviation, and 1.96 is the critical value of a standard normal distribution corresponding to a confidence level of 95%. The term $1.96\sigma_{\bar{X}}$ is known as the margin of error and this equation states that the interval estimates of population mean, calculated using sample mean and standard deviation, with a confidence level of 95%, It where n is the sample size, should be noted that this is only true under certain assumptions. Specifically, this interval estimate assumes that the data is normally distributed and that the sample size is large enough (typically $n > 30$) for the central limit theorem to apply. Usually, we turn the previous statement around and state that based on our sample mean and standard deviation, we are 95% confident that the population mean lies within the following interval.

$$(\bar{X} - 1.96s_{\bar{X}}) \le \bar{X} \le (\bar{X} + 1.96s_{\bar{X}})$$

This is all based on the standard normal distribution. For example, the 95% confidence interval is 1.960, and here we see that the range from −1.96 to +1.96 includes 95% of all values (Figure 1.4.1, left graph). Applying this to our sample reveals the distribution shown in the right graph (Figure 1.4.1).

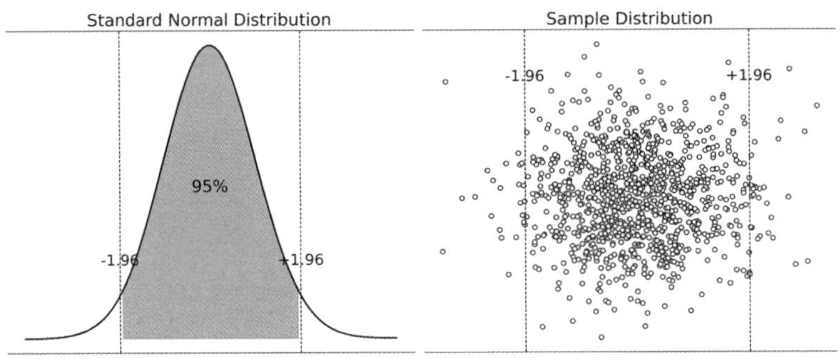

Figure 1.4.1. Understanding 95% confidence intervals with standard normal distribution. The interval from −1.96 to +1.96 SD includes 95% of all values.

The confidence interval is based on the mean and standard deviation. Its formula is

$$X = Zs\sqrt{n}$$

where X is the mean, Z is the Z value from the table below, s is the standard deviation, and n is the number of observations (Table 1.4.1).

[Table 1.4.1] Confidence Interval with the *Z* Score

Confidence Interval	80%	85%	90%	95%	99%	99.5%	99.9%
Z	1.282	1.440	1.645	1.960	2.576	2.807	3.291

+ **Numerical Practice with the Confidence Interval:** For example, Confidence Interval, the degree of conservativeness (1 – 100) was measured among students (Table 1.4.2).

[Table 1.4.2] Sample Data Set

Sample	N	$\bar{X}$	S	$s_{\bar{X}}$	95% *CI*
Sample 1	4	45	10	5	$35 \le \mu \le 55$
Sample 2	25	45	10	2	$41 \le \mu \le 49$
Sample 3	100	45	10	1	$43 \le \mu \le 47$

Let us assume there are nine students who took a research method course last year were randomly selected. The mean and standard deviation of the final scores of these nine students were 80 and 15. What is the 95% confidence interval for the population mean?

$$(\bar{X} - 1.96s_{\bar{X}}) \leq \bar{X} \leq (\bar{X} + 1.96s_{\bar{X}}), \qquad s_{\bar{X}} = \frac{s}{\sqrt{n}}$$

Answer: $N = 9, \bar{X} = 80, SD = 15, CI = 95\%$

$$\left(80 - 1.96 * \frac{15}{\sqrt{9}}\right) \leq \bar{X} \leq \left(80 + 1.96 * \frac{15}{\sqrt{9}}\right)$$

$$(80 - 1.96 * 5) \leq \bar{X} \leq (80 + 1.96 * 5)$$

$$(80 - 9.8) \leq \bar{X} \leq (80 + 9.8)$$

$$70.2 \leq \bar{X} \leq 89.8$$

The mean lies between 70.2 and 89.8.

Numerical Practice with the Sample Mean and Probability

Question: The body weight of weightlifters is distributed (100 kg, 10^2). If we select one weightlifter from this group, what is the probability that the weight of this athlete is greater than 105 kg?

Answer: $\bar{X} = 100$, SD = 10, X = 105

$$Z = \frac{(X - \bar{X})}{s}, \qquad Z = \frac{(105 - 100)}{10} = 0.5 = 30.85\%$$

If we select 16 weightlifters from the group, what is the probability that the mean weight of these 16 athletes will be greater than 105?

$$Z = \frac{(X - \bar{X})}{\frac{s}{\sqrt{n}}} \quad Z = \frac{(105 - 100)}{\frac{10}{\sqrt{16}}} = \frac{5.0}{2.5} = 2.0 = 2.28\%$$

2. Sampling for Inference

Where does probability come in? Although inference is key, we are only able to make inferences from data that are generated by models. Probability is used to generate data from a given model. Consider the following:

Observed/sample data → model → probability/data generation → inference

We are going to create a model that allows us to generate random data, all based on the idea of a simple coin flip (coin flips are classic examples that help break down distributions). What is the likelihood that any given flip returns heads or tails? Answer is simply almost half/half.

Let us create an experiment to simulate a coin flip. The first parameter is the number of pulls or runs. Run the experiment as many times as you want, and you will see that approximately 50% of the time the function returns 1 and the other 50% of the time it returns 0 (in this example, let us call everything equal to or above .5 heads).

If you run the experiment 10 times, you may have the following results:

Output: [1 1 1 1 0 0 1 1 1 1]

You get heads 80% of the time. If you run the experiment repeatedly, you will begin to approximate heads 50% of the time.

Output: [0 0 0 0 1 1 0 1 1 1 1 0 0 1 0 0 0 0 0 1 1 0 1 1 1 0 0 0 1 1 0 0 1 1 1 1 0 1 1 0 1 0 0 0 1 1 1 0 1 0]

You should also run it with greater and greater numbers of flips.

```
[In]
flips = (10000)
mean(flips == 1)
print(mean)
```

→

```
[Out]
0.5006
```

As we implemented, the code above generates a list of 10,000 flips of a coin and calculates the mean of the number of times the coin landed on heads. Almost 50% of the time you get heads. Now let us change it up a bit: Do 1,000,000 draws of our experiment, but this time each test will constitute 10 flips with a 50% likelihood of flipping five heads.

What you see constitutes five flips approximately 25% of the time. This metric is known as the density of the binomial at that point. Simulation helps us answer questions about distribution and its behavior. There is a lot to understand, but it is hoped that this has proven useful as a primer on distribution for basic statistical inference.

3. Hypothesis Testing

A hypothesis is an assumption, an idea that is proposed for the sake of argument so that it can be tested to see whether it might be right. In the scientific method, the hypothesis is constructed before any applicable research has been done.

You ask a question, read up on what has been studied before, and then form a hypothesis. A hypothesis is a statement that predicts a particular result (Mahon & Caramazza, 2008).

+ Null and Alternative Hypotheses: Make a hypothesis and examine whether it is true using statistics. Because it is challenging to prove that a particular hypothesis is true, we usually prove that a reverse hypothesis (null hypothesis) is false. Thus, the only alternative (the original hypothesis) is true.

- Null hypothesis (or statistical hypothesis): a hypothesis of no difference or no relationship

$$H_0 = \mu_E = \mu_C$$

The H_0 is commonly used to indicate the Null hypothesis. In this case, the null hypothesis is that there is no significant difference between specified populations, any observed difference being due to sampling or experimental error.

- Alternative hypothesis: a hypothesis of a difference or relationship

$$H_1 = \mu_E \neq \mu_C$$

The H_1 is also commonly used to indicate the alternative hypothesis. In this case, the alternative hypothesis is that there is a significant difference between the mean of population μ_E and the mean of population μ_C. It states that the two-population means are not equal $\mu_E \neq \mu_C$.

+ Type I Error, Type II Error, and Statistical Power: Hypothesis testing consists of examining null and alternative hypotheses, typically denoted H_0 and H_1, respectively (Raftery et al., 1995).

The null hypothesis is a statement that no difference or effect is expected. Thus, if H_0 is rejected, then you can say that there is no significant difference in the tested data. H_1 is the complement of H_0; that is, if H_1 is accepted, then you can say that there is a significant difference in the tested data.

Statistical power: rejecting H_0 when H_0 is false (probability)

$$Power = 1 - \beta$$

		Truth in the world	
		H_0 is true	H_0 is false
Conclusion	Reject H_0	Type I error $P = \alpha$	Correct conclusion $P = 1 - \beta$
	Fail to reject H_0	Correct conclusion $Power = 1 - \alpha$	Type II error $P = \beta$

Accepting or rejecting the null and alternative hypotheses results in errors. There are two types of errors: type I and type II. Type I error occurs when a researcher rejects the null hypothesis but that hypothesis is true. The probability of type I error is called the significance level. Type II error occurs when a researcher fails to reject the null hypothesis but that hypothesis is false. In practice, type II error is more serious than type I error (i.e., research involving drugs).

Because of type II error, statistical power was created. Statistical power is the probability $(1-\beta)$ of rejecting the null hypothesis when it is false, and this null hypothesis should be rejected to avoid type II error.

These two types of error occur when we do not predict accurately and some of the data are misanalysed.

- Type I error: This error occurs when the actual value is 0 but it is predicted to be 1. In statistical terms, this error reveals the level of significance (i.e., alpha). The value of alpha is chosen by us, so the error depends on what value we choose.
- Type II error: This error occurs when the actual value is 1 but it is predicted to be 0. In statistical terms, this error is called *beta*, and it is mainly dependent on the sample size (n) and variance (sigma).

Therefore, we need to keep the statistical power high, as the higher the statistical power, the fewer type II errors we can expect. An analysis of statistical power (i.e., a power analysis) can be done either before or after data have been collected. Statistical power usually depends on the following:

- Significance level: α
- Level of confidence = $1-\alpha$ (sometimes use percentage)
- Type I error: rejecting H_0 when H_0 is true (probability) → we can reduce type I error by lowering α

The error of rejecting a true hypothesis is type I or alpha (α) error. This is because until the hypothesis is rejected, the probability of having made an error equals the significance (α) level.

- Why do we not use $\alpha = .001$?
- Type II error (β): failing to reject H_0 when H_0 is false (probability)

The error of failing to reject a false hypothesis is type II or beta (β) error. In any one statistical test, there is a possibility of making only one type because the researcher either rejects or fails to reject the hypothesis. In general, reducing the risk of one type of error increases the risk of the other type of error.

To minimize these errors, we must maintain a high statistical power, which depends on factors such as the chosen significance level and sample size. Balancing the risks of Type I and Type II errors is essential in ensuring accurate and reliable statistical analysis within any academic research.

Brief Example of the Hypothesis and Test

Hypothesis: The weight of Austrian weightlifters is different from that of Korean weightlifters.

- Null hypothesis (H_0): The weight of Austrian weightlifters is the same as that of Korean weightlifters.
- Alternative hypothesis (H_1): The weight of Austrian weightlifters is different from that of Korean weightlifters.

From 16 randomly selected Austrian weightlifters, we obtain a mean of 105 and a standard deviation of 10. If the null hypothesis is true, what is the probability that the mean of the 16 athletes is equal to or greater than 105?

$$Z = \frac{(X - \bar{X})}{\frac{s}{\sqrt{n}}}, \qquad Z = \frac{(105 - 100)}{\frac{10}{\sqrt{16}}} = \frac{5.0}{2.5} = 2.0 = 2.28\%$$

Answer: The probability is 2 (2.28%).

Remarks: As we have seen, descriptive statistics includes summarizing and describing data, but inferential statistics is drawing inferences and conclusions about a population based on sample data. A potent method for extracting conclusions and inferences about a population from a sample of data is inferential statistics. When estimating population parameters, such as the mean or percentage, sampling for inference is the process of choosing a representative subset of the population. Inferential statistics, where we utilize sample data to evaluate a claim about a population parameter, must include hypothesis testing. An alternative hypothesis, or the hypothesis we want to test, must be specified together with a null hypothesis, which is the presumption that there is no difference between the population parameter and the sample statistic. The chance of receiving the observed sample data under the null hypothesis is then calculated using statistical tests. We reject the null hypothesis in favor of the alternative hypothesis if the likelihood is sufficiently low. To make data-driven judgments and to infer meaningful conclusions from empirical information, inferential statistics is widely utilized.

Supplemental Information 1.1: Understanding Research

Research is an organized method of finding answers to questions. It is systematic because it is a process broken up into clear steps that lead to conclusions (Kothari, 2004). If government, business, institutions, labor, organizations, and society in general are to function efficiently and effectively, the decisions they make must be based on valid and reliable information and thorough analysis. The search for this information is referred to as the research process. There may be an existing body of evidence (prior research, studies, etc.) we can make use of. If there is not, there is a need for research.

What is research: Research is a way of finding new knowledge-seeking satisfactory explanations for whatever needs explaining (Blaxter et al., 2010). Research can be systematic, controlled, empirical, naturalistic, and so on. It is guided by theory and hypotheses about the presumed relations among phenomena (gathering information without interpretation or mere copying of information is not research).

The ultimate goals of research are to describe and predict measurable and testable data, gradually adding to the accumulation of human knowledge.

1. Types of Research

+ Classified by Purpose

Basic: Basic research involves collecting data to enhance knowledge (Powell, 2006). The primary motivation is knowledge expansion. This is noncommercial research that does not facilitate the creation or invention of anything. One example is performing an experiment to determine a simple fact.

Applied: Applied research focuses on analyzing and solving real-world problems (Hedrick et al., 1993). This type of research helps solve practical problems using scientific methods (cf. action research). Studies play an essential role in addressing issues that impact the overall well-being of humans. One example is finding a specific cure for a disease.

+ Classified by Data Type

Quantitative: Quantitative research is a structured way of collecting data and analyzing them to make conclusions. Unlike qualitative research, quantitative research uses computational and statistical processes to collect and analyze data (Gelo et al., 2008). Quantitative research is all about numbers. It involves a larger population: more people mean more data. With more data to analyze, we can obtain more accurate results. This type of research uses closed-ended questions because researchers are typically looking to gather statistical data.

Qualitative: Qualitative research is a process of inquiry. It helps create an in-depth understanding of problems or issues in their natural settings. This nonstatistical research is heavily dependent on the experience of the researchers and the questions used to probe the sample. The sample size is usually restricted to 6 ~ 10 people. Open-ended questions are asked in a manner that encourages answers that lead to another question or group of questions. The purpose of asking open questions is to gather as much information as possible from the sample.

+ Classified by Process (Research Method)

Descriptive: Descriptive research focuses on expanding knowledge of current issues by collecting data (Grove et al., 2012). Descriptive studies are used to describe the behavior of a sample population. In a descriptive study, only one variable is required. The three main purposes of descriptive research are to describe, explain, and validate the findings. For example, a study may be conducted to determine whether top-level managers in the 21st century have a moral right to receive huge sums of money from their companies.

Experimental: Experimental research adheres strictly to the scientific design (Singh, 2021). It includes a hypothesis; a variable that can be manipulated by the researcher; and variables that can be measured, calculated, and compared. Most important, experimental research is conducted in a controlled environment. The researcher collects data, and the results either support or reject the hypothesis. This method of research is also referred to as hypothesis testing or deductive research.

Quasi-experimental: Quasi-experimental research is similar to traditional experimental designs or randomized controlled trials but specifically lacks the element of random assignment to a treatment or control condition (Thyer, 2012). Instead, quasi-experimental designs typically allow the researcher to control assignment to the treatment condition using some criterion other than random assignment (e.g., an eligibility cutoff).

Ex post facto/causal-comparative: Whereas causal research is experimental research designed to compare groups in a more natural way, causal-comparative research attempts to identify causes or consequences of differences in a nonexperimental setting (Schenker & Rumrill, 2004). These differences already exist, and their impact on the outcome is determined by comparing groups. Causal-comparative designs can have different focuses: (a) exploring effects, (b) exploring causes, or (c) exploring consequences.

Case/field study: A case study is a type of qualitative research in which one or multiple cases with a specific context are investigated (Feagin et al., 1991). The case study focuses on organizations, events, and activities within their particular context, aiming to link theory to practice. Field research involves studying phenomena in their natural setting. Researchers who advocate for field research contend that natural behavior cannot be captured from the outside. Field research is particularly useful when researchers want to study the natural habits of their targets from the point of view of those who are being observed.

Historical: Historical researchers study the meaning of past events in an attempt to interpret the facts and explain the cause of the events and their effect on present circumstances (Amenta, 2003). In doing so, they rely on primary historical data (direct accounts of events, archival data, official documents, personal records, and records of eyewitnesses) and less frequently on secondary historical data (information from persons who did not witness the event; information from textbooks, newspapers, encyclopedias).

Qualitative: Qualitative research focuses on understanding a research query as a humanistic or idealistic approach (Silverman, 2020). Although quantitative methods are more reliable because they are based on data and techniques that can be studied objectively and replicated by other researchers, qualitative methods are used to understand people's beliefs, experiences, attitudes, behavior, and interactions. Quantitative methods generate nonnumerical data.

2. Steps in the Traditional Research Process

The following steps outline a simple and effective strategy for any research in general (Anfara et al., 2002). Depending on your experience with a particular subject and the challenges you encounter along the way, feel free to rearrange these steps.

+ **Problem Formulation:** Formulating the problem focuses on uncovering the nature and boundaries of a situation or question. In defining the issue or problem, the researcher should consider the purpose of the study, the relevant background information, what information is needed, and how it will be used in decision making.

+ **Literature Review:** A literature review is a survey of scholarly research on a specific topic. It provides an overview of current knowledge, allowing you to identify relevant theories, methods, and gaps in the existing research. Conducting a literature review involves collecting, evaluating, and analyzing publications (such as books and journal articles) that relate to your research question.

+ **Problem Specification, Hypothesis Formulation:** A problem specification is a clear and precise statement of questions or issues that are to be investigated. According to Sekaran and Bougie (2019), a research problem is 'any situation where a gap exists between actual and desired ideal estate.' A hypothesis is an assumption about the relationship between the variables. It is the expected outcome of the research based on existing research. It shows the direction of the research.

+ **Identification of the Independent and Dependent Variables:** There are two critical variables in every experiment: the independent variable and the dependent variable.

- Independent variable: what the scientist changes or what changes on its own
- Dependent variable: what is being studied or measured

+ **Selection of a Research Design:** The research design or study design is the most crucial aspect of the research. It encompasses the plan and procedures that will enable you to achieve your goal. A research design gives you direction and helps you avoid taking the wrong path. It defines the study type, hypothesis, objectives, research methodology, and data collection as well as the analysis techniques.

+ **Population Sampling:** Population sampling is the process of selecting a subset of subjects who are representative of the entire population. The sample must be of sufficient size to warrant statistical analysis. Population sampling is done to save time, money, and effort while conducting the research.

+ **Data Collection:** Data collection is the process of gathering quantitative and qualitative information on specific variables to evaluate outcomes or glean actionable insights. Good data collection requires a transparent process to ensure that the data collected are clean, consistent, and reliable.

+ **Data Analysis:** Data analysis is the process of inspecting, cleaning, transforming, and modeling data to discover useful information, inform conclusions, and support decision making. There are many approaches to data analysis, and different techniques are used in the business, science, and social science domains.

+ **Interpretation (Drawing a Conclusion):** Drawing conclusions involves using information that is implied or inferred to make meaning out of what is not clearly stated. Writers give readers hints or clues that help them read between the lines to gather all of the information.

+ **Generalization (Expanding Theory):** A generalization is a broad statement about a group of people or things. It states something they have in common. Scientific generalization is based on evidence of a causal connection (e.g., the generalization 'All crows are black' is based on the causal relationship between being a crow and being black). Empirical generalization is based on experience (e.g., 'All red mangoes are sweet').

3. The Role of Theory in Research

A theory provides significant guidelines for conducting research by pointing to areas that are most likely to be fruitful (i.e., areas in which meaningful relationships among variables are likely to be found). If the researcher studies variables that are not related, the research will be sterile no matter how meticulous the observations and inferences (Venable, 2006).

A theory is a systematic and general explanation of a particular group of phenomena → proved theory → law. It provides the researcher with a specific viewpoint, which goes a long way toward helping him or her inquire into the relationships between particular variables selected from among an almost infinite array.

The importance of theory in scientific research is crucial because it serves as the basis for generalizing research findings and serves as the beginning point for selecting a research challenge. Research that tries to analyze and build upon the flawed idea starts with theory as its foundation. According to Kuhn et al. (1977) in 'Objectivity, Value Judgment, and Theory Choice,' a good theory should have a number of essential qualities. These qualities include being concise, thorough, logical, and clear. They also include having empirical backing and being falsifiable. The reliability, validity, and generalizability of research findings are improved by having a strong theoretical framework.

+ **Accurate**: The term 'accuracy' describes the consistency between the outcomes of previous experiments and observations made within a theory's area and the conclusions that may be drawn from them. A good theory should show precise predictions and outcomes that match what has been observed.

+ **Consistent**: The term 'consistency' refers to the theory's internal coherence as well as its compatibility with other widely accepted theories that address the same element of nature. Consistent theories do not contradict one another and work well with other pertinent hypotheses.

+ **Broad**: The term 'broad' describes the extent to which the implications of a theory extend beyond the specific facts, rules, or sub theories it was initially intended to explain. A wide theory offers a thorough knowledge of the phenomenon under consideration and has far-reaching ramifications.

+ **Simple**: A theory's 'simplicity' refers to its capacity to organize and make sense of complicated occurrences. A straightforward theory offers a cogent justification for singular findings and aids in making sense of baffling evidence.

+ **Fruitful**: The term 'fruitfulness' describes a theory's ability to provide fresh insights and discoveries in the field of inquiry. A useful hypothesis motivates additional research and unearths previously unrecognized new occurrences or linkages. A good theory need to be wide, simple, and productive, as well as accurate and consistent.

4. Validity

Validity is an indication of how sound your research is. It applies to both the design and the methods of your analysis (Heale & Twycross, 2015). When your data collection is valid, then your findings truly represent the phenomenon you claim to measure. Valid claims are substantial claims.

Validity is a main concern in research. Any research can be affected by different kinds of factors that, while extraneous to the concerns of the research, can invalidate the findings. Controlling all possible factors that threaten the validity of the research is the primary responsibility of every good researcher.

+ **Internal Validity** is affected by flaws within the study itself, such as not controlling the major variables (a design problem) or problems with the research instrument (a data collection problem). Findings can be said to be internally invalid because they may have been affected by factors other than those thought to have caused them, or because the interpretation of the data by the researcher is not clearly supportable. Here are some factors that affect internal validity:

- Subject variability
- The size of the population
- The time given for the data collection or experimental treatment
- History
- Attrition
- Maturation
- Instrument/task sensitivity

+ **External Validity** is the extent to which you can generalize your findings to a larger group or other contexts. If your research lacks external validity, the results cannot be applied to contexts other than the one in which you carried out your research. For example, if your subjects are all males from one ethnic group, your findings might not apply to females or other ethnic groups. Or if you conducted your research in a highly controlled laboratory environment, your conclusions may not faithfully represent what might happen in the real world.

Findings can be said to be externally invalid because they cannot be extended or applied to contexts outside those in which the research took place. Here are seven essential factors that affect external validity:

- Characteristics of the population (subjects)
- Interaction between subject selection and research
- Descriptive explicitness of the independent variable
- Effects of the research environment
- Researcher or experimenter effects
- The methodology used for data collection
- Effects of time

5. Reliability

In statistics, reliability is the overall consistency of a measure. It produces similar results under consistent conditions (replicability of research): If an independent researcher can replicate a study and come to the same conclusion, then the reliability of the study is high. For research to be reliable, it must be valid. If a study is not reliable, it cannot be valid.

- If research is valid, it is reliable (O)
- If research is reliable, it is valid (X)

Validity and reliability are two interrelated but completely different concepts (Carmines & Zeller, 1979).

- Validity: Does the test measure what it is supposed to?
- Reliability: Does the test consistently give the same result under the same conditions?

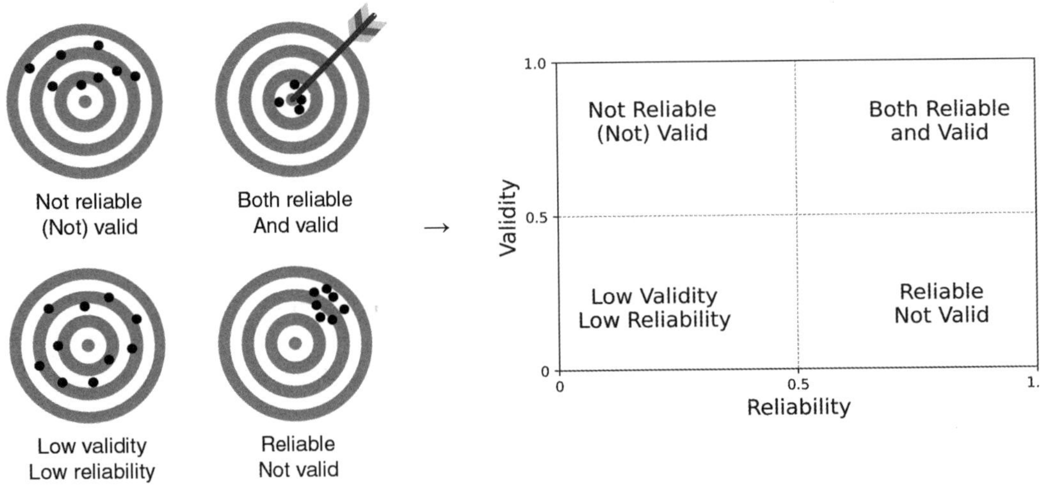

Figure S1.1.1. Schematic drawing of the validity and reliability

Reliability is an essential component of validity but on its own is not a sufficient measure of validity (Figure S1.1.1). A test can be reliable but not valid, but a test cannot be valid yet unreliable. An extremely unreliable test is mostly not valid either; it merely is not measuring what it is meant to.

6. The Research Problem and Hypothesis

Research problems can come from a higher authority in the field, discussion, practical issues, experience, or a literature review. Factors to consider in identifying a research problem are significance, originality, research ability, and feasibility (Buelow, 2006).

+ Examples of Research Problems

- The effect of physical activity on cancer prevention
- Work environment and injury among nurses
- The effect of social activity on the self-esteem of a specific population
- The prevalence of overweight and obesity in preschool children
- Effective means of increasing social activity in elderly

A hypothesis is a conjecture (proposition) about the relationship between two or more variables. There are substantive (research) hypotheses and statistical hypotheses.

The characteristics of a (reasonable) hypothesis are as follows. A hypothesis should reflect the theory behind it and provide direction for the research. It should contain a (directional) expectation(s) about the results of the research.

+ **Examples of Research Hypotheses**

- Physical activity prevents cancer (or people who participate in physical activity are less likely to have the disease).
- A good work environment reduces the risk for injury among nurses.
- Participating in a team enhances self-esteem in the wheelchair-bound population.
- An increasing number of preschool children are obese.
- Marathoners have greater cardiovascular capacity than soccer players.

+ **Operational Definition:** Connecting (by defining) research variable(s) with actual variable(s) that are used in research; selecting a certain variable(s) that is to be included in research. For example:

- Cardiovascular fitness: VO_2 max
- Physical activity: the amount of exercise for a week
- Work environment: the quality of an instrument or building
- Self-esteem: self-esteem measured by a questionnaire

7. Literature Reviews and Research Designs (Quantitative)

+ **Literature Reviews:** The main purpose of a literature review is to update the most recent knowledge in an area, to help identify and specify a research problem and hypothesis. Also, it can be used to examine what has (and has not) been done in terms of research design. A variety of sources can be reviewed, including journal articles, books, theses and dissertations, websites, and others. When reviewing the literature, one needs to be thorough and critical, asking 'Is it logical? Valid? Well written?'

+ **Quantitative Research Designs:** A research design includes procedures for creating and measuring variable(s), collecting data, and analyzing data (Fischer et al., 2014). Things to consider in research design are bias, extraneous variables, and appropriateness of the statistical models used. To control variance of extraneous/confounding variables, we use randomization and statistical control, and can also build factors as an independent variable(s) and hold factors constant (see the Supplement 1.2 for more detail).

Statistical Model, Part 1

Identification of the *t* Test, Independent *t* Test, Independent *t* Test Unequal *n* Two-Tailed, Paired *t* Test, Factors Affecting the *t* Test

The *t* Test, a widely utilized statistical method, is employed to assess the significance of differences between groups (Kruschke, 2013). Essentially, it determines whether the observed disparities in means or averages could have occurred by chance alone. In simpler terms, the *t* Test is a statistical approach that compares two means, either between groups or in repeated measures.

1. Identification of the *t* Test

The *t* Test was initially developed by William Sealy Gosset (1908), a chemist employed at the Guinness brewery in Dublin, Ireland. Gosset's expertise in both biochemistry and statistics was instrumental in the development of the *t* Test, which he devised as a cost-effective means to evaluate and monitor the quality of stout within the context of the brewery's industrial processes.

+ ***t* Distribution:** *t* distributions comprise a family of distributions that are the sampling distributions for many statistics. Each *t* distribution is determined by a degree of freedom. Below, see a simple example;

[In]

```
x = np.arange(-4,4,0.1)
firstT = stats.t.pdf(x,1)
secondT = stats.t.pdf(x,2)
thirdT = stats.t.pdf(x,5)
fourthT = stats.t.pdf(x,30)
plt.plot(x, firstT, x, secondT, x,
thirdT, x, fourthT)
plt.legend({'t with df1':firstT,'t
with df2':secondT,'t with df3':thirdT,
'standard':fourthT})
```

→

[Out]

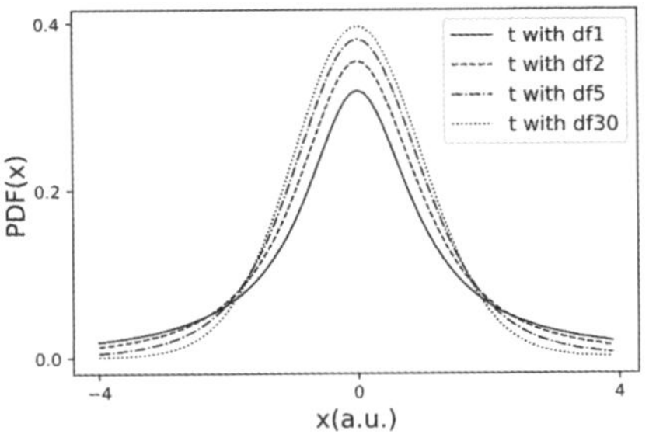

Figure 1.5.1. *t* distribution, used for the *t* Test

This is a continuous distribution characterized by a single parameter called the degrees of freedom. Here, we used 'stats.t.pdf' function, which is part of the 'scypy.stats' library, to create the *t* distribution and plots four probability density for the Student's *t* distribution (Figure 1.5.1).

+ **Types of *t* Test:** As previously mentioned, the *t* Test is a statistical method employed to compare the means of two groups. There are three distinct types of tests: one-sample, independent-samples, and paired-samples *t* Tests.

The one-sample *t* Test is used to compare a sample mean with a known population mean or another pertinent fixed value. The independent-samples *t* Test is employed to compare the means of two separate groups. In contrast, the paired-samples *t* Test is utilized when comparing the means of two repeated measurements for the same participants. For example, in a dual-treatment randomized block design, this test might be used to compare scores obtained from the same subjects at different time points or between matched samples. The *t* Test effectively highlights the difference between means and medians.

+ **Calculating the *t* Statistics:** When the difference between two means is divided by the standard deviation of the mean difference ($s_{\bar{X}1-\bar{X}2}$), the resulting value is *t*. This *t* follows a normal distribution only when the sample size (n) is very large. For each sample size or degrees of freedom (df), there is a unique *t* distribution.

- Independent *t* test: $df = (n_1 - 1) + (n_2 - 1) = n_1 + n_2 - 2$
- Paired (dependent) *t* test: $df = n - 1$

$$Test\ statistic = \frac{score - mean}{standard\ deviation\ of\ the\ score}$$

A large *t* score tells you that the groups are different. A small *t* score tells you that the groups are similar.

$$Z_x = \frac{x - \mu}{\sigma_x}, \qquad Z_{\bar{X}} = \frac{\bar{X} - \mu}{\sigma_{\bar{X}}} = \frac{\bar{X} - \mu}{\sigma/\sqrt{n}}$$

Here, the equation represents how many standard deviations an observation or data point is from the mean of a dataset. It is also known as standard score, of a single variable x is calculated by subtracting the population mean μ from the value of x, and then dividing by the population standard deviation σ_x. The sample mean of a variable x, represented by $\bar{X}$, is calculated by subtracting the population mean μ from the sample mean $\bar{X}$, and then dividing by the population standard deviation divided by the square root of sample size $\sigma/\sqrt{n}$.

$$Z_{\bar{X}_1-\bar{X}_2} = \frac{(\bar{X}_1 - \bar{X}_2) - (\mu_1 - \mu_2)}{\sigma_{\bar{X}_1-\bar{X}_2}}, \qquad t = \frac{(\bar{X}_1 - \bar{X}_2) - (\mu_1 - \mu_2)}{s_{\bar{X}_1-\bar{X}_2}}, \qquad t = \frac{(\bar{X}_1 - \bar{X}_2)}{s_{\bar{X}_1-\bar{X}_2}}$$

These equations are used to determine the probability of obtaining a sample mean difference as extreme or more extreme than the one observed, assuming the null hypothesis is true and the population mean difference is zero.

The first equation represents the Z-score of the difference of means of two samples. It is calculated by subtracting the population mean difference $\mu_1 - \mu_2$ from the sample mean difference $\bar{X}_1 - \bar{X}_2$ and then dividing by the population standard deviation of the difference of means $\sigma_{\bar{X}_1-\bar{X}_2}$. The second equation is the *t* score (or *t*-statistic) for the difference of means of two samples when the population standard deviation is unknown.

The *t* score is calculated by subtracting the population mean difference $\mu_1 - \mu_2$ from the sample mean difference $\bar{X}_1 - \bar{X}_2$ and then dividing by the sample standard deviation of the difference of means $s_{\bar{X}_1-\bar{X}_2}$. The third equation is a simplified version, assuming that the population mean difference is zero.

+ **Standard Error of Difference and *t* Equations (Independent *t* Test):** The *t* score is a ratio of the difference between two groups and the difference within the groups. Again, the larger the *t* score, the greater the difference between the groups. The smaller the *t* score, the greater the similarity between the groups.

$$s_{\bar{X}_1-\bar{X}_2} = \sqrt{\frac{s_p^2}{n_1} + \frac{s_p^2}{n_2}}, \qquad s_p^2 = \frac{(n_1 - 1)s_1^2 + (n_2 - 1)s_2^2}{n_1 + n_2 - 2}$$

$$t = (\bar{X}_1 - \bar{X}_2)/s_{\bar{X}_1-\bar{X}_2}, \qquad t = (\bar{X}_1 - \bar{X}_2)\Big/\sqrt{\frac{s_p^2}{n_1} + \frac{s_p^2}{n_2}}$$

In these equations, the *t* score is calculated by subtracting the sample mean difference $\bar{X}_1 - \bar{X}_2$ and then dividing by the standard error of the mean difference $\sqrt{\frac{s_p^2}{n_1} + \frac{s_p^2}{n_2}}$, where s_p^2 is the pooled standard deviation and n_1 and n_2 are the sample sizes of group 1 and group 2, respectively.

This formula accounts for the fact that the sample variances are different by calculating a pooled estimate of the population variance, which is used to calculate the standard error.

$$t = (\bar{X}_1 - \bar{X}_2)\Big/\sqrt{\frac{s_1^2}{n_1} + \frac{s_2^2}{n_2}}$$

Here, the s_1^2 and s_2^2 are the sample variances for groups 1 and 2, respectively. The *t* statistic allows for inference on the difference of means of two independent samples when the variances of the two populations are not equal. When you run a *t* Test, the bigger the *t*, the more likely it is that the results will be repeatable.

2. Independent *t* Test (Hypothesis Testing)

The *t* Test is employed to analyze the means of two populations through statistical examination (Cohen, 1992). A two-sample *t* Test is commonly used in cases of small sample sizes, and it serves to assess the difference between the samples when the variances of two normal distributions are unknown.

As previously discussed, there are three main types of *t* Tests: (a) the independent-samples *t* Test, which compares the means of two distinct groups; (b) the paired-samples *t* Test, which compares the means of a single group at different time points (e.g., 1 year apart); and (c) the one-sample *t* Test, which tests the mean of a single group against a known mean.

+ Conducting an Independent *t* Test (Assumptions and Example): Before diving into the example, it is important to note the assumptions required for a *t* Test.

These include: the observations being independent of each other, the dependent variable being normally distributed in the population, the dependent variable being measured on an interval or ratio scale, the sample being randomly selected or assigned, and the variances of the dependent variable between the two groups being the same in the population. With these assumptions in mind, let us walk through an independent-samples *t* Test step-by-step:

- Research hypothesis: The heart rate of students who drink one or more cups of coffee per day is higher than that of students who do not drink coffee. → Coffee increases heart rate.

- Statistical hypothesis: Alternative hypothesis (H_1), Null hypothesis (H_0)

$$H_1: \mu_C > \mu_{NC}$$

$$H_0: \mu_C \leq \mu_{NC}$$

- Significance level: $\alpha = 0.05$

[Table 1.5.1] Data Set for *t* Test

	Coffee Group	No Coffee Group
Mean	77.42	69.67
SD	6.56	6.39
N	12	12

Table 1.5.1 shows a comparison between two groups: the Coffee Group and the No Coffee Group, with their respective mean values, standard deviations (*SD*), and sample sizes (*N*).

Calculation: $\bar{X}_1 = 77.42, \bar{X}_2 = 69.67, s_1 = 6.56, s_2 = 6.39, n_1 = 12, n_2 = 12$

$$t = (\bar{X}_1 - \bar{X}_2) / \sqrt{\frac{s_1^2}{n_1} + \frac{s_2^2}{n_2}}, \qquad t = \frac{(77.42 - 69.67)}{\sqrt{\frac{6.56^2}{12} + \frac{6.39^2}{12}}} = \frac{7.75}{2.64} = 2.93$$

- Critical t: $df = n_1 + n_2 - 2 = 22$
  In the critical t table $df = 22, \alpha = 0.05$, one-tailed test → $t_{.05;22} = 1.717$

Statistical conclusion: Because the calculated t is larger than the critical t, we reject H_0 ($\mu_C \leq \mu_{NC}$) and accept $H_1(\mu_C > \mu_{NC})$. Or, it is very unlikely to obtain the calculated t or larger t ($p = 0.008$), so we reject H_0.

[Table 1.5.2] Group Statistics

	Group	*N*	Mean	*SD*	Standard error of the mean
Heart rate	Coffee	12	77.4167	6.55686	1.89280
	No coffee	12	69.6667	6.38654	1.84363

Table 1.5.2 provides group statistics for heart rate in the Coffee and No Coffee groups, including sample size (N), mean, standard deviation (*SD*), and standard error of the mean.

[Table 1.5.3] Independent *t* Test

	t test for equality of means				
	t	df	Sig. (two-tailed)	Mean difference	Standard error difference
Heart rate	2.933	22	.008	7.75	2.642
	2.933	21.99	.008	7.75	2.642

Table 1.5.3 presents the results of an independent *t* Test comparing the means of the two groups, showing the *t* value, degrees of freedom (df), significance level (two-tailed), mean difference, and standard error difference.

Research conclusion: Students who drink one or more cups of coffee per day have a higher heart rate than those who do not.

3. Independent *t* Test (Unequal Sample Sizes)

The independent *t* Test is a statistical method used to compare the means of two independent groups. In cases where the sample sizes (n) of the two groups being compared are unequal (also called Two-tailed hypothesis testing), this kind of *t* Test is still valid.

The *t* Test is often used in hypothesis testing to determine if there is a significant difference between the means of two groups. The two-tailed alternative hypothesis allows for the possibility that the means of the two groups may differ in either direction (higher or lower) rather than just one direction. The independent *t* Test can provide valuable insights for decision-making.

Let us walk through an example with unequal sample sizes and a two-tailed hypothesis:

Research hypothesis: The academic average of students who drink more than two cups of coffee per day is different from that of students who drink no coffee. → Coffee increases students' academic average.

Statistical hypothesis

- $H_1: \mu_C > \mu_{NC}$ Alternative hypothesis
- $H_0: \mu_C \leq \mu_{NC}$ Null hypothesis

Significance level: $\alpha = 0.05$

[Table 1.5.4] Data Set for Independent *t* Test

	Coffee Group	No Coffee Group
Mean	84.7%	77.9%
SD	3.4	7.9
N	7	17

Table 1.5.4 displays the data set for an independent t-test comparing the Coffee and No Coffee groups, including the mean (as a percentage), standard deviation (*SD*), and sample size (*N*) for each group.

Calculation

$$t = \frac{(\bar{X}_1 - \bar{X}_2)}{\sqrt{\frac{s_p^2}{n_1} + \frac{s_p^2}{n_2}}},\ s_p^2 = s_p^2 = \frac{(n_1-1)s_1^2 + (n_2-1)s_2^2}{n_1+n_2-2} = \frac{(7-1)3.4^2 + (17-1)7.9^2}{7+17-2} = \frac{1067.9}{22} = 48.54$$

$$t = 84.7 - 77.9/\sqrt{\frac{48.54}{7} + \frac{48.54}{17}} = 6.8/3.13 = 2.16$$

- Critical *t*: $df = n_1 + n_2 - 2 = 22$
  In the critical *t* table $df = 22, \alpha = 0.05$, two-tailed test → $t_{.05;22} = 2.074$

Statistical conclusion: Because the calculated *t* is larger than the critical *t*, we reject H_0 ($\mu_C \leq \mu_{NC}$) and accept $H_1(\mu_C > \mu_{NC})$. Or, because the probability of obtaining the calculated *t* or larger *t* value is very low ($p = 0.0419$), we reject H_0.

Research conclusion: The academic average of students who drink more than 2 cups of coffee per day is different from that of students who drink no coffee.

+ **Exercise Independent *t* Test:** In this exercise, we will practice conducting an independent *t* Test using data from an experiment where a researcher hypothesized that people who jog have higher bone density than people who swim. The data set for this exercise is presented in Table 1.5.5, which compares the mean, variance, and sample size (n) of Joggers and Swimmers.

[Table 1.5.5] Data Set for Independent *t* Test Exercise

	Joggers	Swimmers
Mean	56.5	50.0
Variance	7.74	8.36
n	8	12

Table 1.5.5 presents the data set for an independent t-test comparing Joggers and Swimmers, displaying the mean, variance, and sample size (n) for each group.

Research hypothesis: People who jog have higher bone density than people who swim.

Statistical hypothesis

- H_1: $\mu_C > \mu_{NC}$ Alternative hypothesis
- H_0: $\mu_C \leq \mu_{NC}$ Null hypothesis

Significance level: $\alpha = 0.05$

Conduct necessary calculations

$$t = \frac{(\bar{X}_1 - \bar{X}_2)}{\sqrt{\frac{s_p^2}{n_1} + \frac{s_p^2}{n_2}}}, \quad s_p^2 = s_p^2 = \frac{(n_1 - 1)s_1^2 + (n_2 - 1)s_2^2}{n_1 + n_2 - 2} = \frac{(8-1)7.74^2 + (12-1)8.36^2}{8 + 12 - 2} = \frac{419.35 + 768.78}{18} = 66$$

The equation calculates the *t* statistic for an independent *t* Test by comparing the difference in means between two groups, taking into account their pooled variance and respective sample sizes.

$$t = 56.5 - \frac{50}{\sqrt{\frac{66}{8} + \frac{66}{12}}} = \sqrt{8.25 + 5.5} = \sqrt{13.75}$$

$$t = 6.5/3.7 = 1.75$$

The equation calculates the *t* statistic value by finding the difference between the means of the two groups and dividing it by the square root of the pooled variance divided by their respective sample sizes, resulting in a *t* value of 1.75.

Critical *t*: $df = n_1 + n_2 - 2 = 18$
In the critical *t* table $df = 20, \alpha = 0.05$, two-tailed test → 2.101

Statistical conclusion: Because the calculated *t* is larger than the critical *t*, we reject H_0 ($\mu_C \leq \mu_{NC}$) and accept H_1 ($\mu_C > \mu_{NC}$).

Research conclusion: The bone density of people who jog is not different from that of people who swim.

4. Paired *t* Test (Analyzing Dependent or Linked Observations)

The paired *t* Test, also known as the correlated samples *t* Test, is a statistical method used to compare two sets of related or dependent observations. This test is employed when examining the same variable measured twice on the same participants to determine if there is a significant difference between the two means.

The formula for the paired *t* Test is as follows:

$$t = (\bar{X}_1 - \bar{X}_2)/s_{\bar{X}_1-\bar{X}_2},\quad S_{\bar{X}_1-\bar{X}_2} = \sqrt{\frac{s_1^2 + s_2^2 - 2rs_1s_2}{n}}$$

Here, $\bar{X}_1$ and $\bar{X}_2$ are the means of the two sets of observations, and $s_{\bar{X}_1-\bar{X}_2}$ is the standard deviation of the difference between the two means, respectively. Therefore, s_1 and s_2 are the standard deviations of the two sets of observations, r is the correlation between the two sets of observations, and n is the number of observations in each set, is used to calculate the standard deviation of the difference between the means. A stronger correlation between the two sets of data will lead to a larger *t* value, hence the value of r can have an impact on the *t* value.

Simply enough, the paired *t* Test is a useful method for determining whether two linked sets of observations have significant differences. It accounts for the correlation between the two sets of observations and evaluates if the observed differences are more likely due to chance or are statistically significant.

+ Paired *t* Test Example (Comparing VO_2 Max before and after Exercise): In this example, we demonstrate the use of the paired *t* Test to compare the VO_2 max of a group of subjects before and after exercise. A researcher would use a paired *t* Test for this purpose, as it compares the means of two related sets of observations.

[Table 1.5.6] Data Set for Paired *t* Test

Subject	Pretest	Posttest	Difference	
1	6	9	–3	
2	12	14	–2	
3	18	20	–2	
Mean	12	14.3	–2.3	
SD	6.0	5.5	0.577	$r = 0.998$

Table 1.5.6 presents a dataset for a paired *t* Test, comparing pretest and posttest scores of three subjects, and showing that the posttest scores increased by an average of 2.3 points with a strong correlation (r = 0.998) between pretest and posttest scores.

Statistical hypothesis

- $H_1: \mu_1 - \mu_2 \neq 0$ Alternative hypothesis
- $H_0: \mu_1 - \mu_2 = 0$ Null hypothesis

Calculation

$$t = \frac{12.0 - 14.3}{\sqrt{\frac{6^2 + 5.5^2 - 2(.9986)(6)(5.5)}{3}}} = \frac{-2.3}{\sqrt{.111}} = \frac{-2.3}{.333} = -6.99$$

The equation calculates the *t* statistic for a paired *t* Test comparing pretest and posttest scores, resulting in a value of -6.99, which indicates a significant difference between the two sets of scores. If the difference between the means is statistically significant, this *t* value may be used to compare against a crucial *t* value from a *t* distribution table.

- critical *t*: $df = n - 1 = 2$
- $t_{.01/2;2} = 9.925$, or $t_{.05/2;2} = 4.303$

In conclusion, we reject the null hypothesis (H_0) as the calculated *t* value is larger than the critical *t* value. This indicates that there is a treatment effect, suggesting that the exercise intervention has led to an increase in VO2 max for the subjects.

+ *Z* and *t* Tests: A Comparison (Which One to Use): *Z* and *t* Tests are both used for hypothesis testing (Savin, 1984). In this section, we discuss the math and assumptions behind each of these tests, assuming you understand the basics of running a hypothesis test.

One-Sample *Z* Test: The *Z* Test is used to determine whether a sample comes from a specific population. In other words, does the sample differ from what we would expect from the given population? When we conduct a *Z* Test, we first need a sample. In a *Z* Test, our sample is generally large (greater than or equal to 30). We also need to ensure that our sample is randomly selected and that the selection process is independent. Finally, we need to know the population's standard deviation.

Once these parameters are met, we can start running the *Z* Test.

To begin, we calculate the *Z* statistic. The *Z* statistic is the standardization of the sample mean to the standard normal distribution. You will recall that a standard normal distribution has a mean of 0 and a standard deviation of 1. Thus, if your *Z* statistic is 0.75, it is .75 SD above the average. The formula for calculating your *Z* statistic is as follows:

$$Z = \frac{\bar{X} - \mu}{\frac{\sigma}{\sqrt{n}}}, \qquad \bar{X} = sample\ mean;\ \mu = population\ mean;$$

$$\sigma = population\ SD;\quad n = smaple\ size$$

Calculating the *Z* for a Sample: Once you have the *Z* statistic, look at the *Z* table. *Z* tables are measurements of the area under the curve to the left. To use the table, find the appropriate *Z* statistic, which will give you a decimal value, which will lead to p.

To find p, you need to consider the alternative hypothesis (see the Supplement Data for more detail about Statistical Tables).

- If your alternative hypothesis states that the sample mean is greater than the population mean, subtract this value from 1 to get the p.
- If your alternative hypothesis states that the sample mean is less than the population mean, use this value as your p.
- If your alternative hypothesis states that the sample mean is not equal to the population mean, use either of the methods above (whichever is more appropriate for your calculated *Z* statistic) and multiply by 2.

Alternatively, you can forgo multiplying the p and halve your alpha. If your p is less than your alpha, then you can reject the null hypothesis and state that there is a statistical difference. If not, you fail to reject the null hypothesis and cannot say that the sample is different from the population.

One-Sample *t* Test: The *t* Test is used to determine whether two samples are different from each other. One of these samples can be the population. However, we use the *t* Test in place of the *Z* Test if the population's standard deviation is unknown.

There are a lot of similar assumptions to the *Z* Test. The sample must be random and independently selected as well as drawn from a normal distribution. The values should be numerical and continuous. The sample size does not necessarily have to be large.

We then calculate the *t* statistic. This looks very similar to the *Z* statistic except for the denominator, in which we use the standard deviation of the sample instead of the population. The other value we need to calculate is the degrees of freedom, which is n – 1.

$$t = \frac{\bar{x} - \mu}{s_{\bar{x}}}, \qquad s_{\bar{x}} = \frac{s}{\sqrt{n}}, \qquad \bar{X} = sample\ mean;$$
$$\mu = Proposed\ constant\ for\ the\ population\ mean;$$
$$s = sample\ SD;\ n = sample\ size;$$
$$s_{\bar{x}} = Estimated\ standard\ error\ of\ the\ mean\ (\frac{s}{sqrt(n)})$$

We then use a *t* distribution table to find the critical value. To use the table, you must know the alpha and degrees of freedom. This table gives the area under one tail. If your *t* statistic is less than this critical value, then you fail to reject the null hypothesis. If your critical value is greater than the critical value, then you can reject the null hypothesis and conclude that there is a significant difference between the sample and the population (see the Supplement Data (Appendix) for more detail about Statistical Tables).

Z and *t* Tests are essential tests for measuring whether a difference between a sample and a population is significant. Although the formulas are similar, the choice of test often relies on the sample size and whether you know the population's standard deviation. If you know the population's standard deviation and the sample size is large enough, then you want to use a *Z* Test. If you do not know the population's standard deviation, you want to use a *t* Test. In addition, regardless of whether you know the population's standard deviation, if your sample size is small, you want to use a *t* Test.

+ **One-Sample *t* Test in Programming** (Ensuring Consistency in Latte Espresso Content): Let us practice with a hypothetical scenario. A coffee shop relocates to Italy and wants to ensure that all lattes have a consistent amount of espresso. The owners believe that each latte should contain an average of 4 oz of espresso. If this is not the case, they must adjust the espresso quantity accordingly. In this example, a random sample of 25 lattes has a mean espresso content of 4.6 oz and a standard deviation of 0.22 oz. We will use an alpha of 0.05 and perform a one-sample t-test to compare the sample mean against the expected population mean of 4 oz.

Solution: Two-Tailed One-Sample *t* Test

State the hypotheses:

- Null hypothesis (H_0): There is no significant difference between the sample mean (M) of espresso in latte and the population mean (μ). Thus, $M = \mu = 4$ oz.
- Alternative hypothesis (H_1): There is a significant difference between the sample mean (M) and the population mean (μ). Thus, the sample mean of espresso in latte is not equal to 4 oz.

The purpose of the one-sample *t* Test is to determine whether the null hypothesis should be rejected, given the sample data. Here we perform a two-tailed (or two-sided) one-sample *t* Test. Notice we are performing a two-tailed *t* Test here because we are not trying to determine whether the mean value is less than or greater than any given value.

Let us solve this simple example using programming. Write the code given below in your platform (i.e., Jupyter notebook). H_0: The mean amount of espresso in latte is not different from the population (mean is 4 oz). H_1: The mean amount of espresso in latte is significantly different from the population (mean is not 4 oz).

```
[In]
import numpy as np
sample_mean = 4.6
population_mean = 4
sample_size = 25
degree_of_freedom = 25-1
standard_deviation = 0.22
print('Degree Of Freedom is : ',
degree_of_freedom)
```

→

```
[Out]
Degree Of Freedom is :  24
```

The code calculates the *t* Test statistic by subtracting the population mean from the sample mean, and dividing the result by the standard error of the mean.

```
[In]
Ttest = (sample_mean - population_mean) /
(standard_deviation/np.sqrt(sample_size))

print('T test value is : ', Ttest))
```

→

```
[Out]
T test value is : 13.6
```

The critical value is depending on the degrees of freedom, which is determined by the sample size and the number of estimates of variance in the sample. Then we can calculate the *p* value. Write the code below in your notebook. We will use the 'scipy stats' function to calculate the *p* value as follows:

```
[In]
import scipy.stats as stats
p = stats.t.cdf(Ttest, df = 24)
pvalue = stats.t.sf(np.abs(Ttest), 24)*2
print("p is:", p)
print("pvalue is:", pvalue)
```

The 'stats.t.cdf' function calculates the cumulative distribution function (CDF) of the *t* distribution at a given value of the *t* Test statistic. And we are performing a two-sided test to find the final *p* value.

```
[In]
pval = (1 - p)*2

if pvalue > 0.05:
 print('The mean of espresso in latte is
not different from population Mean (Fail To
Reject H0)')
else:

 print('The mean of espresso in latte is
significantly different from population
Mean (Reject H0) ')
```

→

```
[Out]
put:  p is:
0.9999999999995757

pvalue is: 8.48542925590331e
-13
```

The mean amount of espresso in latte is significantly different from the population mean (Reject H_0).

Accepting or rejecting the null hypothesis using the *p* value: Per our calculation above, the *p* value is 8.485, which is less than 0.05. (In the majority of analyses, an alpha of 0.05 is used and accepted as the cutoff for significance.)

Hence, we can reject our null hypothesis H_0 to conclude the following: The mean amount of espresso in latte is not different from the population mean.

5. Factors Affecting Power in a *t* Test

Several factors affect the power of a statistical test. Some are under the control of the experimenter, whereas others are not. The following example is used to illustrate the various factors.

$$t = (\bar{X}_1 - \bar{X}_2) / \sqrt{\frac{s_1^2}{n_1} + \frac{s_2^2}{n_2}}$$

+ **Standard Deviation Value (s^2)**: Power is higher when the standard deviation is small than when it is large. For all values of N, power is higher for a standard deviation of 10 than for a standard deviation of 15 (except, of course, when $N = 0$). Experimenters can sometimes control the standard deviation by sampling from a homogeneous population of subjects, by reducing random measurement error, and by making sure the experimental procedures are applied very consistently.

+ **Significance Level (α)**: There is a tradeoff between the significance level and the power. The more stringent (lower) the significance level, the lower the power. Naturally, the stronger the evidence needed to reject the null hypothesis, the lower the chance that the null hypothesis will be rejected.

+ **Sample Size (*N*)**: The larger the sample size, the higher the power. Because the sample size is typically under an experimenter's control, increasing the sample size is one way to increase power. However, it is sometimes difficult or expensive to use a large sample.

+ **One-tailed or Two-tailed Test**: Power is higher with a one-tailed test than with a two-tailed test as long as the hypothesized direction is correct. A one-tailed test at the 0.05 level has the same power as a two-tailed test at the 0.10 level. A one-tailed test, in effect, raises the significance level.

+ **Effect Size Difference ($\mu 1 - \mu 2$):** Naturally, the larger the effect size, the more likely it is that an experiment will reveal a significant effect. This shows the impact of increasing the difference between the mean specified by the null hypothesis (75) and the population mean μ for standard deviations of 10 and 15.

p Value Basics

What is the p value? It is the probability that you will obtain a test result given an actual distribution (Westfall & Young, 1993). Or in an A/B test set, it is the probability that you measure something, like a company's average order, given an initial hypothesis, like 'I believe the average order is $170.'

The p indicates a certain confidence level. You often hear people say, 'I am 90% confident I will get that job' or something like this involving a confidence level (p). We saw in the discussion on the central limit theorem that if we draw samples enough times, the distribution of the sample means will be normal with a mean approaching the population mean. A natural question to ask is how likely it would be to observe a particular sample mean.

Let us look at an example. Assume that the average order of a company's customers is $170. That is, our population mean is $170 (and the standard deviation is $5). Our hypothesis is that our entire customer base has an average order of $170.

Now we want to test our hypothesis. In general, we cannot test our entire population; thus, we resort to testing many smaller samples of the population. If we draw 10,000 samples, then the distribution of those sample means will look like a normal curve (i.e., a normal distribution). The entire customer base (of millions of people) has an average order of $170. If we picked smaller samples of people at a time, repeated this sample taking 10,000 times, and plotted the sample means, they would look like the bell curve (=normal distribution). The sample means would range from $150 to $190. You can see that any given sample's mean could range anywhere from $150 to $190.

Then let us say we took a sample group of people to get their mean order, and it turned out to be $183. We want to know the probability that this could happen. In statistics, we ask the following: What is the probability that a sample mean can be $183 or above given the hypothesis and a population mean of $170? We can calculate this probability by counting all of the numbers greater than 183 and dividing it by the total number of samples draws, which is 10,000.

Use the following code:

```
[In]
pvalue_101(170.0, 5.0, 10000, 183.0)
```

The percentage of numbers greater than 183.0 is 0.35%. This is a tiny percentage, but it is not 0%. It would be wrong for you to reject the hypothesis that the population mean is $170 because we derived this sample mean from that population distribution.

You could also ask the following: What is the probability of getting a sample mean that is more than $13 different from the population mean of $170? That is, what is the probability of getting a sample mean that is less than $157 or more than $183? The percentage of numbers beyond the population mean of 170.0 by ±13.0 is 0.77%.

This is about double the percentage that the sample mean could be just only larger than $183. This is because a normal distribution is symmetrical around the mean. It is essential to understand this small but non-zero probability. Even if we draw samples from the same population, there is a non-zero chance that the sample mean will vary by quite a lot from our population mean. If we run the A/B test for just one day, which is the equivalent of drawing one sample, we cannot determine the population mean.

What we can do is estimate the probability that we would get this sample mean given this population. Now, instead, assume you wanted to know the reverse: how far from the population 95% of the sample means lie. That is when the 68–95–99.7 rule comes in handy.

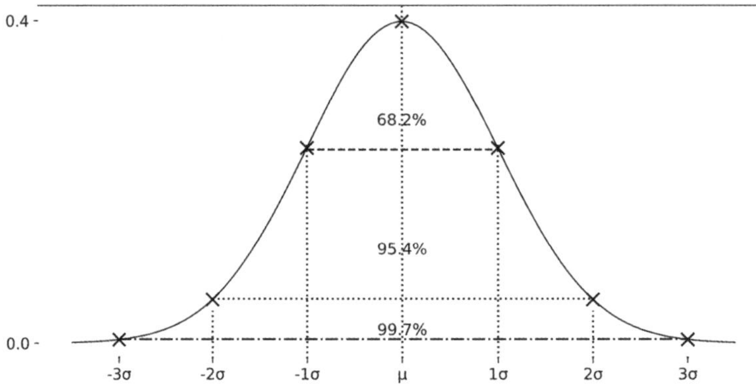

Figure 1.5.2. Population distribution

In Figure 1.5.2, we are 68.2% confident that if we draw many random samples, the sample means will be between $\mu \pm \sigma$. In our case, 170 ± 5 (i.e., 68.2% of the time) the sample means will be between \$165 and \$175. Moreover, 95.4% of the time our sample means will be between $\mu \pm 2\sigma$ (i.e., 95.4% of the time, our sample means will be between \$160 and \$180). p is closely related to this rule. It measures the probability that a sample mean will be a certain value or more given the population mean and standard deviation.

In sum, p gives us the probability of observing what was observed if a hypothesis is true. It does not tell us the probability that the null hypothesis is true. In this example, a p of 0.35% will give the probability that we get a sample mean that is more than \$183 given the hypothesis that the population mean is \$170. A p of 0.77% will give the probability that we get a sample mean that is more than \$183 or less than \$157 given the hypothesis that the population mean is \$170. It does not give us the probability of the hypothesis being true. In fact, it would be dangerous to reject the hypothesis that the population mean is \$170, because we got the sample mean from a population mean of \$170.

Remarks: To draw conclusions regarding the differences between groups or situations, statistical models like the *t* Test are often utilized in a variety of study domains. The paired *t* Test is helpful for comparing two linked groups, whereas the independent *t* Test is frequently employed when comparing two independent groups. To account for different sample sizes or variances, the *t* Test can be adjusted. When using this model to analyze data, careful evaluation and the right changes are crucial. A number of factors, including sample size, variance, and distribution, can dramatically alter the validity and power of the *t* Test.

[1.6]

Statistical Model, Part 2

Identification of ANOVA, Two-Way ANOVA, Comparison of One-Way and Two-Way ANOVA

Analysis of variance (ANOVA) is the generalization (expansion) of the *t* Test and a statistical method that compares three or more means (Kao & Green, 2008). It is a test of a hypothesis that is appropriate to compare means of a continuous variable in two or more independent comparison groups.

1. Identification of ANOVA

ANOVA is used to compare more than two population means (in some clinical trials, there are more than two comparison groups), for example, yields of crops from multiple varieties of seed. The concept dates back to around 1920 and a British statistician named Ronald Fisher, who was analyzing massive amounts of data on crop yields (Fisher, 1992).

Fisher used ANOVA to prove there was a significant difference in the average weight of potatoes when different fertilizers were used. This vital tool enables researchers to conduct tests simultaneously. ANOVA extends the independent-samples procedure discussed previously, which is used when there are precisely two independent groups to compare, to two or more independent groups.

+ ***F* Distribution:** The *F* distribution is a continuous probability distribution that is commonly used in Analysis of Variance (ANOVA) tests.

[In]

```
from scipy import stats
x = np.arange(0, 5, 0.1)
firstF = stats.f.pdf(x, 1, 1)
secondF = stats.f.pdf(x, 2, 2)
thirdF = stats.f.pdf(x, 5, 5)
fourthF = stats.f.pdf(x, 30, 30)
for y, ls in zip([firstF, secondF, thirdF,
fourthF], line_styles):
    plt.plot(x, y, linestyle=ls)
```

→

[Out]

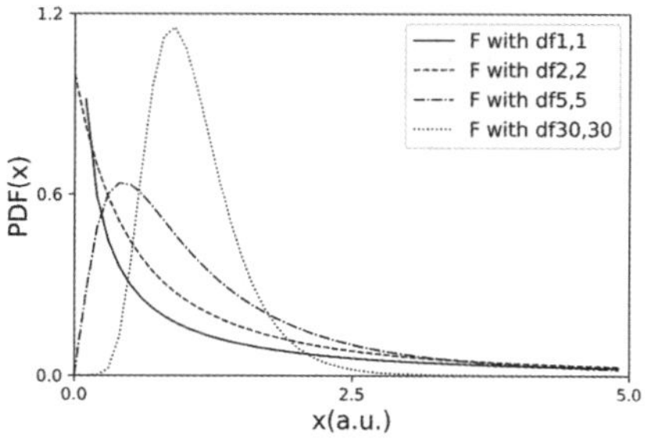

Figure 1.6.1. *F* distribution, used for ANOVA test

In Figure 1.6.1, we plot the *F* distribution in Python using the 'scipy' library. This is a continuous distribution characterized by two parameters called the numerator degrees of freedom ($df1$) and denominator degrees of freedom ($df2$). Because there are more than two groups, computing the test statistic is more involved. The test statistic must consider the sample sizes, sample means, and sample standard deviations in each of the comparison groups. The hypotheses of interest in ANOVA are as follows:

- H_0: $i_1 = i_2 = i_3 \ldots = i_k$
- H_1: Means are not all equal.
- k : the number of independent comparison groups

If you are examining the means of three groups, it might be tempting to perform three separate group-to-group comparisons, but this would be incorrect because each of these comparisons would fail to consider the total data. This would increase your likelihood of incorrectly concluding that there are statistically significant differences, as each comparison would add to the probability of type I error. ANOVA avoids these problems by asking a more global question (i.e., whether there are significant differences among the groups) without addressing differences between any two groups in particular (although there are additional tests that can do this if the ANOVA indicates that there are differences among the groups).

+ **Calculating the *F* Statistics and Analysis:** The fundamental strategy behind ANOVA is to examine within-group and between-group variability. An *F* statistic is calculated by dividing the difference between the group means (variance) by the difference between subjects within groups (variance). This statistic follows the *F* distribution:

$$F = MS_{between}/MS_{within}$$

- $F > 1.0$: there is a difference between groups (i.e., there is a treatment effect).
- $F \leq 1.0$: there is no difference between groups (i.e., there is no treatment effect).

Let us take a simple example of ANOVA and presents the results, highlighting the significance of the statistics.

- Exercise (none, moderate, or strong) influences resting heart rate is the topic of the study's arbitrary generated research problem.

[Table 1.6.1] Data Set for ANOVA Example

	No exercise	Moderate exercise	Strong exercise
P1	77	69	60
P2	75	68	65
P3	73	65	61
P4	78	74	70
P5	79	73	71
P6	74	76	69
Mean	76.00	70.83	66.00

Table 1.6.1 presents data for an ANOVA example comparing the means of three groups: no exercise, moderate exercise, and strong exercise, with the mean scores decreasing as exercise intensity increases.

[Table 1.6.2] Results for ANOVA Example

	Sum of square	df	Mean of square	F	p
$MS_{between}$	300.111	2	150.056	9.923	.002
MS_{within}	226.833	15	15.122		
Total	526.944	17			

Table 1.6.2 displays the results of the ANOVA example mentioned earlier, highlighting the sum of squares, degrees of freedom, mean squares, F statistic, and p value, which indicates a significant difference among the three exercise groups.

[Table 1.6.3] Results for the Dependent Variable: Heart Rate

G2	G2	Mean difference	Standard error	p
1.00 No exercise	2.00 Moderate exercise	5.16667	2.24516	.086
	3.00 Strong exercise	10.00000*	2.24516	.001
2.00 Moderate exercise	2.00 No exercise	–5.16667	2.24516	.086
	3.00 Strong exercise	4.83333	2.24516	.113
3.00 Strong exercise	2.00 No exercise	–10.00000*	2.24516	.001
	3.00 Moderate exercise	–4.83333	2.24516	.113

In the Table 1.6.3, the 'Mean difference' column for each comparison contains the average difference between the two groups for that comparison. The standard error of the mean difference is displayed in the 'Standard error' column. Each comparison's p value is displayed in the p column. The p values represent the chances of finding a difference as significant as the one that was seen, supposing that the null hypothesis—that there is no difference between the groups—is correct. Typically, statistical significance is shown by a p value less than 0.05. The difference between the 'No exercise' and 'Strong exercise' categories in this table is statistically significant, as shown by the p value of 0.001, which is less than 0.05.

Overall, this data indicates that there is (not) a statistically (in)significant difference in heart rate between the 'Strong exercise' group and the 'No exercise' group. In ANOVA, the F statistic is calculated to analyze the data and make inferences about the population. The F statistic is similar to other commonly used statistics, such as the Z and t statistics

$$F = \frac{Treatment\ Differences + Random\ Differences}{Random\ Differences}$$

Here, the F statistic is calculated by dividing the variance between groups (treatment differences) by the variance within groups (random differences). The resulting value is then compared to a critical value from the F distribution which describes the variability of the F statistic under the null hypothesis. If there are no treatment differences (i.e., if there is no actual effect), we expect F to be 1. If there are treatment differences, we expect F to be greater than 1. The F statistic has a one-tailed distribution, much like how the Z and t statistics have their distributions.

+ **One-Way ANOVA Numeric (Anti-Anxiety Medication):** Researchers are testing the efficacy of a new anti-anxiety medication. To do this, they conduct a One-Way ANOVA study where they split participants into three independent groups based on the dosage they receive (0 mg, 50 mg, and 100 mg). The participants then rate their anxiety levels on a scale from 1 to 10.

Are there any differences between the three conditions at alpha = 0.05? Table 1.6.4 presents a data set comparing three groups with different dosages (0 mg, 50 mg, and 100 mg) and their respective values, which appear to generally decrease with increasing dosage.

[Table 1.6.4] Data Set for ANOVA Exercise

0 mg	50 mg	100 mg
9	7	4
8	6	3
7	6	2
7	7	3
7	8	4
9	7	3
8	6	2

To achieve this, let us define Null and Alternative hypotheses, state an alpha, calculate degrees of freedom (maximum number of logically independent values calculated by subtracting one from the number of items within the data sample), state a decision rule, calculate a test statistic, state the results, and state a conclusion as follows.

- $H_0: \mu_{0mg} = \mu_{50mg} = \mu_{100mg}$
- H_1: not all means are equal
- State Alpha: $\alpha = 0.05$

The null hypothesis (H_0) claims that the population means for three groups, μ 0mg, μ 50mg, and μ 100mg, are equal to each other. The competing hypothesis (H_1) contends that at least one of these population means does not equal the other means. Since the significance level (α) for this hypothesis test is set at 0.05, we reject the null hypothesis and conclude that the population means are not equal if the test results show a significance level of less than 0.05.

Calculate Degrees of Freedom: $N = 21, n = 7$

$$df_{Between} = a - 1 = 3 - 1 = 2$$
$$df_{Within} = N - a = 21 - 3 = 18$$
$$df_{Total} = N - 1 = 21 - 1 = 20$$

In the equations above, the $df_{Between}$ represents the degrees of freedom between groups which calculated by subtracting 1 from the number of groups. The df_{Within} denotes the degrees of freedom within groups which calculated by subtracting the number of groups from the total number of observations (N).

And the df_{Total} represents the total degrees of freedom which calculated by subtracting 1 from the total number of observations (N). Thus, if there are 21 observations in total, there will be 20 degrees of freedom.

State Decision Rule: To look up the critical value, we need to use two different degrees of freedom.

$$df_{Between} = a - 1 = 3 - 1 = 2$$
$$df_{Within} = N - a = 21 - 3 = 18$$

(2, 18: See the Supplement Data for more detail about Statistical Tables). If F is greater than 3.5546, reject the null hypothesis.

[Table 1.6.5] Calculate Test Statistic for ANOVA Exercise

	SS	df	MS	F
Between		2		
Within		18		
Total		20		

In Table 1.6.5, the degrees of freedom are used to determine the variance of a particular sample or group.

$$SS_{Between} = \frac{\Sigma(\Sigma a_i)^2}{n} - \frac{T^2}{N}, \quad SS_{Between} = \frac{57^2 + 47^2 + 21^2}{7} - \frac{125^2}{21} = 98.67$$

$$0 \text{ mg}_{\text{ Group}}: 9 + 8 + 7 + 8 + 8 + 9 + 8 = 57$$
$$50 \text{ mg}_{\text{ Group}}: 7 + 6 + 6 + 7 + 8 + 7 + 6 = 47$$
$$100 \text{ mg}_{\text{ Group}}: 4 + 3 + 2 + 3 + 4 + 3 + 2 = 21$$

$$SS_{Within} = \sum Y^2 - \frac{\Sigma(\Sigma a_i)^2}{n}, \quad SS_{Within} = 853 - \frac{57^2 + 47^2 + 21^2}{7} 10.29$$

$$\sum Y^2 = 9^2 + 8^2 + 7^2 + 8^2 + 8^2 + 9^2 + 8^2 + 7^2 + 6^2 + 6^2 + 7^2 + 8^2 + 7^2 + 6^2 + 4^2 + 3^2 + 2^2 + 3^2 + 4^2 + 3^2 + 2^2 = 853$$

The given equations calculate the between-group sum of squares ($SS_{Between}$), within-group sum of squares (SS_{Within}), and total sum of squares (SS_{Total}) for an ANOVA exercise using three groups (0 mg, 50 mg, and 100 mg).

$$SS_{Total} = \sum Y^2 - \frac{T^2}{N}, \quad SS_{Total} = 853 - \frac{125^2}{21} = 108.95$$

The results show that $SS_{Between}$ is 98.67, SS_{Within} is 10.29, and SS_{Total} is 108.95. And the table below (Table 1.6.6) lists the findings of a two-way analysis of variance (ANOVA) that compares the means of two or more groups.

[Table 1.6.6] Test Results for ANOVA Exercise

$F = \frac{MS_{between}}{MS_{within}}$	$MS_{between} = \frac{98.67}{2} 49.34$	$MS_{within} = \frac{10.29}{18} 0.57$

$$F = MS_{between}/MS_{within} = 49.34/0.57 = 86.56$$

	SS	df	MS	F
Between	98.67	2	49.34	86.56
Within	10.29	18	0.57	
Total	108.95	20		

The first column, 'SS' displays the residuals and groups' sum of squares. Degrees of freedom (or df) for each phrase are shown in the second column. The mean squares, which are the total of squares divided by the degrees of freedom, are shown in the third column, 'MS.' The F ratio, shown in the fourth column with the letter 'F' is determined by dividing the mean square for the groups by the mean square for the residuals. The F ratio in this instance is 86.56, which shows that the group means deviate significantly from one another.

State Results: If F is greater than 3.5546, reject the null hypothesis. $F = 86.56$. Reject the null hypothesis.

State Conclusion: The three conditions had significantly different effects on anxiety,

$$F(2, 18) = 86{,}56, p < 0.05$$

The results of the ANOVA test on the effect of exercise on anxiety levels showed that there is a significant difference between the three conditions. The F ratio was calculated to be 86.56, which is greater than the critical value of 3.5546, indicating that the null hypothesis should be rejected. The conclusion is that the three conditions (0 mg, 50 mg, and 100 mg) had a significant effect on anxiety levels, with a significance level of $p < 0.05$.

+ **Practice with ANOVA in Programming:** The main idea behind ANOVA is to test whether there is a statistically significant difference in measured means between subgroups of a sample. To demonstrate ANOVA using actual data, let us look at the prices of wine.

The data for this example is from Kaggle's Wine Review data set (see the Supplement data), filtered down to a subset of 3,000 reviews of wines from the United States, Italy, and France (1,000 per country). We are going to use ANOVA to test whether there is a statistically significant difference in the average price among countries.

First let us import the necessary libraries and modules and import the data.

```
[In]
import pandas as pd
import seaborn as sns
import numpy as np
from scipy.stats import f, norm
import matplotlib.pyplot as plt
larger_df = pd.read_csv('data… /winemag-data_first150k.csv')
larger_df.head()
```

↓

[Out]

	Unnamed: 0	country	description	designation	points	price	province	region_1	region_2	variety	winery
0	0	US	This tremendous 100% varietal wine hails from ...	Martha's Vineyard	96	235.0	California	Napa Valley	Napa	Cabernet Sauvignon	Heitz
1	1	Spain	Ripe aromas of fig, blackberry and cassis are ...	Carodorum Selección Especial Reserva	96	110.0	Northern Spain	Toro	NaN	Tinta de Toro	Bodega Carmen Rodríguez
2	2	US	Mac Watson honors the memory of a wine once ma...	Special Selected Late Harvest	96	90.0	California	Knights Valley	Sonoma	Sauvignon Blanc	Macauley
3	3	US	This spent 20 months in 30% new French oak, an...	Reserve	96	65.0	Oregon	Willamette Valley	Willamette Valley	Pinot Noir	Ponzi
4	4	France	This is the top wine from La Bégude, named aft...	La Brûlade	95	66.0	Provence	Bandol	NaN	Provence red blend	Domaine de la Bégude

We read in a CSV file called 'winemag-dat_first150k.' and stores the data in a 'larger_df'. The 'head' method o the 'larger_df' displays the first few rows of data. By default, it will display as below but we can specify a different number of rows by passing an integer as an argument to the 'head' method. Then, plot the price distributions.

```
[In]
for name in ['US', 'Italy',
'France']:
sns.distplot(wine_df[wine_df.country=
=name].price,
kde_kws=dict(linewidth=3))

plt.show()
```

→

[Out]

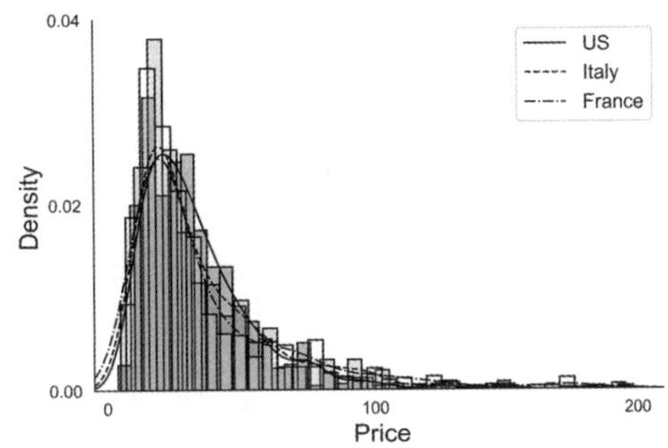

Figure 1.6.2. Price distribution by country

The 'sns.distplot' creates a distribution of the 'price', column for a subset of the 'wine_df' that is filtered by country. The 'kde_kws' argument is for specifying the width of the kernel density estimate line which is set to 3 in this case. The results (Figure 1.6.2) show three similar distributions, with the majority of prices around $25 a bottle. If we take the mean for each country, we see that France has the most expensive wine on average, followed by Italy and then the United States. However, is this a genuine difference in price among countries, or could it just be due to sample variation?

This is where ANOVA comes into play.

```
[In]
wine_df.groupby('country').mean()['price']
```

→

```
[Out]
country    price
France     37.242
Italy      35.286
US         33.776
```

Now we calculate the means and variances for the statistics.

```
[In]
mean_france = wine_df[wine_df.country == 'France']['price'].mean()
mean_italy = wine_df[wine_df.country == 'Italy']['price'].mean()
mean_us = wine_df[wine_df.country == 'US']['price'].mean()
mean_overall = wine_df['price'].mean()
var_france = wine_df[wine_df.country == 'France']['price'].var()
var_italy = wine_df[wine_df.country == 'Italy']['price'].var()
var_us = wine_df[wine_df.country == 'US']['price'].var()
```

We then compute the sum of squares between

```
[In]
SSB = 1000*(mean_overall - mean_france)**2 + \
1000*(mean_overall - mean_italy)**2 + \
1000*(mean_overall - mean_us)**2
print('Sum of Squares Between:
{0:.6g}'.format(SSB))
```

→

```
[Out]
Sum of Squares Between: 60
39.73
```

And the sum of squares within

```
[In]
SSW = 1000*var_france + \
      1000*var_italy + \
      1000*var_us
print('Sum of Squares Within: {0:.9g}'.format(SSW))
```

→

```
[Out]
Sum of Squares Within: 222
6089.55
```

The degrees of freedom

```
[In]
dof_SSB = 2
dof_SSW = 3000-3
print(dof_SSB)
print(dof_SSW)
```

→

```
[Out]
Degrees of Freedom for SSB
: 2
Degrees of Freedom for SSW
: 2997
```

F Statistics

```
[In]
f_stat = (SSB/dof_SSB) / (SSW/dof_SSW)
print('F-Statistic: {0:.6g}'.format(f_stat))
```

→

```
[Out]
F-Statistic: 4.06567
```

Finally, compare the *p* value and significance level to decide whether to reject the hypothesis we made (null or alternative), automating the ANOVA test.

[In]

```
import statsmodels.api as sm
from statsmodels.formula.api
import ols
model = ols('price ~ country',
data=wine_df).fit()
anova = sm.stats.anova_lm(model,
typ=2)
anova
```

[Out]

	sum_sq	df	F	PR(>F)
country	6.039731e+03	2.0	4.069736	0.017176
Residual	2.223863e+06	2997.0	NaN	NaN

This code does all the calculations and produces a summary table complete with an *F* statistic and *p* value. We used 'ols' function stands for ordinary least squares to fit a linear model to the data. The 'fit' is for model object to fit the model to the data. The 'anava_lm' from the 'sm.stats' module specifies the type of test to perform. The result showed an *F* statistic value of 4.06567 and a *p* value, which can be used to determine the significance level and reject or accept the hypothesis.

Degrees of Freedom?

'Degrees of Freedom' in statistics represents the count of independent observations in a sample, crucial for determining the reliability of statistical tests and estimates.

How do degrees of freedom work? Degrees of freedom quantify the changeable independent variables in a statistical study, providing a measure of data inclusion in the sample or test. For instance, examining the heights of 100 adult males in a population gives us 100 degrees of freedom, each representing a unique data point.

What makes degrees of freedom crucial? Its significance lies in affecting the precision and reliability of sample estimates and statistical tests. A sample size of 100 men instead of 10, while studying male heights, will yield a more dependable mean and variance, due to the larger degrees of freedom.

Degrees of freedom shape the probability distribution for the test statistic in various hypothesis tests like *t* Tests, *F* tests, and chi-square tests, ultimately affecting the computed *p* values.

How do you figure out degrees of freedom? The computation of degrees of freedom is typically formula-based and varies with sample and test type. It is generally the count of data observations minus the number of parameters estimated in the analysis. For instance, for sample mean, degrees of freedom equal sample size minus one.

$$Degrees\ of\ Freedom(DoF) = N(observations) - p(parameters\ to\ be\ estimated)$$

Considering a statistical model's degrees of freedom is paramount to ensure accurate and valid results, as it influences the model's ability to fit the data correctly.

2. Two-Way ANOVA

A two-way ANOVA compares the differences in means between groups that have been split on two independent variables (called *factors*). The primary goal of a two-way ANOVA is to determine whether there is an interaction between two independent variables on the dependent variable (Manochehr, 2006).

For example, you could use two-way ANOVA to determine whether there is an interaction between gender and education level on test heart rate among university students, where gender (male/female) and exercise (none/moderate/strong) are your independent variables. You may want to determine whether there is an interaction between physical activity and gender on blood cholesterol concentration in children, where physical activity (low/moderate/high) and gender (male/female) are your independent variables and cholesterol concentration is your dependent variable.

- Used when there are two groupings (independent variable)
- Research problem: Resting heart rate varies by gender and exercise level.
- Example of the two-way ANOVA results

[Table 1.6.7] Data Set for Two-way ANOVA

	No exercise	Moderate exercise	Strong exercise	
9 males	75.00	67.33	62.00	68.11
9 females	77.00	74.33	70.00	73.78
	76.00	70.83	66.00	

Table 1.6.7 presents a data set for a two-way ANOVA, comparing the effects of exercise level (no exercise, moderate exercise, and strong exercise) and gender (males and females) on a dependent variable. We notice that males have a lower average score across all exercise levels compared to females, and the average score decreases as the exercise intensity increases for both genders.

[Table 1.6.8] Results for Two-way ANOVA; Dependent Variable (Heart Rate)

Source	SS	*df*	MS	*F*	*p*
G1	144.500	1	144.500	33.779	.000
G2	300.111	2	150.056	35.078	.000
G1 × G2	31.000	2	15.500	3.623	.059
Error	51.333	12	4.728		
Total (Corr)			.903 (adjusted = .962)		

Notice the table (Table 1.6.8) presents the findings of a two-way analysis of variance (ANOVA) that looked at how two independent factors (G1 and G2) affected the dependent variable (heart rate). The findings show that G1 and G2 both significantly affected heart rate (p .05), with G1 having a bigger impact (F = 33.779, p .000) than G2's (F = 35.078, p .000). Significant (F = 3.623, p =.059) was also the G1 and G2 interaction. The variables had a strong correlation (r =.903, adjusted r^2 =.962). The interaction term in a two-way ANOVA tells you whether the effect of one of your independent variables on the dependent variable is the same for all values of your other independent variable (and vice versa) (Figure 1.6.3).

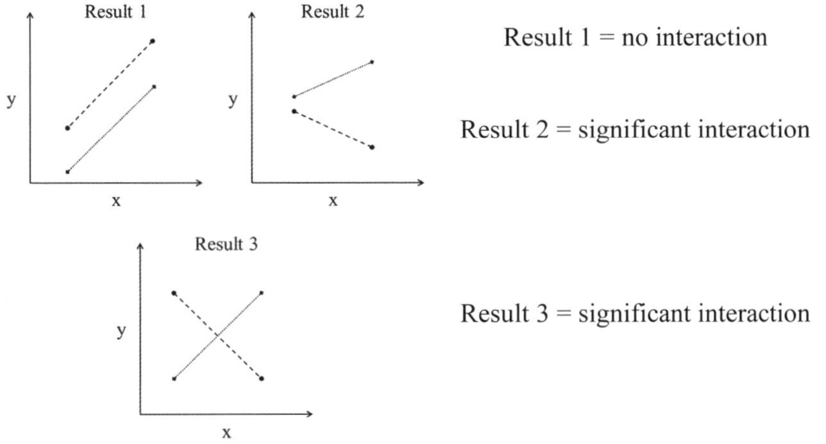

Figure 1.6.3. Schematic representation of the two-way ANOVA

For example, is the effect of gender (male/female) on the test variable influenced by exercise level? Figure 1.6.3 illustrates a schematic representation of the two-way ANOVA results, depicting two graphs with significant interaction effects and one graph showing no interaction between the factors being analyzed. As a matter of fact, if a statistically significant interaction is found, you need to determine whether there are any simple main effects and, if there are, what these effects are.

+ **Performing (Two-Way) ANOVA Numeric:** First, the primary principle behind ANOVA is to test the difference among population means by assessing the ratio of within-group variation to between-group variation.

$$F = s_b^2 / s_w^2$$

Within-group variance is due to random unexplained disturbance, whereas between-group variance is caused by different treatments. In ANOVA, we test the null hypothesis (H_0), wherein all population means are the same, or the alternative hypothesis (H_1), wherein at least one population mean is different. Let us create factors and levels as below and test it.

[Table 1.6.9] Created Factors, Levels, and Organized Data Set

0 mg	50 mg	100 mg
9	7	4
8	6	3
7	6	2
7	7	3
7	8	4
9	7	3
8	6	2

Day 1	Day 2	Day 3
9	7	4
8	6	3
7	6	2
7	7	3
7	8	4
9	7	3
8	6	2

	0 mg	50 mg	100 mg
Men	9	7	4
	8	6	3
	7	6	2
Women	7	7	3
	7	8	4
	9	7	3

Notice in Table 1.6.9, set of left side represents Factor (variables Dosage, Levels: 0 mg, 50 mg, 100 mg). Middle side denotes data set for exercise. And the table of right side denotes organized data set for exercise.

In ANOVA, it is assumed that the sample is drawn from a normally distributed population and that the population variance is equal. Error in one case is independent of error in another (outliers have been removed from the data set). Homogeneity of variance by population variance in different levels of each independent variable is equal.

Main Effects in ANOVA: Pretend we are comparing the test scores of people who have received medication (100 mg dosage group) and people who have not received the medication (0 mg dosage group). The 0 mg condition has a mean of 60, whereas the 100 mg condition has a mean of 80. This can be represented in a graph like the Figure 1.6.4' left side:

- $H_0: \mu_{A1} = \mu_{A2} = \mu_{A3}$
- H_1: not all means are equal

Interaction Effects in ANOVA: Here we have a factorial ANOVA with two factors: dosage (0 mg and 100 mg) and gender (men and women). In the 0 mg dosage condition, men have a mean of 60 whereas women have a mean of 80. In the 100 mg dosage condition, men have a mean of 80 whereas women have a mean of 50. This can be represented in a graph like the Figure 1.6.4's right side:

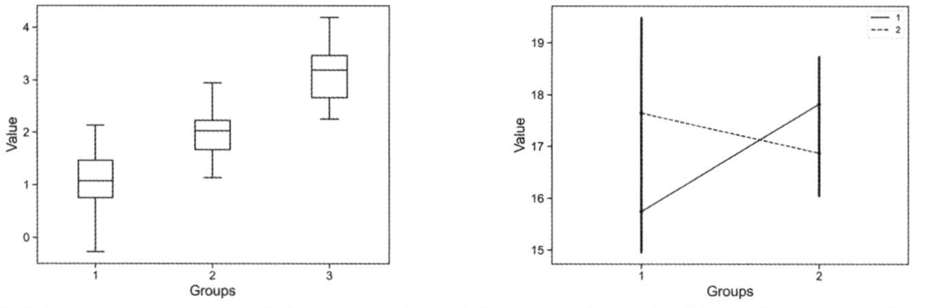

Figure 1.6.4. Representation of the actual data driven result. Left side = the main effects in an ANOVA, Right side = the interaction effects in an ANOVA result (interaction present)

Hypotheses for ANOVA Exercise

$H_0: \mu_{A1} = \mu_{A2} = \mu_{A3}$ H_1: not all means are equal	$H_0: \mu_{B1} = \mu_{B2} = \mu_{B3}$ H_1: not all means are equal	H_0: interaction absent H_1: interaction present

In the 'Hypotheses' above, the null hypothesis for the interaction term states that the effect of one independent variable is the same across all levels of the other independent variable. The alternative hypothesis states that the effect of one independent variable is not the same across all levels of the other independent variable, including the presence of an interaction between the two variables.

3. A Comparison of One-Way and Two-Way ANOVA

Let us clarify some more between the two models. The one-way ANOVA and two-way ANOVA are statistical techniques for comparing group means. In contrast to two-way ANOVA, which looks at the connection between two independent categorical variables and a dependent variable, one-way ANOVA concentrates on the association between a single categorical variable and a continuous dependent variable.

+ **One-way ANOVA:** A hypothesis test in which only one categorical variable or single factor is taken into consideration. With the help of the *F* distribution, it enables us to compare the means of three or more samples. The null hypothesis is 'All population means should be equal,' whereas the alternative hypothesis is 'There should be a difference in at least one mean.'

+ **Two-way ANOVA:** Examines the effect of two independent factors on a dependent variable. Also studies the interrelationship between independent variables influencing a dependent variable, if any.

The tests' outcomes shed light on whether or not the means of various groups are comparable or noticeably different (Table 1.6.10).

[Table 1.6.10] A Comparison of One-Way and Two-Way ANOVA

	One-way ANOVA	Two-way ANOVA
Assumptions	- The populations from which the samples are drawn are approximately normally distributed, and have the same variance. - The samples drawn from the different populations are random and independent.	- The populations from which the samples are drawn are approximately normally distributed. The categorical independent group should have the same size of the population (or samples). - There are two or more categorical independent groups in two factors. The dependent variables can be continuous.
Applications	- Gender as a categorical variable impacting state sales of ecosites (e.g., Amazon). - Different levels of blood pressure in three populations. - Measuring glycogen content for multiple samples of heart, liver, kidney, lung, and so on.	- Analyzing the test scores of a class based on gender and age. - Here test score is the dependent variable and gender and age are the independent variables. - Measuring the response to three different drugs in both men and women. Drug treatment is one factor and gender is the other.

We can use these tests to study any interrelationships between independent factors that might affect the dependent variable as well as the effect of independent variables on a dependent variable.

Choosing the Appropriate Hypothesis Testing: *t* Test, ANOVA, and Chi-Square

***t* Test:** This applies whenever a continuous variable and a categorical variable are involved. We employ the one-sample *t* Test when examining a single continuous numeric variable to determine if the mean is above or below a certain value. The two-sample *t* Test is used in this situation where there are two levels, such as when examining the relationship between height and gender to see whether gender affects average age.

ANOVA: Now imagine that you are utilizing a categorical variable that has more than two levels in comparison to a continuous variable. When this happens, use ANOVA. For instance, because Grade includes more than two levels, ANOVA should be used rather than a *t* Test to determine the impact of Attendance on the grade.

Chi-Square: When attempting to determine if two category variables are related to one another, a Chi-square test is utilized. For instance, we run a Chi-square test on our dataset to see whether there is a relationship between grade and gender or to determine whether grade is influenced by gender (see Statistical Model Part 5 for more detail).

* Please remember that before doing any test, we should first assume a null and alternate hypothesis before drawing conclusions based on the *p* value.

Remarks: As we taught, when comparing two groups, the *t* Test is used, and when comparing three or more groups, the ANOVA is utilized. Both models are susceptible to changes in sample size, alpha level, and effect size as well as assumptions about the homogeneity of variance and normality of the data. ANOVA statistical models are particularly useful for examining differences between three or more groups or conditions. ANOVA with two independent variables on a dependent variable is an extension of ANOVA with one independent variable. Depending on the research objective and experimental setup, it is possible to compare one-way versus two-way ANOVA to decide whether model is better suited for a particular dataset. The proper identification and application of ANOVA can increase the reliability and validity of statistical studies and help in the right interpretation of data.

Supplemental Information 1.2: Experimental Research

Experimental research is research that includes an experiment. It contains one or more experimental treatments (experimental variables), a control condition, and dependent variables (Mutz & Pemantle, 2015).

+ **Experimental Design:** A well-designed experiment requires a detailed plan that outlines the identification and manipulation of the independent variables, the method of data collection, and the measurement of the dependent variable. The goal of an experimental design is to control for extraneous variables, ensure adequate comparison, and ensure the validity and reliability of the data collected.

Example: The objective of this experiment is to examine the impact of different forms of exercise on body fat percentage [X = Treatment, O = Observation]. Three groups will be formed, with each group performing a different type of exercise:

G1: Running		G1:	X1	O1
G2: Walking	→	G2:	X2	O2
G3: Weight training		G3:	X3	O3

The following factors should be considered when designing a successful experiment: adequate control, valid comparison, sensitive data collection, reliable dependent variables, the absence of extraneous variables, lack of confounding, parsimony, naturalness, and representativeness of the subjects, variables, and environment.

+ **Experimental Validity:** The validity of an experiment refers to the extent to which it accurately measures what it is intended to measure. Internal and external validity are two critical aspects of experimental validity, and construct and statistical conclusion validity can also be considered important criteria for a well-designed experiment.

- Internal Validity: Internal validity concerns the cause-and-effect relationship between the independent and dependent variables.
- External Validity: External validity assesses the generalizability of the results beyond the specific conditions of the experiment.
- Construct Validity: Construct validity involves ensuring that the concepts and constructs being studied are clearly defined and accurately measured.
- Statistical Conclusion Validity: Statistical conclusion validity deals with the accuracy of the decision about whether the experimental and control groups are significantly different.

+ **Various Designs:** In general, research design is a preconceived plan for conducting an experiment. To be specific, design is the structure by which variables are positioned, arranged, or built into the test. Such arrangements have certain meanings, which are explored as follows:

- Pretest: test (measurement) before the experimental treatment

- Posttest: test (analysis) after the experimental treatment
- Posttest only, no control group

G: X O

- Posttest only, control group

G1: X1 O1
G2: X2 O2
G3: — O3

- Pretest-posttest: pretest as a control condition

G: O1 X O2

- Pretest-posttest control group

G1: O1 X O2
G2: O3 — O4
⋮ ⋮ ⋮ ⋮

- Solomon four-group design

G1: O1 X O2
G2: O3 — O4
G3: — X O5
G4: — — O6

- Factorial design: A design with two or more independent grouping variables. This design allows one to examine the interaction effect. Example: The effect of exercise (three levels) and gender (two levels) on the change in body fat.

G_{M1}: X1 O1
G_{M2}: X2 O2
G_{M3}: — O3
G_{F1}: X1 O4
G_{F2}: X2 O5
G_{F3}: — O6

- Repeated measures design: The same subject is measured more than once on the dependent variable.

G1: O1 X O2 O3 O4
G2: O5 — O6 O7 O8
⋮ ⋮ ⋮ ⋮ ⋮ ⋮

Notice that creating an appropriate design and adequately implementing it require careful planning with simultaneous attention to a variety of details, but they provide the best safeguard against errors.

Practical Example of Experimental Research

Title: Water-Based Exercise for Cardiovascular Fitness in People with Chronic Stroke: A Randomized Controlled Trial.

Background: Cardiovascular disease is the leading cause of death in people with stroke. Exercise helps to prevent this; it is essential to develop different exercise protocols for stroke patients. Treadmill training, bicycle ergometer training, and circuit training improve cardiovascular fitness. Water-based exercise has potential. In fact, 8~12 weeks of water-based training has been shown to improve cardiovascular fitness and strength in healthy older adults. No studies have been done with stroke patients.

Purpose: To evaluate the effects of an 8-week water-based exercise program (experimental group) and an upper extremity function program (control group) on the cardiovascular fitness of people with stroke in a community setting.

Participants: A volunteer sample of 12 community-dwelling people ages 51~72 with mild to moderate residual motor deficits due to stroke.

Experimental treatment (intervention): One hour of group exercise three times per week for 8 weeks. The experimental group exercised in chest-deep water at targeted heart rates. The control group performed arm and hand exercises while sitting.

Outcomes (dependent variable): VO_2 max, gait speed, muscle strength

Design: Randomized controlled trial, 12 subjects were randomly assigned to two groups (n = 7 in the experimental group, n = 5 [initially 6] in the control group). Dependent variables were measured before and after the treatment.

Statistical analysis: independent t Test, two-way analysis of variance, $\alpha = .05$

[Table S1.2.1] Results for Experimental Research Example

	Experimental group		Control group	
	Pretest	Posttest	Pretest	Posttest
VO_2 max	17.3	21.2	17.1	17.6
Gait speed	0.99	1.18	1.01	1.04
Strength	2.75	2.97	2.12	1.96

Table S1.2.1 presents the pretest and posttest results for both the experimental and control groups in an experimental research study, comparing VO_2 max, gait speed, and strength outcomes.

Conclusion: A water-based group exercise program may be an effective means of promoting fitness in people with stroke.

Supplemental Information 1.3: Quasi-Experimental Research

We use quasi-experimental designs when we have less control over possible extraneous variables (Asgari & Baptista Nunes, 2011). This is research that resembles experimental research but is not true experimental research (the prefix *quasi* means 'resembling').

Although the independent variable is manipulated, participants are not randomly assigned to conditions or orders of conditions. These designs are used when random sampling and random assignment is not possible; intact groups are used. Thus, quasi-experimental designs lack internal validity (compared to actual experiments; i.e., differential selection).

What should we do about this? We have to establish the equivalence of groups before any experimental treatment. The limitations have to be clearly identified.

+ **Example Designs:** There are various experimental designs that can be employed to test a hypothesis. Here are a few examples:

Posttest Only with a Nonequivalent Control Group: In this design, two groups are formed - one group receives the treatment (X) and the other group serves as the control group. The dependent variable is measured only after the treatment has been administered (O1 and O2).

G1: X O1
G2: — O2

Pretest-Posttest with a Nonequivalent Control Group: This design is similar to the posttest only design, but the dependent variable is measured both before and after the treatment has been administered.

G1: O1 X O2
G2: O3 — O4

What if there is a difference in the pretest scores? In the event of a significant difference in pretest scores between the two groups, the validity of the experiment may be compromised.

One potential solution is to match the groups on the pretest score to control for any preexisting differences. Another option is to use a pretest-posttest design with a randomized control group, where the groups are randomly assigned to receive the treatment or serve as the control group.

+ **Single-Subject Experimental Design:** This type of experimental design is used to study the impact of a specific treatment or intervention on a single subject. It is also known as a single-case experimental design, single-system design, time-series design, or small-N design. For example;

Research Problem: The goal of this study is to determine the effectiveness of walking in reducing pain after disc surgery.

Subject: The subject of this study is a 60-year-old patient who is suffering from severe back pain following disc surgery.

Dependent Variable: The dependent variable in this study is the reported pain of the subject, which will be measured on a scale ranging from 0 to 10.

Independent Variable (Treatment): The independent variable, or treatment, in this study is a daily 1-mile walk.

- Results

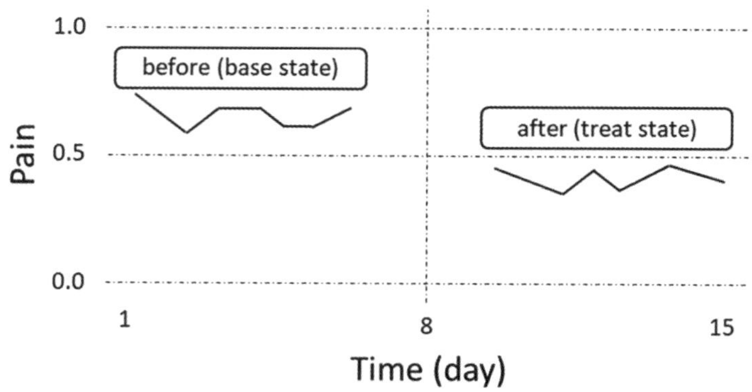

Figure S1.3.1. Representation of the single-subject design

The results of the study will provide insights into the impact of walking on reducing pain after the surgery. Characteristics: N = 1, systematic manipulation of an independent variable, repeated measures include several phases: baseline (control) phase(s) [A], treatment phase(s) [B].

+ **When and Why We Use a Single-subject Design:** This design is utilized in research when it is challenging to obtain a large number of subjects or when the subjects are not homogeneous. Additionally, when the research focus is on individual-level rather than group-level phenomena, a single-subject design may be appropriate.

In single-subject design studies, the dependent variable is typically a measure of frequency, duration, or magnitude, such as the rate of correct performance, the duration of exercise, or the level of pain.

The baseline (Control) phase (A) is a period during which no treatment is administered. Measurements are taken until stability is demonstrated. The treatment (Intervention) phase (B) is the period during which the intervention is administered. Ideally, this phase should be of the same length as the baseline phase and stability is desirable.

+ **Various Designs:** Basic design (A-B design) is one baseline phase (A) and one treatment phase (B). This design is limited control over threats to internal validity → and it can be solved by replication (more phases, subjects, and/or interventions).

- Withdrawal (reversal) designs (A-B-A design)

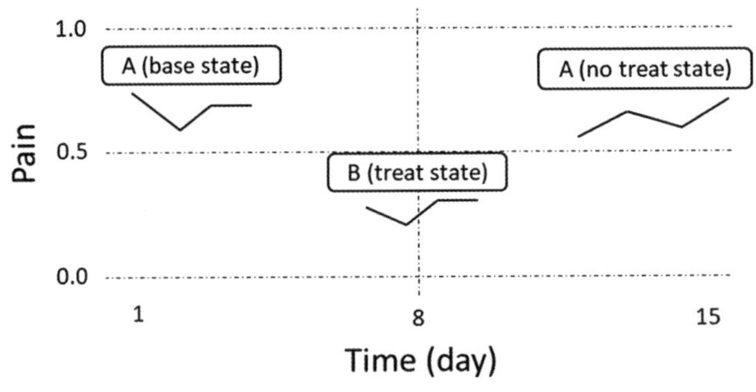

Figure S1.3.2. Representation of the withdrawal (reversal) design

Treatment is withdrawn or reversed and there is better control over internal validity. Problem: If the effect is irreversible (e.g., if there is a learning effect), it is unethical to withdraw treatment.

- A-B-A-B design

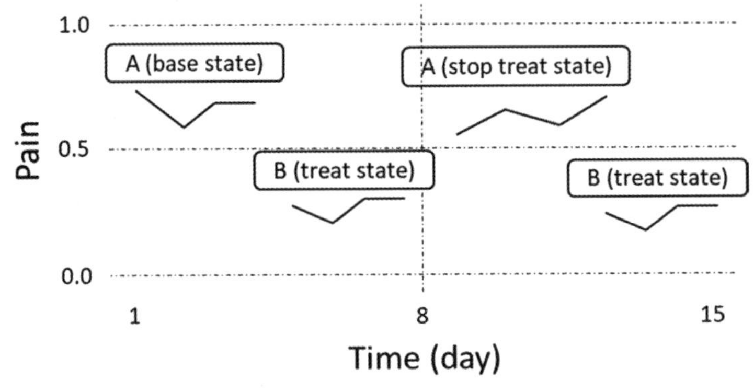

Figure S1.3.3. Representation of the A-B-A-B design

Used when a subject needs an immediate treatment.

Multiple-baseline design: replication of several subjects, dependent variables, situations.

+ **Pros and Cons of the Single-subject Design:** The single-subject design has both advantages and disadvantages.

This design has a number of advantages, including its systematic experimental approach, which enables researchers to look at the causal relationship between an independent variable and a dependent variable, its singular subject focus, its adaptability, which allows researchers to change the design as the study goes along, and its practicality, as the study's results can be used in practice right away. This design also takes the dependent variable's trend of change into account.

- Strengths: Systematic experimental approach → examine the causal relationship. Focused on an individual subject. Flexible → can modify the design as the study progresses. Practical → findings can be applied directly to practice. Many consider the trend of change.
- Weaknesses: Low internal validity: statistical regression and selection bias in particular are concerns. Low external validity.

Single-subject designs, however, can have certain drawbacks. The low internal validity of this design, which may be influenced by elements like statistical regression and selection bias, is one of its main shortcomings. Due to the possibility that the findings will not apply to bigger populations, this approach may also have low external validity. This is because the study only looked at one participant, and it is possible that the findings would not apply to other people or communities.

Other Quantitative Research

What is nonexperimental quantitative research? It is research that includes the collection and analysis of quantitative data but does not manipulate an independent variable (Johnson, 2001).

- Less control over variables and other research conditions.
- Includes many variables simultaneously.
- Often the purpose of a study is to describe facts or examine noncausal relationships between variables.

Names used to indicate this kind of research: ex-post facto research, causal-comparative research, survey research, correlational research, descriptive research (Canning et al., 2004).

+ **Ex-post Facto Research:** Research that examines variables (and relationships between variables) from retrospectively collected data. Various types of data, such as already existing data and self-reported data, can be used.

+ **Causal-comparative Research:** Ex post facto research that explores a possible cause-and-effect relationship with data collected without any manipulation. Often independent and dependent variables can be identified.

+ **Example:** Descriptive Study

- Title: 'Prevalence of Overweight and Obesity in a Provincial Population of Canadian Preschool Children'

- Data: The height and weight of 4,161 children born in 1997 and living in Newfoundland and Labrador were collected (the actual study was conducted in 2000).
- Evaluation: Body mass index (kg/m^2) was calculated and compared to age-specific criteria to determine the prevalence of being overweight (> 90th percentile), obese (> 95th percentile), or normal.
- Results: In all, 25.6% of the children were overweight or obese, and there was no significant difference between boys and girls.
- Interpretation: A high proportion of children are overweight or obese. Prevention measures should begin before the age of 3 years.

+ **Survey Research:** Surveys are often used to measure a large number of variables from a large number of people within a short period. In general, there are two kinds of survey questions: one measuring facts from subjects' simple recall and one measuring psychological attributes (e.g., attitudes, opinions, perceptions).

- **Survey:** In the development process of a survey questionnaire, researchers should first identify the constructs to be measured, create relevant questions and answer options, assess content validity, conduct a pilot test and make revisions accordingly, gather data from a large sample, and evaluate the validity of the instrument.

- **Survey Administration:** It can be conducted through various methods, including group-to-person surveys, mail surveys, interview surveys, telephone interview surveys, and web-based surveys. However, some challenges to consider when conducting surveys are nonresponse and maintaining confidentiality of respondents' information.

+ **A study Examining Change (comparing different times):** A study investigating change over time, such as a longitudinal study, can involve various approaches, including trend studies, cohort studies, and panel studies, while cross-sectional studies focus on comparing different groups at a specific point in time.

Supplemental Information 1.4: Qualitative Research

Qualitative research is used to investigate complex phenomena in depth. Priori assumptions and conclusions (i.e., the directional research hypothesis in quantitative research) are avoided (Newman et al., 1998).

+ **General Characteristics:** Like quantitative research, qualitative research encompasses a range of methods, but there are some common characteristics that are unique to qualitative research. These include:

- No Manipulation or Intervention: In qualitative research, there is no manipulation or intervention of the subjects or environment. The focus is on understanding and describing the experiences and perspectives of the subjects in their natural setting.
- Conducted in Natural Settings: Qualitative research is mostly conducted in the subjects' natural environment, rather than in a laboratory or controlled setting.
- Less Structured: Qualitative research is less structured than quantitative research, allowing for a more flexible and exploratory approach.
- Flexible: Qualitative research is flexible, allowing the researcher to adapt and modify the research design and methods as the research progresses.
- Researcher Dependent: Qualitative research is often dependent on the skills, knowledge, and experience of the researcher. The researcher plays a critical role in determining the research design, data collection methods, and data analysis.

+ **Components of the Research Design** (working design): This is the preliminary plan that starts and identifies the research; decisions are made regarding the following.

- Subjects to be studied (Who?)
- Site (Where?)
- Schedule (When and for how long?)
- Possible information needed (What about?)

Working hypothesis: Qualitive research uses inductive inquiry without any preconceived theories or hypothesis. Although there are no hypothesis statements, foreshadowed problems, research problems, and questions come in at this point, which may be reviewed as the data collection and analysis proceed. For example:

- Does work culture play any role in injuries among caregivers in hospitals? If it does, what are the critical elements?

Data collection: A host of issues must be dealt with. The researcher needs to gain access to data formally and decide on a particular role; will the researcher be a participant-observer or simply an observer?

Notice the most commonly used methods of data collection include:

- Methods: interview, observation, compilation and review of documents
- Interactive data collection: data are collected through interaction between the researcher and subjects (e.g., interviews).
- Noninteractive data collection: data are obtained without interaction between the researcher and subjects (e.g., observation, review of documents).

Analysis and interpretation: Because qualitative research usually tests working hypotheses with unanticipated results, data analysis and interpretation run together with the data collection.

- The purpose is to accurately describe and explain a phenomenon.
- Organization, categorization, synthesis: reduction of data.

Because there will be a lot of information from field notes or interviews, they need to be reduced by organization, and comparisons can be made based on the working hypotheses or initial theories.

+ **Case Study:** A case study is a detailed investigation of a case, person, organization, specific event, and so on. Widely used in qualitative study. The most commonly used designs for case study research are historical organizational studies and observational studies.

+ **Validity:** Reliability and validity may cause more difficulty in qualitative research versus quantitative research. In general, qualitative research lacks internal validity. To enhance internal validity, include multiple observers (interviewers) and verify results and conclusions from two or more sources. Broad generalization (to other situations) is typically not a concern (External validity). The application of the results to other situations is left to the reader. The reader is more concerned with comparability and translatability (e.g., whether there is adequate theory and proper procedure, how well the results are documented).

Some aspects, such as a well-organized procedure including logical analysis of results, and logically well-documented descriptions can enhance validity.

+ **Example:** Work organizational culture and injuries of caregivers.

- Method: interviews of caregivers working at eight hospitals: four high-injury hospitals and four low-injury hospitals
- Research questions: Does work culture play a role in injuries among caregivers in hospitals? If it does, what are the essential elements?
- Issues mentioned: bad equipment, too many patients (staff shortage), schedule, doctor–nurse relationship, bad organization of work, teamwork, leader (supervisor) of a team

Epidemiological Research

What is epidemiology? Epidemiology is the study of the distribution and determinants of disease frequency in human populations (MacMahon & Pugh, 1970). Physical activity epidemiology is the study of the relationship between physical activity and health (disease).

+ Types of Epidemiological Studies

Descriptive study: A study describing the general characteristics of the distribution of disease by person, place, or time.

Case-control study: A study comparing a case group (that has the disease) and a control group (that does not have the disease) on possible determinants of the disease.

Example: 'Lung Cancer and Cigarette Smoking in Women: A Multicenter Case-Control Study in Europe' (Agudo et al., 2000)

- Case group: 1,556 patients with lung cancer in Europe
- Control group: 2,450 randomly selected (age-matched) subjects without cancer
- Researchers interviewed the case and control groups on smoking history and other possible determinants (drinking, medication use, etc.).

[Table S1.4.1] Results for Epidemiological Studies

	Cases	Controls
Ever-smokers	1,096 (*a*)	849 (*b*)
Never-smokers	460 (*c*)	1,601 (*d*)
Percentage of smokers	70.4%	34.7%

The table S1.4.1 presents results from epidemiological studies, showing a higher percentage of smokers among cases (70.4%) compared to controls (34.7%).

Odds ratio = $(a/b)/(b/d) = ad/bc = (1{,}096 \times 1{,}601)/(849 \times 460) = 4.49$
Smokers have an approximately 4.49 times higher risk of having lung cancer.

Conclusion: Smoking is a critical determinant of lung cancer.

Cohort study (follow-up study): a study in which observations are made of a specific group (can be a population or a sample), typically over a period of time

- Retrospective cohort study
- Prospective cohort study

Intervention study = experimental study = clinical trial: The characteristics of an intervention study are the same as those of an experimental study. Intervention = experimental treatment.

[1.7]

Statistical Model, Part 3

Overview of Correlation, Basic Assumptions about Correlation, Facts about Correlation

There are two major types of analysis. One is used with nominal or categorical independent variables (i.e., ANOVA), and the other is used with continuous independent variables (i.e., correlation). Continuous variables can take on many values (in theory an infinite number of values).

Examples of continuous variables include height and weight, scores on tests like Scholastic Assessment Test, and measures of sentiment (such as job satisfaction). The most straightforward correlation and regression models assume linear relations and continuous independent and dependent variables.

1. Overview of Correlation

Correlation means 'co-relation' or the degree to which two variables go together (Pearson, 1920). Statistics can show the relationship between two continuous variables, such as exercise frequency and life span or caffeine intake and blood pressure.

- One variable can cause or depend on the value of another variable.
- One variable can be weakly associated with another variable.
- Two variables can depend on a third unknown variable.

It can be useful in data analysis and modeling to better understand the relationships between variables. The statistical relationship between two variables is referred to as their correlation.

$$r = \sum \frac{Z_x Z_y}{N-1}, \qquad (or) r = \frac{cov_{xy}}{s_x x_y} = \frac{\frac{\sum (x-\bar{x})(y-\bar{y})}{N-1}}{s_x x_y}$$

Here, Z_x and Z_y are the standardized values of the two variables (x and y). N is the sample size, $\bar{x}$ and $\bar{y}$ are the means of the two variables. cov_{xy} is the covariance of the two variables, and s_x and x_y are the standard deviations of the two variables.

A linear correlation is a straight line.

- A correlation can be positive, which means that both variables move in the same direction, or negative, which means that when one's value increases, the other's value decreases.
- A correlation can also be neutral or 0, which means that the variables are unrelated (Figure 1.7.1).

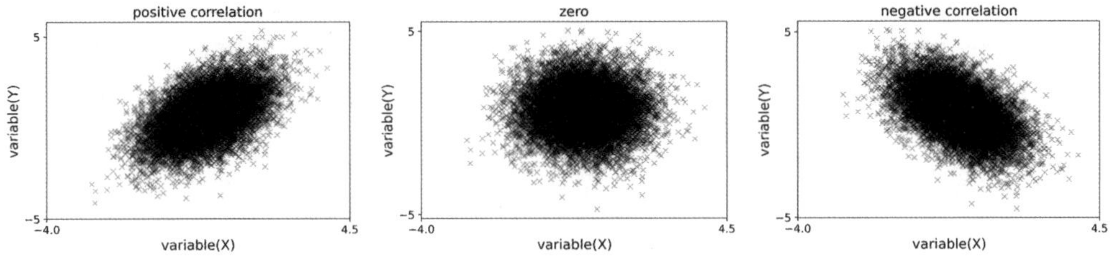

Figure 1.7.1. Prototype of various correlations

The plot on the left shows a positive correlation; both variables change in the same direction. The plot in the middle denotes a neutral correlation; there is no relationship between the changes in the variables. The plot on the right shows a negative correlation; the two variables change in opposite directions.

The structure of the relationship may be known (e.g., it may be linear), or we may have no idea whether a relationship exists between two variables or what form it may take. Depending on what is known about the relationship between and distribution of the variables, different correlation scores can be calculated.

2. Basic Assumptions about Correlation

A correlation coefficient is a number that summarizes the direction and degree (closeness) of the linear relationship between two variables. It is also known as the Pearson product-moment correlation coefficient.

- Range: $-1.00 \leq r \leq 1.00$
- Shows linear relationship only
- Dose not represent a causal relationship
- Often used to represent validity and reliability

The population value is called ρ (rho), the sample value is called r, the coefficient of determination is called r^2.

- r^2 represents the percentage of the variance shared by two variables
- The correlation coefficient can take a value between –1 and +1
- The sign (+ or –) of the correlation affects its interpretation

+ The Statistical Significance of Correlation Coefficients: Correlation coefficients have a probability (*p*), which shows the probability that the relationship between the two variables is equal to 0 (null hypotheses; no relationship).

- Strong correlations have low *p* because the probability that they have no relationship is very low.

[Table 1.7.1] Example of Correlation Results

		Exam 1 score	Exam 2 score	Age of respondent	Sex of respondent
Exam 1 score	Pearson correlation	1	.610(**)	.705(**)	.469(*)
	Sig. (2-tailed)		.004	.001	.037
	N	20	20	20	20
Exam 1 score	Pearson correlation	.610(**)	1	.839(**)	.766(*)
	Sig. (2-tailed)	.004		.000	.000
	N	20	20	20	20
Age of respondent	Pearson correlation	.705(**)	.839(**)	1	.709(**)
	Sig. (2-tailed)	.001	.000		.000
	N	20	20	20	20
Sex of respondent	Pearson correlation	.489(*)	.766(**)	.709(**)	1
	Sig. (2-tailed)	.037	.000	.000	
	N	20	20	20	20

In scientific research as represented in Table 1.7.1, correlations are typically considered statistically significant if *p* is less than 0.05, but the researcher has the liberty to decide the *p* at which he or she will consider the relationship significant. The value of *p* at which a correlation is considered statistically significant is called the *alpha level* and must be reported.

- Notation for *p*: Sig. (2-tailed). In the previous example, $r = 0.62$ and $p = 0.03$. The *p* of 0.03 is less than the acceptable alpha level of 0.05, which means that the correlation is statistically significant.

Four things must be reported to describe a relationship, as follows:

- First, the correlation coefficient gauges how closely two variables are related to one another.
- Second, the correlation coefficient's sign tells us how the connection is going to go. The variables move in the same direction when it is positive; in the opposite direction when it is negative.
- Third, when two variables are correlated linearly, it is measured by the Pearson correlation coefficient. For the correlation coefficient to be calculated, the connection must be linear.
- Finally, the *p* value is employed in order to ascertain whether the association is

statistically significant. The link is deemed statistically significant if the p value is below a specific cutoff (often 0.05).

The images above (Figure 1.7.2) illustrate what relationships might look like at different degrees of strength (for different values of r).

- When $r = 0$, the points have no direction (middle plot), the shape is almost round, and no line fits to the points on the graph.

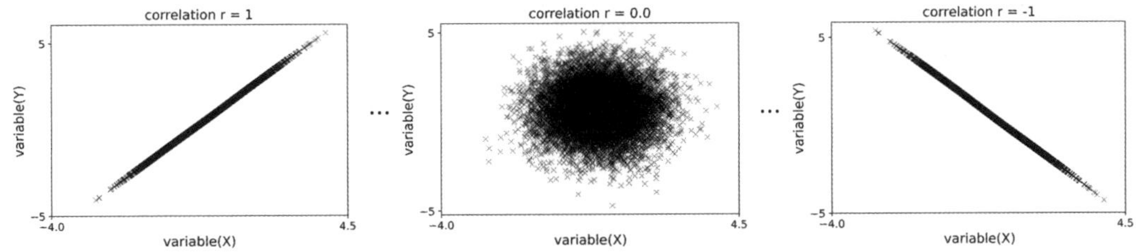

Figure 1.7.2. Simulated random variables with different correlations

As the correlation coefficient increases (or decreases), the observations group closer together in a linear shape. The line is difficult to detect when the relationship is weak but becomes more apparent as relationships become stronger (left and right plots).

Variance and Covariance

In analysis, the concepts of variance and covariance are crucial.

Variance: Mathematically speaking, the variance is defined as the average of the squared differences from the mean. Here is the mathematical representation of variance (Figure 1.7.3):

Population variance

$$\sigma^2 = \frac{\sum_{i=1}^{N}(x_i - \mu)^2}{N}$$

σ^2=population variance
x_i=value of element
μ=population mean
N=population size

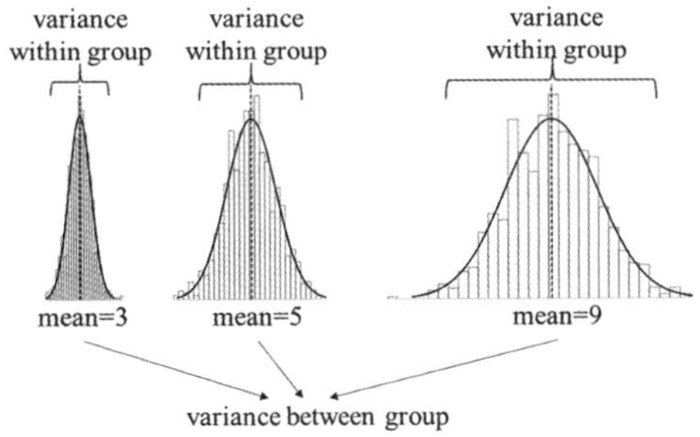

Sample variance

$$s^2 = \frac{\sum_{i=1}^{n}(x_i - \bar{x})^2}{n - 1}$$

x^2=sample variance
x_i=value of element
$\bar{x}$=sample mean
n=sample size

Figure 1.7.3. Mathematical representation of the variance

Thus, the variance measures the spread between numbers in a data set.

Covariance: Covariance is a measure of the joint variability of two random variables, which is defined as the expected value (or mean) of the product of their deviations from their individual expected values. Covariance helps us measure (and understand) the linear relationships between variables (Wildt et al., 1978).

Let us use an example to better understand this. Suppose we have two variables, height and weight, and we want to measure the relationship between them. Before we do that, however, are we sure there is in fact a relationship between them? How do we know whether any sort of relationship exists?

One crude way to understand this is to check. For example, we can see whether height increases as weight increases, or height decreases as weight increases. Refer to the graph below (Figure 1.7.4):

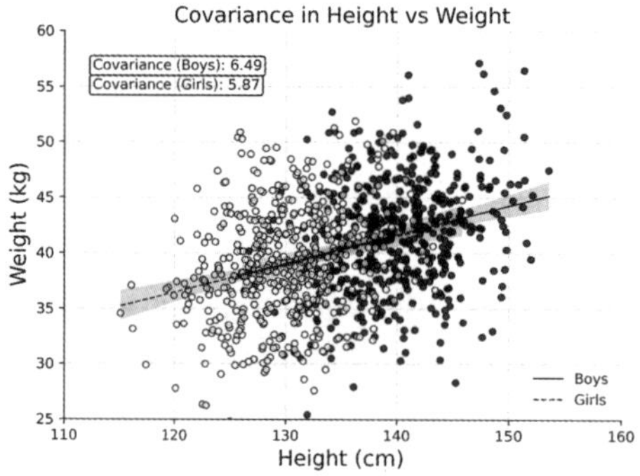

$cov(x,y) = \frac{1}{n}\sum_{i=1}^{n}(x-\bar{x})(y-\bar{y})$
$cov(x,y)$ = cov between x and y variables
x & y = members of x and y variables
$\bar{x}$ & $\bar{y}$ = mean of x and y variables
n = number of members

Figure 1.7.4. Mathematical representation of covariance

In the graph, it is very clear that as height increases, weight also increases irrespective of whether the person is a girl or a boy. When observed (using a scatter plot), this gives us an idea about the linear relationship between the variables.

- $cov(x,y)$ = positive; if x increases, y also increases
- $cov(x,y)$ = negative; if x increases, y decreases

Covariance is represented in its mathematical form in the formula on the right; the above two observations are based on this formula for covariance.

Covariance analysis in programming: Some variables may show a linear relationship. This relationship is consistently additive across two data samples. The relationship can be summarized between two variables. It is calculated as the average of the product of the values of each sample when the values have been centered (had their means subtracted). Sample covariance is calculated as follows:

$$cov(x,y) = sum\big(x - mean(x)\big) * \big(y - mean(y)\big) * 1/(n-1)$$

The use of the mean in the calculation suggests the need for each data sample to have a Gaussian or Gaussian-like distribution. The sign of the covariance can reflect whether the two variables change in the same direction (positive) or in different directions (negative). The magnitude of the covariance is not easily interpreted. A covariance of 0 indicates that both variables are entirely independent.

The *cov()* NumPy function can be used to calculate a covariance matrix between two or more variables.

```
[In]
Covariance = cov(data1, data2)
```

The diagonal of the matrix contains the covariance between each variable and itself. The other values in the matrix represent the covariance between the two variables; in this case, the remaining two values are the same, given that we are calculating the covariance for only two variables.

We can calculate the covariance matrix for the two variables in our test problem. The complete example is listed below.

Import libraries and modules.

```
[In]
From numpy.random import randn
from numpy.random import seed
from numpy import cov
```

Seed the random number generator and prepare the data.

```
[In]
Seed(1)
data1 = 20 * randn(1000) + 100
data2 = data1 + (10 * randn(1000) + 50)
```

Calculate the covariance matrix.

```
[In]
Covariance = cov(data1,
data2)
print(covariance)
```

→

```
[Out]
[[385.33297729    389.7545618]
[389.7545628     500.38006058]]
```

The 'cov' function takes two arguments; 'data1' and 'data2' which we want to calculate and the function returns the covariance matrix.

The covariance and the covariance matrix are widely used in statistics and multivariate analysis to characterize the relationships between two or more variables. Running the example calculates and prints the covariance matrix. Because the data set was contrived, with each variable drawn from a Gaussian distribution and the variables correlated linearly, covariance is reasonable for describing the relationship.

The covariance between the two variables is 389.75. We can see that it is positive, which suggests that the variables change in the same direction, as we would expect.

A problem with covariance as a statistical tool alone is that it is challenging to interpret. This leads us to Pearson's correlation coefficient.

3. Facts about Correlation

The order of variables in a correlation is not essential. Correlations provide evidence of association, not causation. The r has no units and does not change when the units of measure of x, y, or both are changed. A positive r indicates a positive association between the variables, and a negative r indicates a negative association.

- The r is always a number between -1 and 1.

Here we look at one score for variables that have a Gaussian distribution and a linear relationship and another that does not assume a distribution and will report on any monotonic (increasing or decreasing) relationship. First, let us import libraries and modules such as 'From numpy import mean', 'from numpy import std', 'from numpy.random import randn', 'from numpy.random import seed', 'from matplotlib import pyplot'.

Before we look at correlation methods, we need to define a data set we will use to test the methods. We generate 1,000 samples of two variables with a strong positive correlation. The first variable is random numbers drawn from a Gaussian distribution with a mean of 100 and a standard deviation of 20. The second variable is valued from the first variable with Gaussian noise added, with a mean of 50 and a standard deviation of 10.

We use the *randn()* function to generate random Gaussian values with a mean of 0 and a standard deviation of 1, then multiply the results by our own standard deviation and add the mean to shift the values into the preferred range. The pseudorandom number generator is seeded to ensure that we get the same sample of numbers each time the code is run.

Seed the random number generator and prepare the data as follows:

```
[In]
seed(1)
data1 = 20 * randn(1000) + 100
data2 = data1 + (10 * randn(1000) + 50)
```

When it summarized, the output will be as follows:

```
[In]
print('data1: mean=%.3f stdv=%.3f' %
(mean(data1), std(data1)))
print('data2: mean=%.3f stdv=%.3f' %
(mean(data2), std(data2)))
```

→

```
[Out]
data1: mean=100.776
stdv=19.620
data2: mean=151.050
stdv=22.358
```

Running the example first prints the mean and standard deviation for each variable: A scatter plot of the two variables is created. Because we contrived the data set, we know there is a relationship between the two variables. This is clear when we review the generated scatter plot, in which we can see an increasing trend.

[Out]

```
[In]
Pyplot.scatter(data1,
data2)
pyplot.show()
```

→

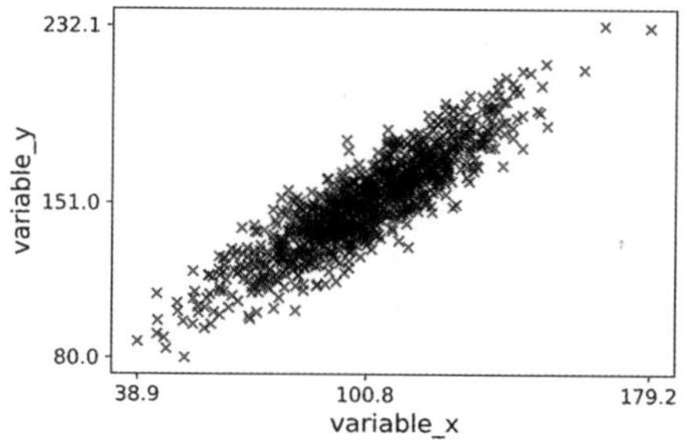

Figure 1.7.5. Results of the correlation - variables x and y (represent each data 1 and 2)

The scatter plot is a useful visualization tool for exploring the relationship between two continuous variables (Figure 1.7.5). The 'scatter' in the code above takes two arguments; 'data1' and 'data2', which are plotted on the x-axis and y-axis, respectively. Each data point represented as a dot on the plot.

+ **Pearson's Correlation:** Pearson's correlation coefficient (named for Karl Pearson) can be used to summarize the strength of the linear relationship between two data samples. Pearson's correlation coefficient is calculated as the covariance of the two variables divided by the product of the standard deviation of each data sample. It normalizes the covariance between the two variables to give an interpretable score.

$$Pearson's\ correlation\ coefficient\ =\ covariance(X,Y)/(stdv(X)\ \times\ stdv(Y))$$

The use of the mean and standard deviation in the calculation suggests the need for the two data samples to have a Gaussian or Gaussian-like distribution. The result of the calculation, the correlation coefficient, can be interpreted to understand the relationship.

The coefficient has a value between -1 (complete negative correlation) and 1 (complete positive correlation). A value of 0 means no correlation. The value must be interpreted, where often values below -0.5 or above 0.5 indicate a strong correlation and values below those values suggest a weaker correlation.

A computer programing (i.e., *pearsonr()* SciPy) function can be used to calculate Pearson's correlation coefficient for two data samples of the same length. We can calculate the correlation between the two variables in our test problem. The complete example is listed below.

Import libraries and modules.

```
[In]
From numpy.random import randn
from numpy.random import seed
from scipy.stats import pearsonr
```

Seed the random number generator and prepare the data.

```
[In]
Seed(1)
data1 = 20 * randn(1000) + 100
data2 = data1 + (10 * randn(1000) + 50)
```

Calculate Pearson's correlation.

```
[In]
Corr, _  = pearsonr(data1, data2)
prtin('Pearsons correlation: %.3f' %
corr)
```

→

```
[Out]
Pearson's correlation: 0.888
```

Running the example calculates and prints Pearson's correlation coefficient. We can see that the two variables are positively correlated and that the correlation is 0.8. This suggests a high correlation (i.e., a value above 0.5 and close to 1.0). Pearson's correlation coefficient can be used to evaluate the relationship between more than two variables. This can be done by calculating a matrix of the relationships between each pair of variables in a data set. The result is a symmetrical matrix called a *correlation matrix* with a value of 1.0 along the diagonal, as each variable always correlates perfectly with itself.

+ **Spearman's Correlation:** Two variables may be related nonlinearly, such that the relationship is stronger or weaker across the distribution of the variables. Furthermore, the two variables may have a non-Gaussian distribution. In this case, Spearman's correlation coefficient (named for Charles Spearman; Spearman, 1961) can be used to summarize the relationship between the two data samples (Hauke & Kossowski, 2011). This test of relationship can also be used if there is a linear relationship between the variables, but it will have slightly less power (e.g., it may result in lower coefficient scores).

As with Pearson's correlation coefficient, values are between −1 (perfectly negative correlation) and 1 (perfectly positive correlation). Instead of calculating the coefficient using covariance and standard deviations for the samples themselves, we can derive these statistics from the relative rank of the values in each sample. This is a common approach used in nonparametric statistics (i.e., statistical methods that do not assume a distribution of the data such as Gaussian).

$$Spearman's\ correlation\ coefficient = covariance(rank(X), rank(Y))/(stdv(rank(X)) * stdv(rank(Y)$$

A linear relationship between the variables is not assumed, although a monotonic relationship is assumed. This is a scientific name for an increasing or decreasing relationship between two variables. If you are unsure of the distribution of and possible relationships between two variables, Spearman's correlation coefficient is an excellent tool to use. The computer programing (i.e., *spearmanr()* SciPy) function can be used to calculate Spearman's correlation coefficient for two data samples of the same length.

We can calculate the correlation between the two variables in our test problem. The complete example is listed below.

Import libraries and modules.

```
[In]
From numpy.random import randn
from numpy.random import seed
from scipy.stats import spearmanr
```

Seed the random number generator and prepare the data.

```
[In]
Seed(1)
data1 = 20 * randn(1000) + 100
data2 = data1 + (10 * randn(1000) + 50)
```

Calculate Spearman's correlation.

```
[In]
Corr, _ = spearmanr(data1, data2)
prtin('Spearmans correlation: %.3f' % corr)
```

→

```
[Out]
Spearman's correlation: 0.872
```

Running the example calculates and prints Spearman's correlation coefficient. We know that the data are Gaussian and that the relationship between the variables is linear. Nevertheless, the nonparametric rank-based approach shows a strong correlation between the variables of 0.8.

Remarks: Correlation analysis stands as an essential method within systems analysis. This technique quantifies the extent and direction of a linear association between pairs of variables, capturing how variations in one are mirrored by changes in the other. Employed chiefly to evaluate the degree and character of the linkage between two variables, correlation analysis underscores a statistical measure of association rather than causality. The presence of a strong correlation does not infer a cause-and-effect relationship between the variables in question. In summary, correlation analysis is indispensable for elucidating the dynamics between variables, offering valuable insights into their directional and strength-related associations. This, in turn, facilitates a deeper understanding of potential cause-and-effect relationships and informs decision-making processes. By accurately delineating the connections between variables, correlation analysis proves instrumental across various fields and applications, enhancing our capacity for insight generation and prediction.

[1.8]

Statistical Model, Part 4

Overview of Simple Regression, Multiple Regression, Logistic Regression

When we want to examine the relationship between a quantitative outcome and a single quantitative explanatory variable, we typically use simple linear regression. The word *simple* means that we are only considering a single explanatory variable. Linear regression is the most widely used type of regression for predicting the value of one dependent variable from the value of one independent variable.

The concept of regression was introduced by Francis Galton to describe a biological phenomenon: Heights of descendants of tall ancestors tend to regress down toward a normal average (Galton, 1989). His term was later extended to a more general statistical context.

1. Overview of Simple Regression

In linear regression, there are usually many different values of the explanatory variable, and we usually assume that values between the observed values are also possible values of the explanatory variables (Sykes, 1993). We postulate a linear relationship between the population mean of the outcome and the value of the explanatory variable (Figure 1.8.1). If we let *Y* be an outcome and *x* be an explanatory variable, then we can express the structural model as follows:

x = *predictor (present in data)*
$\beta 1$ = coefficient (estimated by regression)
$\beta 0$ = intercept (estimated by regression)
Y = *predicated value (calculated from* α, β, *and x)*

→

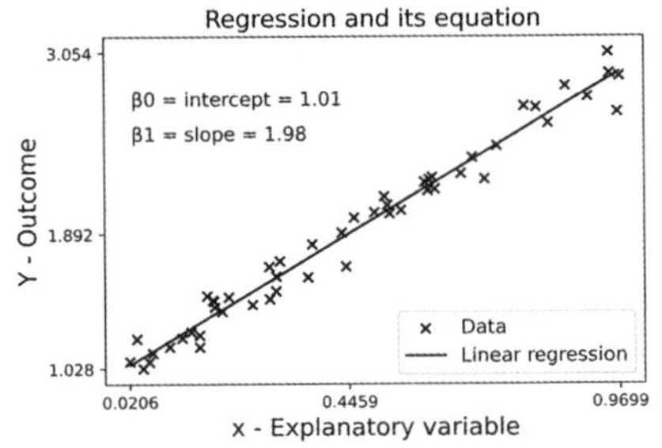

Figure 1.8.1. Schematic representation of regression and its equation

$$E(Y|x) = \beta 0 + \beta 1x$$

The equation $E(Y|x) = \beta 0 + \beta 1x$ represents a linear relationship between two variables, where $E()$ denotes the expected value or population mean, and $(Y|x)$ signifies the possible values of Y given a specific value of x. In this equation, $\beta 0$ (beta zero) refers to the intercept parameter, while $\beta 1$ (beta one) represents the slope parameter. Parameters or parameter estimates used to predict Y from x are commonly known as coefficients. It is typical to replace the 1 subscript in $\beta 1$ with the explanatory variable's name or an abbreviation of it.

+ **Example of a Regression Results Table:** An example of a Regression results table, which predicts one variable based on another, is provided below:

[Table 1.8.1] Simple Regression Results

Model	R	R^2	Adjusted R^2	SE
1	.639[a]	.408	.387	.42961

Model	Unstandardized coefficients		Standardized coefficient	t	Sig.
	B	SE	Beta		
1 (Constant)	7.626	.495		15.408	.000
I_jp8	–.044	.010	–.639	–4.392	.000

* Dependent variable: dash8

Table 1.8.1 presents the results of a regression analysis with 'dash8' as the dependent variable. The table displays the coefficients for each predictor in the model, both unstandardized and standardized, as well as the t statistic and p value for each predictor. The p value helps evaluate the statistical significance of the predictor's association with the dependent variable, while the coefficients indicate the change in the dependent variable for each unit change in the predictor. The results suggest a significant relationship between the dependent variable and both the constant and the predictor 'I_jp8.'

According to the structural model, for each specific value of x, the population mean of Y (over all subjects with that particular value of x for their explanatory variable) can be calculated using the linear expression $\beta 0 + \beta 1x$.

+ **Practice with Simple Linear Regression:** Now let us put the theory into practice with real data. To begin, we need to import the required library and load the csv file (use the supplement file HousePrice.csv).

```
[In]
import pandas as pd
import matplotlib.pyplot as plt
df=pandas.read_csv(' ~ ./Dataset/HousePrice.csv')
```

The csv file has a number of columns. However, we use only two of them to show how simple regression works in a step-by-step process: Price (Older) and Price (New), where older is the x coordinate and new is the y coordinate as follows:

```
[In]
df=df[['Price (Older)', 'Price (New)']]
```

Slice the Price (Older) and Price (New) columns the data frame. And we define the feature list (X) target (Y).

```
[In]
X=df[['Price (Older)']]
Y=df[['Price (New)']]
```

Load the predefined linear Regression model, and divides the data (Train Test Split) set into 75% training and 25% testing data.

```
[In]
from sklearn.model_selection import train_test_split
from sklearn.linear_model import LinearRegression
xTrain,xTest,yTrain,yTest=train_test_split(X,Y)
```

Linear Regression().fit(X, Y)-> puts the X values and Y values into the given function.

```
[In]
Lreg=LinearRegression().fit(xTrain,yTrain)
```

W1 and b (final weight and final bias) give the best fit.

```
W1=Lreg.coef_
b=Lreg.intercept_
```

The 'coef_' represents the coefficients or weights of the model, which are used to predict the response variable based on the predictor variables. The 'intercept_' represents the intercept of the model, which is the value of the response variable when all the predictor variables are 0. Then, let us see the coefficient and intercept in the given function.

```
[In]
print('Coef(W1):',Lreg.coef_)
print('Intercept(W0/b):',Lreg.intercept_)
```

Plot using matplotlib.

```
[In]
import matplotlib.pyplot as plt
plt.scatter(X,Y)
plt.plot(X,W1*X+b,'r-')
plt.show()
```

→

[Out]

Figure 1.8.2. Data-driven regression result

In Figure 1.8.2, the 'plot' function takes three arguments: $X, W1 * X + b$, and $r -$. The first argument is the data for the x-axis, the second argument is the data for the y-axis, and the third argument is a string that specifies the format of the plot. In this case, the plot is red line ($r -$). Finally, we can predict a value as follows:

[In]		[Out]
`print(Lreg.predict(someValue))`	→	X = 141,070, y = 150,589

We used simply a single argument here 'someValue', which is the value for which we want to make a prediction to display the predicted value. The plotted the regression line given input value shows how simple linear regression can be used to predict one variable based on another, providing a practical application for data-driven decision making.

How do Regression, Correlation, and Correlation Coefficient Differ?

This is a frequently asked query in data analysis. The correlation coefficient assesses the intensity and direction of this link, whether it is positive or negative, whereas correlation investigates the straight-line relationship between two variables. Let us evaluate each in isolation.

Correlation: Correlation enables us to comprehend the connection between two variables, such as height and weight, earnings and expenses, and so on. If there is a linear connection between the two variables, it informs us that. Use of a scatter plot, which may show whether two variables are linked or not, is one method of observing correlation. Correlations can be positive or negative, such as those between the cost and demand of items or between illiteracy and poverty in a nation.

Correlation coefficient: The strength and direction of a linear link between two variables are assessed using the correlation coefficient. While a scatter diagram can show us whether or not two variables are connected, the correlation coefficient offers a more accurate assessment of the strength of the association. '*r*' stands for the correlation coefficient, which has a positive and negative value. A number less than 0 denotes a negative relationship, whereas a value of 0 shows no association between the two variables.

Linear regression: Linear regression is a statistical method for examining the relationship between two variables, such as the impact of price increases on demand or changes in advertising spending on company sales. By analyzing the data and fitting a linear equation to it, linear regression makes an attempt to describe the connection between the two variables; one variable is regarded as an independent variable, and the other is regarded as a dependent variable. By way of illustration, it is possible to model an individual's weight to their height, with the help of which weight may be predicted using a certain height value.

How regression differs from correlation: Regression analysis varies from correlation in that it tries to describe the link between the two variables by fitting a linear equation to the data observed, rather than just recognizing it. Regression analysis primary objective is to utilize the connection between the variables to generate hypotheses or estimates. Regression analysis can be used, for instance, to predict or estimate the values of one variable from the other if two variables have a significant correlation. This helps with future planning in a variety of areas, like forecasting rainfall, product prices, or the number of births.

2. Multiple Regression

Predicting (explaining) one variable from more than one variable. Linear regression is also widely used for predicting the value of one dependent variable from the values of two or more independent variables. When there are two or more independent variables, the technique is called *multiple regression* (Tranmer & Elliot, 2008) (Figure 1.8.3).

State the research hypothesis
State the null hypothesis
Gather the data
Fit the data into the regression
Visualize whether the model fits the data

→

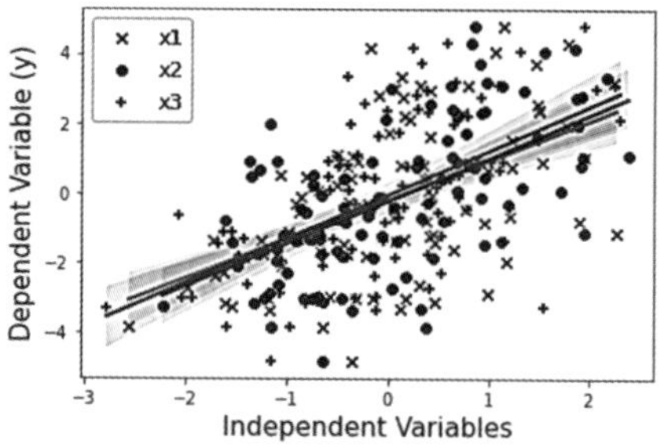

Figure 1.8.3. Schematic representation of the multiple regression characteristics

Similar procedures apply to both simple regression and multiple regression analyses:

- First, investigate each variable independently to see if it is normally distributed, and collect measurements of central tendency and dispersion, as well as frequency distributions and graphs.
- Second, look at the interactions between each independent variable and the dependent variable: To find out if the two variables are linearly connected, compute the correlation coefficient and make a scatter plot.
- Third, examine the correlation between independent variables: Compile the correlation coefficients for each independent variable to see whether there is an excessive amount of correlation between them.
- Fourth, calculate the regression equation using the data, and then perform statistical significance tests and association tests on each coefficient individually as well as on the entire regression equation.
- Finally, test your hypotheses by accepting or rejecting both the null hypothesis and the research hypothesis. In terms of the correlation between the independent and dependent variables, discuss the real-world implications of the findings.

Let us comprehend what makes up an equation for multiple regression. We will delve further into the elements of the multiple regression equation and how they help us understand how the independent and dependent variables relate to one another.

$$Y = \alpha + b_1X_1 + b_2X_2 + b_3X_3$$

where Y is the value of the dependent variable, or what is being predicted or explained; α (alpha) is the constant or intercept; b_1 is the slope (beta coefficient) for X_1; X_1 is the first independent variable to explain the variance in Y; b_2 is the slope (beta coefficient) for X_2; X_2 is the second independent variable to explain the variance in Y; b_3 is the slope (beta coefficient) for X_3; and X_3 is the third independent variable to explain the variance in Y.

+ Example Result Tables (Multiple Linear Regression): Predicting a single variable based on multiple explanatory variables. In Table (1.8.2) below, 'SE' is the standard error of coefficient. R^2 is the proportion of the variance in the values of the dependent variable (Y) explained by all independent variables (X) in the equation. Sometimes this is reported as the adjusted R^2 when a correction has been made to reflect the number of variables in the equation.

[Table 1.8.2] Regression Results: Multiple Regression

Model	*R*	R^2	Adjusted R^2	SE
1	.684[a]	.468	.429	.41463

Model	Unstandardized coefficients		Standardized coefficient	*T*	Sig.
	B	SE	Beta		
1 (Constant)	7.540	.480		15.703	.000
I_jp8	-.039	.010	-.563	-3.834	.001
amhng8	-.008	.005	-.257	-1.750	.092

* Dependent variable: dash8

The results of a multiple regression analysis, where 'dash8' is the dependent variable and 'I jp8' and 'amhng8' are the independent variables, are displayed in the table (Table 1.8.2). The values of the coefficients appear in the 'Unstandardized Coefficients' column, whereas the 'Standardized Coefficient' column lists the standardized coefficients. With the t statistic, each coefficient's significance is shown in the 'T' column. The degree of significance for each coefficient is indicated in the 'Sig.' column. When compared to 'I jp8,' the intercept has a significant coefficient of 7.540 (p =.000), while 'I jp8' has a significant coefficient that is negative (– 0.039, p =.001). With a coefficient of – 0.008 (p =.092), the 'amhng8' coefficient is not significant at the.05 level.

In reality, we know that although the equation predicts the true mean of the outcome for any fixed value of the explanatory variable, it would be unwise to use extrapolation to make predictions outside of the range of x values that we have available for study. However, it is reasonable to interpolate (i.e., to make predictions for unobserved x values between the observed x values). The structural model is essentially an assumption of linearity, at least within the range of the observed explanatory data. It is essential to realize that the 'linear' in 'linear regression' does not imply that only linear relationships can be studied. It only means that the betas must not be in a transformed form. It is alright to turn x or y, and doing so allows many nonlinear relationships to be represented on a new scale that makes the relationship linear.

+ Practice with Multiple Linear Regression: We implemented a linear regression model to make predictions on the variable, size. Here, we will expand on that a little bit and include a few more variables. Multiple linear regression is used to estimate the relationship between two or more independent variables and one dependent variable.

First, we import the Pandas library and load the csv file (use the supplement file Advertising.csv). In its essence, the library (Pandas) that adds support to Python is allowing us to create multi-dimensional arrays that allow vectorization, matrices, and other high-level functions that can be utilized with the mentioned arrays. Then we save the data set in a data frame. The first column in this data set is reviewing with df.head().

```
[In]
import pandas as pd
df= pd.read_csv('…/Advertising.csv')
df = df.iloc[:,1:len(df)]
df.head()
```

→ [Out]

	TV	Radio	Newspaper	Sales
0	230.1	37.8	69.2	22.1
1	44.5	39.3	45.1	10.4
2	17.2	45.9	69.3	9.3
3	151.5	41.3	58.5	18.5
4	180.8	10.8	58.4	12.9

We get the independent variable x into a data frame. We save the dependent variable Sales as y in a different data frame. As a result of these operations, we separate the dependent and independent variables from each other.

```
[In]
X = df.drop('Sales', axis = 1)
y = df[['Sales']]
y.head()
X.head()
```

→ [Out]

	Sales
0	22.1
1	10.4
2	9.3
3	18.5
4	12.9

	TV	Radio	Newspaper
0	230.1	37.8	69.2
1	44.5	39.3	45.1
2	17.2	45.9	69.3
3	151.5	41.3	58.5
4	180.8	10.8	58.4

Now we run the calculation by set up the multiple linear regression model with Statsmodel. This is a bit of a primitive method. We import the Statsmodel library to install the model. Then we create the 'lm' model object with the ordinary least squares (OLS) method. When we set up a model with Statsmodel, we obtain a model that we can learn more about.

```
[In]
import statsmodels.api as sm
lm = sm.OLS(y, X)
model = lm.fit()

model.summary()
```

Here, the 'OLS' function takes two arguments: y, and X and the code fits the model to the data using the 'fit' method. The 'summary' on the fitted model generates a summary of the results of the model fit including coefficients, standard errors, t values, p values as below. The outcomes of a multiple regression analysis between the dependent variable and three independent variables are shown in the table (Table 1.8.3; TV, Radio, and Newspaper).

[Table 1.8.3] OLS Regression Results

Dep. Variable	Sales	R-squared (uncentered):	0.982
Model	OLS	Adj. R-squared (uncentered):	0.982
Method	Least Squares	F-statistic:	3566.
Date	Sat, 07 Aug 2021	Prob (F-statistic):	2.43e-171
Time	10:57:25	Log-Likelihood:	- 423.54
No. Observations	200	AIC:	853.1
Df Residuals	197	BIC:	863.0
Df Model	3		
Covariance Type	nonrobust		

Note: R^2 is computed without centering (uncentered) because the model does not contain a constant.

	coef	*std err*	*t*	*P>|t|*	[0.025	0.975]
TV	0.0538	0.001	40.507	0.000	0.051	0.056
Radio	0.2222	0.009	23.595	0.000	0.204	0.241
Newspaper	0.0168	0.007	2.517	0.013	0.004	0.030

Note: Standard errors assume that the covariance matrix of the errors is correctly specified.

The dependent variable's change for a unit change in each independent variable, while leaving all other variables constant, is represented by the independent variables' coefficients. The accuracy of the estimate is shown by the coefficients' standard error.

Each coefficient undergoes a rigorous test for statistical significance, employing both the t-value and p-value. A p-value of 0.05 or lower indicates a significant relationship. Furthermore, the confidence interval provides a range within which the true value of the coefficient is likely to fall, established at a 95% confidence level. The results reveal a notable positive influence of both radio and television on the dependent variable.

3. Logistic Regression

Now that we are somewhat knowledgeable about linear regression, we can learn about logistic regression and its use in data analysis.

What is logistic regression? Logistic regression as a general statistical model was originally developed and popularized primarily by Joseph Berkson (1944), who coined the term *logit*. Logistic regression is named for the function at its core, the logistic function, which outputs a categorical value (for binary logistic regression outputs are 0 and 1). Logistic regression is generally used for classification (Menard, 2002).

+ **Differences between Linear and Logistic Regression:** Linear regression is used to determine the value of a continuous dependent variable. The output is the weighted sum of the inputs. Logistic regression is a generalized linear regression in the sense that it does not output the weighted sum of the inputs but instead passes through a sigmoid function that can map any real value between 0 and 1 (Han & Moraga, 1995).

In the case of classification where we want a categorical output, linear regression cannot be used; instead, we must use logistic regression.

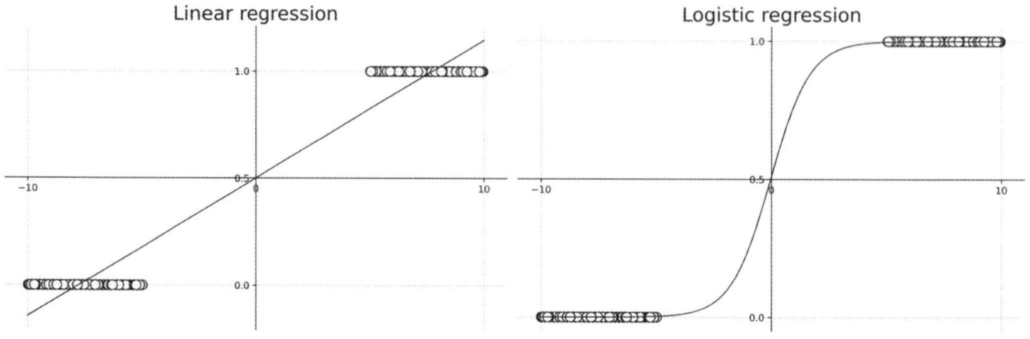

Figure 1.8.4. Schematic representations of linear and logistic regression. Notice the x-axis denotes independent variable, the y-axis represents dependent variable, and the marker of the circles represents data.

Two statistical methods that are often employed in predictive modeling are logistic regression and linear regression (Figure 1.8.4). In contrast to logistic regression, which is used to forecast binary outcomes like whether a student passed or failed an exam, linear regression may be used to forecast continuous values like test scores on a range of 0 to 100. Furthermore, logistic regression generates probability scores that support the model's classifications; these probability scores are then translated to probabilities by using the sigmoid function.

A helpful tool for converting predictions into probabilities in machine learning applications is the sigmoid function, which converts any actual value to a value between 0 and 1 (Figure 1.8.5). The sigmoid function, denoted numerically by '$S(z)$' takes an input value 'z' and transforms it into a probability estimate between 0 and 1. The prediction made by the method is denoted by 'z' for instance, the linear equation '$mx + b$'. The mathematical letter 'e' represents the natural logarithm's base. Logistic regression typically employs the sigmoid function to convert predictions into a probability framework that can be used to determine whether to classify something as true or untrue.

$$S(z) = \frac{1}{1} + e^{-z} \quad \rightarrow$$

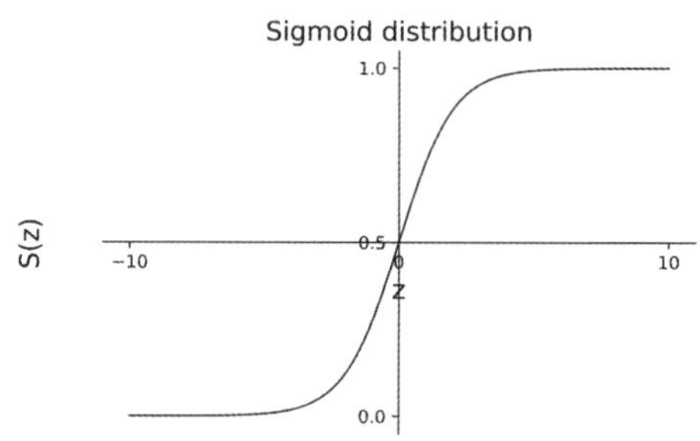

Figure 1.8.5. Representation of the Sigmoid distribution. Notice the vertical axis label is S(z) and the horizontal axis label is z.

Let us make a simple Python coding example

```
[In]
def sigmoid(z):
  return 1.0 / (1 + np.exp(-z))
```

We can use 0.5 as the probability threshold to determine the classes. If the probability is greater than 0.5, we classify it as Class-1($Y = 1$); otherwise, classify as Class-0($Y = 0$).

+ **Decision Boundary:** The output of the logistic regression model is a probability score that quantifies the chance that an observation belongs to a certain class. We must decide on a threshold value that distinguishes between the two groups in order to categorize these probabilities in a binary manner.

A typical threshold value is 0.5, which designates observations as belonging to class 1 if their projected probability score is more than or equal to 0.5, and class 0 if it is lower.

$$p \geq 0.5, class = 1$$

$$\rightarrow$$

$$p < 0.5, class = 0$$

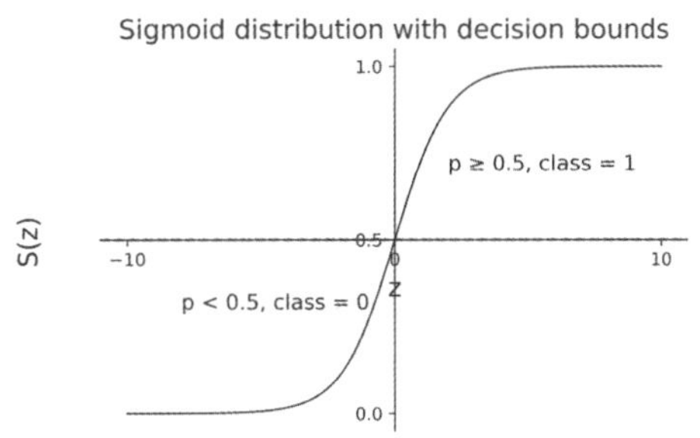

Figure 1.8.6. The sigmoid distribution with decision bounds. Notice the vertical axis label is S(z) and the horizontal axis label is z.

The observation would be categorized as class 1 if the threshold was set at 0.5 and the prediction function returned a value of 0.7, whereas class 0 would be indicated by a prediction of 0.2 (Figure 1.8.6). The class with the highest predicted probability is chosen as the final classification in a logistic regression with several classes.

+ **Making Predictions:** Using our knowledge of sigmoid functions and decision boundaries, we can now write a prediction function. A prediction function in logistic regression returns the probability of our observation being positive, True, or 'Yes.' We call this *class 1* and note it as $P(class = 1)$. As the probability gets closer to 1, our model is more confident that the observation is in class 1. Let us use the same multiple linear regression equation from the linear regression.

$$z = a + b_1X_1 + b_2X_2, \quad X_1 = studied, \quad X_2 = slept$$

This time, however, we transform the output using the sigmoid function to return a probability value between 0 and 1.

$$P(class = 1) = 1/1 + e^{-z}$$

If the model returns .4, it believes there is only a 40% chance of passing. If our decision boundary is .5, we would categorize this observation as 'Fail.'

Simple Python coding example

```
[In]
def predict(features, weights):
  z = np.dot(features, weights)
  return sigmoid(z)
```

The 'features' argument represents the input data for the model, and the 'weights' argument represents the model coefficient. The function calculates the product of the 'features' and 'weights' using the 'dot' and called a 'sigmoid' to map the output of a model to a probability between 0 and 1.

Logistic regression is a statistical method used for binary classification, distinguishing itself from linear regression by predicting probabilities through the sigmoid function. This approach allows for classifying outcomes into two categories based on a probability threshold, usually set at 0.5. It effectively translates linear predictions into a binary decision framework, making it crucial for tasks that require categorical outcomes. Demonstrated through the programing examples, logistic regression proves to be a valuable tool for predictive modeling, enabling clear distinctions between binary classes with practical application.

Remarks: Regression analysis serves as a fundamental tool for predicting the value of a dependent variable based on the values of one or more independent variables. This method illuminates the dynamics between variables, allowing for the forecasting of outcomes. In this framework, the coefficient associated with each independent variable reflects the magnitude of its impact on the dependent variable. Specifically, a positive coefficient indicates a direct relationship, whereas a negative coefficient points to an inverse relationship. While regression analysis offers a pathway to uncover potential causal links, drawing definitive conclusions about causality necessitates further evidence and a robust experimental framework. Regression analysis encompasses various models, including logistic regression for binary outcomes and linear regression for continuous outcomes. Logistic regression employs the sigmoid function to transform predicted values into probabilities, facilitating decision-making by setting a threshold probability to determine outcomes. These analytical techniques are invaluable across a spectrum of systems analysis and decision-making scenarios, enhancing our ability to leverage logistic and regression models for insightful predictions and informed strategies.

[1.9]

Statistical Model, Part 5

Overview of Chi-Square, Chi-Square in Statistics, Chi-Square Test of Independence

The chi-square statistic (χ^2) shows the relationship between two categorical variables (Pearson, 1920; Plackett, 1983). In statistics, there are two types of variables: numerical (countable) variables and nonnumerical (categorical) variables. Chi-square is a single number that tells you the difference between your observed counts and the counts you would expect if there were no relationship at all in the population.

1. Overview of Chi-Square

Chi-square is used to examine the relationship between two nominal or ordinal variables (often the data are frequencies). In chi-square analysis, you must calculate the χ^2 statistic and compare it to a critical value.

$$\chi^2 = \frac{\Sigma(O - E)2}{E}, \qquad where\ O = observed\ value, E = expected\ value$$

Here, the 'O' is the observed frequency of an event, and 'E' is the expected frequency of the event under the statistical model. The sum is taken over all the events being considered.

This is calculated by dividing the sum of the squared differences between the observed and expected frequencies by the expected frequencies. If the calculated statistics is larger than the critical value, the null hypothesis is unlikely to be true.

Let us see a simple example as follows.

+ **Example Result Tables (Chi-square Analysis):** One-way classification table.

[Table 1.9.1] Data Set for Chi-Square Example 1

	Kim's	Park's	Total
Observed	9	23	32
Expected (H_0)	16	16	32

In Table 1.9.1, the observed frequencies of occurrences in Kim's and Park's groups (9 and 23, respectively) notably differ from the expected frequencies under the null hypothesis (H_0), which assumes equal frequencies of 16 for each group.

- Calculation

$\chi^2 = \sum (O - E)^2/E = ((9 - 16)^2/16) + ((23 - 16)^2/16) = 6.13$
$df = k - 1 = 1$, critical value = 3.841

Conclusion: The statistical analysis suggests a preference for Park's group over Kim's, as evidenced by the significant deviation from expected frequencies under the null hypothesis.

[Table 1.9.2] Preferences for Brands of Drink

	Kim's	Park's	Lee's	Cho's	Total
Observed	4	5	8	15	32
Expected (H_0)	8	8	8	8	32

In Table 1.9.2, the observed preferences for brands of drink among Kim's, Park's, Lee's, and Cho's groups (4, 5, 8, and 15, respectively) display substantial deviation from the expected frequencies under the null hypothesis (H_0), which posits equal preferences of 8 for each group.

- Calculation

$\chi^2 = \sum (O - E)^2/E = ((4 - 8)^2/8) + ((5 - 8)^2/8) + (8 - 8)^2/8 + (15 - 8)^2/8 = 9.25$
$df = k - 1 = 3$, critical value = 7.815

Conclusion: People's preferences for the four different drinks are not evenly distributed, indicating that there is variation in their choices.

2. Chi-Square in Statistics (with a Numerical Solution)

Let us detail chi-square in statistics based on the previous fundamentals. There are two types of random variables, and they yield two types of data:

- Numerical
- Categorical

The chi-square statistic (χ^2) is used to investigate whether distributions of categorical variables differ from one another.

- Categorical variables yield data in categories, and
- numerical variables generate data in mathematical form.

Responses to such questions as 'What is your major?' or 'Do you own a car?' are categorical because they yield data such as 'biology' or 'no.'

In contrast, responses to such questions as 'How tall are you?' or 'What is your grade point average?' are numerical. Numerical data can be either discrete or continuous.

[Table 1.9.3] Difference between the Variables

Data type	Question type	Possible responses
Categorical	What is your sex?	Male, Female
Numerical	Discrete: How many cars do you own?	2 or 3
Numerical	Continuous: How tall are you?	172 cm

Table 1.9.3 may help you see the differences between these two variables. Notice that discrete data arise from a counting process, whereas continuous data originate from a measuring process.

The chi-square test is used to test whether there is any relationship between two categorical variables. In other words, it is a test for independence. Therefore, our hypothesis statements are the following:

- H_0: X and Y are independent.
- H_1: X and Y are dependent.

+ **Contingency (2 × 2) Table:** There are several types of chi-square test, and the type you use will depend on the way the data were collected and the hypothesis that is being tested.

We begin with the simplest case of the 2 × 2 contingency table (Table 1.9.4). If we set the 2 × 2 table to the general notation, using the letters a, b, c, and d to denote the contents of the cells, then we have the following table.

[Table 1.9.4] General Notation for a 2 × 2 Contingency Table

	Data type 1	Data type 2	Total
Category 1	A	B	a + b
Category 2	C	D	c + d
Total	a + c	b + d	a + b + c + d

For a 2 × 2 contingency table, the chi-square statistic is calculated with the formula

$$\chi^2 = \frac{(ad - bc)^2(a + b + c + d)}{(a + b)(c + d)(b + d)(a + c)}$$

Note that the four components of the denominator are the four totals from the table columns and rows. Suppose you conducted a drug trial on a group of animals, and you hypothesized that the animals receiving the drug would show increased heart rates compared to those that did not receive the drug. You conduct the study and collect the following data (Table 1.9.5):

- H_0: The proportion of animals whose heart rate increases is independent of drug treatment.
- H_1: The proportion of animals whose heart rate increases is dependent on drug treatment.

[Table 1.9.5] General Notation for a 2 × 2 Contingency Table with Data

	Heart rate increase	No heart rate increase	Total
Treated	36	14	50
Not treated	30	25	55
Total	66	39	105

Applying the formula above we get

$$\chi^2 = \frac{(36 * 25 - 14 * 30)^2(36 + 14 + 30 + 25)}{(36 + 14)(30 + 25)(14 + 25)(36 + 30)}$$

and the result is $\chi^2 = 3.417673$.

Before we can proceed, we need to know how many degrees of freedom we have. When a comparison is made between one sample and another, a simple rule is that the degrees of freedom equals (number of columns minus 1) × (number of rows minus 1) not counting the totals for rows or columns. For our data, this gives

$$(2 - 1) \times (2 - 1) = 1$$

We now have our chi-square statistic ($\chi^2 = 3.418$), our predetermined alpha level of significance (0.05), and our degrees of freedom ($df = 1$). Entering the chi-square distribution table with 1 degree of freedom and reading along the row we find that our χ^2 (3.418) lies between 2.706 and 3.841.

The corresponding probability is between the 0.10 and 0.05 probability levels. That means that p is above 0.05 (it is 0.065). Because a p of 0.065 is greater than the conventionally accepted significance level of 0.05 (i.e., $p > 0.05$), we fail to reject the null hypothesis. In other words, there is no statistically significant difference in the proportion of animals whose heart rate increased.

What would happen if the number of control animals whose heart rate increased dropped to 29 instead of 30 and, consequently, the number of controls whose heart rate did not increase changed from 25 to 26?

+ **Run the Calculation:** Notice that the new χ^2 value is 4.125, and this value exceeds the table value of 3.841 (at 1 df and an alpha level of 0.05). This means that $p < 0.05$ (it is now 0.04), and we reject the null hypothesis in favor of the alternative hypothesis—the heart rate of animals differs by treatment group (Table 1.9.6). When $p < 0.05$, we generally refer to this as a significant difference.

[Table 1.9.6] Probability Level (Calculation for χ^2)

	Probability level (alpha)					
df	0.5	0.1	0.05	0.02	0.01	0.001
1	0.455	2.706	3.841	5.412	6.635	10.827
2	1.386	4.605	5.991	7.824	9.21	13.815
3	2.366	6.251	7.815	9.837	11.345	16.268
4	3.357	7.779	9.488	11.668	13.277	18.465
5	4.351	9.236	11.07	13.388	15.086	20.517

We can also calculate the chi-square of the 2 × 2 contingency table in another way (Table 1.9.7). Let us say we have a random sample of 237 pupils who were asked whether they ever got into trouble at school. We compile the results in the table below.

[Table 1.9.7] Results from the Sample (Calculation for χ^2)

	Trouble	No trouble	Total
Boys	46	71	117
Girls	37	83	120
Total	83	154	237

Because the number of pupils varies by gender, it is hard to compare boys and girls. Therefore, we standardize joint frequencies by dividing counts within each row with the corresponding row total. We also normalize marginal rates by dividing each marginal frequency by the overall total (Table 1.9.8; located in the bottom right cell - 237).

[Table 1.9.8] Converting Observations to Frequencies (Calculation for χ^2)

	Trouble	No trouble	Total
Boys	46/117	71/117	117/237
Girls	37/120	83/120	120/237
Total	83/237	154/237	237

[Table 1.9.9] Results in Frequencies (Calculation for χ^2)

	Trouble	No trouble	Total
Boys	0.39	0.61	0.49
Girls	0.31	0.69	0.51
Total	0.35	0.65	237

This way, joint frequencies become joint (conditional) probabilities or observed probabilities (see sections 2.6 and 2.7 in Part 2 for more detail): they consider two categorical variables.

In Table 1.9.9, for instance, 0.39 is the probability of a boy being in trouble [*P*(boy, trouble)]. Marginal frequencies become marginal probabilities: they take into account only one of the categorical variables. For instance, 0.49 is a probability of being a boy [*P*(boy)], and 0.35 is the probability of being in trouble [*P*(trouble)].

[Table 1.9.10] Marginal Probabilities (Calculation for χ^2)

	Trouble	No trouble	Total
Boys	0.39	0.61	0.49 [*P*(boy)]
Girls	0.31	0.69	0.51 [*P*(girl)]
Total	0.35 [*P*(no trouble)]	0.65 [*P*(no trouble)]	237

The marginal probability computation for chi-square is shown in Table 1.9.10. There are two categories for the data: 'Trouble' and 'No trouble.' The two variables under investigation are 'Boys' and 'Girls'. Each category's probability is expressed as a percentage of the total. Boys had a 0.39 percent chance of falling into the 'Difficulty' group and a 0.61 percent chance of falling into the 'No trouble' category. These odds add up to 0.39 + 0.61, which equals the overall likelihood of being a male, which is 0.49.

Similar to the 'Boys' category, 'Girls' have a 0.31 likelihood of being 'Trouble' and a 0.69 probability of being 'No trouble.' The likelihood of being a female is calculated as the product of these probabilities (0.31 + 0.69), which is equal to 0.51. With *P*(No problem) equal to 0.65 and *P*(Trouble) equal to 0.35, the marginal probabilities for the 'No trouble' and 'Trouble' categories are listed at the bottom of the table. There are 237 sightings in all, according to the data.

Previously, we outlined our hypotheses as follows:

- H_0: *X* and *Y* are independent.
- H_1: *X* and *Y* are dependent.

Probabilities in statistics state that if two events are independent, the following equation is satisfied:

$$P(X,Y) = P(X) \times P(Y)$$

where *X* and *Y* are events. The chi-square test is based on this assumption. Therefore, if it is true (i.e., if *X* and *Y* are independent), the following equation will be satisfied:

$$P(\text{gender}, \text{situation}) = P(\text{gender}) \times P(\text{situation})$$

On the left side of this equation, we see the joint probability, and on the right side of this equation we see two marginal probabilities. Chi-square analysis uses this assumption to calculate expected probabilities (joint probabilities) using the marginal probabilities.

For instance, the expected probability of boys being in trouble is

$$P(\text{boy}) \times P(\text{trouble}) = 0.35 \times 0.49 = 0.17$$

The expected probability of girls being in trouble is

$$P(\text{girl}) \times P(\text{trouble}) = 0.35 \times 0.51 = 0.18$$

and so on. Find the expected probabilities in the parentheses below (Table 1.9.11).

[Table 1.9.11] Finding *P*(*X*, *Y*), *P*(*X*), *P*(*Y*)

	Trouble	No trouble	Total
Boys	0.39 × 0.17	0.61 × 0.32	0.49
Girls	0.31 × 0.18	0.69 × 0.33	0.51
Total	0.35	0.65	237

In other words, when we calculate the expected probabilities, we calculate the probabilities that we should expect if *X* and *Y* are independent variables. This means that if gender and trouble status are independent variables, our expected probability of boys not being in trouble is 0.32.

+ **Measure the Differences between Probabilities**: We should also measure the differences between the actual probabilities (in our tables) and the expected probabilities we have just calculated.

If we see that the difference between the actual probabilities and the probabilities we would expect if two categorical variables are independent is enormous, then our variables are most likely not independent.

Similarly, if the difference is small, then our two variables are most likely independent. The difference between the two probabilities is represented by a value that we can calculate using the following formula:

$$\chi^2 = \sum \frac{(observed - expected)^2}{expected}$$

We just plug in the values from the table above:

$$\chi^2 = \sum \frac{(0.39 - 0.17)^2}{0.17} + \frac{(0.61 - 0.32)^2}{0.32} + \frac{(0.31 - 0.18)^2}{0.18} + \frac{(0.69 - 0.33)^2}{0.33}$$

$$= 1.034$$

The last step is to compare a value with the value in the distribution table to conclude whether you should accept or reject it. The following procedures apply:

- $\chi^2 >$ table value: Accept. You have enough statistical data to conclude that the two variables are dependent.
- $\chi^2 \leq$ table value: Reject. You have enough statistical data to conclude that the two variables are independent.

To get the proper value from the table, we have to know two things: the significance level and the degrees of freedom. In the table, the significance level is at the top, and the degrees of freedom are on the left side.

- The significance level is something you choose yourself. Let us use a significance level of 5% ($\alpha = 0.05$).
- The degrees of freedom are calculated with the following formula:

$$v = (number\ of\ rows - 1) * (number\ of\ columns - 1)$$
$$= (2 - 1) * (2 - 1) = 1$$

With a significance level of 0.05 and 1 df, we have a table value of 3.8415 (Table 1.9.12: first row, third column).

[Table 1.9.12] Significance Level

df	0.5	0.1	0.05	0.02	0.01	0.001
1	0.455	2.706	3.841	5.412	6.635	10.827
2	1.386	4.605	5.991	7.824	9.21	13.815
3	2.366	6.251	7.815	9.837	11.345	16.268
4	3.357	7.779	9.488	11.668	13.277	18.465
5	4.351	9.236	11.07	13.388	15.086	20.517

+ **Probability Level (alpha):** Now we have both χ^2 and table values and can compare the two to make a decision. An estimate from the above calculation $\chi^2 = 1.034$ that is below our table value of 3.8415—that is, any difference between our actual probabilities and the probabilities we would expect if the two variables are independent, that is below 3.8415—means that the two variables are independent or have no relationship. As our χ^2 is less than the table value, we can reject H_1 and conclude that gender and trouble status are not correlated. Thus, here are the steps of a chi-square test: (a) Add marginal frequencies to a contingency table. (b) Convert joint and marginal frequencies into probabilities. (c) Estimate the expected probability for each cell. (d) Calculate χ^2. (e) Compare χ^2 to the table value and make a decision.

- $\chi^2 >$ table value = accept = dependent
- $\chi^2 \leq$ table value = reject = independent

3. Chi-Square Test of Independence

Suppose we want to determine whether there is a statistically significant association between smoking and being a professional athlete. We collect data from 28 persons.

The smoking variable is measured as smoker or nonsmoker, and the professional athlete variable is measured as athlete or nonathlete. Because the two variables of interest are qualitative variables, we need to use a chi-square test of independence.

In this example we use binary variables (i.e., qualitative variables with two levels) for the sake of simplicity, but the chi-square test of independence can also be performed on qualitative variables with more than two levels.

For instance, if the smoking variable had three levels—nonsmoker, moderate smoker, heavy smoker—the steps and the interpretation of the results would be similar. Observed frequencies: Our data are summarized in the contingency table below (Table 1.9.13).

[Table 1.9.13] Data Set for Chi-Square Practice

	Nonsmoker	Smoker	Total
Athlete	14	4	18
Nonathlete	0	10	10
Total	14	14	28

+ **Expected Frequencies:** Remember that for the chi-square test of independence, we need to determine whether the observed counts are significantly different from the counts we would expect if there were no association between the two variables.

We have the observed counts (see the table above), so we now need to compute the expected counts in the case that the variables are independent. These expected frequencies are calculated for each subgroup one by one with the following formula:

$$Expected\ frequencies = \frac{total\ \#\ of\ obs.for\ the\ row * total\ \#\ of\ obs.for\ the\ column}{total\ number\ of\ observations}$$

where obs. is observations. Given our table of observed frequencies above, below is the table of expected frequencies computed for each subgroup (Table 1.9.14).

[Table 1.9.14] Expected Frequencies for Each Subgroup

	Nonsmoker	Smoker	Total
Athlete	(18 × 14)/28 = 9	(18 × 14)/28 = 9	18
Nonathlete	(10 × 14)/28 = 5	(10 × 14)/28 = 5	10
Total	14	14	28

Note that the chi-square test of independence should only be used when the expected frequencies in all groups are equal to or greater than 5. This assumption is met for our example, as the minimum number of expected frequencies is 5.

+ **The Test Statistic:** We have the observed and expected frequencies. We now need to compare these frequencies to determine whether they differ significantly. The difference between the observed and expected frequencies, referred to as the test statistic and denoted χ^2, is computed as follows:

$$\chi^2 = \sum_{i,j} \frac{(O_{ij} - E_{ij})^2}{E_{ij}}$$

where O represents the observed frequencies and E represents the expected frequencies. We use the square of the differences between the observed and expected frequencies to ensure that positive differences do not compensate for negative differences. The formula looks more complicated than it is, so let us illustrate it with our example.

We first compute the difference in each subgroup one by one according to the formula.

- In the subset athlete and nonsmoker: $\frac{(14\text{-}9)^2}{9} = 2.78$
- In the subset nonathlete and nonsmoker: $\frac{(0\text{-}5)^2}{5} = 5$
- In the subset athlete and smoker: $\frac{(4\text{-}9)^2}{9} = 2.78$
- In the subset nonathlete and smoker: $\frac{(10\text{-}5)^2}{5} = 5$

Then we sum them all to obtain the test statistic.

$$\chi^2 = 2.78 + 5 + 2.78 + 5 = 15.56$$

+ **Critical Value:** The test statistic alone is not enough to conclude whether the two variables are independent or dependent. As mentioned previously, this test statistic (which in some sense is the difference between the observed and expected frequencies) must be compared to a critical value to determine whether the difference is large or small. You cannot know whether a test statistic is large or small without comparing it to the critical value.

A critical value is a cut-off value (or two cut-off values in case of a two-tailed test) that constitutes the boundary of the rejection region(s). In other words, critical values divide the scale of your test statistic into the rejection region and non-rejection region. If the test statistic is above the critical value, this means that the probability of observing such a difference between the observed and expected frequencies is low. However, if the test statistic is below the critical value, this means that the probability of observing such a difference is high. If we are likely to observe this difference, we cannot reject the hypothesis that the two variables are independent; otherwise, we can conclude that a relationship exists between the variables.

The critical value can be found in the statistical table of the chi-square distribution and depends on the significance level (α) and the degrees of freedom (df). The significance level is usually set equal to 5%. The degrees of freedom for a chi-square test of independence are found as follows:

$$df = (number\ of\ rows - 1) \times (number\ of\ columns - 1)$$

In our example, the degrees of freedom is (2 – 1) × (2 – 1) = 1, because there are two rows and two columns in the contingency table (totals do not count as rows or columns).

We now have all of the necessary information ($\alpha = 0.05$, $df = 1$) to find the critical value in the chi-square table.

To find the critical value, we need to look at row $df = 1$ and column $\chi^2 = 0.050$ (because $\alpha = 0.05$) in the chart below (Figure 1.9.1 and Table 1.9.15).

The critical value is 3.84146.

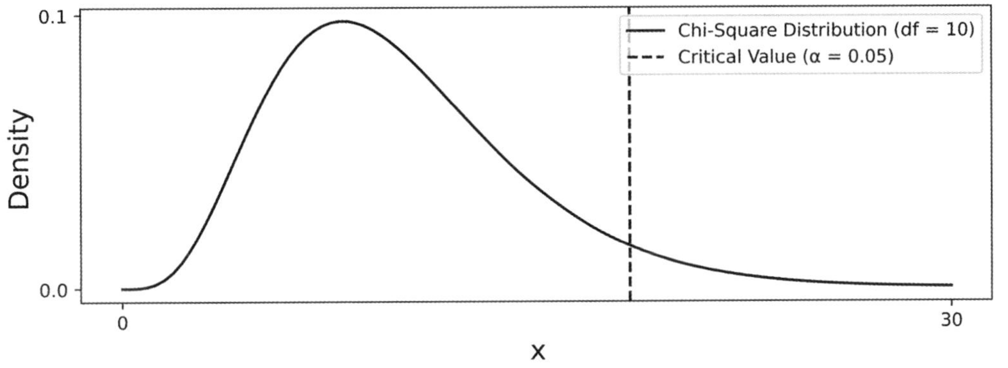

Figure 1.9.1. Percentage point of χ^2 distribution

[Table 1.9.15] Chi-Square Table: Critical Value for Alpha = 5% and *df* = 1

χ^2 =0.100	χ^2 =0.050	χ^2 =0.025	χ^2 =0.010	χ^2 =0.005	*df*
2.70554	3.84146	5.02389	6.63490	7.87944	1
4.60517	5.99147	7.37776	9.21034	10.5966	2
6.25139	7.81473	9.34840	11.3449	12.8381	3
7.77944	9.48773	11.1433	13.2767	14.8602	4

+ **Conclusion and Interpretation:** Now that we have the test statistic and the critical value, we can compare them to check whether the null hypothesis of independence between the variables is rejected or not. In our example,

- test statistic = 15.56 > critical value = 3.84146.

For any statistical test, when the test statistic is larger than the critical value, we can reject the null hypothesis at the specified significance level. In our case, we can therefore reject the null hypothesis of independence between the two categorical variables at the 5% significance level.

This means that there is a significant relationship between smoking and being an athlete. Knowing the value of one variable helps to predict the value of the other variable.

+ **Practice Chi-Square with a Simple Arbitrary Example:** For a practical example of how we run chi-square in programing (i.e., Jupyter notebook), let us suppose we want to test whether there is a statistically significant difference between smokers and non-smokers.

First, we need to generate some sample data to work on it importing a couple of necessary libraries

```
[In]
import pandas as pd
import numpy as np
from scipy.stats import chi2_contingency
import seaborn as sns
import matplotlib.pyplot as plt
```

Sample data

```
[In]
df = pd.DataFrame({'Athlete' : ['Nonathlete', 'Nonathlete', 'Nonathlete', 'Athlete', 'Athlete'] * 10,'isSmoker' : ['Smoker', 'Smoker', 'Non-Smpoker', 'Non-Smpoker', 'Smoker'] * 10})
df.head()
```

→

[Out]

	Athlete	isSmoker
0	Nonathlete	Smoker
1	Nonathlete	Smoker
2	Nonathlete	Non-Smpoker
3	Athlete	Non-Smpoker
4	Athlete	Smoker

The easiest way to run the chi-square test is to convert the data to a contingency table with frequencies.

We will use the crosstab command from Pandas.

```
[In]
contigency=
pd.crosstab(df['Athlete'], df['isSmoker'])
contingency
```

→

[Out]

isSmoker	**Non-Smpoker**	**Smoker**
Athlete		
Athlete	10	10
Nonathlete	10	20

The 'crosstab' function takes two arguments: 'Athlete' and 'isSmoker', that will be used to create the contingency table.

Let us get the percentages by athlete (row).

```
[In]
contigency_pct =
pd.crosstab(df['Athlete'], df['isSmoker'], normalize='index')
contigency_pct
```

→

[Out]

isSmoker	**Non-Smpoker**	**Smoker**
Athlete		
Athlete	0.500000	0.500000
Nonathlete	0.333333	0.666667

If we want the percentages by column then we should write normalize='column', and if we want the total percentage then we should write normalize='all'.

An easy way to visualize the contingency tables is with heat maps (Figure 1.9.2).

```
[In]
plt.figure(figsize=(5,4))
sns.heatmap(contigency, annot=True,
cmap="binary")

plt.xlabel('Smoking', fontsize=15)
plt.ylabel('Athlete', fontsize=15)
plt.grid(True)

plt.show()
```

→ [Out]

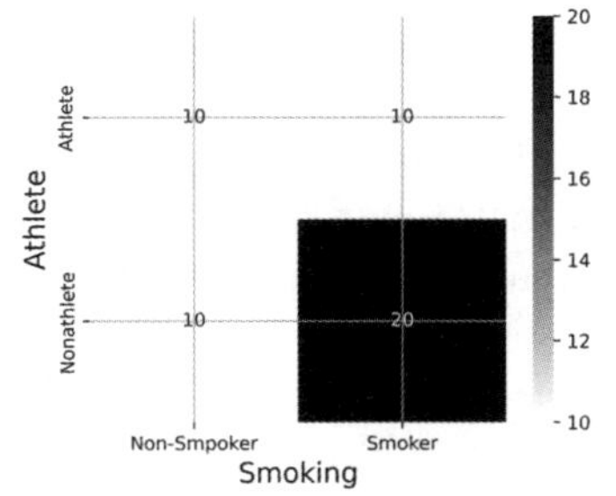

Figure 1.9.2. Chi-square test of contingency tables using heat maps

Now that we have built the contingency table we can pass it to the chi2_contingency function from the scipy package, which returns the test statistic (chi2), the p value of the test (p), degrees of freedom (dof), and the expected frequencies (expected), based on the marginal sums of the table.

```
[In]
# Chi-square test of independence.
c, p, dof, expected =
chi2_contingency(contigency)
# Print the p-value
print(p)n
```

→

```
[Out]
0.3767591178115821
```

The statistic is calculated by comparing the observed frequencies in the contingency table under the null hypothesis. The resulting p value is then used to determine whether the null hypothesis can be rejected. The p value is 37.67%, which means that we do not reject the null hypothesis at the 95% level of confidence. The null hypothesis was that smokers and gender are independent. In this example, the contingency table was 2×2. We could have used a Z test for proportions instead of the chi-square test. Notice that the chi-square test can be extended to $m \times n$ contingency tables.

Remarks: To evaluate if there is a statistically significant correlation between two category variables, the chi-square test is utilized. The significance of the association is then assessed using the resultant chi-square statistic, and the test may be used to either reject or fail to reject the null hypothesis of independence between the two variables. In conclusion, the chi-square test is a procedure that is frequently used in statistics to examine the association between categorical variables and can give important insights into the underlying relationships.

Supplemental Information 1.5: Writing a Research Paper

The structure of a research paper typically comprises the title, abstract, introduction, literature review, method, results, and discussion (Randolph, 2009).

1. General Structures

+ **Title**: The title should succinctly and precisely convey the essence of the research. It is important to strike a balance between brevity and specificity to ensure the title fully encapsulates the study's scope and focus.

+ **Abstract:** Begin with a concise sentence that clearly states the research's objective. Follow this with 3~5 sentences detailing the research methodology, outlining the approaches and techniques employed. Then, in another 3~5 sentences, present the key findings or results, highlighting the most significant data and outcomes. Conclude with 1~2 sentences that provide a brief discussion, summarizing the implications and potential impacts of the research findings.

+ **Introduction:** This section should lay the foundation for the current research, articulating its rationale with a strong grounding in relevant theoretical frameworks and prior studies. It should clearly establish the context and significance of the research, drawing connections to existing literature and identifying the specific gap or question that the study addresses. For instance:

- Exercise (sports, physical education) holds significance.
- The subject of investigation is also important.
- Theory endorses the relevance (or role) of the topic studied in this research.
- There has been some related research conducted.
- Overall, there is a scarcity of research in this area.
- Conducting this research is necessary.

+ **Literature Review:** This section should take about existing research and theories, analyzing studies that have explored similar themes or populations. The review critically assesses the methodologies, findings, and conclusions of these studies, highlighting any consistencies or discrepancies. This evaluation will frame the research problem by pinpointing where your study will contribute new insights or address existing knowledge gaps in this area.

+ **Method:** The methodology section must be sufficiently detailed to allow readers to accurately replicate the study. Start with a comprehensive overview of the overall procedures, then delve into the specific steps taken. Employ figures and tables to illustrate and structure the information for clarity and ease of understanding. Organize the section into subsections such as 'participants', 'Materials', and 'Procedures' to improve the readability and logical flow of the content.

+ **Results:** Employ tables and figures to effectively present and interpret the research findings. Ensure that the presentation remains objective, maintaining a neutral perspective throughout. It is beneficial to include parts of the analysis procedures to contextualize the results.

- Remember, tables should display numerical values or text in an organized format of rows and columns, aiding in clarity and comprehension.
- Figures, encompassing various types of illustrations such as charts, graphs, photographs, drawings, and maps, should be used to visually represent data and enhance understanding.

When incorporating figures or tables from other sources, it is crucial to use proper citations. Be aware that the rules for citing tables and figures may differ, and adhere to the appropriate citation style guidelines.

What to Include in Tables and Figures?

Tables usually show numerical values or textual information and are almost always characterized by a row-column structure (Table S1.5.1). Any type of illustration other than a table is referred to as a figure.

[Table S1.5.1] Sales of Take-Home Ice Cream

	Year ago	Current year	Change (%)
Dollar sales	87.845	91.698	4.4
Package sales	19.984	19.682	-1.5
Aver price per pack ($)	4.40	4.66	6.0

Note: The data are from the Market Information (moving annual total).

A figure may be a chart, graph, photograph, drawing, or any other illustration or contextual depiction. Any type of illustration other than a table is referred to as a figure (Figure S1.5.1).

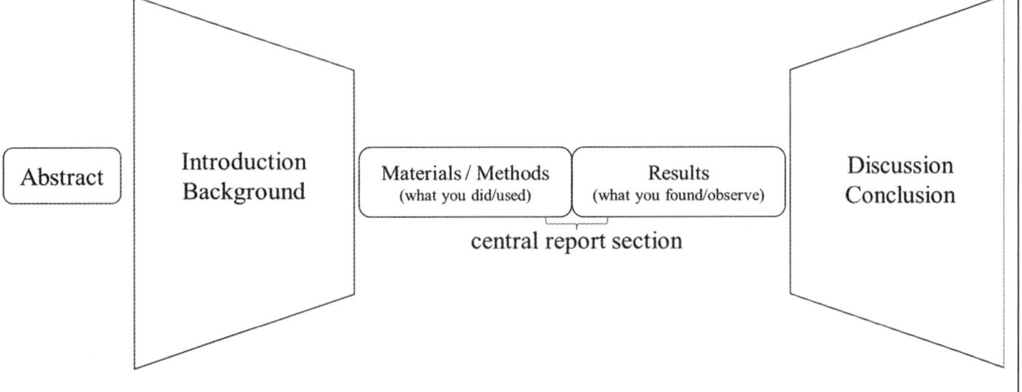

Figure S1.5.1. Schematic of a research article

When you use a figure that has been adapted or copied directly from another source, you need to reference that original source.

+ **Discussion:** In this section, critically analyze whether your results align with existing theoretical frameworks. Ask and answer key questions:

- Do the findings corroborate the theory, or does the theory elucidate your results?
- Compare your outcomes with those of other research studies.
- Investigate whether your findings are consistent or inconsistent with previous research, and explore the reasons behind these agreements or disagreements.
- Organize the discussion starting from the most significant findings, progressively addressing each in detail.
- Summarize the results succinctly, providing clear examples for clarity.

For instance, you might state 'Daily exercise was found to have a detrimental effect on health.' Then, connect this finding to existing literature, noting, 'While the extent of impact varies, this observation partially concurs with previous studies.' Elaborate on possible reasons such as increased fatigue, compromised immune system, etc., to provide a comprehensive analysis.

2. Things to Consider When Writing

In order to provide a more comprehensive explanation and adhere to the principles of professional writing, consider the following:

- Tone: Use a professional tone.
- Clarity: Be clear (think as a reader).
- Effectiveness: Shorter is better, as long as the sentence delivers the message.
- Formatting: Be consistent with font, size, indentation, margins, spacing, the number of lines in a table, and so on.
- Grammar: No mistakes, simple and clear.

Adopting a professional tone, ensure that your writing is clear and concise, conveying complex concepts in an easily understandable manner, while maintaining consistency in font, size, indentation, margins, spacing, and other formatting elements; additionally, use simple and grammatically correct language to enhance the overall readability and effectiveness of your academic work.

3. The Writing Process

To write a research paper, you must work through the writing process to turn your rough ideas into a polished, publishable finished piece (Zamel, 1982). Let us learn more about this necessary process in detail.

+ **Brainstorming**: The writing process starts before you put pen to paper or fingers to keyboard. The first step is brainstorming. Depending on the assignment, you may be given a topic or you may have to create one yourself. Do an internet search for the topic you will be covering to get a stronger grasp on it and all the potential directions your writing can take. Jot down every idea you have while you brainstorm, even if it is only tangentially related to your topic. The goal here is not to create a coherent piece of writing—it is to clear a path for your writing.

Brainstorming is not just about developing a clear topic and set of supporting content to cover; it is about determining the most effective way to present your information to your intended audience. Think about the type of writing you are doing and for whom you are writing. Once you have a clear central theme for your writing and a strong grasp of your supporting arguments, it is time to finesse your brainstorming results into a logical outline.

+ **Preparing to Write**: In this stage, you are taking all the ideas, connections, and conclusions you encountered during your brainstorming session and organizing them into an outline. An outline is basically a skeleton of a finished piece of writing that maps the topics you will cover and where each paragraph fits into the piece. It provides a structure that helps ensure your ideas flow logically and clearly. It can be helpful to look at outline templates online, especially if you have been assigned a type of writing that you have not done before.

+ **Choosing Authoritative Sources**: At this stage, you will also identify which sources to use. With certain types of research, you will need to cite your sources as references. If this is the case for your current research, this stage is the point at which you should familiarize yourself with the applicable style guide and its formatting requirements for citations (i.e., American Psychological Association, Chicago, Modern Language Association, Harvard, Vancouver).

Make sure your chosen sources are appropriate for your writing before you decide to use them. For any research writing, the range of acceptable sources you can use is typically limited to academic articles, government or nonprofit research groups, and, if you are writing a literature review, the literary works you are comparing in your writing. For other kinds of writing, appropriate sources are relevant sources. Thus, working with credible sources gives your research a strong foundation, whereas weak sources undermine the position your writing takes.

+ **Striking the Right Tone**: This is also the stage when you clarify the tone you will use in your work. Usually figuring out the right tone for your writing is easy—if it is a piece of academic writing, it needs a formal tone. If it is a cover letter, your tone should be confident but not arrogant. When you are unsure about the right tone to use or how to achieve it, do an internet search for examples of the type of writing you are doing and familiarize yourself with the structure, vocabulary, and overall tones used.

+ **Writing First Draft:** Now you are ready to write the rough draft. Your goal is to get words on the page, not to churn out something that is ready to publish. Using the outline, you created, start building your draft, sentence by sentence and paragraph by paragraph. Here you should not write your rough draft from start to finish. When you reach a challenging spot in your writing, it is easy to get stuck there and waste a lot of time trying to figure out what to write. Save yourself the time and stress by writing the easiest parts first, then moving on to the tougher spots.

+ **Editing and Revising**: Once you have a completed rough draft, the next step in the writing process is to shape it into a final draft. This is known as editing. As you move through the writing process, you are able to use different kinds of editing. At this stage, you are content editing, line editing, and copy editing. Later you will proofread your work, and, depending on the content you are covering, you might fact-check it as well. In some cases, you are totally on your own with editing. In others, editing your work involves incorporating feedback an editor or instructor left on your first draft. When you are facing the latter scenario, be sure to read the feedback carefully and address or incorporate all of it.

Edit with fresh eyes. Do not jump from writing the first draft to editing it unless you are crunched for time and you absolutely must. By taking time between writing and editing, you are giving yourself some distance from your work. This enables you to view it with fresh eyes and catch mistakes and areas for improvement more easily than you would if you had not created that distance (i.e., with fresh eyes, look for logical inconsistencies or fallacies, inconsistent and inappropriate tone, areas where your writing can be made more concise, opportunities to replace words with more effective synonyms, confusing phrases and sentences).

Pay attention to how effectively your supporting arguments prove and strengthen your research statement. The goal of editing and revising your work is to optimize your writing to make your central theme as clear and powerful as possible. Once you are finished editing, it is time to revise your draft into its final version. This is the process of implementing all the changes you noted during the editing process.

+ **Proofreading Final Draft**: At this stage, you are finished writing, but you are not quite ready to hand in your research. Proofreading is one last look to catch any spelling mistakes, grammar errors, typos, formatting errors, or incorrect structure or syntax. Unless something is egregiously wrong, you are not changing any of your content—you are simply double-checking that everything is grammatically correct. Ideally you have enough time to proofread your work with fresh eyes.

+ **Publishing Finished Work**: Your work is ready to be shared with others. What it means to publish your work depends on the type of research you are doing. If you are the publisher as well as the writer, this stage is essentially you uploading your own research and making it available to others. If you just completed academic research or a commissioned piece for a journal or another outlet to which you are a contributor, this step is when you send it off to your professor or the journal editor. No matter what publishing means for your particular piece, take a moment to celebrate. You wrote something, and now people are going to read it.

[1.10]

Concluding Remarks

Most analysis projects simply involve questions about situations and phenomena (or objectives). If you are really interested in knowing what something is about, you cannot simply believe whoever offers you an answer.

What should we do and which one to use? First and foremost, you need to be clear what you want to do.

- Do you want to validate whether the population mean is correct?
- Do you want to compare groups of data to see whether differences are statistically significant?
- Do you want to validate whether there is a relationship between categorical variables?

Next you need to decide on the proper hypothesis based on your test objective. A null hypothesis (H_0) proposes no significant difference in a set of given observations, and an alternative hypothesis (H_1) proposes otherwise.

To reject a null hypothesis, you must calculate a test statistic. This test statistic is then compared to a critical value. Critical values are the boundaries of the critical region.

- If the test statistic is greater than the critical value, the null hypothesis is rejected.
- If the test statistic is less than or equal to the critical value, the null hypothesis is accepted.

In addition to the test statistic, it is important to look at the p value. The p value is the level of marginal significance within a statistical hypothesis test representing the probability of the occurrence of a given event. A p value less than 0.05 indicates strong evidence against the null hypothesis, so you reject the null hypothesis. The lower a p value, the stronger the evidence. The larger the absolute value of the test statistic, the smaller the p value, and the greater the evidence against the null hypothesis.

Once you have the question and hypothesis settled, you may use the process below as a general guide for deciding which statistical test to run (Figure 1.10.1). It is not comprehensive, but it demonstrates the basic differences among these tests.

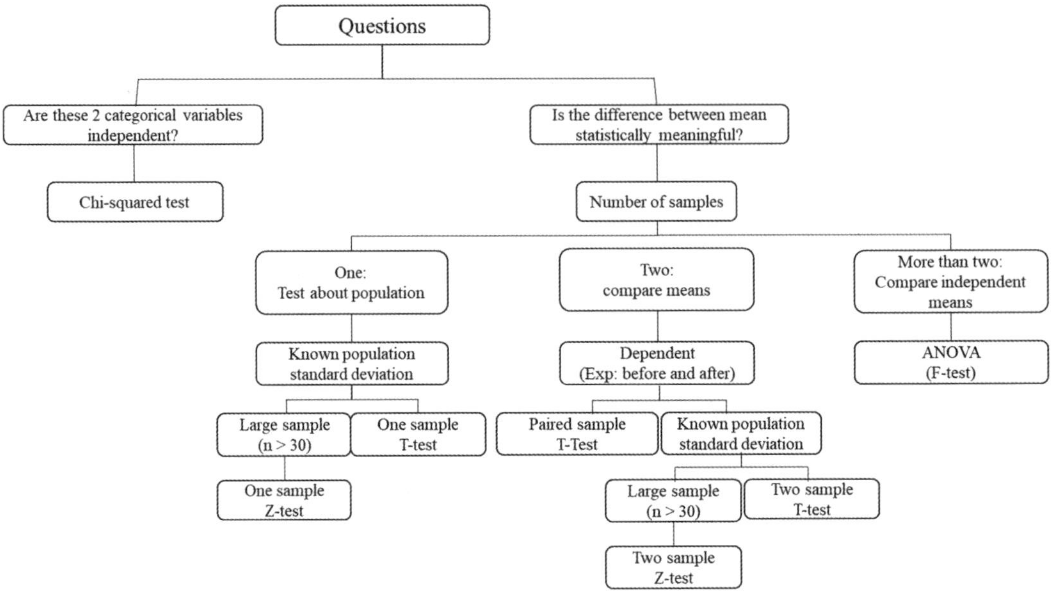

Figure 1.10.1. General guide for deciding which statistical test to run

Finally, all these things are biased and subjective, even if we have a lot of experience and information. Ask whether you can really trust the answer. Asking this question is quite hard, because people do not like thinking critically (Kahneman, 2011), but it is necessary in research. Do not forget that critical thinking is one of the best methods in any investigation.

Reading List 1.1: Applied Statistical (Variance-Based) Research Example with Perceptual Accuracy

Purpose: This study measured whether self-produced movement influences haptic perception ability (experiment 1) as well as the factors associated with levels of influence (experiment 2) in racket sports. **Methods:** The haptic perception accuracy of five male table tennis experts and five male novices was examined under two different conditions (no movement vs. movement). Inferential statistics (analysis of variance, *t* test) and custom-made devices (shock and vibration sensor) were used to determine haptic perception accuracy and its association with expertise. **Results:** The results of this research show that expert-level players acquire higher accuracy with less variability (racket vibration and angle) than novice-level players, especially in their self-produced movement-coupled performance. The important finding from this result is that, in terms of accuracy, skill-associated differences were enlarged during self-produced movement (Park & Kim, 2014).

Background: Perception contributes to action, but action also contributes to perception (Gibson, 1979). Attempting to study one aspect while neglecting the other may fail to reveal both aspects (Turvey, 1977). Although a number of studies have confirmed this haptic ability in various conditions, most experimental designs are based on the premise that only perception facilitates action. They mainly consider the aspect of perception as an independent variable and the interpretation of the pattern of action as a dependent variable. Although few studies have addressed how action facilitates perception (Oudejans et al., 1996; Ranganathan & Carlton, 2007), those measuring the contribution of action may simply be measuring movement or action rather than the influence of self-produced movement considering the main dependent variable. Therefore, we investigated nonvisual haptic perception accuracy, that is, whether self-produced movement influences haptic perception ability in two different conditions (unproduced and produced). We did this while controlling movement as an independent variable and then interpreting the pattern of perception as a dependent variable. We expected that there would be a strong correlation between self-produced movement and expertise as regards determining skill level. In addition, this may be identified as a new developmental investigation of the action task, providing a template with which perception accuracy ability and haptic function can be assessed.

Method: The performance of five highly skilled table tennis experts was measured and compared to that of five novice participants (ages 19–23 years old). All experts played in the highest Korean national league and were considered top players. The novices were five male university students. None of the novices had ever had any regular table tennis technical guidance. Individuals were excluded on the basis of self-reported acute or chronic physical and psychiatric disorders. In addition, all participants had normal mobility of their right arms.

Apparatus and Procedures: A participant was seated at a desk with the right forearm supported to the elbow. During each trial, a table tennis racket was placed on the right-hand side, with the racket divided into 16 compartments arranged into four columns and four rows. This type of division allowed the most comparable size for comparing the ball diameter (40 mm). Participants were instructed to hold the racket on the side (racket hand) occluded by an opaque curtain. In front of the participant (3 m) was a screen displaying an image such that the participant was aware of the regional number of each racket. The task was the same for both groups—judging one hit position of the racket with a ball among the 16 compartments and reporting the position number verbally—differing only in terms of self-produced movement being coupled or uncoupled.

[Table R1.1.1] Design of the experiment

Experimental Design	Independent variable	Dependent variable
Two-way ANOVA	Skill (2-levels) Condition (2-levels)	Performance accuracy (Haptic)

Note: The skill levels (expert vs. novice) task conditions (uncouple vs. coupled)

When the subject was ready, the trial was initiated by the experimenter. In the uncoupled condition, a ping-pong ball was dropped from a height of 30 cm onto the racket by an assistant. In the coupled condition, a ping-pong ball was suspended by a thread and the subject wielded the racket following the beat of a metronome. To control the contact location between the ball and the racket, the experimental assistant managed the area randomly at only 50% of the compartment (50% around the center, 50% around the outside). All wielding motions typically occurred in all three spatial dimensions and matched the same contact speed as the uncoupled condition according to the law of acceleration ($F = ma$).

To control (in advance) and compare (during the trial) the size of the impulse at the moment of contact between the racket and the ball, we used the coefficient of restitution (see equation below), a custom-made piezoelectric accelerometer (shock and vibration sensor; vso300A1, T-nest, Korea), and a software program (LabView 8.5, NI, Korea).

$$e = \sqrt{h_{up}/h_{down}}$$

Here e denotes the coefficient of restitution, h_{up} denotes the rebound height of the ball, and h_{down} represents the falling height of the ball. According to the formula for the related impulse and momentum ($F = mv_f - mv_i$), the impulse is increased by a faster velocity; this can be represented by the coefficient of restitution in the circumstance of a collision between two objects. All participants were given a similar impulse according to the formula with the device: stimulus range (uncoupled: .03~.26 V, coupled: condition .02~.26 V), which was significant during the tasks, $r(160) = -.03$, $p = .55$. We did not use any device to prevent auditory feedback because while performing the pilot trials, the auditory stimulus did not make any difference due to the controlled impulse procedure. Each subject received eight different consecutive trials in each condition block, with trials separated by approximately a 20-second pause and each condition block separated by approximately a 2-minute pause. For each condition block, the participants were instructed to report the compartment number on the screen as quickly and correctly as they could. Prior to each series of critical trials, they were allowed to practice for two trials per task. When they were ready and able to follow instructions, they completed the eight trials with two condition blocks.

Statistical Analysis: We analyzed the data to determine uncoupled and coupled self-produced movement in terms of haptic perception precision and to determine expertise. Two-way analysis of variance was used to determine the uncoupled and coupled haptic perception accuracy associated with expertise. The alpha level for all statistical tests was .05.

Results: The critical dependent measure was the accuracy of the test depending on the two independent variables (skill levels: 2, conditions: 2); only relevant and significant findings are presented.

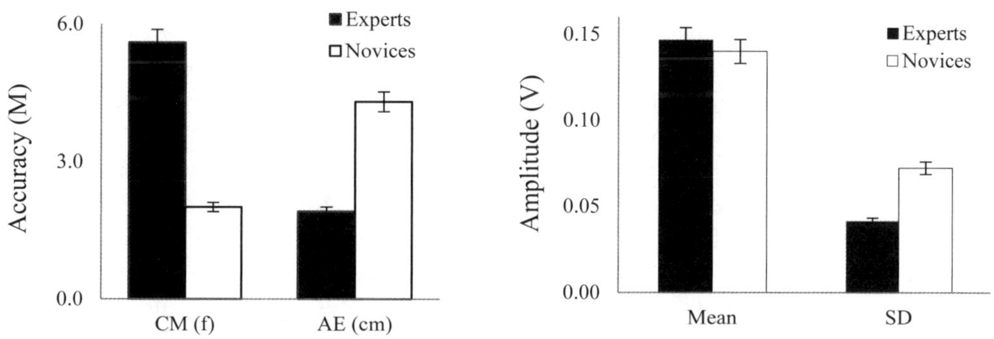

Figure R1.1.1. Comparison of the performance variables in experiment

In Figure R1.1.1, the left plot shows participants' frequency (f) of a correct match (CM: black bar) (+SE), size (cm) of the absolute error ($AE = \sum | \chi_i - T | / n$) between an actual hit position and the perceived designated position of the racket (+SE). The average frequency of matching the contact point by experts was 5.6 (70%) with a pointed error size of 1.9cm, while for novices the matching frequency was 2 (25%) and the pointed error size was 4.3cm. The plot of the right side shows the racket vibration (+SE); also shown are the voltage mean (left bar graph) and standard deviation (right bar graph) at the moment of a strike of the ball with the racket between the two groups.

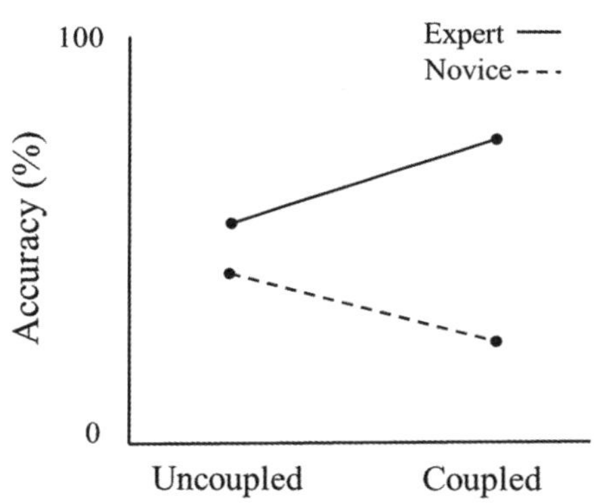

Figure R1.1.2. Mean performance accuracy by task condition and skill level. The average frequencies of matching the contact point were 20.8 (wrong: 19.2) in the uncoupled condition (*M* = 52%) and 28 (wrong: 12) in the coupled condition (*M* = 70%) for the experts and 12.8 (wrong: 28.2) in the uncoupled condition (*M* = 32%) and 8 (wrong: 32) in the coupled condition (*M* = 20%) for novices.

Figure R1.1.2. shows nonvisual haptic accuracy for self-produced movement by skill level. As shown in the figure, the main effect in the conditions was not significant, $F(1, 160) = .12$, $\eta^2 = .001$. The skill level main effect was significant, $F(1, 160) = 22.78$, $\eta^2 = .127$, and the skill level by condition effect on haptic accuracy was significant, $F(1, 160) = 4.19$, $\eta^2 = .026$. These results indicate that although the experts exhibited significantly greater performance accuracy in both conditions compared to the novices, skill-associated differences were intensified during self-produced movement.

From the results, we found that expert-level players acquire higher accuracy than novice-level players, especially in performances involving self-produced (coupled condition) movement. The implication of these results was that the skill-associated differences were exacerbated during self-produced movement. Furthermore, these features stemming from the same component can contribute differently depending on the user's capability.

Throughout the earlier discussions of accuracy and skill levels, many references were made to observations that accuracy is lower in novices. However, most of the data reported are average performance scores over many individuals at different skill levels rather than reflections of change within individuals over a series of trials. In addition to the general responses, performance variability is another indicator of haptic perception integrity, and from these concepts, it is indeed clear that the self-produced accuracy and variability levels of haptic features differ depending on the skill level. Thus, participants become less heterogeneous as regards the key features of a skill with an increase in the skill level. These factors as investigated in this study would be useful criteria to educators and suggest a broader hypothesis for further research into the effects of the haptic accuracy.

Reading List 1.2: Applied Statistical (Regression-Based) Research Example with Executive Function

Purpose: The purpose of this study was to demonstrate the effects of a motor coordination relationship on the executive function of older adults. Bimanual coordination and eye–hand coordination responses were used to evaluate motor coordination. **Methods:** To assess executive function, the Stroop Color-Word Test was used, which is considered an ideal tool for studying typical and atypical executive processing. The methodology in this study established which differences in variables were and were not predictive of different Stroop interference scores to determine the most robust motor coordination features related to executive function. **Results:** The analysis revealed that some significant relationships were detected in light of the variance in Stroop error effects. In particular, discrete bimanual spatial coordination was most strongly related to Stroop error effects. This shows that the asymmetrical spatial coordination pattern can be regarded as a significant contributor to the executive function of the elderly (Park, 2018).

Background: The present study investigated two aspects of motor coordination in line with behavioral findings. The first was bimanual coupling (Franz et al., 1991) in terms of the measured spatial dimension of motor control. The second was an eye–hand response (Fitts & Seeger, 1953) regarding temporal-based reactions to highlight how these types of coordination can be understood as realizations of neural connectivity. To match these movement results to cognitive function, the Stroop Color-Word Test (Stroop, 1935) was used, which is considered an ideal tool for studying typical and atypical executive process (Adleman et al., 2002). We attempted to focus more on the inhibition or overriding of the tendency to produce a more dominant or automatic response (Stuss & Benson, 1986) by assessing an aspect of the cognitive function of the elderly. We expected that these matches (Stroop task and motor coordination) could inform us of the degree of regulation of cognitive functions and actions (Kiefer et al., 1998). This refers to a strong relationship between not only the prefrontal regions of the frontal lobe (Petersen et al., 1988) but also the parietal lobes (Grieve et al., 2007).

Methods: Elderly participants (N = 30: 20 females, 10 male) took part in this cross-sectional experiment. The mean age was 80 years old (SD = 5, range 71–89), and the mean level of education was 11.3 years (SD = 2.77, range 9–16). The mean level of activity on a weekly basis according to leisure and physical exercise was 6.5 hours (SD = 2.02, range 5–10), and the average body mass index was 23.85 kg/m^2 (SD = 1.06). Systolic blood pressure was 140.30 mm Hg (SD = 3. 22), and diastolic blood pressure was 82.38 mm Hg (SD = 2.01). Individuals (N = 6) were eliminated from the study (3 = gave up [two males, one female], 3 were outliers [two males, one female]) using an exclusion criteria score. Thus, 24 sets of data were analyzed in total. All individuals participated voluntarily in the study. Participants were recruited in Seoul (Korea) through the member registry of a college program for the elderly (Golden Well-Being). As part of the selection process, volunteers were screened for medical problems, including health restrictions; individuals whose general motor (cardiovascular, osteoarthritis, and arthrosis) and neurological (brain, Parkinson's disease, and stroke) functions were normal were recruited.

Procedure and Data Analysis: Experimental test sessions tested the simple reaction time, anticipation time, bimanual symmetry drawing, bimanual asymmetry drawing, and Stroop tasks (a modified version of the Stroop task). Every participant was tested in these five individual sessions, which lasted approximately 30 ~ 40 minutes total. The spatial coordination score equals the index of circularity. It was computed for each period of these repetitive drawings. If a participant produced a perfect circle, the ratio of the two axes should have been 1. To consider a symmetry condition and eliminate the dominant hand effect, all participants were instructed with two dual-different pair conditions. The line and circle trajectories of a typical dual-different trial are depicted in Figure R1.2.1.

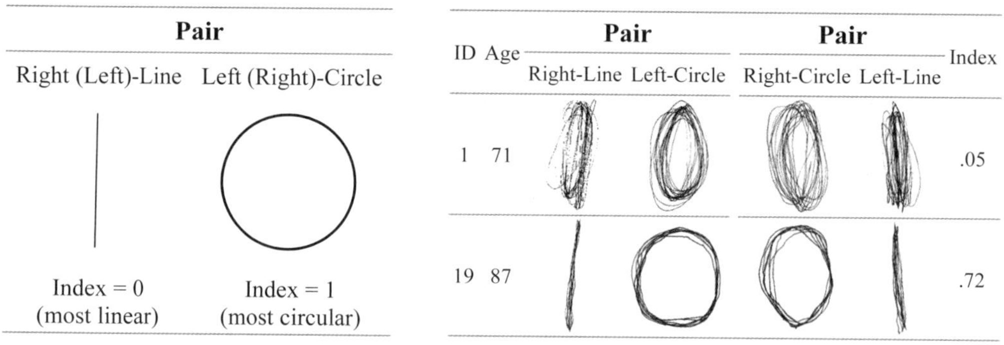

Figure R1.2.1. Index of independence for the dual-different trials. (Left) A perfect independent would have a higher index score (perfect is 1.0), and the worst independent would have a lower index score (the worst is 0.0). (Right) Asymmetrical bimanual drawing displacement in millimeters: the X (major axis) and Y (minor axis) dimensions of lines and circles when subjects performed dual-different tasks for a 20-second trial for each pair condition.

Results: The result showed no remarkable regressions with SEs (age: $R^2 = -.003$, simple reaction time: $R^2 = 0.004$, anticipation time: $R^2 = -.059$, SEE: $R^2 = .133$). Although the regression exhibited causality between spatial coordination and SE, the effect on the coordinative movement appeared to be weak relative to the other variables.

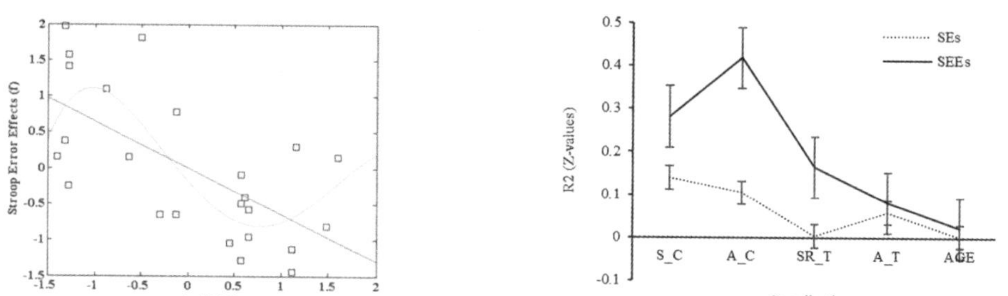

Figure R1.2.2. Regression-based analysis results. Left: Predicted Stroop interference scores with SEE (f: frequency), A_C as the dependent variable indicating asymmetrical bimanual coordination. Includes both linear (straight line) and logarithmic (curved line) regressions, yielding a fourth-degree polynomial fit equation. Right: Normalized (Z scores) significant regression (R^2) for each motor condition, predicting Stroop effects (SEs) and error effects (SEEs). Conditions include S_C (simultaneous bimanual drawing), A_C (asymmetrical bimanual coordination), SR_T (simple reaction time), A_T (anticipation time), and AGE (participant age range). Vertical bars show standard error mean.

Specifically, spatial coordination according to the measurements of the asymmetric bimanual drawings was more closely related to the SEEs (R^2 = –.42) than to the other variables (age: R^2 = –.023, simple reaction time: R^2 = 0.164, anticipation time: R^2 = –.081, simultaneous bimanual drawing: R^2 = 0.281, SE: R^2 = .133). We hypothesize that there is a different effect on executive function when coordinative movement is controlled, especially in SEEs.

Figure R1.2.2. displays a possible generalization of different motor conditions with Stroop interference effects by comparing various parameters of the normalized regression function. The evidence showed that both the temporally dependent responses and the spatially dependent responses incurred weakly significant results from the variable of the SEs. Meanwhile, the SEEs and spatial dependent coordination variables were statistically significant, showing increased drawing performance among the participants.

[Table R1.2.1] Multiple Regression Analyses for the Stroop Test Interference Time

Variable	B	T	Significant t	R^2
Dependent Variable: Stroop Effects				0.246
Age	.177	.196	.452	
Simple reaction time	–.138	–.532	.602	
Anticipation time	–.210	–.751	.463	
Symmetric spatial coordination	–.295	–1.002	.330	
Asymmetric spatial coordination	–.085	–.253	.803	
Dependent Variable: Stroop Error Effects				0.635
Age	.160	.909	.376	
Simple reaction time	.359	2.250	.038*	
Anticipation time	–.388	–2.230	.040*	
Symmetric spatial coordination	–.120	–.576	.572	
Asymmetric spatial coordination	–.475	–2.321	.033*	

*$p < .05$. Note: Using the regression-based approach, we normalized the raw scores of each independent variable into Z scores. The dependent variable was the Stroop interference score.

The analysis (Table R1.2.1) revealed that all variables were weak contributors after the component variables were accounted for in the SEs: SE R^2 = .246, adjusted R^2 = –.020, $F(6, 17)$ = 924, $p > .502$. However, this was interpreted in terms of the variance of the SEEs, R^2 = .635, adjusted R^2 = .506, $F(6, 17)$ = 4,922, $p < .004$, and a significant relationship was detected by a strong coefficient. In particular, asymmetrical spatial coordination was the most significant contributor (SEE $p < .033$). This observation implies that although the significance levels vary for drawing motor ability, asymmetrical bimanual drawing can predict SEEs from degradation. SEs may not have significant negative explanatory power; that is, decreasing the Stroop error index did not significantly reflect increasing motor coordination.

Part 2.

Applied Time Series Analysis

Contents

Part 2. Applied Time Series Analysis

[2.1]

Fundamentals of Time Series Analysis

Overview of Time Series, A Guide to Stochastic Concepts (4 ways)

A time series, typically characterized by data points chronologically organized and recorded at uniform time intervals, often serves as the backbone of applied systems analysis. The time factor functions as the primary independent variable.

The observation frequency in a time series can span a broad range, from high-frequency data recorded every second or minute—such as user visits or mouse clicks—to data observed on an hourly, daily, weekly, monthly, quarterly, or annual basis. This versatility makes time series data pervasive in a multitude of contexts. Essentially, any sequentially observed data over time form a time series (Klein & Klein, 1997). Such data ranges widely, including daily stock prices, monthly sea temperatures, quarterly company sales, and annual corporate profits.

1. Overview of Time Series Analysis

In the realm of applied systems analysis, time series analysis stands as a crucial tool, enabling us to comprehend various elements of a series to generate precise and meaningful forecasts.

+ **Embarking on Forecasting:** Time series analysis provides the foundation for forecast development, with deeper understanding of the data translating into enhanced forecast accuracy (Yaffee & McGee, 2000). Effective problem framing necessitates a clear understanding of forecast utilization, the target audience, and the forecast's organizational role. A forecaster's engagements must span all individuals involved in data collection, database management, and forecast implementation for future planning.

+ **Information Acquisition:** The process invariably requires a blend of statistical data and expertise accrued by data collectors. The challenge often lies in gathering sufficient historical data to establish a robust statistical model. Older data might occasionally lose relevance due to systemic changes in the forecasted subject, prompting reliance on recent data. Yet, a well-constructed statistical model takes evolutionary system changes into account, warranting the preservation of valuable data.

+ **Preliminary Analysis:** The analysis begins with data visualization, prompting several key questions:

- Are there discernible recurring patterns?
- Is there a significant trend?
- Does seasonality play a notable role and are there identifiable cycles?
- Are outliers present, necessitating expert explanation?
- How robust are the relationships among available variables?

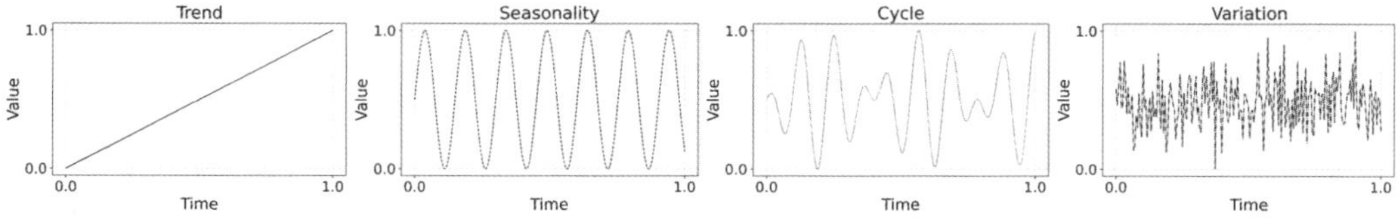

Figure 2.1.1. Time series components analysis. Four plots represent the primary components of time series data: Trend, Seasonality, Cycle, and Variation. It also provides an overview of the relationships among these components in the data.

Figure 2.1.1 presents a time series component analysis, delineating four key elements; trend, seasonality, cycle, and variation. Comprehensive understanding of these components is paramount for effective time series analysis, yielding insights into underlying data patterns and behaviors.

+ **Model Selection and Evaluation:** The ideal model choice hinges on historical data availability, the strength of relationships between forecast and explanatory variables, and forecast application. Model comparison is typical, each being an artificial construct based on assumptions and parameters to be estimated using historical data. Post model selection and parameter estimation, the model is deployed for forecasting. Model performance can be effectively evaluated only after forecast period data is available. Several methods exist for forecast accuracy assessment, along with organizational considerations regarding forecast application.

+ **Uncertainty:** Forecasting is not the sole objective of time series analysis. Other key attributes, such as stationarity, autocorrelation, and master equations, warrant attention. Before progressing to advanced modelling, a firm grounding in forecasting basics is necessary. Humans demonstrate a unique susceptibility to uncertainty (Berkes, 2007). Research suggests that the inability to tolerate uncertainty is central to most anxiety disorders and fuels panic-inducing catastrophic interpretations. In times of global crises (like the COVID-19 pandemic or the 2008 financial crisis), uncertainty disrupts routine behaviors, inducing conflict, and heightened reactivity to negative experiences.

+ **Addressing Uncertainty:** Tackling uncertainty's impacts head-on is essential. In the face of considerable uncertainty, humans are prone to envisage worst-case scenarios. Many studies revealed that worry impairs problem-solving, reducing confidence in problem-solving abilities and deteriorating the quality of solutions. Information overload can also overwhelm an individual's information processing capacity. One should avoid associating uncertainty and instead, strive to reduce uncertainty to a manageable level. To counteract the effects of uncertainty arising from abundant information or numerous random variables, forecasting processes are invaluable. They aid in strategizing to diminish uncertainty, improving decision-making, and fostering a sense of control.

2. A Guide to Stochastic Concepts (4 ways)

Building on the concept of forecasting under uncertainty, we now transition to the realm of stochastic processes, mathematical models that effectively grapple with randomness and unpredictability in our data. A stochastic process is a mathematical object usually defined as a condition or random variables (Heckman, 1987). The word *stochastic* is an adjective that describes something that is randomly determined. It describes a random probability distribution or pattern that may be analyzed statistically but may not be predicted precisely.

[Table 2.1.1] Probability Theory and Stochastic Processes: Basic Concepts

	Discrete state	Continuous state
Discrete time	Shuffle cards Markov chain	Timing of repetitive motor responses, stock prices Iterative maps (e.g., moving average models)
Continuous time	Biological reactions Master equations	Fluctuations in intensity motor responses, stock prices Langevin equation, Fokker-Planck equation Single-person behavior, social dyads (two people), small-group research, group and mass behavior/many-body systems research (networks)

Probability theory and stochastic processes can be represented in four different ways (Table 2.1.1) according to the relations of their states and time (Medhi, 1994).

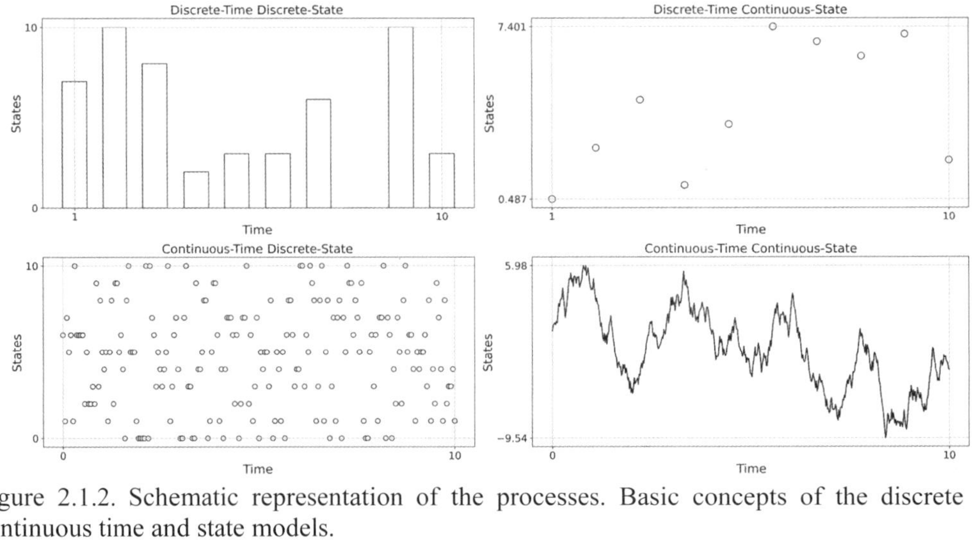

Figure 2.1.2. Schematic representation of the processes. Basic concepts of the discrete vs. continuous time and state models.

The details of each of these are presented throughout Part 2, but let us briefly introduce them here.

In Table 2.1.1, Shuffled card games are examples of discrete time systems, while biological processes are examples of continuous time systems and are described by master equations. Iterative mapping or integrating fluctuations using Langevin or Fokker-Planck equations can be used to represent the time of human repeated motor responses and stock prices for continuous state systems. The models' use in many-body systems study spans individual to collective behavior.

A schematic illustration of the procedures involved in discrete and continuous time and state models is shown in Figure 2.1.2. The graphic demonstrates how these two models' fundamental ideas might be utilized to explain and comprehend complicated systems. The illustration offers guidance by way of visuals and these simple application examples to help guide concepts in this section follow.

+ **Discrete Time and Discrete States (e.g., Card Shuffle):** A stochastic process can be thought of as a sequence of random variables indexed over time, where time can be a discrete set or dense set. For example, we have three cards (A, B, C) defined with six states.

[Table 2.1.2] Shuffle Rule: Take One and Put It to the Right

	1	2	3	4	5	6
States	ABC	ACB	BAC	BCA	CAB	CBA

We can characterize it as below:

- Characterizing the change from one state to another, such as the change from ABC to BCA, is a key component in the study of transitions between states.
- One may figure out, for instance, that $p(i) = 1/6$ is the probability of witnessing a state i denoted as $p(i)$.
- Calculating the probabilities of transition between states is also possible, as shown by the example of $p(1\ \ 4) = 1/3$.
- In the context of Markov chains, Monte Carlo simulation methods are used to represent these transitions.

For example, let us suggest a simple model for the discrete time and discrete states (e.g., emotional dynamics; Isenhower et al., 2012). Participants were asked to report their emotional state. Every 3 minutes, a cued response is requested: report how you feel (analog scale: positive or negative, and to what degree). The recoding periods is 1 day (periods ranged from 14 h to 19 h).

The characteristics of the data are below:

- Discrete time

- Only the qualitative aspect of the response was considered
- Responses were classified as either 'positive' or 'negative' → two states $\{p, n\}$ → discrete states
- Participants made transitions between two emotional states (Markov chain modeling)

The findings suggest that the positive emotional state was more attractive than the negative one.

+ **Discrete Time and Continuous States (e.g., Tapping):** To build stochastic models with continuous state spaces, stochastic processes can be used. They can be seen as a generalization of (Markov) chains because they do not rely on a discrete state space. Let us take a simple experiment as an example:

Tapping with right-hand fingers on a key at a required frequency (Wing and Kristofferson, 1973).

- Continuation paradigm: pacing at the beginning of the trial, then pacing is switched off
- Question: Do participants use feedback? Is this evidence for the feedback system?
- Answer: Moving average model (no feedback); negative correlation is a purely statistical effect.

This experimental model includes discrete time and discrete events (taps), a continuous-state variable (= time intervals), and applies a moving average model (a special case of the stochastic iterative map).

- Application to financial stock market: states = prices and indices, time = days

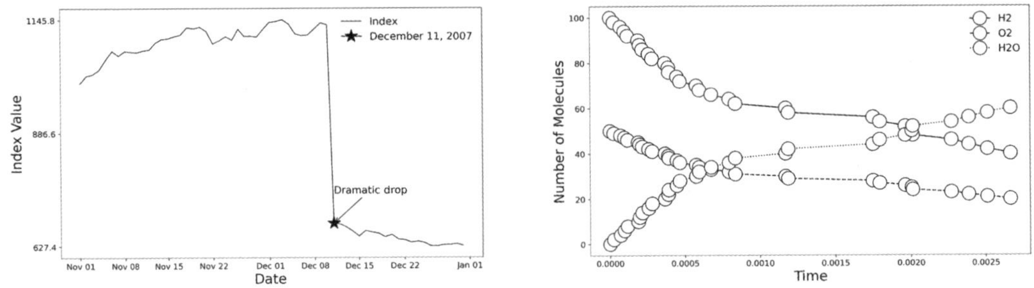

Figure 2.1.3. Stochastic method for the stock market (discrete time and continuous states) and the chemical reactions (continuous time and discrete stats). The simulation of the left side on December 11, 2007, the index drops dramatically (created data from Wikipedia). The simulation of the right side shows a chemical reaction $2H_2 + O_2 \rightarrow 2H_2O$ using the Gillespie algorithm.

The left side of the Figure 2.1.3, alludes to a practical illustration of the idea of discrete time and continuous states. The illustration concentrates on the stock market, particularly the decline in an index that took place on December 11, 2007. The state of the system is recorded at certain points in time, and the time dimension is described as discrete. Since there are only a limited number of potential values for the state of the index at any one moment, it is probable that the time increments in this situation will be daily. A continuous model is used to represent the system's state, in this example, the value of the index.

+ **Continuous Time and Discrete States (e.g., Biochemical Reactions):** For continuous time modeling, the time parameter space is considered continuous.

To be specific, let us take a simple example:

- $2H_2 + O_2 \rightarrow 2H_2O$, Molecules: n_1: H_2, n_2: O_2, n_3: H_2O.

The variables can assume only integer values → discrete states.

- Evolution in time: $n_1(t), n_2(t), n_3(t)$

Thus, the time variable *t* is continuous. This can be also applied by stochastic case called Gillespie algorithm.

In Figure 2.1.3' right side, the variables n_1, n_2, and n_3 are thought to exist in discrete states as the number of molecules can only have an integer value. However, $n_1(t)$, $n_2(t)$, and $n_3(t)$ continually capture the change of these variables across time. The constant time variable t illustrates this. The Gillespie algorithm, a stochastic method for modeling the temporal development of chemical reactions, is one example of a discrete-time and continuous-state methodology that may be used to represent the chemical reaction.

Another choice for modeling systems with continuous-time and discrete-state features is the master equation model. For the continuous-time probability of the number of particles in each state, this model offers an evolution equation. This model may be used to explain the development of cancer states in the field of health research, where the states reflect various phases of the illness and time is continuous.

The master equation may also be used to simulate the emergence of addiction in the context of material misuse, such as local smoking habit. For instance, Selya et al. (2013) used this model to examine the evolution of smoking habit across time. In this instance, time is depicted as a continuous dimension, and the states indicate various levels of smoking behavior.

- Discrete states: Nonsmoker, experimental smoker, dependent smoker
- Transition in continuous time

According to the results of the example case, environment and prevention can affect the rate of transition from experimental smoker to dependent smoker. A time series model allows for model-based data analysis.

+ **Continuous Time and Continuous States (e.g., Collective Phenomena):** For continuous time models with a continuous state space, a differential equation for the rules can be derived.

- States: continuous
- Time: continuous

Generally speaking, the most potent analytical tools for modeling complex systems are continuous-time and continuous-state models. These models help to understand the dynamics at work more thoroughly and may be mathematically examined to determine how they will behave before being put through a simulation.

Several systems, such as stock prices, rotational Brownian motion, light intensity in bistable lasers, temperature in many-body systems, synchronized flashing of fireflies, orientational order in liquid crystals, and swarming behavior, may be described using this method. These models offer a more thorough knowledge of the underlying processes and enable a more accurate description of the system's ongoing change through time.

Systems where the underlying variables change continuously and smoothly across time benefit the most from the usage of continuous-time and continuous-state models.

Discrete-time and continuous-state modeling has also been used, but not that frequently.

- Application of time series modeling in a new or developing research field (small-group research)

Continuous-time and continuous-state modeling of motor control and coordinated movements

- Single person: within-person motor coordination
- Between persons: coordinated movement between persons (small group)
- Group behavior: mass behavior (e.g., synchronization $N > 1{,}000$)

As an illustration of a system that may be represented using continuous-time and continuous-state analysis, consider the bi-manual pendulum movement. In this model, the left and right oscillators' relative phases can change continuously throughout a range of 0° to 180° and are influenced by the strength differences between the two oscillators. Using the programming (i.e., Python) platform, straightforward techniques may be used to examine the continuous-time and continuous-state data.

This platform offers an intuitive user interface for data processing and enables simple display of the outcomes. It is possible to gain a deeper knowledge of complex systems and gain insightful information for a variety of applications by combining continuous-time and continuous-state models with contemporary programming tools, such Jupyter notebook.

First, we need to import libraries

```
[In]
import numpy as np
import pandas as pd
import matplotlib.pyplot as plt
from sklearn.decomposition import PCA
```

And we are going to generate data which has been created by a programing concerning movement (left side and right side) oscillations.

```
[In]
N = 5000
x = np.concatenate([np.random.normal(loc=30, scale=5, size=N),
np.random.normal(loc=20, scale=5, size=N)])
y = np.concatenate([np.random.normal(loc=26, scale=2, size=N),
np.random.normal(loc=25, scale=2, size=N)])
z = np.concatenate([np.random.normal(loc=12, scale=1, size=N),
np.random.normal(loc=10, scale=1, size=N)])
hand = np.concatenate([np.ones(N), np.zeros(N)])  # 1 for left, 0 for right
```

Combine the created data into a 'DataFrame' and convert each oscillator

```
[In]
data = pd.DataFrame({'x': x, 'y': y, 'z': z, 'hand': hand})
data['hand'] = data['hand'].astype(int).astype(str)
```

The code reads in a 'data', and separates the values in the file. In this case, the data is specified as columns that the data are separated into three axes (x, y, z) because the movement data are three dimensional.

Now we create two arrays called 'left' and 'right' from the 'data' set. The 'left' array is intended to contain the values from the x, y, and z columns for the rows where the value in the 'hand' columns is the one. At the same time, the 'right' array is intended to contain the values from the x, y, and z columns for rows where the value in the 'hand' column is the other one.

Then, we create a plot to transform data into data for 5,000time duration.

```
[In]
left  =  data.loc[data['hand']  ==
'1', 'x':'z'].to_numpy()

right = data.loc[data['hand'] ==
'0', 'x':'z'].to_numpy()
```

→

```
[Out]
array([[ 30.643,  26.500,  11.993],
       [ 31.197,  26.519,  11.876],
       [ 31.721,  26.524,  11.754],
       ...,
       [ 20.237,  24.950,  10.092],
       [ 20.429,  25.015,  10.127],
       [ 20.628,  25.078,  10.166]])
```

In Figure 2.1.4, we used principal component analysis to reduce the dimensionality of the 'left_1d' and 'right_1d' arrays and then plotting 1D arrays.

[In]

```
pca = PCA(n_components=1)
left_1d = pca.fit_transform(left)
right_1d = pca.fit_transform(right)

# Reshape the 1D PCA values into 1D arrays
left_1d = left_1d.reshape(-1)
right_1d = right_1d.reshape(-1)

# Generate time array
t = np.arange(left_1d.shape[0])
```

→

[Out]

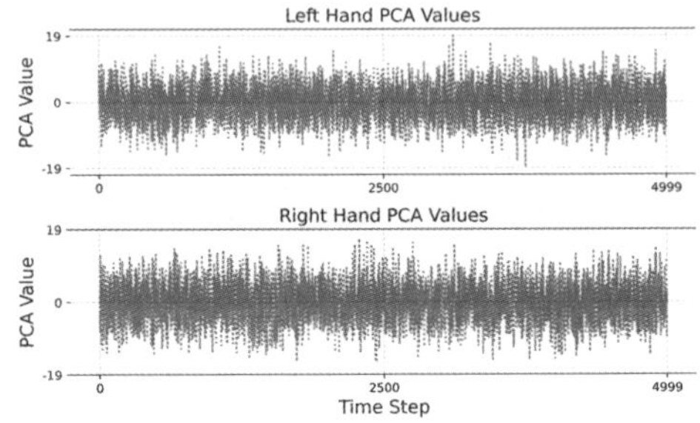

Figure 2.1.4. Output of results using continuous time and state event values. The vertical axis denotes the degree of each hand's pendulum movement (left hand = blue, right hand = green) and the horizontal axis represents time duration cutting 5000, PCA = Principal Component Analysis.

The visualizations represent continuous-time models where the 'Time Step' corresponds to the continuous timeline and the Principal component analysis value corresponds to the continuous state of motion. By adopting continuous-time and continuous-state models, we can capture collective phenomena with greater accuracy, offering deeper insights into the underlying dynamics, and enabling predictive and mathematical analysis of system behavior prior to simulation.

+ **Practice with Continuous Time and Continuous States (A First Look):** Let us assume we have a linear, deterministic model whose evolution is given by

$$\frac{d}{dt}X(t) = -a.X(t)$$

This describes the time evolution of a variable X. The equation states that the rate of change of X with respect to time (dx/dt) is equal to the negative of a constant a multiplied by the value of X at time t. X = observable, T = time, Parameter: $a > 0$ (e.g., a = 0.05), initial condition: $X(0)$ (e.g., $X(0) = 2.0$), and the $dx(t)/dt$ is derivative of X with respect to t. The interpretation of $dx(t)/dt$ can be as below:

Change in X with time t
Slope of function $X(t)$ at time t
Example: if $X(t)$ increases linearly in a time interval around t, then the slope is $\Delta X/\Delta t$.

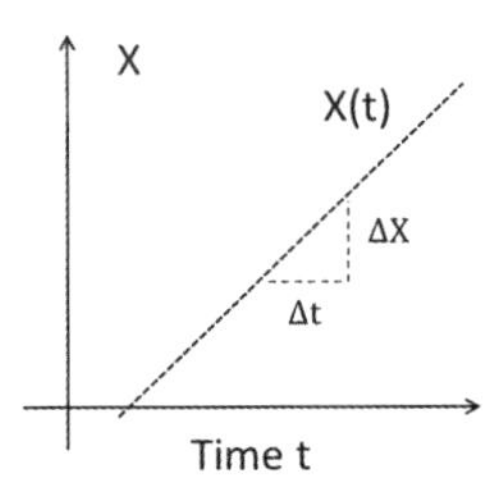

Figure 2.1.5. Schematic representation of the continuous time and states $(dx(t)/dt)$

Following the concepts represented above (Figure 2.1.5), we suggest one simple simulation (a nonlinear deterministic time series model) to show a more vivid image. It describes the evolution of a continuous variable X on a continuous time scale described by t. The evolution equation for the trajectories $X(t)$ of this model is given by

$$\frac{d}{dt}X(t) = -a.[X(t)]^3$$

That is, on the right side of the equation, the variable X is raised to the power of 3. We consider the case $a > 0$. In this case, it can be shown that for any initial condition $X(0)$, the trajectory $X(t)$ converges to the state $X = 0$ over time.

Now, let us make a task overview up: We can determine solutions numerically using a discrete-time model, such as the Euler forward method, using the equation ($\frac{d}{dt}X(t)$).

- We create a programing (i.e., Jupyter notebook) script to obtain a new script that can be used to solve equation ($\frac{d}{dt}X(t)$) numerically via the method given in terms of the discrete-time model.
- Then solve the equation numerically for $a = 0.5$ and initial condition $X(0) = 3$, in the time window $t = 0$ to $t = 10$.
- Finally, compare the numerical solution and the exact solution in the same diagram.

The exact solution for $X(0) > 0$ is given by

$$X_{exact}(t) = \frac{1}{\sqrt{2 \cdot a \cdot t + 1/[X(0)]^2}}$$

This is an exact solution to the following initial value problem

$$\frac{d}{dt}X(t) = -a * X(t), \qquad X(0) = X\_0$$

Here, a and X_0 are constants and $X(t)$ is the unknown function being solved for. Notice it is a function of the time t, the decay constant a, and the initial value X_0. For the solution, it is worth to write the programming code for the precise solution slightly differently. For example, we may use a for-loop and calculate in the for-loop each element of the exact solution separately.

The code works by approximating the derivative on the right-hand side of the equation using the following sets. The value of a to 0.5, the step size dt to 0.001, and the initial value $X(0)$ to 2. And it generates a vector of time steps 'time_vec' from 0 to 10 with a step size of dt. A for loop is then used to iterate over the time steps and approximate the solution oat each time step continues until the final time step is reached.

```
[In]
a = 0.5
delta_t = 0.001
time_vec = np.arange(0, 10, delta_t)
t_vec_length = len(time_vec)
X = np.zeros(t_vec_length)
X[0] = 2

for n in range(1, t_vec_length):
    X[n] = (1 - delta_t * a) * X[n - 1]
```

Let us create plot to see what is going on with the calculation.

```
[In]
import numpy as np
import matplotlib.pyplot as plt

plt.figure(1)
plt.plot(time_vec, X, 'k-', linewidth=5,
alpha=0.3)

Xexact = X[0] * np.exp(-a * time_vec)
plt.plot(time_vec, Xexact, 'k--',
linewidth=1, alpha=0.7)
```

→ [Out]

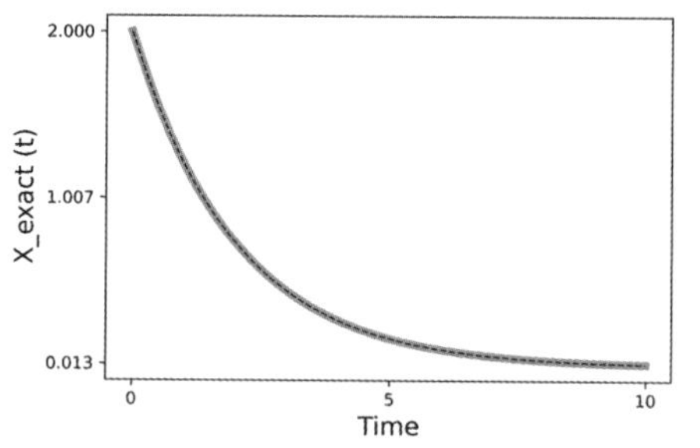

Figure 2.1.6. Graphical representation of the X-exact solution

This is the simple prototype, so feel free to generate the solution (Figure 2.1.6) using the numerical solution and the exact solution for $a = 0.5$ and $X(0) = 3$. Produce a similar figure for the numerical solution and exact solution for $a = 0.5$ and $X(t) = 2$.

Remarks: As briefly mentioned earlier, stochastic concepts are crucial for understanding the time series and systems analysis. These concepts provide insights into system behavior over time, enabling the modeling and examination of random processes. Focusing on four key elements—discrete time, discrete state, continuous time, and continuous state—the study of stochastic ideas involves applying mathematical models to depict the evolution of systems. By exploring the interconnections between these elements, we can gain a deeper understanding of system behavior and make predictions about their future states. Various disciplines, such as physics, biology, economics, and engineering, as well as numerous practical fields, necessitate a grasp of stochastic concepts. Thus, we will delve into further detail and explore various practical applications in the subsequent sections.

[2.2]

Probability Theory, Part 1

System Variables, Relative Frequency, Frequency Based Approaches, Probability State Spaces

In everyday conversation, the term 'probability' encompasses multiple meanings, which play a significant role in the development and application of mathematical theory (Kneale, 1951).

Probability is simply how likely something is to happen. Whenever we are unsure about the outcome of an event, we can talk about the probabilities of certain outcomes—how likely they are. One meaning has to do with the interpretation of probabilities as relative frequencies, for which simple games involving coins, cards, dice, and roulette wheels provide examples. The distinctive feature of games of chance is that the outcome of a given trial cannot be predicted with certainty, although the collective results of a large number of tests display some regularity.

For example, the statement that the probability of getting heads when tossing a coin equals one half, according to the relative frequency interpretation, implies that in a large number of tosses the relative frequency with which heads actually occurs will be approximately one half, although it contains no implication concerning the outcome of any given toss.

There are many similar examples involving groups of people, molecules of a gas, genes, and so on (Schrodinger et al., 1992). Actuarial statements about the life expectancy of persons of a certain age describe the collective experience of a large number of individuals but do not purport to say what will happen to any particular person. Similarly, predictions about the chance of a genetic disease occurring in a child of parents with a known genetic makeup are statements about relative frequencies of occurrence in a large number of cases but are not predictions about a given individual.

This section presents the fundamental mathematical concepts of probability theory, accompanied by examples of applications that have driven their development. Since applications necessitate simplifying assumptions that emphasize certain aspects of a problem while disregarding others, it is beneficial to start with elementary experiments, such as coin tossing or dice rolling. We will then explore how these seemingly trivial investigations connect to crucial scientific inquiries and systematic analysis.

1. System Variables

The term 'variable' refers to a value that may be changed. It can represent a number, a vector, a matrix, a function, the argument of a function, a set, or an element of a set.

System variables are predefined variables (Burton & Zinober, 1986). They control many kinds of behavior and appearance. For example, system variables control the number of random samples used to represent uncertain quantities. There are a few essential concepts that you should understand before we go any further.

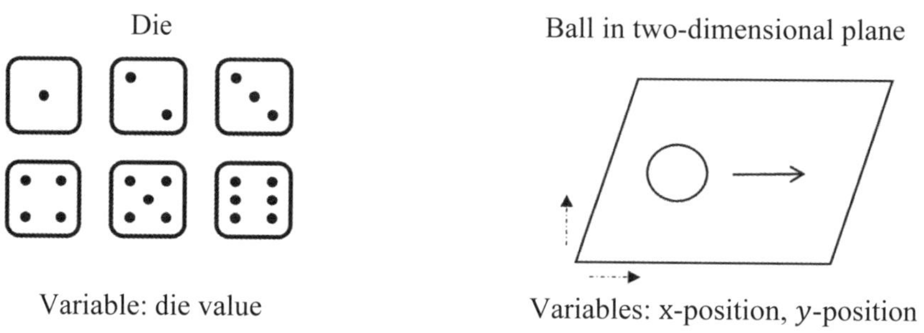

Figure 2.2.1. State variables defined by scales of measurement, discrete (left) and continuous (right)

The probability, which involves random variables and occurrences (realizations), describes the state of a system (e.g., a die value, the position and velocity of a ball; Figure 2.2.1).

+ **Parameter and Variable:** A parameter is a constant value that characterizes a particular aspect or condition of a system or model. Parameters are generally fixed, although they can sometimes be adjusted during model development or calibration. They act as a reference point or guide for calculating other variables. On the other hand, a variable represents a value that can change or vary within a system or model. Variables can be influenced by parameters or other variables and can be classified as discrete, continuous, or state-space, depending on their nature and the system they are part of.

- Discrete variables which are numerical or categorical: numerical (e.g., die value 1, die value 2), categorical (e.g., health status: healthy, preclinical cancer, clinical cancer)
- Continuous (numerical) variables: position, velocity
- State-space (discrete or continuous) variables which represent all possible states of a system, such as the various positions of a ball in two dimensions: the set of all possible states of a system (e.g., die: 1, 2, 3, 4, 5, 6; ball in two dimensions: X, Y in X^2)

Within the state space, coordinates or vectors of coordinates characterize a particular state (e.g., x and y coordinates).

+ **Random Variables and Realizations:** Examples of random variables include the outcome of a die roll, the height of randomly selected students, or the velocity of a molecule in the air at room temperature. Each time an experiment is conducted, a random variable X may assume a specific value, denoted as $X^{(k)}$, where k represents the iteration of the experiment. In this context, $X^{(k)}$ is referred to as a realization of X, as it represents a particular instance or observed value of the random variable X during the experiment.

- A random variable X assumes particular values $X^{(k)}$ each time an experiment is repeated $\rightarrow X^{(k)}$ realization of X.
- A random variable is a variable that has realizations k that in general are not the same (because of the effects of randomness) for all k.

The probability distribution of a random variable describes the possibility of witnessing each of its potential realizations. The realizations k of a random variable might not be consistent across studies due to the effects of randomness.

+ **Deterministic Systems:** A deterministic system is a conceptual doctrine of determinism applied to a system to understand everything that has occurred and will occur in the system based on the physical outcomes of causality (Bishop & Bishop, 1989). In a deterministic system, every action (or cause) produces a reaction or effect, and every reaction in turn becomes the cause of subsequent reactions. The totality of these cascading events can theoretically show precisely how the system will exist at any moment in time.

In mathematics, computer science, physics, and data analysis, a deterministic system is a system in which no randomness is involved in the development of future states (Skyttner, 2005). To understand this concept, start with a relatively small system. Visualize a set of three dominoes lined up in a row, with each domino less than a domino's length away from its neighbors. Once the first domino has toppled, the third domino will topple because the second will topple upon contact with the first domino.

A deterministic model always produces the same output from a given starting condition or initial state. Thus, it is a system whose state is described by a state variable for which all realizations are the same $X^{(i)}$ for all i, k (see Supplemental Information 2.2's Deterministic and Stochastic for more detail).

2. Relative Frequency

Frequency is the number of occurrences of a repeating event per unit of time. It is also referred to as temporal frequency, which emphasizes the contrast with spatial frequency and angular frequency. Frequency is measured in hertz, which is equal to one event per second.

+ **Notation**: Consider data set $X^{(k)}$, comprising elements $\{1, 1, 3, 5, 2, 1, 4, 6, 5, 6\}$.

Within this, we denote subsets of indices w corresponding to a particular state j, for example, $I\_1 = \{1, 2, 6\}$ and $I\_2 = \{5\}$, where I_1 and I_2 represent these unique subsets.

$$I_j = \{w | X^w = j\}$$

Frequency of state j j

$$h_j = \sum_{k \in I_j} 1 \quad \rightarrow \quad h_j = \sum_{k=1,\ldots,N; X^{(k)}=j} 1$$

Here, we defined the index set as the set of all indices k such that $X^{(k)} = j$, where $X^{(k)}$ is some function for which the value of $X^{(k)}$ is equal to j. The Greek letter sigma $\sum$ is used to denote a summation, and the symbol $\in$ represents membership in a set.

Outcome: State j (discrete state space)

- Collection (or set) of states = $I = \{j_1, \ldots, j_m\}$ (discrete state space)
- Interval or subspace (continuous state space)

Probability of an outcome = proportion (percentage) of occurrences of that outcome in the statistical ensemble

$$P_j = \frac{1}{N} \sum_{k=1,\ldots,N; X^{(k)}=j} 1, \qquad P(X \in [a, b]) = \frac{1}{N} \sum_{k=1,\ldots,N; X^{(k)} \in [a,b]} 1,$$

$$P(j \in I) = \frac{1}{N} \sum_{k=1,\ldots,N; X^{(k)} \in j} 1$$

The equation is defined as the fraction of indices k that satisfy condition which is equal to the sum of the values 1 over the index set $X^{(k)} = j$ divided by the total number of indices N. Let us implement step-by-step coding example of the frequency: histogram—Single-step method for frequencies—hist(data, nbins):

```
[In]
import numpy as np
import matplotlib.pyplot as plt
data = np.random.normal(size=100)
def hist(data, nbins=10):
    fig, ax = plt.subplots()
    min_val = np.min(data)
    mid_val = np.median(data)
    max_val = np.max(data)
    ax.hist(data, nbins)
```

→

[Out]

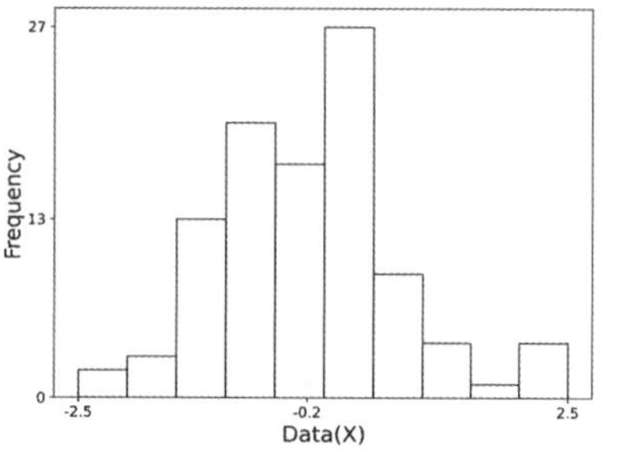

Figure 2.2.2-1. Representation of the frequency (using the generated random data)

We created data from the '`random.normal`' commad into a variable 'data'. The 'nbins' is set to 10 specified the number of bins to use in the histogram (Figure 2.2.2-1). The 'hist' function is then used to plot for the 'data' variable.

+ **Probability and Relative Frequency:** Probability is computed from an ensemble. This is an infinitely large set of realizations and not a random variable. In the probability, relative frequency is the ratio of the number of outcomes in which a specified event occurs to the total number of trials. Relative frequency is computed from a sample. This is a finite number of observations and a random variable.

[Table 2.2.1] Frequency Table (Stress)

Level	Rating = j	Frequency = h_j	Relative frequencies = f_j	Percent (%)
Highest	10	14	14/151	9.3
	9	15	15/151	9.9
	8	26	26/151	17.2
	7	31	31/151	20.5
	6	13	13/151	8.6
⋮	5	18	18/151	11.9
	4	16	16/151	10.6
	3	12	12/151	7.9
	2	3	3/151	2.0
	1	1	1/151	0.7
lowest	0	2	2/151	1.3
Total	11	151	1	99.9

In Table 2.2.1, a frequency analysis of stress levels (ranging from 0 to 10) reveals that the majority of the 151 observations were concentrated around the moderate to high ratings, with the largest proportion (20.5%) reporting a stress level of 7, followed by 17.2% at level 8, and the least frequent (0.7%) at level 1, cumulatively accounting for approximately 99.9% of the sample.

For $N \to \infty$ the relative frequency f_j corresponds to the probability P_j. Frequency (left side of the equation) and its relative frequency (right side of the equation)

$$h_j = \sum_{k=1,\ldots,N;X^{(k)}=j} 1 \qquad \to \qquad f_j = \frac{h_j}{N}$$

The equation simply states that f_j is equal to h_j divided by N, which means theat the f_j is a fraction whose numerator is h_j and whose denominator is N.

- Probability (left side of the equation) and its alternative notation (right side of the equation)

$$P_j = \lim_{N\to\infty} \frac{1}{N} \sum_{k=1,\ldots,N;X^{(k)}=j} 1 \qquad \to \qquad P_j = \lim_{N\to\infty} f_j$$

Here, the equation defines a quantity P_j as the limit of the quantity f_j as the parameter N approaches infinity. A key property of probability is that it is a tool for managing uncertainty. Whenever an event is neither certain (probability = 1) nor impossible (probability = 0), we are facing an uncertain situation; hence, we need to attribute to our event a likelihood of occurrence, which is probability. Thus, probability P is a dimensionless variable (semi-positive: $P \geq 0$). It must be less than or equal to 1: $P \leq 1$, where the probability of the outcome (event) is 1.

- State-space Ω: $P(X \in \Omega) = 1$
- Discrete sate space: States j = 1, …, M → $P(X \in \{1, \dots, M\}) = 1$
- Continuous state space defined by the interval [A, B] → $P(X \in [A, B]) = 1$
- Implication: normalization condition → $\sum_{j=1}^{M} P_j = 1$

Step-by-step coding example of the frequency histogram: Using program, the data is binned into a predetermined number of intervals (nbins), and the relative frequency of each bin is calculated by dividing the frequency of the bin by the total number of data points (f = counts/length(data)). The relative frequencies that result are then shown against the bin centers using the bar() function. For more complicated data sets, an alternative is to utilize the sophisticated histc() method.

```
[In]
data = np.loadtxt('data set_stress.txt')
nbins = 11

counts, bins = np.histogram(data,
bins=nbins)
f = counts / len(data)

min_val = np.min(bins) - 1
mid_val = (np.max(bins) + np.min(bins)) /
2
max_val = np.max(bins) + 1

ax.bar(bins[:-1], f)
```

→ [Out]

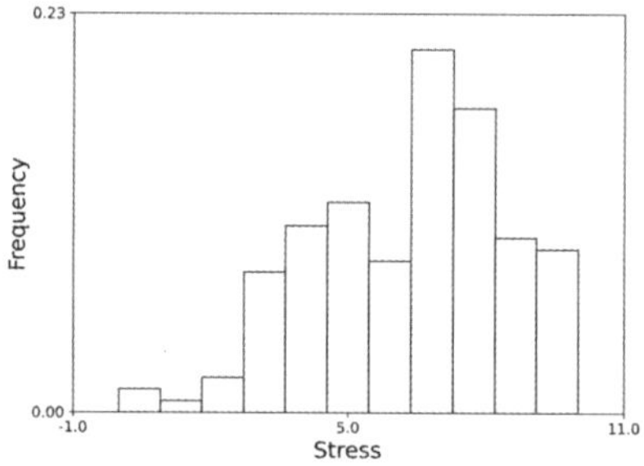

Figure 2.2.2-2. Representation of the relative frequency (using the Supplement data set of Stress.txt)

The plot (Figure 2.2.2-2) represents the fraction of data points in each bin. To be specific, we compute the histogram of the data in the 'data' variable. The 'nbins' are passed to the function to specify the number of bins and the width of the bars. The function returns the bin counts in the 'counts' variable and the bin edges in the variable. And the 'f' is defined as the ratio of 'counts' to the length of the 'data' array.

+ **Practice Histograms:** Data exploration is a critical step in every science project, and it usually starts with looking at the distribution of single variables. This is where histograms shine. A histogram is a graph that shows the frequency of numerical data using rectangles. The height of a rectangle (the vertical axis) represents the distribution frequency of a variable (the amount, or how often that variable appears).

Histograms are great for visualizing the distribution of columns, which helps to understand important aspects of data. When working with histograms, we almost always end up adjusting the bin width, which is a critical parameter that determines how much and what kind of information we can extract (or include) from the plot.

We can quickly find the optimal bin width by creating a simple histogram via computer programming (Jupyter Notebook). Notice that although we show a simple plot below, we can apply the logic to any plotting library, such as Matplotlib.

Import the libraries and input defined some values.

```
[In]
import matplotlib.pyplot as plt
import numpy as np


mu = 0
sigma = 2
x = np.random.normal(mu, sigma, size=10000)
```

Create histograms reflecting the values.

```
[In]
bin_size = 10 # apply with different bin_size like 100, 1000 etc.,
plt.hist(x, bin_size, density=True, histtype='bar')
plt.show()
```

We took several optional arguments in the code 'hist', which includes 'bin_size', 'density', and 'histtype'. The 'bin_size' specifies the size of the bins to use in the histogram. In this case, we set to 10 which data will be divided into bins of size 10. The 'density' specifies whether the histogram should be normalized to the probability density function of the data. Because we set to True, the plot will show normalized such that area under the curve equal to 1. Finally, the 'histtype' is for plot.

In this case, we set to 'bar' which that the histogram will be plotted as below.

```
[Out]
```

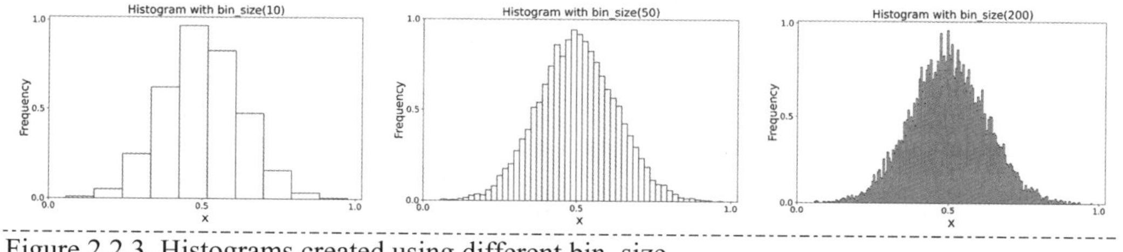

Figure 2.2.3. Histograms created using different bin_size

The width of each bin in a histogram may be changed by examining how frequently events occur (see Figure 2.2.3). We can create a rudimentary blueprint for creating a histogram by combining these component parts. This graphical form is a useful tool for examining certain columns in a data collection, especially when we are curious about the distribution of values.

3. Frequency-Based Approaches

Here we discuss the three approaches to probability: the classical approach, the empirical approach, and the subjective approach. Each approach provides different interpretations of the concept and starts with different assumptions (Fine, 2014).

+ **Classical Approach:** This approach traces back to the field in which probability was first systematically used: gambling (flipping coins, tossing dice, etc.). Gambling problems are characterized by random experiments that have n possible outcomes, all equally likely to occur. Because no one outcome is more or less likely to occur than the other ones, they are said to be symmetrical. The idea behind the classical approach is that given a collection of k elements out of n (where $0 \leq k \leq n$), the probability of the occurrence of event E represented by that collection is equal to

$$k/n$$

Imagine you are tossing dice and want to predict the probability of the following collection of outcomes (Figure 2.2.4):

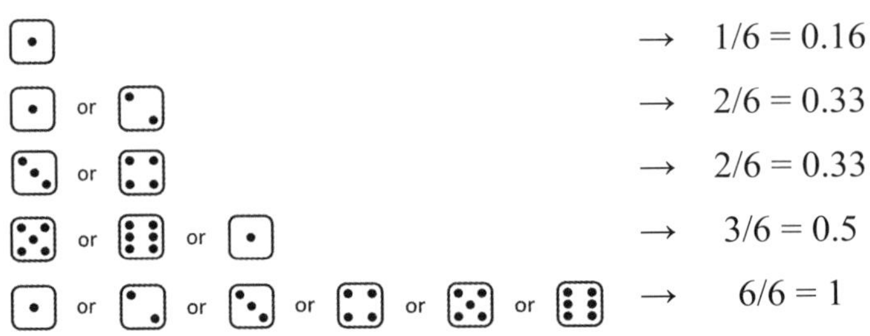

Figure 2.2.4. Collection of dice outcomes

We know that there are six possible outcomes. The event '1' is 1 out of 6 results; hence, its probability is 1/6. Similarly, the event '5 or 6 or 1' (i.e., the event in which one of those numbers turns up) represents 3 outcomes out of 6; hence, the probability is 3/6 = 0.5.

The classical approach is intuitive; nevertheless, it suffers from some pitfalls. The assumption of symmetry is far too strong and unrealistic. Imagine you want to know the probability of the event 'Tomorrow we will have a car accident.' There are two possible outcomes of this scenario: having a car accident and not having a car accident. Given k

= having a car accident, the probability of that event is 1/2, which, besides being a bit worrying, is not representative of the real likelihood of the event.

In this approach, there is no space for information, which is strictly related to probability. Let us think about the previous example of the die. Imagine you are told that this die is loaded, and instead of having a 1 face, it has two 6 faces (so the faces are 2, 3, 4, 5, 6, 6). Given this information, what probability would you attribute to the event '1'? Because it is impossible to roll a 1 with this die, the probability is equal to 0, not 1/6. Hence, probability depends on the available information (this becomes clearer in the discussion of the subjective approach).

Practice with Classical Probability

Classical probability is defined as the ratio of the number of favorable outcomes to the total number of possible outcomes. For example, let us think about the probability of getting heads in a coin toss.

$$possible\ outcom = \{H,T\}, favorable\ outcome = \{H\}, probability\ of\ heads = 1/2$$

+ **Empirical Approach:** This approach was formally introduced in the field of natural science, where the assumption of symmetry fails. Instead, this approach is based on the idea that several experiments can be run under certain conditions considered to be equivalent. Each test results in either success or failure. Hence, given n random experiments run under equivalent conditions, we define the frequency of success (which is an event E) as follows:

$$f_n(E) = \frac{number\ of\ successes\ in\ n\ experiments}{n}$$

If we consider the empirical law of change, which states that the more n increases, the more stable the frequency becomes, we can conclude that the limit of that frequency, for $n \to$ infinity, exists and is equal to the probability of the event 'success':

$$P(E) = \lim_{n \to \infty} f_n(E)$$

Let us compare the empirical and classical approaches using the following example. Imagine you want to know the probability of a coin toss coming up heads. You start with the classical approach: Because there are two possible outcomes (heads and tails), the probability of getting heads is 1/2 = 0.5.

Now you decide to use the empirical approach, and you start by tossing your coin 100 times. In these 100 tosses, you get 55 heads and 45 tails. Hence, the frequency of the event 'heads' is 55/100 = 0.55, and this approximates the probability of the event 'heads.' As you can see, you obtained two different probabilities (0.5 vs. 0.55) for the same event. The critical difference is the role of information: After 100 experiments, you gathered empirical evidence that heads occurred more often than tails. It might be that your coin is not perfect, and you can consider this information when formulating your conclusions. This approach is not lacking in criticism though. Again, there is one big assumption, which is the convergence property of the frequency, whose limit might not exist. Repeating the experiments under equivalent conditions might not be possible. Some events are extremely rare, and it is impossible to run many simulations of them (think about extreme natural events like tsunamis).

Practice with Empirical Probability

Empirical probability is defined as the ratio of the number of times a favorable outcome occurs to the total number of times an experiment is performed. For example, we toss a coin three times and get the following results (Table 2.2.2).

[Table 2.2.2] Coin Results

Experiment	1	2	3
Result	Heads	heads	Tails

$$probability\ of\ heads = 2/3, probabilityof\ tails = 1/3$$

+ **Subjective Approach:** The subjective approach is the most intuitive definition of probability. According to this approach, the probability of an event is the degree of belief a person attaches to that event based on his or her available information. This reasoning holds only under the assumption of rationality, which assumes that people act coherently. Let us consider a more specific definition. Imagine a lottery in which you can win an amount of money equal to S if event E occurs. To participate, you have to buy one ticket. What is the price you would be willing to pay to participate in this lottery? If you indicate that price as $\pi(E,S)$, the probability of event E is given by

$$P(E) = \pi(E,S)/S$$

Imagine you want to predict the probability that your favorite football team will win the game tomorrow. You can participate in a lottery in which, if the team wins, you obtain a prize of $1,000; otherwise, you gain nothing. What is the price you would be willing to pay to participate? Let us say you are very confident in your team, and you are willing to pay $700. The probability that your team will win the game tomorrow is

$$P('win') = \pi('win', 1000)/1000 = 700/1000 = 0.7$$

This approach does not garner serious criticism, as it resolves some of the pitfalls of the previous methods (like the impossibility of repeating experiments under equivalent conditions because of the uniqueness of many events) and at the same time does not contradict other theories. Indeed, the evaluator who must decide the price of the lottery is not prevented from running experiments. Compute the frequency of successes and use this information to propose a price. What in other approaches is a rule in the subjective approach is an option.

Practice with Subjective Probability

Subjective probability is based on an individual's own experience or personal judgment. For example, the probability of a reader understanding the definition of probability is 0.85. Let 's' be the sample space. a and b are events that belong to the sample space.

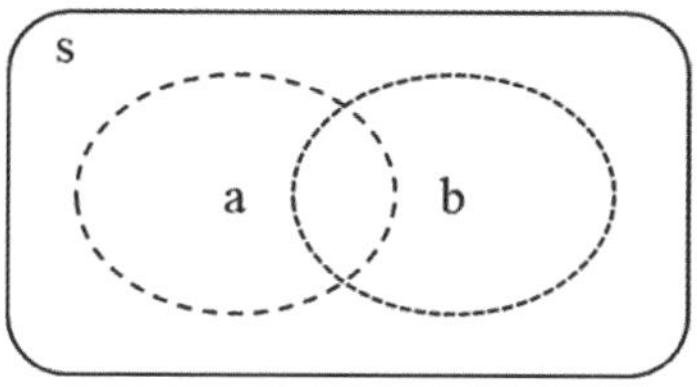

There are three axioms of probability:
$0 \le p(a) \le 1$, $0 \le p(b) \le 1$
$p(s) = 1$
p(aUb) = '(a) + p(b) – p(ab)

Problem: Our family is planning to visit our grandparents on either May 30 or May 31. We believe the chance of rain on May 30 is 30%, on May 31 is 40%, on both May 30 and May 31 is 20%, and on either May 30 or May 31 is 60%.

Question: What is wrong with our beliefs?
Solution (a = rain on May 30, b = rain on May 31): p(a) = 0.3, p(b) = 0.4, p(ab) = 0.2, p(aUb) = 0.6

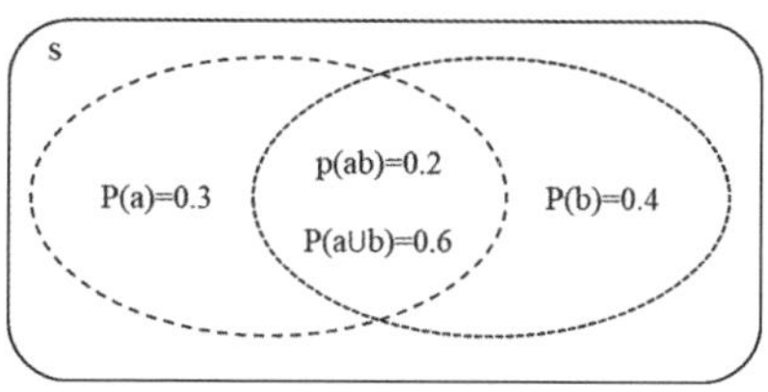

However,
p(aUb) = p(a) + p(b) – p(ab)
p(aUb) = 0.3 + 0.4 – 0.2
p(aUb) = 0.5,
which is not equal to 0.6

Question: How does our family need to modify our belief to be consistent with the axioms of probability?
Solution: p(a) = 0.3, p(b) = 0.4, p(ab) = 0.1, p(aUb) = 0.6.

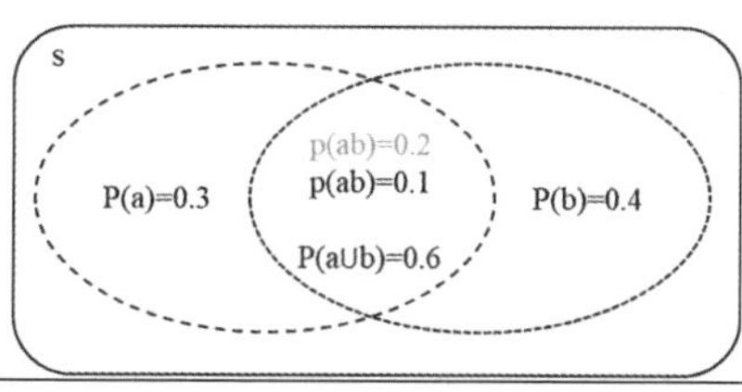

However,
p(aUb) = p(a) + p(b) – p(ab)
p(aUb) = 0.3 + 0.4 – 0.1
p(aUb) = 0.6

4. Probability State Spaces (Discrete and Continuous)

A probability distribution is a statistical function that connects each possible outcome of a random variable in a random process with its respective probability of occurrence (Grimmett & Stirzaker, 2020).

In this section, we will briefly explore probability distributions through examples such as rolling two dice and examining continuous bounded state spaces.

+ **Discrete (Finite) State Spaces:** Set of probabilities $P_1, \ldots, P_m$ (e.g., illustration by bar graph), for example, probabilities of counts 2, 3, …, 12 obtained from rolling two dice

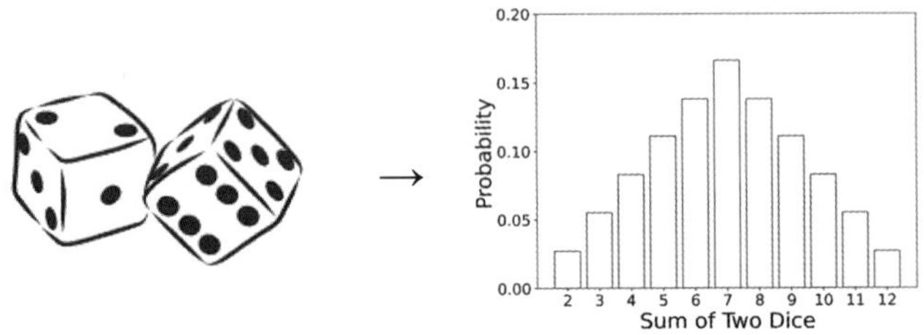

Figure 2.2.5. Probability distribution of totals of two dice (analytical, numerical)

[Table 2.2.3] Numerical Representation of Rolling Two Dice

Values	2	3	4	5	6	7	8	9	10	11	12	x_j
Possible outcomes obtained from the rolling of two dice	11	12	13	14	15	16	26	36	46	56	66	
		21	22	23	24	25	35	45	55	65		
			31	32	33	34	44	54	64			
				41	42	43	53	63				
					51	52	62					
						61						
Frequency	1	2	3	4	5	6	5	4	3	2	1	$h_j = \sum_{k=1,\ldots,N; X^{(k)}=j} 1 = 36$
Relative frequency	1/36	2/36	3/36	4/36	5/36	6/36	5/36	4/36	3/36	2/36	1/36	$f_j = \frac{h_j}{N}$
Probability	0.027	0.055	0.083	0.111	0.138	0.166	0.138	0.111	0.083	0.055	0.027	$P_j = \lim_{N \to \infty} f_j$

Figure 2.2.5 and Table 2.2.3 demonstrate the probability distribution of the sum of values obtained from rolling two dice. The table shows the various possible outcomes, their frequencies, relative frequencies, and probabilities for each sum value ranging from 2 to 12. Calculate relative frequencies f_j (approximate the limiting case $N \to \infty$ by choosing a large N)

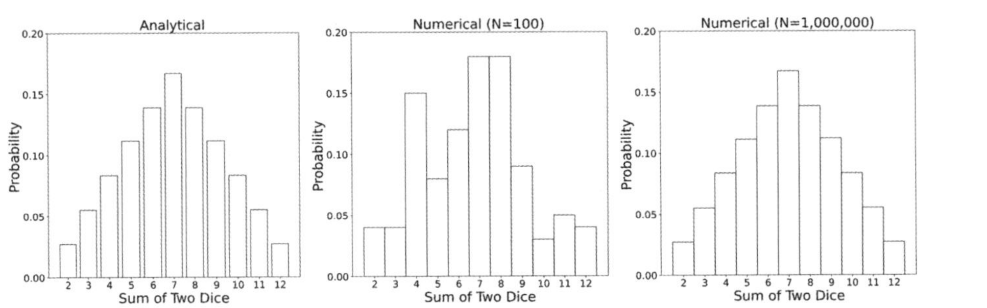

Figure 2.2.6. Simulation results: Analytical (left), numerical (N = 100; middle), numerical (N = 1,000,000; right)

To estimate relative frequencies f_j, a large N is chosen to approximate the limiting case as N approaches infinity ($N \to \infty$), with Figure 2.2.6 showcasing simulation results by comparing analytical outcomes (left) to numerical results for N = 100 (middle) and N = 1,000,000 (right).

+ Continuous (Bounded) State Space [*A*, *B*]: In a continuous bounded state space with an interval [A, B], a grid with bin size is utilized, and the probabilities, p_j, correspond to $p(X \in [a\ ,b])$, with intervals $a_j\ = A + (j - i)\Delta x$ and $b_j\ =\ a_j{+}\Delta x$. The distribution is represented by the set of probabilities as below.

$$p_j = p\big(X \in \left[a_j\ ,b_j\right]\big) \rightarrow p(x,\Delta x) = p(X \in [x, x +\ \Delta x]),$$

$$[a_1\ ,b_1] =\ [A, A +\ \Delta x],\ \ [a_2\ ,b_2] =\ [A +\ \Delta x, A + 2\Delta x]$$

The equations defined two quantities p_j and $p(x,\Delta x)$ as probabilities of certain events occurring. The first line of the equation defines p_j as the probability of the event $\left(X \in \left[a_j\ ,b_j\right]\right)$ where X is a random variable and a_j and b_j are constants. This represents the occurrence of the value of X belonging to the interval $\left[a_j\ ,b_j\right]$. The next line of the equation defines $p(x,\Delta x)$ as the probability of the event $X \in [x, x +\ \Delta x]$ where X and Δx are constants. This represents the occurrence of the value of X belonging to the interval $[x, x +\ \Delta x]$. Example: particle in gravitational field

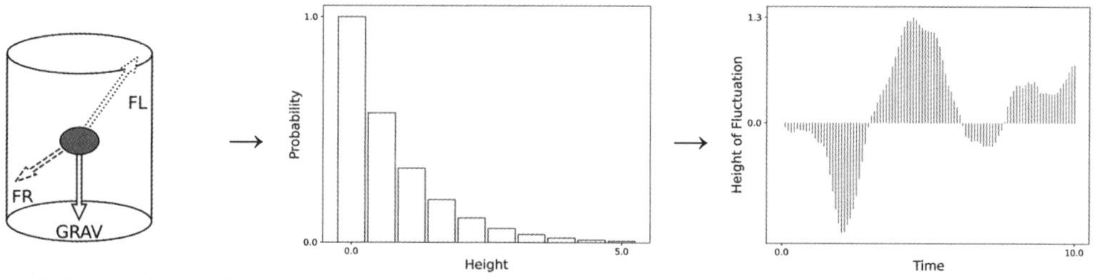

Figure 2.2.7. Schematic representation of the continuous state. Comparative analysis of an Exponential Probability Distribution reflecting particle height probabilities (left), and a Langevin Equation Simulation (right) demonstrating the time-dependent height fluctuations in a particle subjected to fluctuating forces (FL), damping forces (FR), and gravity (GRAV), with each force distinctly represented in the simulation.

In Figure 2.2.7, a schematic representation of the continuous state (exponential function) is illustrated, showcasing a particle in a gravitational field subject to fluctuating force (FL), friction or damping force (FR), and gravity (GRAV).

Remarks: Part 1 of the probability theory course introduces key ideas such system variables, frequency, relative frequency, and probability distribution. We may comprehend the likelihood of events happening in a certain system better by comprehending these notions. Histograms are only a couple of the frequency-based techniques that may be used to explore data sets and comprehend their probability distribution. These skills are necessary for many applications, including as data analysis and decision-making in any industries. The next section takes a deeper dive into continuous distribution functions with more detail.

[2.3]

Probability Theory, Part 2

Random Variables, Distributions and Densities, Various Discrete and Continuous Distributions

Although variables and distributions are a fundamental concept, as covered in Part 1.2, their context can make them challenging to grasp, especially from a probability time series perspective with various applications.

A random variable is simply a function that converts outcomes into numerical values. While this might be obvious to statisticians, it can be difficult for individuals from other fields, such as systemic analysis, as it requires a solid understanding of the core ideas. However, mastering variables and distribution is essential for comprehending advanced concepts like Gaussian processes and Bayesian statistics, which heavily rely on or are based on them (Van Belle, 2011).

1. Random Variables

A random variable is a variable whose value is unknown or a function that assigns values to each of an experiment's outcomes (Rubin, 2005). Contrary to our beliefs about randomness, possible values from a function are not equally likely. In other words, a random variable is one whose values depend on the outcomes of a random phenomenon.

Random variables depend on functions that use randomness but do not necessarily output uniform randomness. When we talk about a random variable, usually denoted X, its final value is unknown. The random variable is X, not the output of X, with possible outcomes and associated probabilities. Random variables generally fall into two categories: discrete and continuous. Let us start conceptually with a simple notation from the previous section because it is more in line with how we as humans view the world.

Notation: Probability = P in [0, 1]

- Probability function (cumulative probability) = F in [0, 1]
- Probability density = $P \geq 0$
- Coordinate = x (lowercase)
- Random variable = X (capital)

Outcomes: State = j (discrete state space)

- Collection (or set) of states: $I = \{j_1, \ldots, j_m\}$ (discrete state space)
- Interval or subspace (continuous state space)

Probability of an outcome: proportion (percentage) of occurrences of that outcome in the statistical ensemble;

- Discrete state: $P_j = \sum_{k=1,\ldots,N; X^{(k)}=j} 1$
- Continuous state: $P(X \in [a,b] = \sum_{k=1,\ldots,N; X^{(k)} \in [a,b]} 1$

The two equations are related by the fact that they define probabilities of events involving the same function $X^{(k)}$, but the events are expressed in different ways. The first one (discrete state) defines the probability being equal to a specific value j while the second one (continuous state) defines the probability belonging to a specific interval $[a,b]$.

+ **Probabilities for Discrete and Continuous Variables:** Understanding these variables can help us see patterns in otherwise random-looking data.

Let us discuss in program-ming terms the two main types of random variables—discrete and continuous—and how to plot probability for each.

Discrete random variables: When the image (or range) of X is countable, the random variable is a discrete random variable, and its distribution can be described by a probability for each value in the image of X. A random variable is discrete if it can only take a finite number of values. Less formally, it can be thought of as a model for the set of possible outcomes of any single experiment that asks a yes–no question.

Flipping a coin is a discrete event because the result can only be heads or tails. This follows a Bernoulli distribution, with only two possible outcomes and a single coin flip at a time.

Let us generate data (NumPy) to model this (Figure 2.3.1).

```
[In]
import numpy as np
import matplotlib.pyplot as plt
n = 1    # number of trials
p = 0.5  # probability of success
sample = np.random.binomial(n, p,
100)
plt.hist(sample, bins=10)
```

→

[Out]

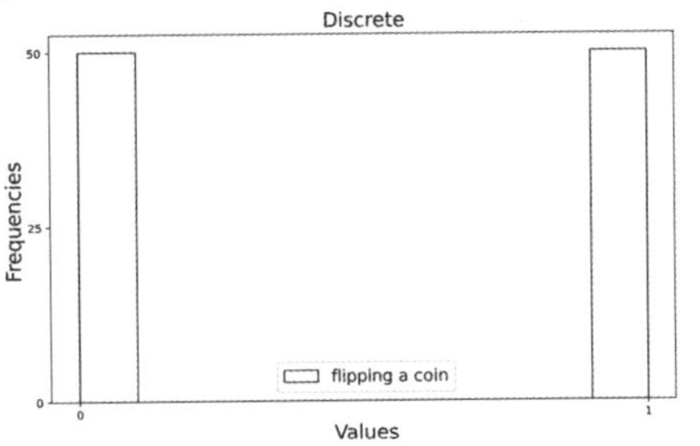

Figure 2.3.1. Output of using discrete event values (flipping a coin)

Also, weight rounded to the nearest pound is a discrete variable because there are discrete buckets at 1 *lb* intervals the weight can fall into (use the Supplement data set of 'weight-height.csv'). Unrounded weight is continuous, so we will come back to this example again in the next section on continuous random variables.

```
[In]
import pandas as pd
df = pd.read_csv('weight-height.csv')

# Convert values to integers to round them.
df['Height'] = df['Height'].astype(int)
df['Weight'] = df['Weight'].astype(int)
df.head(3)
```

→ [Out]

	Gender	Height	Weight
0	Male	73	241
1	Male	68	162
2	Male	74	212

Following our examination of the discrete buckets in the second example of discrete variable simulation, let us turn our attention to continuous random variables.

Continuous random variables: If the image is uncountably infinite, then X is a continuous random variable. The output can be an infinite number of values within a range. Although the combined probability under the curve is equal to 1, it is impossible to calculate the probability for any individual point—it is infinitesimally small. It is possible to calculate the probability for a range of x values under the curve.

Continuous random variables have a probability density function (PDF). The PDF defines a curve that will vary depending on the distribution rather than lists the probability of each possible output. Let us come back to our example about weight. This time, the weight is not rounded.

Why is weight continuous? Because in the real world, weight does not fall neatly into 1 *lb* intervals. A person can weigh 150 *lb* when standing on a scale. However, if we zoomed in to the molecular level, they may actually weigh 150.0000001 *lb*.

In general, the steps for the continuous random variables include:

- Plot the sample data on a histogram
- Find a well-known distribution that the sample appears to follow
- Calculate the parameters required to generate the distribution from the sample
- Generate and plot the PDF on top of your histogram

Let us do this with our weight example from above. Import the data set (use the same Supplement data set 'weight-height.csv') and save it in the same directory as your notebook.

```
[In]
Import pandas as pd
df = pd.read_csv('weight-height.csv')
```

Inspect the shape of the data.

```
[In]
df.head(3)
df = df[df['Gender']=='Female']
df.head(3)
```

→

[Out]

	Gender	Height	Weight
5000	Female	58	102
5001	Female	65	141
5002	Female	63	131

Next, we remove the males to make our lives easier. Including both men and women would result in a bimodal distribution (two peaks instead of one), which would complicate our calculation. Plot the remaining weight data.

```
[In]
import numpy as np
import matplotlib.pyplot as plt

weights = df['Weight'].tolist()
plt.hist(weights, bins=20)
```

→

[Out]

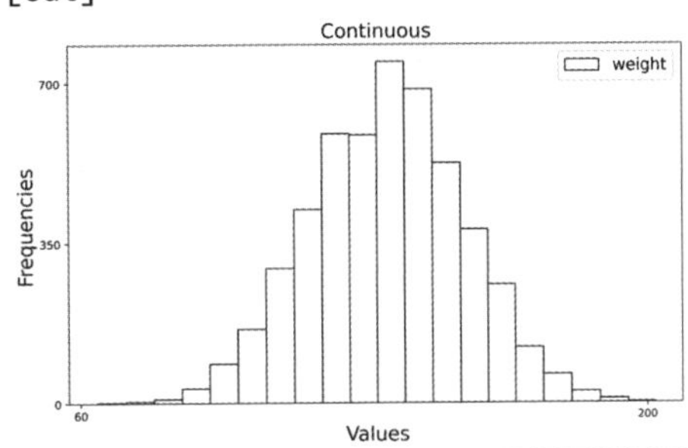

Figure 2.3.2. Output of the data using the continuous variables (frequency)

What common distribution does this look like? Figure 2.3.2's example of continuous variables (frequency) outputs data that looks to follow a normal distribution.

2. Distributions and Densities (Why Do We Need Them?)

A random variable can hold a different set of values from the outcome of any random process. However, it cannot capture the probability of getting those different values. The probability distribution helps to create a clear picture of all possible sets of values with their respective probabilities of occurrence in any random process (Figure 2.3.3).

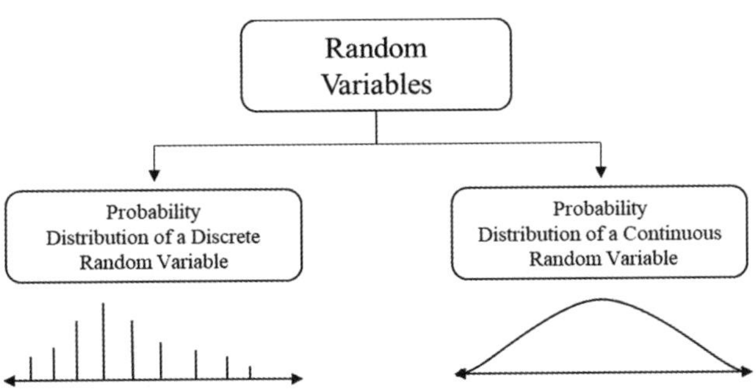

Figure 2.3.3. Schematic representation of the distributions of discrete (left) and continuous (right) random variables

If a random variable can take only a finite set of values (i.e., it is a discrete random variable), then its probability distribution is called a probability mass function (PMF). The probability distribution of a discrete random variable is the list of values of different outcomes and their respective probabilities. In other words, for a discrete random variable X, the PMF $P(X)$ is given as $P(X) = P(X = x)$. If X, a discrete random variable, takes different values X_1, X_2, X_3 ..., then

$$\sum_{i=1}^{\infty} P(X_i) = 1, \quad \text{and} \quad 0 \leq p(xi) \leq 1$$

+ **Example:** If X is a random variable associated with the rolling of a six-sided fair die, then the PMF of X is given as follows:

[Table 2.3.1] Rolling a Six-Sided Fair Die

X	$X = 1$	$X = 2$	$X = 3$	$X = 4$	$X = 5$	$X = 6$
$P(X = x)$	1/6	1/6	1/6	1/6	1/6	1/6

For example, when considering a random variable X associated with rolling a six-sided fair die, the probability mass function (PMF) of X is demonstrated in Table 2.3.1, where each outcome ($X = 1$ to $X = 6$) has an equal probability of occurrence at 1/6.

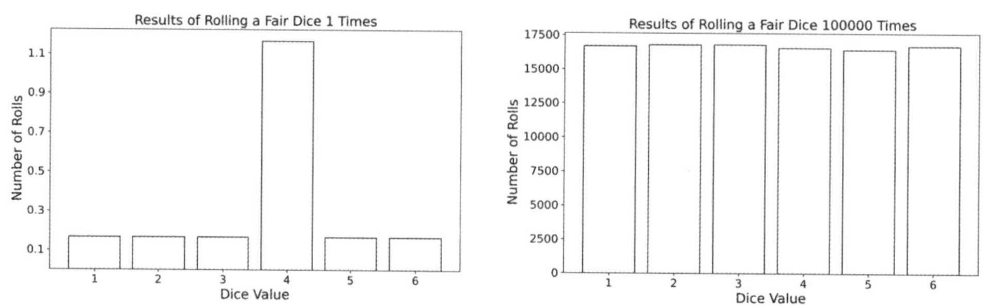

Figure 2.3.4. Rolling a six-sided fair die. The plot of the left side denotes the results of rolling a fair dice 1 times. The plot of the right side denotes the results of rolling a facie dice 100,000 times.

We used random module to select one of the six values (1/6) in the dice value (X) for each dice roll. The results are stored in a dice counts which keeps track of how many times each value was rolled (Figure 2.3.4). Random variables play a vital role in probability distributions and also serve as the base for probability distributions (Manski, 2003). A random variable gives an idea of the underlying probability distribution by showing all possible values it can take along with the likelihood of those values.

+ **Practice for the Probability Distribution:** Let us compute a normal distribution using the same data set in more technical depth.

We imported the 'mean' and 'std' functions from the 'numpy' module (`from numpy import mean,std`), and calculated the mean and standard deviation for a list of weights.

```
[In]
from numpy import mean,std
mean = mean(weights)
std = std(weights)
print(mean, std)
```

→

```
[Out]
Mean: 135.86009300746835
SD: 19.020565463416645
```

Then we use these parameters to generate a normal distribution.

```
[In]
from scipy.stats import norm
distribution = norm(mean, std)
min_weight = min(weights)
max_weight = max(weights)
values = list(range(int(min_weight), int(max_weight)))
```

The ‘norm’ function is being used to create a normal distribution with the specified mean and standard deviation. Then, a list of values is being created using the ‘range’ function which generates a sequence of integers from the min weight to the max weight. Let us visualize the sample distribution and the probability density function you created while also generating probabilities for that range.

```
[In]
probabilities = [distribution.pdf(v)
for v in values]
from matplotlib import pyplot
pyplot.hist(weights, bins=20,
density=True)
pyplot.plot(values, probabilities)
```

→

[Out]

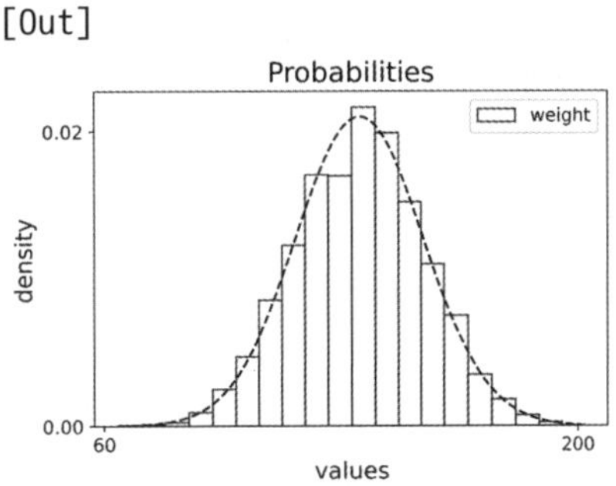

Figure 2.3.5. Created output with the continuous variable example (density)

We intentionally chose a sample that followed a normal distribution to simplify the process (Figure 2.3.5). However, we are well advised to know the different common types of distributions and the parameters required to generate them. Conceptually, here we discussed the fundamental distributions.

Use grid with bin size Δx
Choose Δx such that $M \times \Delta x = \mathrm{B} - \mathrm{A}$
Probabilities $P_j = P(X \in [a_j, b_j])$
Intervals $a_j = \mathrm{A} + (j-1) \times \Delta x$, $b_j = a_j + \Delta x$
Distribution = set of probabilities $P_1, \ldots, P_m$

→

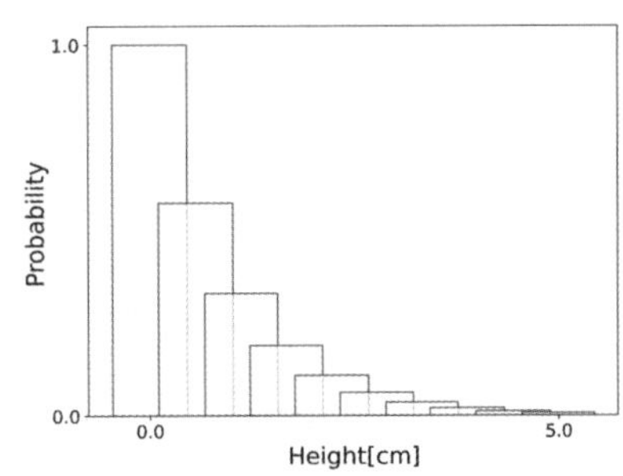

Figure 2.3.6. Probability distribution: Continuous bounded state space with interval $[A, B]$

Continuous bounded state space (Figure 2.3.6) which state space is interval $[A, B]$.

$$P_j = P(X \in [a_j, b_j]) = P(x, \Delta x) = P(X \in [x, x + \Delta x])$$

Continuous bounded state spaces (bin size): where x is the size of the subinterval or bin size in a continuous bounded state space, can be used to express the likelihood of an event happening in a subinterval $[a_j, b_j]$. Figure 2.3.6 shows the probability distribution for a continuous limited state space with [A, B] as its interval. A density function is used to characterize the distribution shown below, which reflects the data distribution regardless of the size of the bins.

+ **Probability Densities:** In probability, a probability density is the relationship between observations and their probability. And a probability density function, or density of a continuous random variable, is a function whose value at any given sample in the sample space can be interpreted as providing a relative likelihood that the value of the random variable would be close to that sample.

For a discrete random variable X that takes on a finite or countably infinite number of possible values, we can determine $P(X = x)$ for all possible values of X and call it the probability function. For continuous random variables, the probability that X takes on any particular value x is 0. That is, finding $P(X = x)$ for a continuous random variable X is not going to work. Instead, we need to find the probability that X falls in interval (a, b); that is, we need to find $P(a < X < b)$. We do that using a probability density function. How many realizations per measurement unit?

- For example, density (x) = how many realizations per centimeter at height x?

For construction of the density function, we need to divide the probability by the bin size and make the bin size smaller and smaller as below (the right side of the Figure 2.3.7).

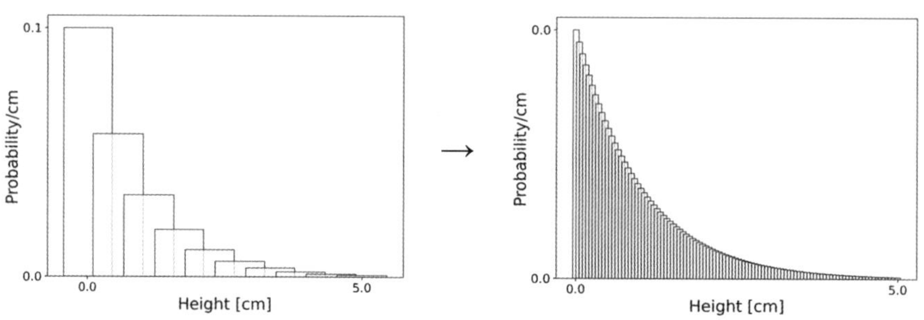

Figure 2.3.7. Density function with bin sizes

Unlike a discrete random variable, a continuous random variable holds different values from an interval of real numbers. Hence, it is challenging to sum these uncountable values like you can for discrete random variables. Therefore, an integral over the set of values is done. The probability distribution of a continuous random variable is called a probability density function (PDF: Parzen, 1962). Given the probability function $P(x)$

for a random variable x, the probability that x belongs to A, where A is an interval, is calculated by integrating $P(x)$ over the set A:

$$P(x \in A) = \int AP(x)dx$$

where $0 \leq p(x) \leq 1$ for all x and $\int P(x)dx = 1$. The probability is calculated by integrating the probability density function of the random variable over the set A. The integral represents the area under the curve of the probability over the set A. It is used to describe the probability distribution of a continuous random variable which would give us the probability x takes on a value within the range.

+ **Example:** A clock stops at any random time during the day. Let x be the time (hours plus fractions of hours) at which the clock stops. The PDF for A is

$$f(x) = \{\begin{matrix} 1/24 & 0 \leq x \leq 24 \\ 0 & otherwise \end{matrix} \quad \longrightarrow$$

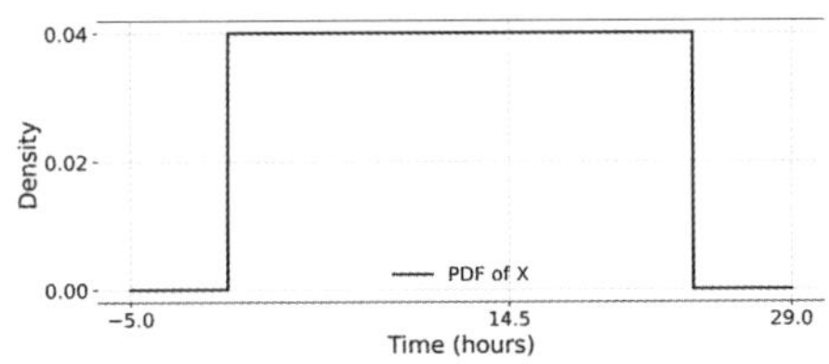

Figure 2.3.8. Representation of the probability density function (for the above example)

Corresponding to the distribution function, the density function is denoted $P(x)$. For any value of x, the function gives the probability of event $X \leq x$. Therefore, in Figure 2.3.8, if $f(x)$ is the probability function of x, then the definitions of the probability density P

$$P(x) = \lim_{\Delta x \to 0} \frac{p(x, \Delta x)}{\Delta x} = \lim_{\Delta x \to 0} \frac{p(x \in [x, x + \Delta x])}{\Delta x}, \qquad Semi-positive: P \geq 0$$

The probability function F can be conceptualized as follow [$F(x) = p(X < x)$, $P(x, \Delta x) = F(x + \Delta x) - F(x)$]. The equation represents the probability that a continuous variable X takes on a value within a small interval around x. It is calculated as the difference between the cumulative distribution function at $x + \Delta x$ and the cumulative distribution function $F(x)$ at x.

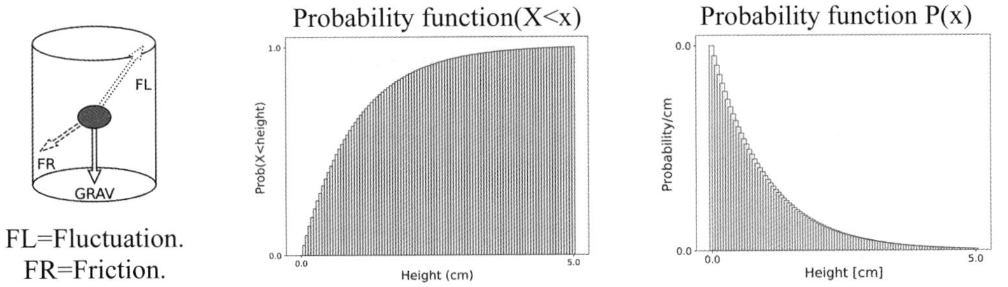

Figure 2.3.9. Schematic representation of different probability functions

$$P(x) = \lim_{\Delta x \to 0} \frac{p(x, \Delta x)}{\Delta x} = \frac{d}{dx} F(x) = \frac{d}{dx} P(X < x)$$

Then, if we use an integral relationship between p and P (Figure 2.3.9), it would be as follows:

$$P(x) = \frac{d}{dx} P(X < x), \qquad P(X \in [a, b]) = P(X < b) - P(X < a),$$

$$P(X \in [a, b]) = \int_a^b P(x)dx$$

The integral $\int_a^b P(x)dx$ represents the area under the curve of the probability density function over the range $[a, b]$. The variable X takes on a value within the range $[a, b]$ is equal to the area under the curve of the probability density function over that rage.

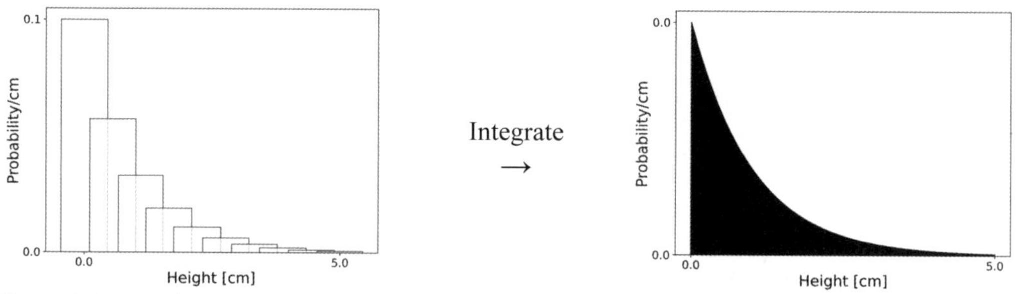

Figure 2.3.10. Schematic representation of the relationship between functions

In Figure 2.3.10, notice that the density P in general is not dimensionless (e.g., density of particle position $\rightarrow$ $[P] = 1/m$). Density P does not need to be less than 1. The normalization condition (when a sure event has a probability of 1) is

$$X \in \Omega = \int_\Omega P(x)dx$$

Frequently seen in the scientific literature: $P(x)dx$ = the probability that the system is in the infinitesimal small interval $[x, x + dx]$

+ **Empirical Probability Densities from Data:** From the fundamentals mentioned above, the probability density function can be determined as below:

- For Stochastic model: analytical methods, numeric
- For data: Nonparametric $\rightarrow$ kernel density, etc. Maximum Likelihood, etc.

To compute the probability density function, one can use a big N for large sample sizes, N, and a small x for short interval sizes, $x0$.

- $N \rightarrow \infty$? Use large N, $dx \rightarrow 0$? Use small dx

An appropriate technique for observables x specified on the real line is the kernel density estimator. As a matter of fact, we may use programing to mimic the kernel density estimate.

First, we need to read the data and command of the density function (use the Supplement dataset 'dataset_velocity.txt').

```
[In]
data = np.loadtxt('dataset_velocity.txt')
```

We used density function from the 'sns' toolbox in programming to estimate the probability density function of a dataset of velocity.txt values. This method takes a vector of data values as input and returns the estimated density corresponding values of x of xout.

Then, let us create the output

```
[In]
# Generate the density plot
sns.kdeplot(data, color='k',
linewidth=1) # black solid line

# Calculate the density values
density = gaussian_kde(data)
x_vals = np.linspace(min(data),
max(data), 1000)
y_vals = density(x_vals)
```

→

[Out]

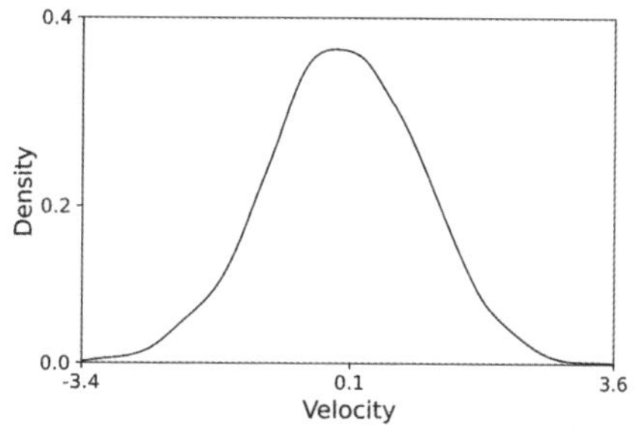

Figure 2.3.11. Kernel density estimate from 5,000 data points: Stationary velocity (from Brownian motion).

The density estimator is good for an observable X defined on a real line (Figure 2.3.11). Further, for density with error bands, we can compute density functions for all trials and compute the mean density function.

Also, we can compute at every grid point the standard deviation of the values from the single-trial density functions. First, we need to parameterize the density function (damping constant; gamma = friction coefficient).

```
[In]
gamma = 1.0
deltat = 0.01
sigma = (deltat)**0.5
nsubjects = 10
```

In the code above, the 'gamma' is a scalar value representing a parameter, the 'delta' is a scalar value representing the time step, the 'sigma' is a scalar value calculated as the square root of 'deltat' which is commonly used as a measure of the standard deviation or volatility of some process. And the 'nsubjects' is an integer value representing the number of subjects in this experiment.

Let us initialize with loop taking only final 5000 data points

```
[In]
import numpy as np
nsubjects = 10
deltat = 0.01
sigma = np.sqrt(deltat)
gamma = 1.0

# Initialize the arrays
v = np.zeros((nsubjects, 10000))
vstationary = np.zeros((nsubjects, 5000))
# Populate the v array
for k in range(nsubjects):
    v[k, 0] = 0
    for n in range(9999):
        e = np.random.normal(0, sigma)
        v[k, n+1] = v[k, n] - deltat * gamma * v[k, n] + e
# Populate the vstationary array
for k in range(nsubjects):
    vstationary[k, :] = v[k, 5000:10000]
xout = np.arange(-3, 3.1, 0.1)
xoutdim = len(xout)
pdf = np.zeros((nsubjects, xoutdim))
```

We used the 'normal' function from the 'stats' toolbox to generate random values form a normal distribution. Here, we create an array 'v' with dimensions 'nsubjects' by 10,000 and initializes all elements to zero. We use a loop to simulate the 'nsubjects' over 10,000 time steps. At each step, it updates the velocity of each subject according to the equation that involves 'gamma' and 'deltat', and creates an array 'pdf' with dimensions 'nsubjects' by the length of 'xout' initialized all elements to zero. Then, let us create the output

```
[In]
xout = np.arange(-3, 3.1, 0.1)
xoutdim = len(xout)
pdf = np.zeros((nsubjects, xoutdim))
for k in range(nsubjects):
    density = gaussian_kde(vstationary[k,
:])
    pdf[k, :] = density(xout)
    plt.plot(xout, pdf[k, :], color='k')
```

[Out]

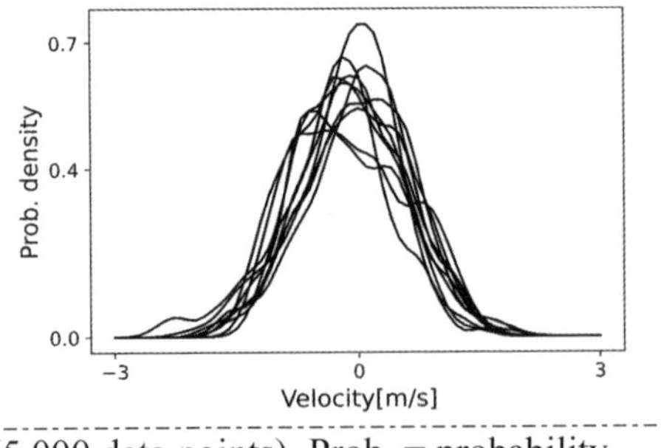

Figure 2.3.12. Density with error bands: Friction data (5,000 data points), Prob. = probability

The density with error bars for the friction data, which consists of 5000 data points, is shown in Figure 2.3.12. Probability is referred to as Prob.

The sample mean probability density function are also included in the theoretical probability density shown in the image (sample mean pdf and error bands).

```
[In]
theopdf = norm.pdf(xout, 0, np.sqrt(0.5))

# Compute mean and standard deviation
mpdf = np.mean(pdf, axis=0)
sdpdf = np.std(pdf, axis=0)

# Plot mean pdf, and one standard deviation
plt.plot(xout, mpdf, '-')
plt.plot(xout, mpdf + sdpdf)
plt.plot(xout, mpdf - sdpdf)
plt.axis([min(xout), max(xout), 0, max(mpdf
+ sdpdf) * 1.1])
```

→ [Out]

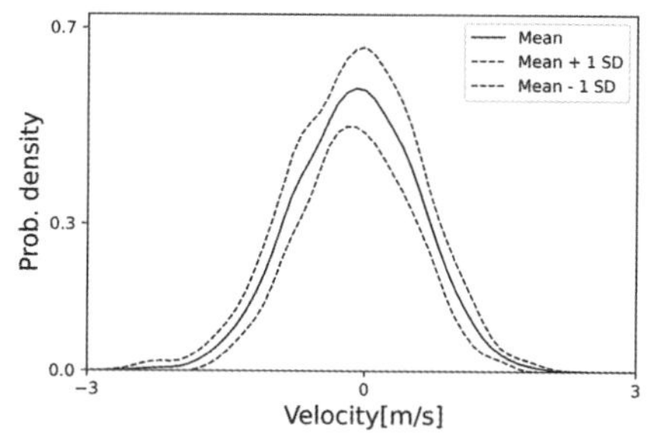

Figure 2.3.13. Density with error bands

The mean probability density is shown in the plot's upper error band plus one standard deviation, while the mean probability density is shown in the plot's lower error band minus one standard deviation. Probability is meant in this context by the prefix Prob (Figure 2.3.13). We use the 'axis' function to provide the upper and lower bounds of the x and y axes on the display. The y-axis limits are set to 1.1 times the maximum value of 'mdpf+sdpdf,' while the x-axis limits are set to the lowest and maximum values of 'xout.'

+ **Practice with Probability Density:** Imagine five participants who perform bi-manual anti-phase coordination movements, and here, we use computer-generated data. For each participant, the relative phase time series is recorded, and the probability density is estimated. The five probability densities are then averaged.

Task: Calculate and plot an estimated probability density for a sample of participants. Use five data files (use the Supplement data set 'phasedataS1, …, phaedataS5') that contain computer-generated data from hypothetical participants asked to perform in-phase coordination movements. Calculate the probability density in the range from –20 degrees to 20 degrees. We may proceed as follows:

2a. Work with the command window

- Define 21 grid points from –20 to 20 at intervals of 2 with command = [-20:2:20].
- Load the data points in the file phasedataS1.txt to a variable (e.g., *d*).
- Estimate the probability density at the grid points x-out with command density.
- Print the 21 values of the estimated density in the command window P'.

- Copy them into Excel.
- Repeat this procedure for all five participants.

2b. Work in Excel: Average the probability density values of the five participants at each of the 21 grid points.

[Table 2.3.2] Data Set for Probability Density

S1	S2	S3	S4	S5	Average	SD	SD+	SD–
0.003	0.002	0.003	0.003	0.005	0.003	0.001	0.004	0.002
0.004	0.004	0.006	0.005	0.007	0.005	0.001	0.007	0.004
0.006	0.007	0.009	0.008	0.010	0.008	0.001	0.009	0.006
0.011	0.011	0.012	0.011	0.013	0.011	0.001	0.012	0.010
0.015	0.017	0.015	0.016	0.018	0.016	0.001	0.017	0.015
0.021	0.023	0.021	0.022	0.023	0.022	0.001	0.023	0.021
0.028	0.029	0.028	0.028	0.030	0.029	0.001	0.030	0.028
0.034	0.037	0.036	0.036	0.038	0.036	0.001	0.037	0.035
0.042	0.044	0.040	0.040	0.043	0.042	0.002	0.043	0.040
0.049	0.048	0.043	0.041	0.045	0.045	0.003	0.048	0.042
0.051	0.050	0.044	0.042	0.046	0.047	0.004	0.050	0.043
0.049	0.048	0.043	0.043	0.045	0.045	0.003	0.048	0.043
0.045	0.046	0.040	0.042	0.041	0.043	0.002	0.045	0.040
0.040	0.040	0.036	0.038	0.034	0.038	0.003	0.040	0.035
0.034	0.031	0.031	0.034	0.025	0.031	0.003	0.034	0.028
0.025	0.023	0.027	0.028	0.020	0.024	0.003	0.028	0.021
0.017	0.015	0.022	0.020	0.016	0.018	0.003	0.021	0.015
0.011	0.010	0.015	0.014	0.012	0.012	0.002	0.015	0.010
0.006	0.006	0.011	0.010	0.009	0.008	0.002	0.010	0.006
0.003	0.004	0.007	0.007	0.006	0.005	0.002	0.007	0.004
0.002	0.002	0.004	0.004	0.004	0.003	0.001	0.004	0.002

Table 2.3.2 presents the probability density values for five participants at 21 grid points, with the average, standard deviation (*SD*), and the *SD* + and *SD* – values calculated for each point, showcasing variability across the participants.

2c. Work with script: Write a script with three lines to plot the average estimated probability density.

```
[In]
Xout=[-20:2:20];
average=[here you enter the 21 values from Excel];
plot(xout,averageP)
```

In the code above, the 'xout' is an array containing sequence of values from −20 to 20 with a step of 2, and 'average' is an array containing the 21 values from the data Excel.

2d. Optional task: In the figure (below), the dashed lines are 1 *SD* above and 1 *SD* below the mean estimated probability density. Calculate these lines for the data sets and plot them in the same figure as the mean estimated probability density. Import data (use your data set), define the empirical density function

```
[In]
df = pd.read_csv('data.csv')
```

And create output

```
[In]
xout = np.linspace(-20, 20, 21)
average = df['Average'].values
plt.plot(xout, average, 'k-', alpha=0.7)
density = gaussian_kde(df['Average'])
# Generate values for density function
x_density =
np.linspace(min(df['Average']),
max(df['Average']), 1000)
y_density = density(x_density)
# Plot the density function
plt.plot(x_density, y_density, 'k:',
alpha=0.7)
plt.axis([min(xout), max(xout), 0,
max(y_density)*1.1])
```

→ [Out]

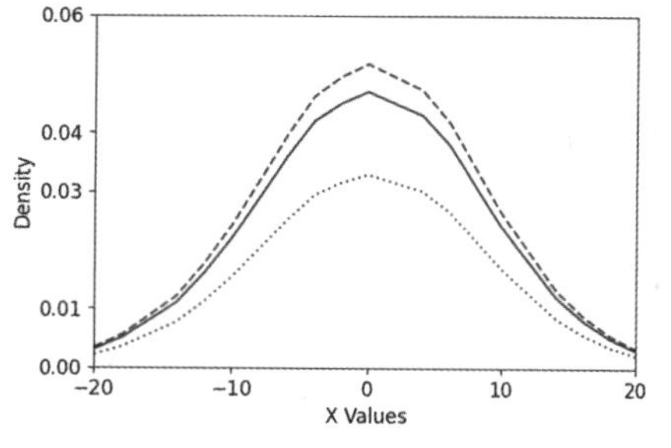

Figure 2.3.14. Data-estimated probability density with standard deviation

The 'figure' function is used to create a new plot and it takes an optional integer argument that specifies the figure number. The 'set' function is for the property values of an object. In this case (Figure 2.3.14), we are using it to set of the current axes to 25 by the property. The 'xout' is an array containing the x-axis values and 'density' is an array containing the y-aixs values specifying the line width and color using the 'linewidgth' respectively.

3. Various (Discrete and Continuous) Distributions

The section builds upon our previous practices, where we learned to create figures and set their properties. This time, we use these techniques to elaborate on the concepts of probability distributions.

Through figures displaying histograms and probability functions, we gain a better understanding of the behavior of these distributions

```
[In]
import scipy.stats as stats
import numpy as np
%matplotlib inline
import matplotlib.pyplot as plt
```

The stats module (i.e., SciPy) offers many distributions, so you are bound to find what you are looking for. They are organized into three categories: continuous, discrete, and multivariate. In this section, we focus on two continuous distributions and two discrete distributions.

+ **Continuous Distributions:** We start with arguably the most famous distribution: the normal (Gaussian) distribution. We create an instance of the 'stats.norm' class and initialize this object to have a mean of 100 and standard deviation of 15.

```
[In]
normal=stats.norm(loc=100, scale=15)
```

Next we generate a random sample by calling method .rvs(size=10) on this object to generate 10 random numbers.

```
[In]
normal.rvs(size=10,random_state=143)
```

→

```
[Out]
array([ 87.50268488, 118.56536514,
62.86403442, 84.42053641, 114.91967271,
102.67337385, 92.6638058, 93.36867073,
96.22587266, 92.97029531])
```

Random_state =143 is just so that the results are reproducible. If you include this, the result will always be the same. If you want to repeat and get different results, just omit this so that your code reads normal.rvs(size=10).

Now let us generate 1,000 random numbers from this distribution and make a histogram (Figure 2.3.15).

```
[In]
normal_sample =
normal.rvs(size=1000,random_state=12
3)
plt.hist(normal_sample,density=True,
label="Sample")

plt.show()
```

→

[Out]

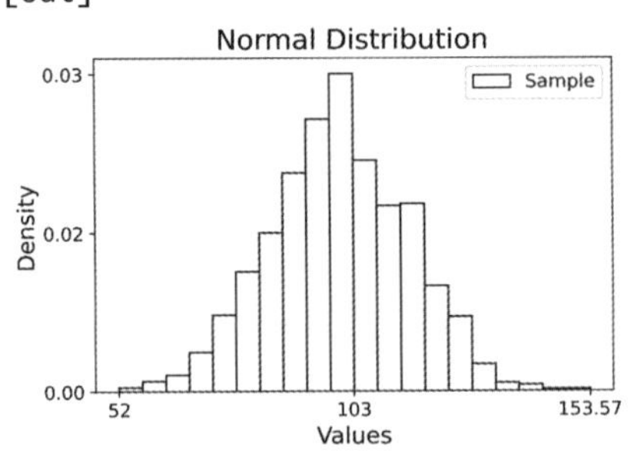

Figure 2.3.15. A histogram of 1,000 random numbers (generated from a normal distribution)

Focus on the PDF using the same (distribution) object we created, we can access the PDF via the method .pdf. The first argument can be a scalar of an array of values.

```
[In]
print(normal.pdf(120))
print(normal.pdf([120,130,140]))
```

→

```
[Out]
0.010934004978399577
[0.010934   0.0035994  0.00075973]
```

Next let us see how to plot the PDF. First, create some values for the horizontal axis. We use np.linspace(start,stop,num) to generate equally num spaced numbers between start and end. The relevant values for our case seem to be between 50 and 150: t.np.linspace (50,150,100). These will serve as the horizontal (x) values for the plot. To get the vertical (y) values, we can compute the PDF on these values: normal.pdf(t).

Below is the code that puts this all together and gives us a plot of the normal PDF (Figure 2.3.16).

```
[In]
t = np.linspace(50,150,100)
plt.plot(t,normal.pdf(t),label="PDF")

plt.show()
```

[Out]

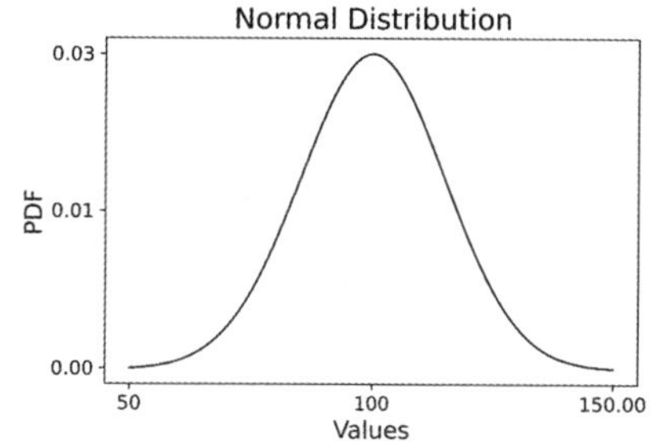

Figure 2.3.16. A normal probability density function (PDF)

Finally, we can overlay this on top of the histogram to see how well the random number generator did (Figure 2.3.17).

```
[In]
normal_sample =
normal.rvs(size=1000,random_state=123)
plt.hist(normal_sample,
density=True,label='Sample')
plt.plot(t,normal.pdf(t), label='PDF')
plt.show()
```

[Out]

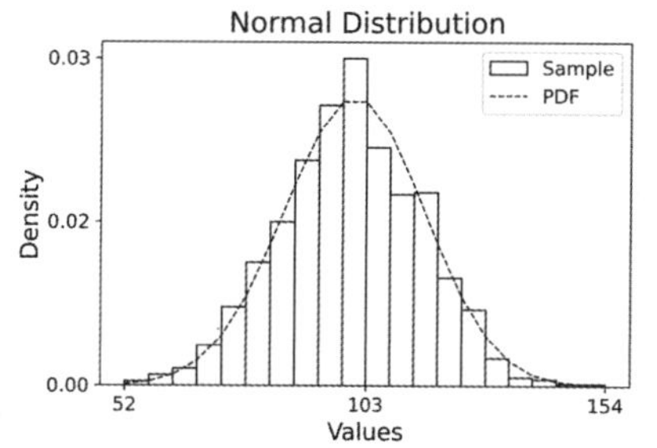

Figure 2.3.17. Histogram of a sample from the normal distribution and the corresponding plot of the PDF. PDF = probability density function

Another popular distribution is the beta distribution. This is also a continuous distribution, and it is used to model random variables between 0 and 1.

For instance, if your random variable is the probability of success, then this might be a good distribution to use.

The beta distribution takes two parameters, a and b. Therefore, to create a beta distribution object in SciPy, we put beta=stats.beta(a=2,b=5).

The code below generates a random sample from this distribution.

```
[In]
beta=stats.beta(a=2,b=5)
beta.rvs(size=5,random_state=133)
```

```
[Out]
array([0.13580999, 0.55628319,
0.39061464, 0.51549706, 0.3439727 ])
```

Like before, let us get 1,000 numbers from this distribution, make a histogram, and overlay it on the PDF (Figure 2.3.18).

[In]

```
beta_sample=beta.rvs(size=1000,random_state=133)

t=np.linspace(0,1,100)

plt.hist(beta_sample,density=True,label="Sample")

plt.plot(t,beta.pdf(t),label="PDF")
plt.show()
```

→

[Out]

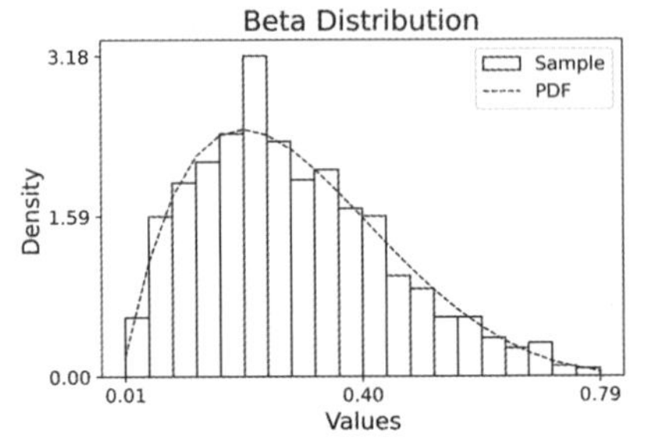

Figure 2.3.18. Histogram of a sample from the beta distribution and a plot of the corresponding PDF. PDF = probability density function

+ **Discrete Distributions:** Sometimes we need to work with discrete distributions. Here we look at two: binomial and Poisson. The binomial distribution is used to model the number of successes with a repeated trial where each trial can either result in success or failure.

The probability of success remains the same from one trial to another, the trials are independent of one another, and the number of trials is fixed in advanced. A typical example is a fixed number of coin tosses.

Here is how we get binomial distribution objects in SciPy and generate some random numbers:

[In]

```
binomial=stats.binom(n=10,p=0.4)
binomial.rvs(size=10,random_state=143)
```

→

[Out]

```
array([5, 3, 4, 4, 4, 6, 1, 1, 3, 2])
```

Remember, the discrete distribution does not have a PDF. The corresponding function is the probability mass function (PMF). In SciPy we can access this function for distribution objects as follows: binomial.pmf.

Just like for the PDF, we can compute the PMF of a single number or a list of numbers.

[In]

```
binomial.pmf(2)
binomial.pmf([2,3,4])
```

→

[Out]

```
0.12093235199999994
[0.12093235 0.21499085 0.25082266]
```

Next let us plot the PMF. Because it is discrete, we use plt.scatter instead of plt.plot. Also, the PMF is 0 except on the support of this distribution, which happens to be 0, 1, 2, …, 10. Instead of np.linspace, which is used to generate numbers from a continuous range, we use np.arange(start,stop,step=1) to generate values to follow a discrete set.

For us, this is

```
[In]
t=np.arange(0,11)
t
```

→

```
[Out]
array([ 0,  1,  2,  3,  4,  5,  6,
7,  8,  9, 10])
```

Notice how we have to use np.arange(0,11) instead of np.arange(0,10), as it does not return the last number. Now, using these values, we can plot the PMF of our binomial distribution (Figure 2.3.19).

```
[In]
plt.scatter(t,binomial.pmf(t),label="PM
F")
plt.title("Binomial Distribution")

plt.show()
```

→

[Out]

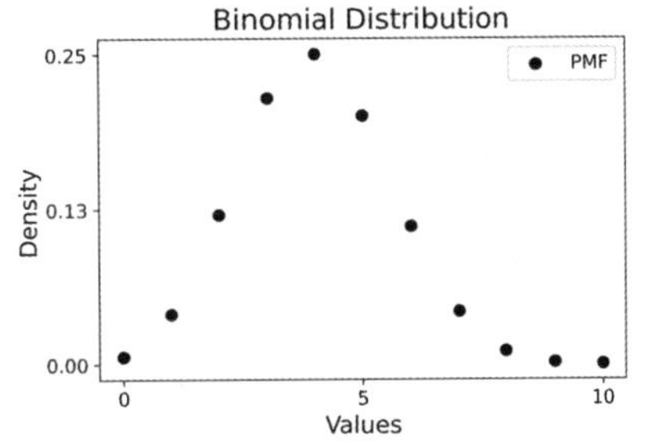

Figure 2.3.19. A plot of the PMF of a binomial distribution. PMF = probability mass function

Now we generate 1,000 random numbers from this distribution.

```
[In]
binomial_sample=binomial.rvs(size=1000,
random_state=143)
plt.hist(binomial_sample,bins=11,range=
(0,11),density=True,

align='left',rwidth=0.75,label="Sample"
)

plt.show()
```

→

[Out]

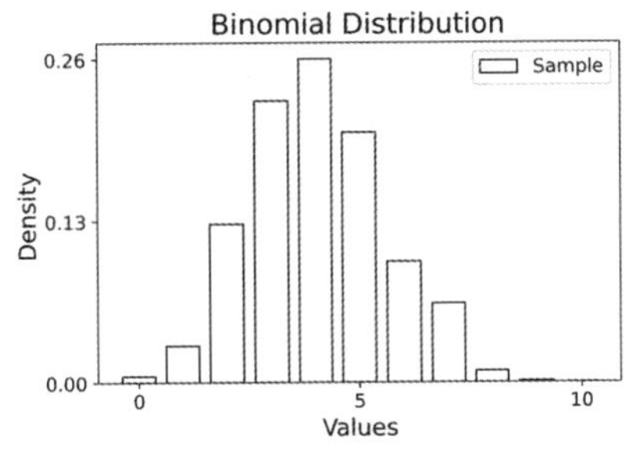

Figure 2.3.20. A histogram of 1,000 random numbers from a binomial distribution

The command line of 'binomial_sample=binomial.rvs(size=1000, random_state=143)' is used to create a histogram (Figure 2.3.20). Because we are dealing with discrete numbers, we have to add some options to control how the histogram is calculated.

The bins=11 to make sure we have 11 bins, one for each number 0, 1, …, 10 range=(0,11) to make sure the respective classes are [0,1), [1,2), …, [10,11) align='left' to make sure that the bar is aligned with the PMF (you will see why later when we overlay everything). rwidth=0.75 so the bars of the histogram are 75% of the class width.

This will make sure the bars do not touch to communicate that we are working with discrete data. alpha=0.75 so that the bars are semitransparent to allow the overlay to look nice.

Finally, here is the code that puts the PMF and the histogram together (Figure 2.3.21):

[In]

```
binomial_sample=binomial.rvs(size=1000,rand
om_state=143)
t = np.arange(0,11)
plt.hist(binomial_sample,bins=11,range=(0,1
1),density=True,
align='left',rwidth=0.75,label="Sample")
plt.scatter(t,binomial.pmf(t),label="PMF)
plt.show()
```

→

[Out]

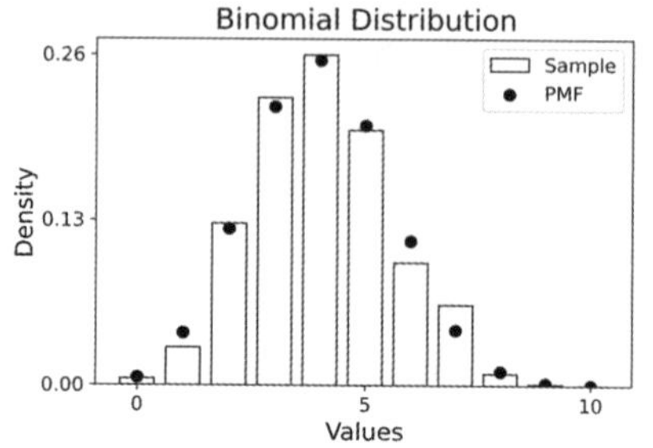

Figure 2.3.21. Histogram of a random sample from the binomial distribution and a plot of the associated PMF. PMF = probability mass function

Understanding these various probability distributions are important because they can be used to model different types of data. They will allow us to select the most appropriate tool for a given data set and to make most relevant inferences based on the simulation results.

Remarks: Probability Theory Part 2 explored a number of key ideas, including discrete and continuous random variables, probability distributions and densities, practice with probability densities, and numerous discrete and continuous distributions. These ideas serve as a basis for comprehending and using probability theory in a number of disciplines, including programing, and machine learning. The probability connected to various occurrences may be better understood and better judgments based on system analysis can be made by comprehending these notions.

[2.4]

Probability Theory, Part 3

Empirical Probability, Empirical Probability Densities, Moment and Expectation Values

Building on the foundation of probability, which quantifies the likelihood of an event occurring, we explore its significance in various random processes, such as flipping a coin, rolling a die, or drawing a playing card from a deck. For a random variable x, $p(x)$ is a function that assigns a probability to all values of x. The probability of a specific event A for a random variable x is denoted $p(x = A)$, or simply $p(A)$. Probability is calculated as the number of desired outcomes divided by the total possible outcomes, where all outcomes are equally likely.

$$p(A) = (number\ of\ desired\ outcomes)/(total\ number\ of\ possible\ outcomes)$$

This concept is easily grasped when considering an empirical random variable like the roll of a die. With a solid understanding of probability for a single random variable, we can now delve into its application within an empirical context.

1. Empirical Probability

The empirical probability, relative frequency, or experimental probability of an event is the ratio of the number of outcomes in which the event occurs to the total number of trials, not in a theoretical sample space but in an actual experiment.

In a more general sense, the empirical probability estimates probabilities from experience and observation (Hertwig & Pleskac, 2010). Given an event A in a sample space, the relative frequency of A is the ratio m/n, with m the number of outcomes in which event A occurs and n the total number of outcomes of the experiment. Empirical probabilities are based on how likely it is that an event will occur; thus, they are always estimate. Let us look at an example that relates empirical probability to the critical concept of expected value. The topics we are going to consider in this chapter are symmetry, modality, tails, and kurtosis.

Note that symmetry and modality also apply to distributions of discrete numerical state variables (e.g., the value of two dice = 2, 3, …, 12).

+ **Symmetry:** Symmetry is a sense of harmony, proportion, and balance. It refers to something that is invariant under some transformations, including translation, reflection, rotation, or scaling. A symmetrical distribution occurs when the values of variables appear at regular frequencies; often the mean, median, and mode all occur at the same point (Figure 2.4.1). If a line were drawn dissecting the middle of the graph, it would reveal two sides that mirror one other.

- Symmetrical: Equal numbers of realizations are above and below an appropriately defined (symmetry) point (or axis).
- Asymmetrical: The condition of symmetry is violated.

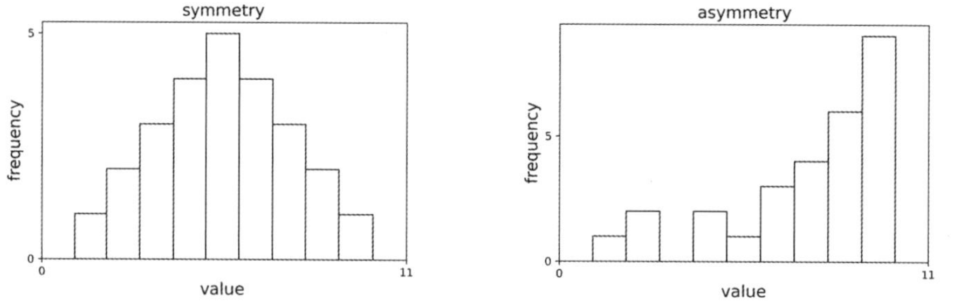

Figure 2.4.1. Distributions of data: Symmetry (left), asymmetry (right)

- A special case of asymmetry: skewed (one side is more spread out than the other, like a tail)

+ **Skewed Distributions:** A skewed distribution is neither symmetric nor normal because the data values trail off more sharply on one side than on the other.

- Direction of the skewness: The side with the more sparsely distributed data (i.e., the side that looks like a tail) gives the name.
- Quantitative measure: skewness (real number)

A dataset's distribution may be explained in terms of its skewness, which gauges how asymmetric the distribution is.

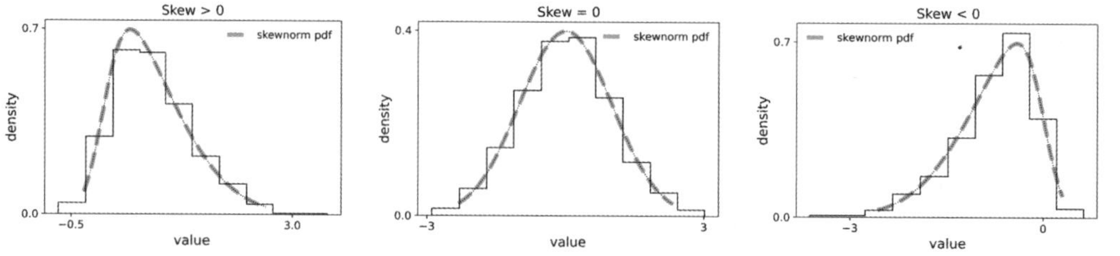

Figure 2.4.2. Distributions of data: Positive skewness (left: skewness > 0), middle = normal distribution (middle), negative skewness (right: skewness < 0)

- Symmetrical distribution: skewness = 0, Skewed distribution: skewness ≠ 0

In Figure 2.4.2, the skewness is 0 if the distribution is symmetric. The skewness, on the other hand, is non-zero if the distribution is skewed. If the skewness is high (skewness > 0), the distribution's tail extends to the right, and if it is negative (skewness < 0), the distribution's tail extends to the left. The left plot in the accompanying figure displays positive skewness, the middle plot a normal distribution, and the right plot displays negative skewness.

How Do We Capture Skewness?

In a symmetrical graph, all measures of central tendency are in the middle. However, if we have an asymmetrical distribution, how do we determine the extent of the asymmetry? Notice how the measures of central tendency tend to spread when the normal distribution is distorted.

When the tail is to the left the data are left (negatively) skewed, and when the tail is to the right, they are right (positively) skewed. It is intuitive to think that the greater the skewness, the further apart these measures will be. Thus, let us jump to the formula for skewness now:

$$Skewness = Mean - Mode/Standard\ Deviation$$

Dividing by the standard deviation enables a relative comparison of distributions on the same standard scale. Because it is not recommended that we calculate mode for small data sets, to arrive at a more robust formula for skewness we replace the mode with a calculation derived from the median and the mean.

$$Mode = 3(median) - 2(Mean)$$

For approximately skewed distributions, replacing the mode in the formula for skewness gives

$$Skewness = 3(Mean - Median)/Standard\ Deviation$$

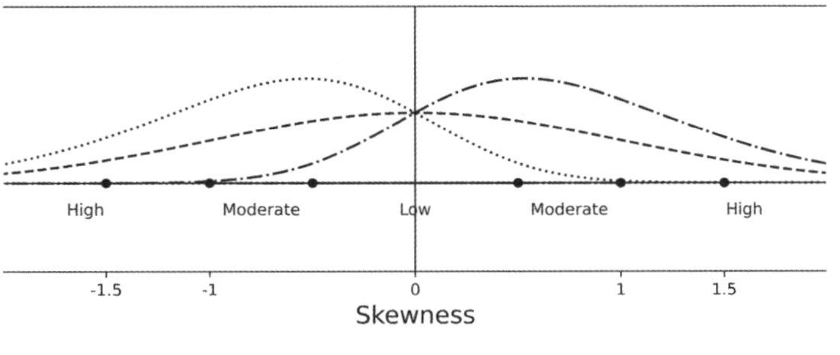

Figure 2.4.3. Scale of skewness. Low = −0.5, 0.5. Moderate = −1, −0.5 U 0.5, 1. High = −1 & beyond U 1 & beyond

Skewness is defined as 3(Mean-Median)/Standard Deviation in the equation for skewness. There are three levels of skewness: low (−1 to −0.5 or 0.5 to 1), moderate (−1 to −0.5 or 0.5 to 1), and extreme (less than −1 or greater than 1). Figure 2.4.3, which displays the scale of skewness, demonstrates this. Thus, we can conclude that the horizontal push-or-pull distortion of a normal distribution curve is captured by the skewness measure.

+ **Modality:** The modality of a distribution is determined by the number of peaks it contains (i.e., How many maxima are there?). Most distributions have only one peak, but it is possible that you will encounter distributions with two or more peaks.

- Unimodal: having one peak
- Bimodal: having two peaks
- Multimodal: having two or more peaks

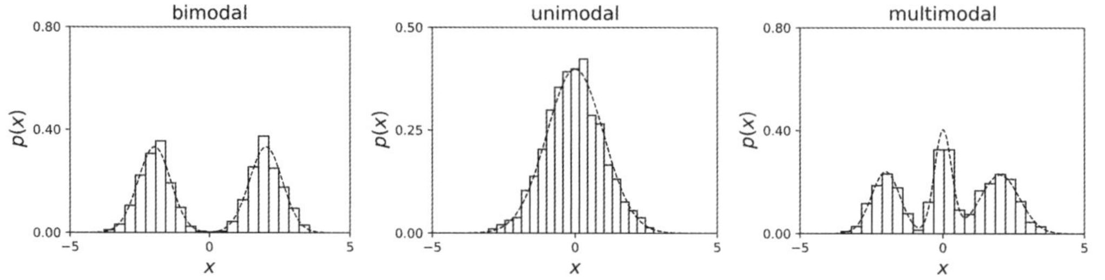

Figure 2.4.4. A summary of modality and symmetry. Bimodal and symmetrical (left), unimodal, multimodal and asymmetrical (right)

Figure 2.4.4 explains the link between a distribution's modality and symmetry. Depending on the quantity and placement of the peaks, the data distribution might take on various forms. A bimodal distribution features two peaks, compared to a unimodal distribution's single peak. Multimodal distributions have two or more peaks.

With examples of four different distribution types, including unimodal and symmetrical, bimodal and symmetrical, unimodal and asymmetrical (right skewed), and bimodal and asymmetrical,

+ **Kurtosis:** Kurtosis is a measure of the combined weight of a distribution's tails relative to the center of the distribution. It is the degree to which the tails of a distribution are heavy or light compared to a normal distribution (Figure 2.4.5).

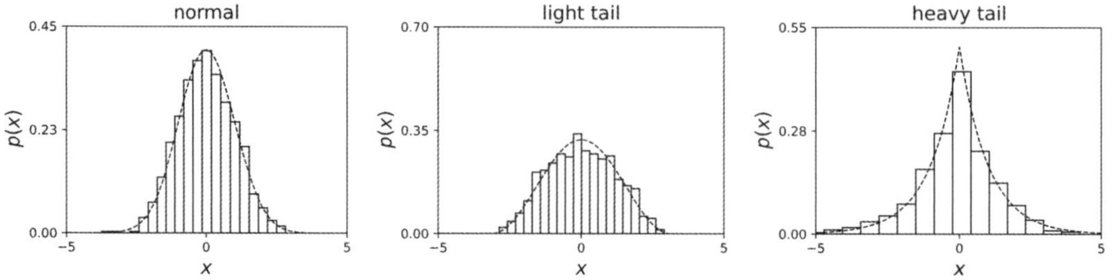

Figure 2.4.5. Kurtosis: normally distributed data (left), light tail (middle), and heavy tail (right)

- Heavy tails: relatively many values at the extreme ends (e.g., power-law distributions)
- Light tails: relatively few values at the extreme ends

Kurtosis ranges from 1 to infinity. As the kurtosis measure for a normal distribution is 3, we can calculate excess kurtosis by keeping the reference 0 for a normal distribution. Now excess kurtosis will range from – 2 to infinity.

$$Excess\ Kurtosis = Kurtosis - 3$$

A probability distribution's excess kurtosis, which ranges from −2 to infinity, is a measurement of how tail-heavy it is. Excess kurtosis for a normal distribution is (3 – 3 = 0). And the lowest value for excess kurtosis is when kurtosis is 1 (1 – 3 = – 2).

We can figure it out by taking 3 out of a distribution’s kurtosis. The excess kurtosis for a normal distribution is 0, and its lowest value, which is −2 when kurtosis is 1, corresponds to a kurtosis of 1.

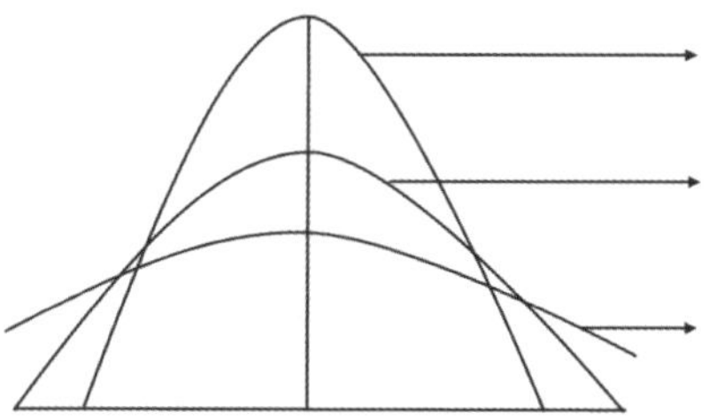

Figure 2.4.6. Schematic representation of kurtosis

We can assume that vertical push-or-pull distortion is captured by the kurtosis measure (Figure 2.4.6). Also, outliers dominate the kurtosis effect, which has its roots in the fourth-order moment-based formula.

2. Empirical Probability Densities

Normal density is a probability and statistical concept widely used in scientific studies for its many benefits (Gelman et al., 1995). It is simple. Its mean, median, and mode have the same value, and it can be defined with just two parameters: mean and variance. It also has important mathematical implications like the central limit theorem (see section 1.5 Central limit theorem in Part 1).

Real-world data sets do not always follow a normal distribution. They are often skewed, which makes the results of our statistical analyses invalid. In this section, we are going to look at the characteristics of normal distributions and lognormal distributions.

+ **Characteristics of a Normal Distribution:** unimodal, symmetrical (bell shaped), two parameters: μ and σ^2

$$P(x) = \frac{1}{\sqrt{2\pi\sigma^2}} exp\left\{-\frac{(x-\mu)^2}{2\sigma^2}\right\}$$

Here, the x is a real number, μ is the mean of the distribution, σ is the standard deviation of the distribution, and π is the mathematical constant pi (approximately equal to 3.14159). It represents a continuous probability that is defiend by a symmetrical bell-shpaed curve. The mean μ of the distribution is the center of the curve and determines the location fo the distribution, and the standard deviation σ is a measure of the dispersion of the distribution and determines the shape of the curve.

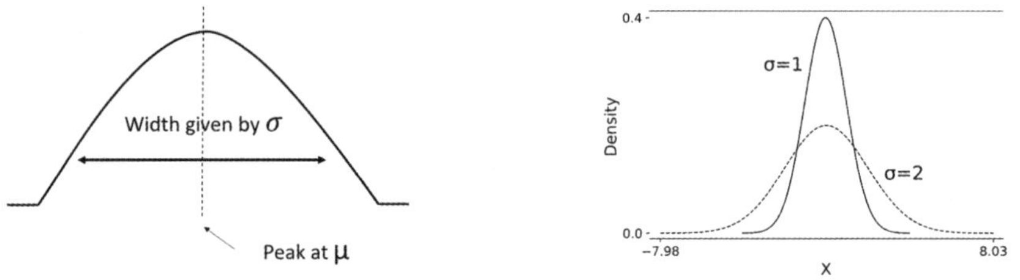

Figure 2.4.7. A normal distribution with parameters. The parameters are μ = mean and σ^2 = variance (width given by σ). Prob. = probability

As we learned, the Gaussian distribution, normal distribution, normal probability density, and Gaussian probability density are other names for the normal distribution (Figure 2.4.7). The standard normal distribution, which has a mean of 0 and a standard deviation of 1, is a specific instance of the normal distribution.

Standard normal distribution (mean $\mu = 0$ and standard deviation $\sigma = 1$)

Consider, as an illustration, the Boltzmann distribution of linear forces with quadratic potential energy, where $P(x)$ denotes the probability density function, $U(x)$ denotes potential energy, K_B denotes the Boltzmann constant, and T denotes temperature. The probability density function is as follows:

- Boltzmann distribution of linear forces: quadratic potential energy.

$$P(x) = \frac{1}{z} exp\left\{-\frac{U(x)}{K_B T}\right\}, \qquad U(x) = a\frac{(x - x_0)^2}{2}$$

In this case, the Boltzmann distribution is a specific example of the normal distribution, demonstrating their relationship. The Boltzmann distribution of kinetic energy is distribution of the velocity of Brownian particles as follows;

$$P(x) = \frac{1}{z} exp\left\{-\frac{E_{kin}(v)}{K_B T}\right\}, \qquad E_{kin} = m\frac{v^2}{2} \qquad \rightarrow$$

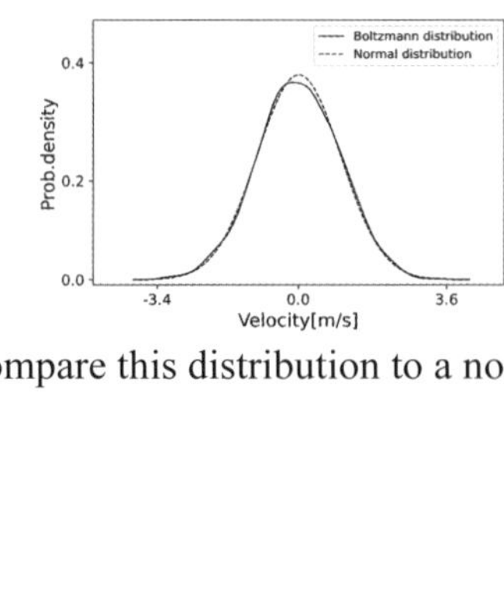

Figure 2.4.8. Stationary velocity (from Brownian motion). Compare this distribution to a normal distribution (dotted line). Prob. = probability

This example highlights the application of the Boltzmann distribution to model the velocity distribution of Brownian particles, which are particles undergoing random motion due to collisions with surrounding molecules. Here, the exponential term that establishes the form of the probability density function is $exp\left\{-\frac{E_{kin}(v)}{K_B T}\right\}$. $E_{kin}(v)$ denotes the kinetic energy of a particle traveling at speed v, which is computed as $m\frac{v^2}{2}$, where m is the particle's mass. T is the system's temperature, and K_B is the Boltzmann constant. The Boltzmann distribution is a specific instance of this (Figure 2.4.8); that explains the distribution of particle speeds in a gas, is frequently referred to as the equation.

- For example: Central limit theorem (simplified)

The Central Limit Theorem (simplified) says that the observable X has a normal distribution when it is the average of several non-correlated distributions or disturbances. The membrane potential U, as described in Abeles (1991), is one example of how this theorem may be used in practice.

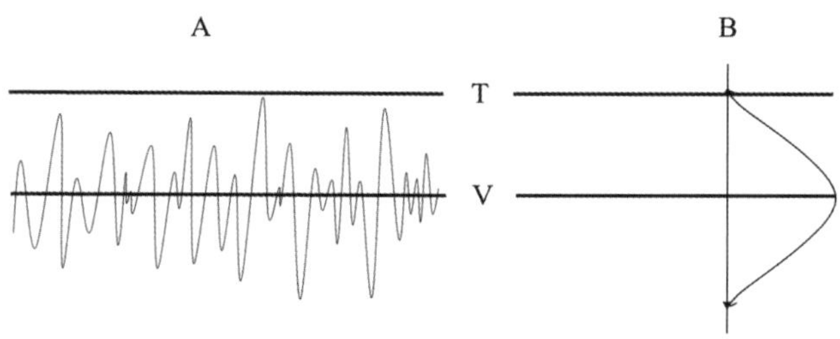

Figure 2.4.9. Normally distributed potential (noncorrelated inputs from many dendrites). A is the sum of 100,000 simulated synaptic potentials per second arriving at random times. B is the probability density function of the potentials of part A lying on its side.

We can see a regularly distributed potential in Figure 2.4.9 that is the consequence of non-correlated inputs from several dendrites. Part A specifically displays the total of 100,000 simulated synaptic potentials per second.

+ **Characteristics of a Lognormal Distribution** (defined on $x \geq 0$): Unimodal, skewed to the right, two parameters μ and σ, does not correspond to the mean and standard deviation of X

$$P(x) = \frac{1}{x\sqrt{2\pi\sigma^2}} exp\left\{-\frac{(\log(x) - \mu)^2}{2\sigma^2}\right\}$$

Here, the lognormal distribution is a continuous probability that is defined by a skewed curve. The function 'log' represents tne natural logarithm of x which is the base-e logarithm of x. The natural logarithm is defiend as the inverse of the expoenetial function. For its parameterization and lognormal computation, we can simulate the (log)density as shown below:

```
[In]
mu=2;
sigma=1;
xout = np.linspace(lognorm.ppf(0.01, sigma, scale=np.exp(mu)),
                   lognorm.ppf(0.99, sigma, scale=np.exp(mu)), 100)
```

Compute lognormal distribution

```
[In]
P = lognorm.pdf(xout, sigma, scale=np.exp(mu))
```

With parameter which defined above, the 'lognpdf' function is a built-in function and returns the value of the lognormal probability density function at each element of the array 'xout' which is by the mean 'mu' and standard deviation 'sigma' of the distribution. And create the output

```
[In]
P = gaussian(xout, mu, sigma)
plt.plot(xout, P)

# setting the axes limits
plt.xlim([min(xout),
max(xout)])
plt.ylim([0, max(P)*1.1])
```

→

[Out]

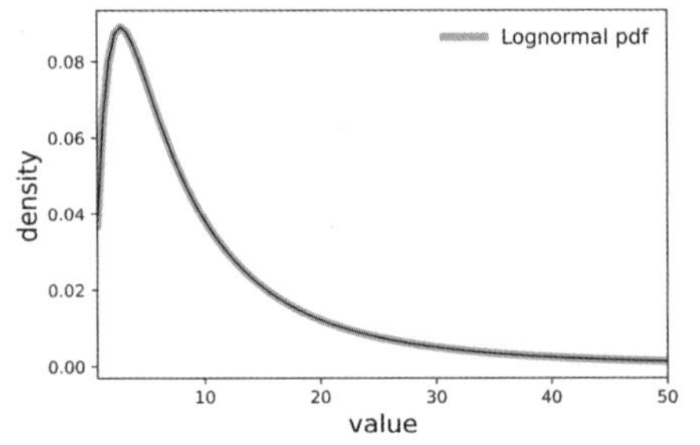

Figure 2.4.10. A lognormal distribution. The parameters are $\mu = 2$ and $\sigma = 1$. Prob. = probability

In Figure 2.4.10, the 'axis' function is used to set the axis limits of the current axes to minimum and maximum values in 'xout', and the y-axis limits to 0 and the maximum value in p multiplied by 1.1.

+ **Lognormal and Normal Distributions:** If X has a lognormal distribution, and the log-transformed variable $Z = \log(x)$, then Z has a normal distribution. The parameters μ and σ of a lognormal distribution are the mean and standard deviation of the transformed variable Z.

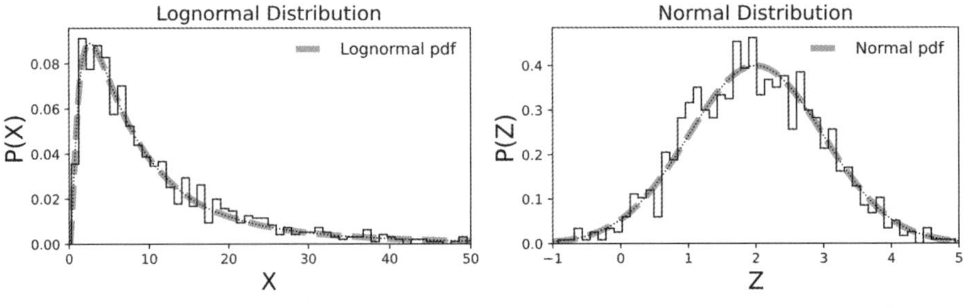

Figure 2.4.11. Lognormal (left) and normal (right) distributions. The parameters are $\mu = 2$ and $\sigma = 1$.

The values of the parameters and in a lognormal distribution correspond to the converted variable Z's mean and standard deviation. The lognormal distribution is presented on the left and the normal distribution is shown on the right in Figure 2.4.11, which shows a visual comparison between the two. The variables are = 2 and = 1 for this specific lognormal distribution.

Log Transformation Interpretation

Logarithms are essential tools in systematical modeling and analysis (Packard et al., 2011). A logarithm can be defined with respect to a base (b) where the base b logarithm of X is equal to y because X equals b to the power of y ($\log(X) = y$ because of $X = b^y$). We can take any positive number as the base of the logarithm, but the most commonly used bases are base 2 and base 10.

- Base 2: the base 2 logarithm of 8 is 3 because $2^3 = 8$
- Base 10: the base 10 logarithm of 100 is 2 because $10^2 = 100$

Natural log—the base of the natural log is the mathematical constant e, or Euler's number, which is equal to 2.718282. Thus, the natural log of 7.389 is 2 because $e^2 = 7.389$.

- Logarithm formula: $y = log_b(x),\ b^y = x$

What Is Log Transformation? Log transformation is a data transformation method that replaces each variable x with a $\log(x)$. The choice of the logarithm base is usually left up to the analyst, and it depends on the purposes of the statistical modeling. Here we focus on the natural log transformation.

- The natural log is denoted: ln.

When original continuous data do not follow a bell curve, we can log transform them to make them as normal as possible so that the results of the statistical analysis of these data become more valid. In other words, the log transformation reduces or removes the skewness of the original data. The important caveat here is that the original data have to follow or approximately follow a lognormal distribution. Otherwise, the log transformation will not work. Applying log transformation in programming is very simple.

First, we have to install and import NumPy, the fundamental package for scientific computing with Python. After that, we just have to apply the natural log transformation function of NumPy (numpy.log or np.log) to the values we want to log transform, as follows:

```
[In]
import numpy as np
x = [1, 2, 3, 4, 5]     →
y = np.log(x)
print(y)
```

```
[Out]
array([0.    , 0.69314718,
1.09861229,   1.38629436,
1.60943791])
```

The log function calculates the natural logarithm of each element in the input array and prints the resulting array y.

+ **Example (Kernel Density Estimation):** A study investigates the duration of psychiatric treatment, focusing on the distribution of treatment durations and the appropriate methods to analyze and represent such data (Silverman, 1986). In this context, kernel density estimation can be a useful technique to estimate the probability density function of the treatment durations. However, when applying kernel density estimation, it may fail to account for the fact that negative treatment durations are impossible. Consequently, the kernel density estimation may predict a non-zero probability for negative durations, which is incorrect.

- Density 'falls' prob to find negative durations > 0
- Set the smoothing parameter per hand, histogram (middle plot), maximum likelihood estimation fit to given distribution (right plot)

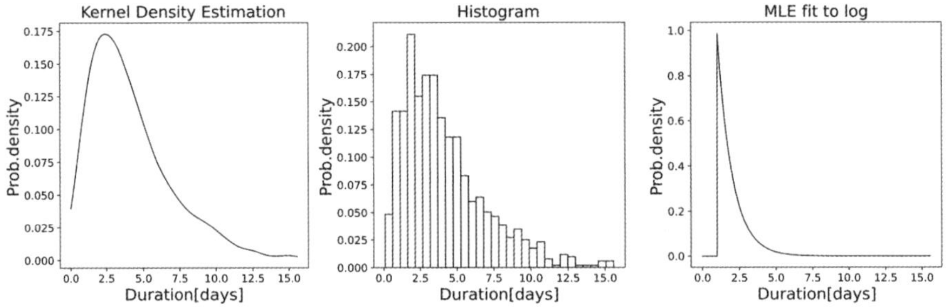

Figure 2.4.12. Schematic representation of the smoothing parameter

In Figure 2.4.12, the plot of the left side shows naïve density estimate of the probability density function (PDF), the plot of the middle side shows the probability density function represented as a histogram, and the plot of the right side shows the probability density function as determined by fitting the maximum likelihood estimation (see Supplement 2.3 for more detail) to the lognormal (Prob. = probability).

3. Moments and Expectation Values

What is the moment and expectation value in probability? To understand the features of a probability distribution, we looked at the features that could be summarized, including skewness and kurtosis. Expected value is another typical measure. The n-th moment of a distribution (or set of data) about a number is the expected value of the n-th power of the deviations about that number. All of these measures are examples of a mathematical quantity called a moment. Let us assume the random variable we are interested in is X. The moments are the expected values of X: $E(X)$, $E(X^2)$, $E(X^3)$, and so on.

- $E(X)$ = the first moment
- $E(X^2)$ = the second moment
- $E(X^3)$ = the third moment

  …
- $E(X^n)$ = the n-th moment

These are (real) numbers computed from a mathematical model. It can be an infinitely large data set, probability distribution (discrete-state variable), or probability density (continuous-state variable) describing certain aspects of a distribution. For example, the mean is the center of a distribution of values, and variance is variability in the values of a distribution.

+ **Moments for Discrete and Continuous Numerical Variables:** This does not make sense for categorical variables.

[Table 2.4.1] Moments and Expectation Values for Rolling Two Dice

Values	2	3	4	5	6	7	8	9	10	11	12	x_j
Possible outcomes obtained from the rolling of two dice	11	12	13	14	15	16	26	36	46	56	66	
		21	22	23	24	25	35	45	55	65		
			31	32	33	34	44	54	64			
				41	42	43	53	63				
					51	52	62					
						61						
Frequency	1	2	3	4	5	6	5	4	3	2	1	$h_j=\sum_{k=1,\dots,N;X^{(k)}=j} 1 = 36$
Relative frequency	1/36	2/36	3/36	4/36	5/36	6/36	5/36	4/36	3/36	2/36	1/36	$f_j = \frac{h_j}{N}$
Probability	0.027	0.055	0.083	0.111	0.138	0.166	0.138	0.111	0.083	0.055	0.027	$P_j = \lim_{N\to\infty} f_j$
Moment	2/36	6/36	12/36	20/36	30/36	42/36	40/36	36/36	30/36	22/36	12/36	$x_j p_j$

Table 2.4.1 presents the moments and expectation values for rolling two dice, illustrating the possible outcomes, their corresponding frequencies, relative frequencies, and probabilities, with the moment calculated as the product of the value and its probability.

Mean (first moment) → n-th moment

$$m = \frac{1}{N} \sum_{k=1,\dots,N} X^{(k)}, \qquad N \to \infty \qquad \longrightarrow \qquad m_n = \frac{1}{N}\sum_{k=1}^{N} \left(X^{(k)}\right)^n, \qquad N \to \infty$$

The formula involves an infinite sum of the k-th power of the variable X '$X^{(k)}$', raised to the power of n, and divided by N. The symbol $\sum_{k=1}^{N}()$ denotes a sum, and the under-script and over-script indicate that the sum is taken over the range of k from 1 to N. The $N \to \infty$ represents that the limit is taken an N approaches infinity.

+ **Moments Computed from Probability Distributions:** Discrete numerical state space: $X \in \{x_1, x_2, \dots, x_M\}$

- Mean

$$m = \lim_{N\to\infty} \frac{1}{N} \sum_{k=1,\dots,N} X^{(k)} = \sum_{j=1,\dots,M} x_j p_j$$

- By analogy: nth moment and expectation values

$$m = \lim_{N\to\infty} \frac{1}{N} \sum_{k=1,\dots,N} [X^{(k)}]^n = \sum_{j=1,\dots,M} [x_j]^n p_j$$

Here, the sum is taken over the range of j from 1 to M '$1, \dots, M$', and the term being summed is the n-th power of $[x_j]^n$, mutipled by p_j.

$$m = \lim_{N\to\infty} \frac{1}{N} \sum_{k=1,\dots,N} X^{(k)} = \sum_{j=1,\dots,11} x_j p_j$$

Also, the moments can be computed from probability densities as given below:

Mean		nth moment
$m = \lim_{N\to\infty} \frac{1}{N} \sum_{k=1,\dots,N} X^{(k)} = \int_{\Omega} xP(x)dx$	$\longrightarrow$	$m_n = \langle X^n \rangle = \int_{\Omega} x^n P(x)dx$

Here, the formula involves an integral over the domain Ω of the function x^n multipled by the probability density function $P(x)$. The probability density functions that describes the probability of a random variable taking on a particular value, and the integral is calculated by taking the product of x^n and $P(x)$ for all values of x in the domain Ω. Then it summed up all these products over the entire domain and the result of the integral is the n-th moment of the random variable x. For example: show that mean value m is given by parameter μ

$$P(x) = \frac{1}{\sqrt{2\pi\sigma^2}} exp\left\{-\frac{(x-\mu)^2}{2\sigma^2}\right\}$$

That is, calculate: M=$\int_{-\infty}^{\infty} xP(x)dx$

+ **Practice with Moment Values:** With respect to the logic denoted above, let us start this practice by defining the mean as below.

$$m = \lim_{n\to\infty} m_e(N), \quad m_e(N) = \frac{1}{N}\sum_{k=1,\dots,N} X^{(k)}$$

Task (moment values): where $X^{(k)}$ are realizations (i.e., statistically independent scores) from a statistical ensemble. The average value m_e can be interpreted as a sample average. The figure below illustrates the limiting process. It shows sample means m_e taken from a normal distribution with parameters $\mu = 7$ and $\sigma = 1$ for samples of increasing size. The sample size used in the figure is shown in the Table 2.4.2.

[Table 2.4.2] Sample Size for Practice

N	10	10^1.5 32	10^2 100	10^2.5 316	10^3 1,000	10^3.5 3,162	10^4 10,000	10^4.5 31,623	10^5 100,000	10^5.5 316,228	10^6 1,000,000
m_e											

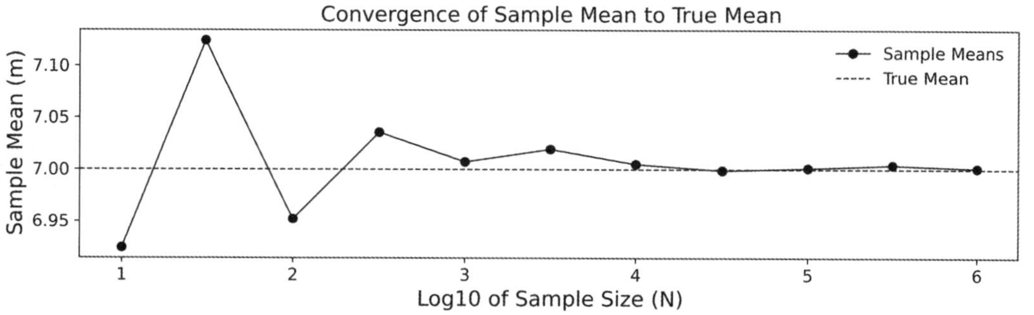

Figure 2.4.13. Simulation of the convergence case (scaled by log)

Figure 2.4.13 suggests that in the limiting case when N goes to infinity, the sample mean converges to the value 7 (i.e., we have $m = \mu = 7$).

Problem 1a: To demonstrate convergence for normally distributed data with a mean (μ) of 11 and a standard deviation (σ) of 2, generate samples with varying sample sizes, compute the mean for each sample, and report the values in a table. Create a plot similar to the referenced figure. You can follow the step-by-step instructions provided or devise an alternative approach. Utilize program to generate a dataset named 'dataset.txt' containing normally distributed scores with the specified parameters $\mu = 7$, $\sigma = 2$, and $N = 10$. Generate data with a normal distribution

```
[In]
mu=0;
sigma=1;
samplesize=500;
data = np.random.normal(mu, sigma, samplesize));
```

The 'mu' is the mean of the normal distribution which is set to 0. The 'sigma' is the standard deviation of the normal distribution which is set to 1. The 'samplesize' is the number of random numbers that will be generated which is set to 500. And the generated data is normalized by the 'random.normal' function which is likely a function from a statistical argument.

```
[In]
d = np.loadtxt('dataset.txt')
mean_d = np.mean(d)
xout = np.arange(1, 6.5, 0.5)

# generating 11 random values
mestim = np.random.rand(11)
plt.plot(xout, mestim, marker='o',
linestyle='-')
```

→

[Out]

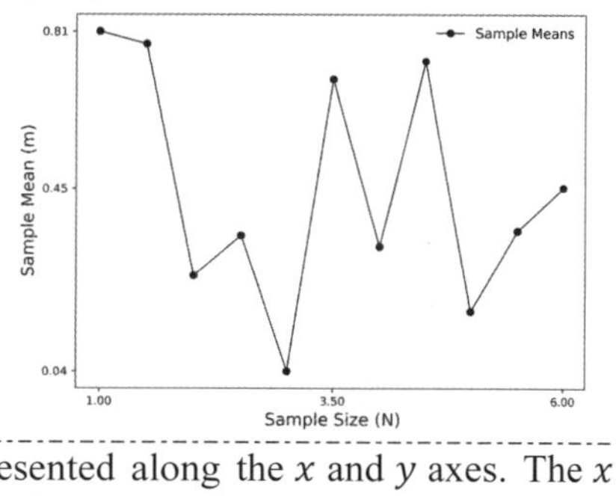

Figure 2.4.14. Relationship between two variables, represented along the x and y axes. The x axis ranges from 1 to 6 with an interval of 0.5, while the y axis presents the corresponding random values.

We road these data in the variable 'd' and compute the sample mean via the command; record the result in the table, and repeat this procedure for all *N*, create a result with programing defined 'xout.' Defined vector that contains the sample means, creating a plot. The Figure 2.4.14 visually present the randomness of the generated values. It does not convey any practical context due to its randomness, but serves as a fundamental demonstration of how to generate and plot a simple random data sequence.

Problem 1b: Compute the means of scores originating from a distribution with parameters $\mu = 1.10$ and standard deviation $\sigma = 1.18$. Similar to Problem 1a, generate lognormally distributed data instead of normally distributed data. Follow the step-by-step instructions provided earlier, utilizing the generated data to produce tables and figures. Determine the relationship between the mean (m) and the parameters μ and σ. Calculate m and compare the result to the numerical findings (Table 2.4.3 and 4).

[Table 2.4.3] Solution to Problem 1a

P1	*N*	10	32	100	316	1,000	3,162	10,000	31,623	100,000	316,228	1,000,000
	Me	7.163	7.415	7.262	7.024	6.986	6.987	7.006	7.015	6.996	7.001	7.001

[Table 2.4.4] Solution to Problem 1b

P2	*N*	10	32	100	316	1,000	3,162	10,000	31,623	100,000	316,228	1,000,000
	Me	7.784	5.880	5.966	6.268	6.183	6.045	5.989	6.116	6.062	6.016	6.018

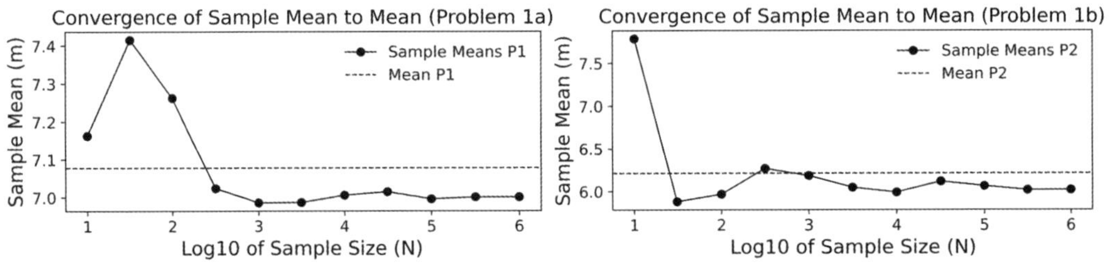

Figure 2.4.15. Solution of the convergence case (scaled by log). The plot of the left side shows the convergence of sample means to the overall mean for Problem 1a. The plot of the right side illustrates the convergence of sample means to the overall mean for Problem 1b.

The data related to Problem 1a show that when *N* becomes large, the means converge to a certain number. That is, they do not fluctuate anymore. They become stable. For Problem 1b, the mean m and variance v of a lognormal random variable are functions of μ and σ that can be calculated with the function $\mathrm{m} = \exp(1.10 + \frac{1.18^2}{2})$. The data also show that the sample mean decreases as the sample size increases.

+ **Expectation Values:** Given a function g of the random variable X: $g(X)$. Which function value do we expect to observe on average?

- Answer = expectation value
- Notation: mathematics = $E[g(X)]$, physics = $< g(X) >$

- Example: first moment (mean): $g(X) = X \rightarrow m = E[X] = <X>$

Given a function g of the random process X, $g(X)$, how do we calculate $E[g(X)]$?

- Answer

$$E[g] = \frac{1}{N}\{g(X^{(1)}) + g(X^{(2)}) + \cdots + g(X^{(N)})\}, \quad N \rightarrow \infty$$

This is a mathematical expression for calculating the expected value of a function 'g' based on a sequence of random variables $(X^{(1)}), (X^{(2)}), \ldots (X^{(N)})$. The answer of the formula involves a sum of the function 'g' applied to each element of the sequence 'N' divided by the number of elememts in the sequence '$(X^{(1)}), (X^{(2)}), \ldots (X^{(N)})$' The symbol '$N \rightarrow \infty$' indicates that the limit is taken as N approaches infinity. Notation of the expectation value of function g is;

$$E[g(X)] = \langle g(X) \rangle = \lim_{N \to \infty} \frac{1}{N} \sum_{k=1,\ldots,N} g(X^{(k)})X^{(k)}$$

Variance ($\sigma^2 = \frac{1}{N}\sum_{k=1}^{N}(X^{(k)} - m)^2$) and standard deviation ($\sigma = \sqrt{\frac{1}{N}\sum_{k=1}^{N}(X^{(k)} - m)^2}$), expectation value (expected squared deviation from mean) as follow;

$$\sigma^2 = \langle (X - m)^2 \rangle, \; m = \langle X \rangle$$

Here, the equation involves the variance of X and the expected value (mean of X). The variance is calculated as the expected value of the squared difference between X and the mean m, and the mean m is calculated as the expected value of X. Moments and Expectation Values Computed from Probability Densities: nth moment $m_n = \langle X^n \rangle = \int_\Omega x^n P(x)dx, \quad \Omega = state\ space$, by analogy: expectation value of function g

$$\langle g \rangle = \lim_{N \to \infty} \frac{1}{N} \sum_{k=1,\ldots,N} g(X^{(k)}) = \int_\Omega g(x)P(x)dx$$

+ **Practice with Expectation Values:** The mean and the variance are measures of central tendency and variability for continuous variables defined on a real line. They may also be applied under certain circumstances to periodic variables such as angular variables. In general, circular statistics should be applied to periodic variables to determine the center and the variability of probability densities (distributions) and samples.

Task: Expectation values (cluster phase and cluster amplitude of circular statistics).

Among several measures, we focus here on cluster phase θ and cluster amplitude r. The cluster phase describes the center of a distribution of angles. The cluster amplitude can assume values between 0 and 1. Low values indicate high variability, whereas high values indicate low variability. For $r = 1$, all angles have precisely the same value. The cluster amplitude and cluster phase are defined by means of the expectation value of a complex function.

$$c = E[\exp(i * X)], \qquad r = |c|, \qquad \theta = angle(c)$$

The complex function is the exponential function of iX where i is the imaginary unit (i.e., the square root of -1). The amplitude r is the absolute value of the complex number c, where the phase is the angle of the complex number c. You can solve the problem described below using calculus. We now describe how to determine cluster amplitude and cluster phase using real expectation values. This is the part that should be solved (working with complex numbers is optional). The cluster amplitude and cluster phase can be computed from expectation values of the sine and cosine functions like

$$a = E[\cos(X)], \qquad b = E[\sin(X)], \qquad r = \sqrt{a^2 + b^2}, \qquad \tan(\theta) = \frac{b}{a}$$

The last relation must be inverted using the arctan function to obtain the cluster phase. To obtain the cluster phase and amplitude from a sample, we must replace the two expectation values (a, b) with the respective sample averages.

Problem: Let us determine cluster amplitudes r and cluster phases θ for the data sets HW3phasedataA.txt and HW3phasedataB.txt. They are identical to HW2phasedataA.txt and HW2phasedataB.txt except that the required phase has not shifted the scores in the HW3 data sets. Compute for each of the two data sets the sample amplitude and sample cluster phase. Compare your results to the results obtained using HW2 (use the Supplement data file).

You may proceed as follows: Load HW3phasedataA.txt into program and store the scores into the variable (vector)

```
[In]
d=load(…);
```

The angular data are given in degrees, and we need them in rad. Therefore, we put

```
[In]
d=d/360*(2*pi);
```

For each score X^k in d we want to know the sine function of that score, i.e., $\cos(X^k)$. Therefore, we compute a vector

```
[In]
cosvec=cos(d);
```

You may have a look at the first 10 function values using ‘cosvec(1:10)’; the parameter ‘a’ is just the average of all function values stored in the vector.

Therefore, we put

```
[In]
a=mean(cosvec);
```

Now, we compute the parameter 'b' in the same way. Using a single step, we put

```
[In]
b=mean(sin(d));
```

Finally, we compute r using the 'sqrt' function and the cluster phase using the atan2 function like

```
[In]
r=sqrt(a*a+b*b)
theta=atan2(b,a)/pi*180
```

When you write down r and theta, repeat the procedure with the data set HW3phase-datatB.txt. Then compare the findings to the results obtained using HW2.

[Table 2.4.5] Solution of Practice (Expectation Values)

	HW3phasedataA	HW3phasedataB	HW2phasedataA	HW2phasedataB
R	0.9831	0.9903	0.9831	0.9903
Theta	179.2475	- 0.5763	- 0.7525	- 0.5763

The cluster amplitude of the two data sets seems to be variable, and the cluster phase appears to be both anti-phase and in-phase. In Table 2.4.5, when it comes to the HW3 data, the A data look anti-phase, as theta is 179, whereas the B data have a theta of – 0.57. Regarding the HW2 data, the data seem to have been shifted by the required phase, as most values are identical to the HW3 values. The exception is the HW2_A theta (– 0.75), which looks like the shifted value (180 – 179.2475 = 0.7525). The moments and expectation values provide insight into the features of a probability distribution, helping to better understand the underlying characteristics of the actual data set.

Remarks: Probability Theory Part 3 explored a number of crucial ideas, including empirical probability, empirical probability densities, moments, and expectation values. While moment and expectation values are statistical measurements that are used to describe characteristics of a distribution, empirical probability and probability densities are generated through seeing and counting real-world occurrences. Understanding these ideas is important in many different domains, including as data science, finance, and engineering, where probability theory is frequently utilized to make predictions and judgments based on existing data.

[2.5]

Probability Theory, Part 4

Exponential Probability Density, Information Theory, Maximum Likelihood Estimation

Probability theory allows us to make uncertain statements and reason in the presence of uncertainty. Logic gives us an idea about deterministic things, and probability theory extends that logic to uncertain things.

Systematical methods are classically divided into supervised and unsupervised algorithms. Unsupervised algorithms have more uncertainty, but even supervised algorithms are mathematically uncertain. Information theory allows us to quantify the amount of uncertainty in a probability distribution (Howard, 1966). You will remember that probability distributions are mathematical functions that provide the probabilities of the occurrence of different possible outcomes in a given experiment. They come in many different shapes with different characteristics as defined by a mean, standard deviation, skewness, and kurtosis. In this section, we discuss exponential probability, information theory, and maximum likelihood with their respective parameters.

1. Exponential Probability Density

The exponential distribution is a widely used continuous distribution and the exponential function is an essential function in mathematics and analysis. It is often used to model the time that elapses between events. To create an exponential function, we let the independent variable (X) be the exponent. Characteristics of the exponent are as follows: data $x \geq 0$, unimodal, peak at 0, skewed to the right, one parameter that corresponds to mean μ. These features are defined in several famous functions as follows.

+ **Exponential Distribution:** The exponential distribution, as previously mentioned, can be considered a type of Boltzmann distribution or measure. It is used in statistical mechanics to calculate the probability of a system being in a specific state, based on the energy of that state and the system's variable like temperature (Jaynes, 1957).

$$P(x) = \frac{1}{\mu} exp\left\{-\frac{x}{\mu}\right\}$$

Here, the probability is defined as the inverse of the mean μ of the distribution multiplied by the exponential function $exp\left\{-\frac{x}{\mu}\right\}$. As the x increase, the probability of x taking on that value decrease exponentially and the mean μ of the distribution determines the rate at which the probability decrease as x increase.

Consider a particle system in thermodynamic equilibrium. The system contains several mass points with positions, velocities, and interactions (spring-like forces, gravity, etc.).

[Table 2.5.1] Generalization of Particle Systems

Newtonian mechanics	→ Statistical mechanics	→ Phenomenological thermodynamics → Second Law: Entropy increases or constant → Maximum entropy principle of canonical systems → Boltzmann distribution

Table 2.5.1 outlines the transition from Newtonian mechanics through statistical mechanics to phenomenological thermodynamics. The Second Law, the Boltzmann distribution, and the maximization of entropy in canonical systems are some of the main ideas that are highlighted in this sequence. Overall, the information in this table gives a helpful picture of how particle systems have evolved and the underlying theories behind them.

+ **Maximum Entropy Principle (Canonical Systems):** The exponential distribution, as discussed earlier, helps describe the statistical mechanics of probability for systems like canonical systems, which are in thermodynamic equilibrium, composed of many particles interacting with an environment E and maintaining a constant temperature T. When considering the Maximum Entropy Principle, the exponential distribution plays a crucial role in determining the likelihood of a system being in a particular state, based on its energy and temperature, while exhibiting maximal disorder (measured as entropy) and satisfying the constraint.

$$Entropy = -\{p_1 \, log \, p_1 + p_2 \, log \, p_2 + \ldots + p_M \, log \, p_M\} = -\sum_{j=1,\ldots,M} p_j \, log \, p_j$$

Here, the formula involves a sum over the range of j from $1 \; to \; M \; (1, \ldots, M)$ of the product of the probability p_j and the logartithmof $log \; p_j$. The logarithm is taken to the base 2 as indicated by the symbol log. Taking the product of the p_j and $log \; p_j$, the entropy is calculated for each value of the discrete random variable x, and then summing up all these products over all possible values of x.

$$Entropy = -\int_{\Omega} P(x) \log P(x) \, dx$$

The formula involves an integral over the domain Ω of the function $P(x)$ multiplied by the logarithm of $\log P(x)$. The integral is calculated by taking the product of $P(x)$ and the logarithm of $\log P(x)$ for all values of x in the domain Ω, and then summing up all these products over the entire domain.

The maximum entropy principle, which is followed by a canonical system in thermodynamic equilibrium at temperature T, results in an energy probability density with an exponential distribution.

- Maximum entropy principle → energy probability density = exponential distribution
- Exponentially decaying energy distribution = Boltzmann distribution

The exponential distribution is the name given to the Boltzmann exponentially decreasing energy distribution. In order to characterize the behavior of systems at equilibrium, the distribution, a fundamental idea in statistical mechanics, is often utilized in many disciplines, including physics, chemistry, and engineering.

$$P(E) = \frac{1}{T} exp\left\{-\frac{E}{T}\right\}$$

Let us specify of the (exponential) density using a programing simulation. For parameterize

```
[In]
mu=2
xout = np.arange(0, 15.1, 0.1)
```

The 'mu' is defined as 2, and 'xout' is defined as an array of values ranging from 0 to 15, in increments of 0.1. Then compute the exponential distribution

```
[In]
pdf = expon.pdf(xout, scale=mu)
```

The function 'expon.pdf' is a statistical toolbox that calculates the probability density function of an exponential distribution.

Then, create the output (Figure 2.5.1)

```
[In]
figure
ax.plot(xout, pdf)

ax.set_xticks([min(xout), np.mean(xout),
max(xout)])
ax.set_yticks([min(pdf), np.mean(pdf),
max(pdf)])
```

→

[Out]

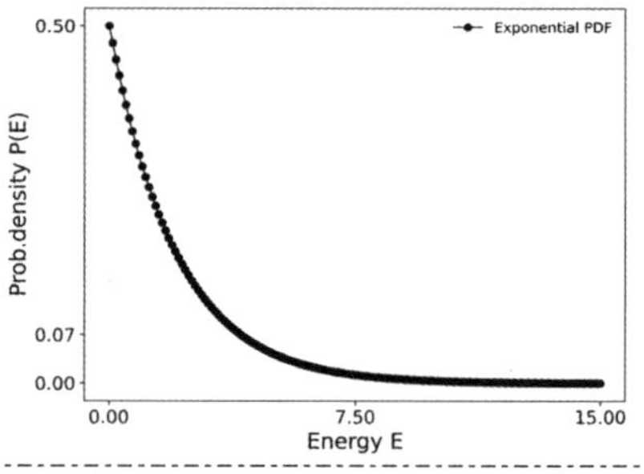

Figure 2.5.1. Boltzmann distribution of exponentially decaying energy: No degeneration of energy levels. Prob. = probability

+ **Brownian Particles and Maxwell-Boltzmann Distribution (Velocities):** Connecting this to the concepts of exponential distribution and maximum entropy, we can further explore the behavior of Brownian particles and their velocities. Brownian particles possess kinetic energy $E = Mv^2/2$, and their Boltzmann distribution $P(E)$ corresponds to a normal distribution $P(v)$ of velocities, known as the Maxwell-Boltzmann distribution.

$$P(E) = \frac{1}{T} exp\left\{-\frac{E}{T}\right\}, \qquad E = m\frac{v^2}{2}, \qquad P(v) = \sqrt{\frac{m}{2\pi T}} exp\left\{-\frac{mv^2}{2T}\right\}$$

The probability is defined as the square root of the ratio of the mass m of the system to the product of 2 and π multiplied by the exponential function $exp\left\{-\frac{mv^2}{2T}\right\}$. It represents as v increase, the probability of v taking on that value decrease expoenetially. And the mass m of the system and the temperature T determine the rate at which the probability decrese as v increase.

Let us specify the Maxwell-Boltzmann distribution using a programing simulation. To load data set (use the Supplement data set 'dataset_velocity.txt')

```
[In]
data = np.loadtxt('dataset_velocity.txt')
```

For its empirical density function

```
[In]
kde = gaussian_kde(data)
```

Alternatively, if you want to use your own grid

```
[In]
xout = np.arange(-10, 10.1, 0.1)
pdf = kde(xout)
```

The function 'kde' is a statistical toolbox that estimates the probability density function (pdf) of a data using a kernel density estimation (kde) method. The kernel density estimation is a non-parametric method for the probability density function of a random variable. Then, create its output

```
[In]
ax.plot(xout, pdf)

ax.set_xticks([min(xout), np.mean(xout),
max(xout)])
ax.set_yticks([min(pdf), np.mean(pdf),
max(pdf)])
```

→

[Out]

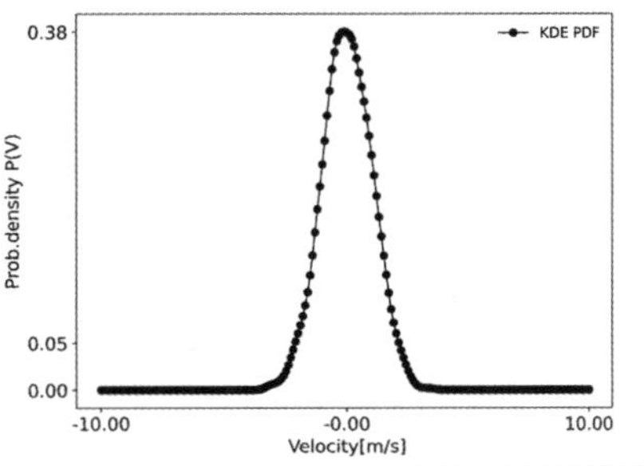

Figure 2.5.2. Output of the Maxwell-Boltzmann distribution of velocities

The Figure 2.5.2 shown above depicts a probability density function with a line and modifies the font size of the plot axis. The plot's axis limitations are specified by the 'axis' command. The output of the Maxwell-Boltzmann distribution of velocities is depicted in the ensuing image.

Distribution and Brownian Motion

We have now discussed kinetic theory to the point that you should understand the force of the particles against pressure, their average kinetic energy (i.e., temperature), their various excitations, and the likely velocity of a particle chosen at random (by probability distribution). This leads to a number of interesting behaviors, such as the Brownian motion of larger particles suspended in liquids or gases (Mörters & Peres, 2010).

Concept: The discovery of Brownian motion is attributed to the botanist Robert Brown. In 1827, Brown noticed the irregular motion of pollen particles suspended in water and was able to rule out the motion being due to the pollen being alive by repeating the experiment using suspensions of dust in water. This is due to the kinetic motion of particles. Imagine a large beach ball 20 m in diameter. This ball is so big that many people in a crowd can hit it all at once (Figure 2.5.3).

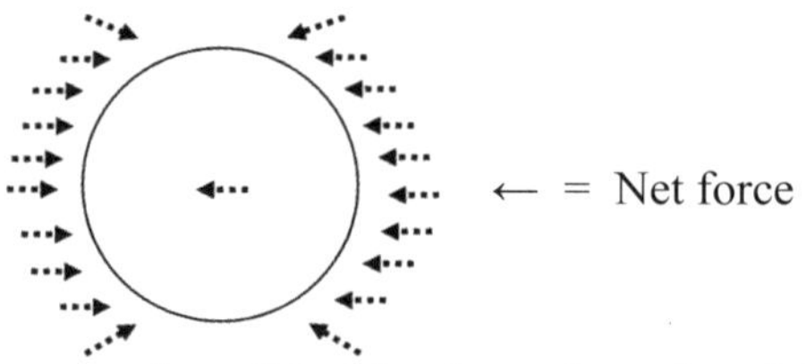

Figure 2.5.3. Schematic representation of the kinetic motion of particles

Application: Consider the force on it, when 20 people might be pushing it to the left, and 21 people might be pushing it to the right. The forces to the left and right are almost balanced, but there is a small net force to the right, so the ball will make a small movement right. Next time it might go another direction. The motion will be even more random now: One person might want to send it a certain direction, but on average this will be cancelled out by all of the other directions people are pushing it in, so at each point the net force on the ball will be random, and the strength and direction of the net force will depend entirely on the balance of all of the little pushes it receives. Brownian motion is essential when a lot of small elements and structures are immersed in place (i.e., water). It is even used by many natural (i.e., biological) entities to move things around.

2. Information Theory

Building on the principles of statistical mechanics and the behavior of particle systems, information theory emerged as a groundbreaking field in the 20th century. Claude Shannon developed information theory, which has grown into a robust branch of mathematics that has influenced various scientific fields, including statistics, biology, behavioral science, neuroscience, and statistical mechanics (Shannon, 1948).

The techniques used in information theory are probabilistic, and some view information theory as a branch of probability theory.

- A system with M discrete states $k = 1, \ldots, M$
- Occupation probabilities $p_1, p_2, \ldots$
- Entropy: reinterpretation as Shannon information measure
- Information gains due to observation

+ **Entropy:** Entropy is a metric for how chaotic or unpredictable a system is. For instance, the entropy of the system is initially low in a box holding two initially distinct particles (i.e., gases). But when the particles mix together over time, the system's entropy rises (Figure 2.5.4).

Figure 2.5.4. Representation of low and high entropy

The Second Law of Thermodynamics states that in the absence of an external force acting on the system, neither the entropy nor the degree of disorder can ever decrease in an isolated system. Consider, for example, a coin toss. You toss the coin four times and the events come up [tails, heads, heads, tails]. If you (or a machine-learning algorithm) were to predict the next coin flip, you would be unable to predict the outcome with any certainty—the system contains high entropy. However, a weighted coin with events [tails, tails, tails, tails] has very low entropy, and given the current information we can almost definitively say that the next outcome will be tails.

Entropy about One Possible Case

Let us take a simple example. We have a fair coin and flip it once. There are two possible outcomes (heads or tails), so we need at least one bit (a computer bit takes the values 0 or 1) to represent all possible outcomes. The probability of getting heads is 1/2. When we think about what happens if we flip the coin twice, there are four possible outcomes, and we need at least two bits 00, 01, 10, 11) to represent them. The probability of both flips being heads is 1/4. Next, we ask the key question: How much information does it take to represent the event of getting one head in a row ($p = 1/2$)? If we let bits be our unit of information, it takes one bit of information, as we just discussed. What about two heads in a row ($p = 1/4$)? Two bits. Extending the pattern, three heads in a row (p = 1/8) would take three bits, and n heads in a row [$p = (1/2)^n$] would take n bits. From this pattern we can see a relationship:

$$I(number\ of\ bits) = \log\_2(1/p)$$

This is the most widely used equation for information, and it is called *Shannon's information* in honor of Claude Shannon, who discovered it. Now we have come up with a definition for information. How surprising an event is or, more mathematically, the number of bits needed to represent the event. We might also be interested in the average information (or average number of bits) over all possible events. This is called the *entropy*. Going back to our coin example, let us consider the two-flip case. There are four possibilities, all equally likely. Thus, the entropy is just the information about one possible case, which has a probability of 25%: log_2(4) = 2.

+ **Entropy = Information Gain:** It is possible to think about entropy as a way to gauge how unpredictable or unknowable a system is. Entropy pertains to how much uncertainty there is in a user's activities in the context of Fitts' Law (1954).

- Large lack of knowledge → high entropy
- Complete knowledge → entropy = 0

The system's entropy is high when there is a significant knowledge gap, and the opposite is also true. The entropy of the system is zero when there is total knowledge. Entropy is a fundamental idea that may be applied to represent the degree of uncertainty or predictability in a number of systems.

$$Information\ (gain) = -\{p_1 \log p_1 + p_2 \log p_2 + \ldots + p_M \log p_M\} = -\sum_{j=1,\ldots,M} p_j \log p_j$$

Why is S a measure of information gain? We know the probabilities P_k. We do not know in which state the system is. We observe the state and in doing so gain information (S = quantitative measure of information gain). High entropy means low information gain, and low entropy means high information gain. Information gain can be thought of as the purity in a system or the amount of clean knowledge available in the system (Figure 2.5.5).

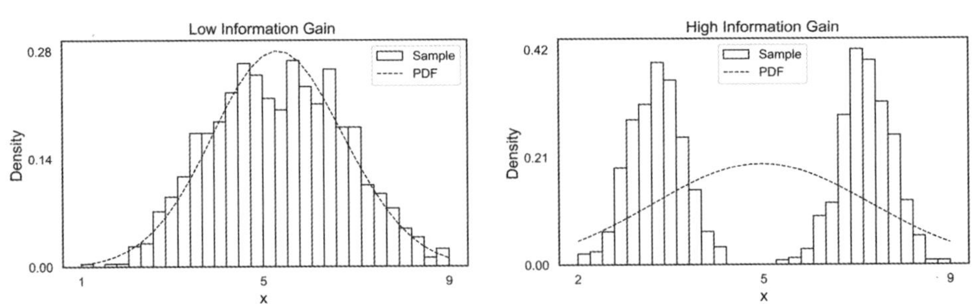

Figure 2.5.5. Schematic comparison of low and high information gain

Although entropy is not always the best loss function—especially when the target function p is not clearly defined—it is generally performance enhancing, which explains its presence everywhere.

$$Information\ (gain) = -\int_{\Omega} P(x) \log P(x)\, dx$$

When entropy is used in data analysis, its core components—uncertainty and probability—are best represented through information gain. Entropy is explicit about dealing with the unknown, which is much desired in model building. When models are optimized in terms of entropy, they are able to wander through plains of unpredictability with an elevated sense of knowledge and purpose.

Let us consider three possible states, for example, the final course grades A, B, and C of students who will participate in an experiment that you conduct. You know the distribution of grades. You do not know which student has which grade. One student enters your lab. You ask, 'What was your grade?' The student says, 'You figure it out.'

All students have As: no information gain {$S = 0$ (PA = 1, PB = 0, PC = 0)}.

[In]

```
from scipy.stats import entropy
entropy([3/3, 0, 0])

Output: 0.0
```

→

[Out]

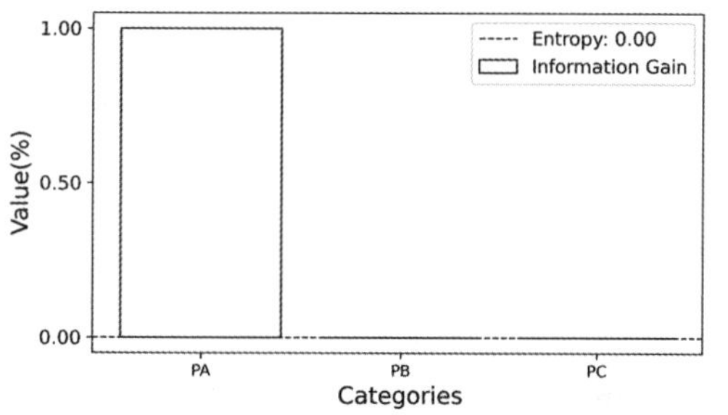

Figure 2.5.6. Simulation of information gain 1

The code calculates the entropy of a discrete random variable based on its probability function (Figure 2.5.6). It is called with a single argument, a list [3/3, 0, 0] which means that the random variable has only one possible value, and the value is 1 (3/3).

Half the students have As and half have Bs: you gain information about whether the student is an A or B student {Figure 2.5.7; medium information gain $S = 0.70$ (PA = 0.5, PB = 0.5, PC = 0)}.

[In]

```
from scipy.stats import entropy
entropy([1.5/3, 1.5/3, 0/3])

Output: 0.6931471805599453
```

→

[Out]

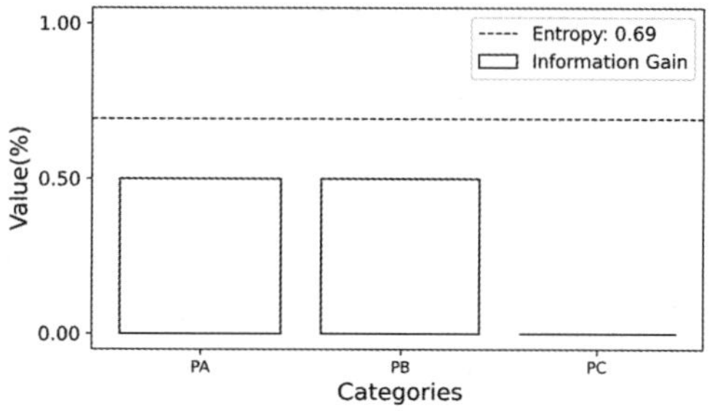

Figure 2.5.7. Simulation of information gain 2

As, Bs, and Cs are uniformly distributed: you gain information about whether the student is an A, B, or C student {Figure 2.5.8; large information gain $S = 1.1$ (PA = 0.33, PB = 0.33, PC = 0.33)}.

```
[In]
from scipy.stats import entropy
entropy([1/3, 1/3, 1/3])

Output: 1.0
```

[Out]

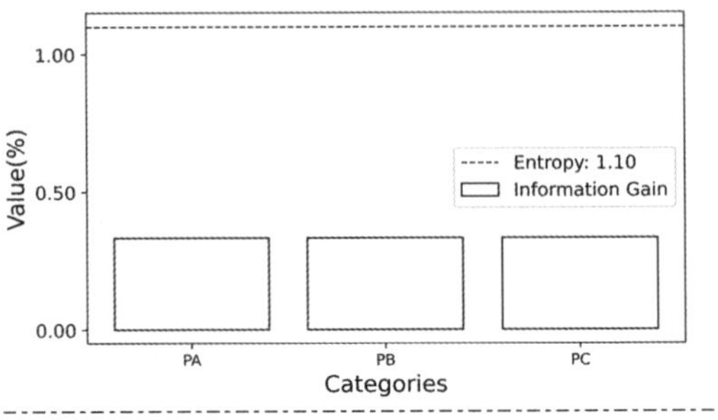

Figure 2.5.8. Simulation of information gain 3

+ **Information Gain and Probability:** Information gain was developed by Claude Shannon (1948) while working at the U.S. telephone company Bell Labs. A foundational concept of information gain is the quantification of the amount of information in things like events, random variables, and distributions. Quantifying the amount of information requires the use of probabilities, hence the relationship between information theory and probability.

We can calculate the amount of information in an event using the probability of the event, calculating it for a discrete event x as follows:

$$Information(x) = -\log(p(x))$$

where log() is the base 2 logarithm and $p(x)$ is the probability of the event x. The calculation of information is often written as $h(x)$, for example

$$h(x) = -\log(p(x))$$

The negative sign ensures that the result is always positive or zero. Information will be zero when the probability of an event is 1.0 or the event is certain (i.e., there is no surprise).

Let us explore the information in a single roll of a fair six-sided die (e.g., the information in rolling a 6). We know the probability of rolling any number is 1/6, which is a smaller number than 1/2 for a coin flip; therefore, we would expect more surprise or a larger amount of information.

Import the library for calculation and enter the probability of the die event.

```
[In]
from math import log2
p = 1.0 / 6.0
```

Then simply calculate the information for the event and see the results.

[In]
```
h = -log2(p)
print('p(x)=%.3f, information: %.3f bits' % (p, h))
```
→

[Out]
```
p(x)=0.167, information: 2.585 bits
```

To take this further, we can calculate information for probabilities between 0 and 1 and plot it all. We can then create a plot of probability versus information. We would expect the plot to curve downward from low probabilities with high information to high probabilities with low information. The complete example is shown below.

First let us import the libraries and enter the list of probability values.

[In]
```
from math import log2
from matplotlib import pyplot

probs = [0.1, 0.2, 0.3, 0.4, 0.5, 0.6, 0.7, 0.8, 0.9, 1.0]  # list of probabilities
```

Then we calculate the information and plot the results.

[In]
```
info = [-log2(p) for p in probs]

# plot probability vs information
pyplot.plot(probs, info,
marker='o', markersize='10')
pyplot.show()
```
→

[Out]

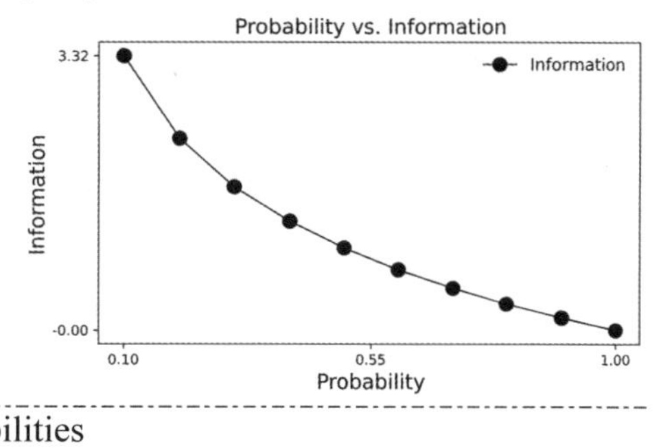

Figure 2.5.9. Output of information gain and probabilities

The code creates a new list by iterating over the elements in the 'probs' list and applying the function '−log2(p)' to each element (Figure 2.5.9). As we can see, the result shows a list of value that are negations of the base-2 logarithms of the elements in the '' list.

+ **Estimating Entropy with Random:** The system in life sciences that maximizes information gain while satisfying certain constraints. The benefit of an information theory approach is that it does not need to meet the condition of thermodynamic equilibrium.

- Discrete state spaces: $j = 1, \ldots, M$

$$Entropy = -\{p_1 \log p_1 + p_2 \log p_2 + \ldots + p_M \log p_M\} = -\sum_{j=1}^{M} p_j \log p_j$$

- Continuous state spaces

$$Entropy = -\int_{\Omega} P(x)\log P(x)\,dx$$

To estimate entropy from a sample of scores using coding, follow these steps: First, define the bins (e.g., [−20, −19], [−19, −18], ..., [18, 19], [19, 20]). Second, create a frequency distribution in programing to obtain the frequencies. Then, calculate the estimated probabilities by dividing the frequency by the sample size N to obtain the relative frequencies. For this simulation, let us parameterize with initial conditions.

```
[In]
import numpy as np
deltat = 0.06
sigma = np.sqrt(deltat)
N = 100
x = np.zeros((N, 1000))
timepoints = np.arange(0, deltat*1000, deltat)
```

Here, the variable 'deltat' is defined and set to 0.06, the variable 'sigma' is defined as the square root of 'deltat', the variable N is defined and set to 100, the variable x is defined as array with dimensions N by 1000, and the variable 'timepoints' is defined as a list of equally spaced values ranging from 0 to 'deltat ∗ (1000)' with a step size of 'deltat'. Then we calculate by loop over realizations.

```
[In]
from scipy.stats import norm
for k in range(N):
    x[k, 0] = 0
    for n in range(999):
        # Python indexing starts at 0, so 1000-1 becomes 999
        e = norm.rvs(loc=0, scale=sigma)
        xnew = x[k, n] + e
        if xnew > 20:
            diff = xnew - 20
            xnew = 20 - diff
        if xnew < -20:
            diff = xnew + 20
            xnew = -20 - diff
        x[k, n+1] = xnew
```

The code generates a sequence of random numbers with a normal distribution, with a mean of 0 and a standard deviation of 'sigma' stored in the array x which has N rows and 1000 columns.

For each value of k from 1 to N, the code initializes the first element in the k-th row of x to 0, then generates 1000 random numbers using the 'norm' function which is random numbers with a normal distribution.

Now we display the output (Figure 2.5.10).

[In]

```
import matplotlib.pyplot as plt
fig, ax = plt.subplots()
ax.tick_params(axis='both',
which='major', labelsize=20)

for k in range(N):
    ax.plot(timepoints, x[k, :])
plt.show()
```

→

[Out]

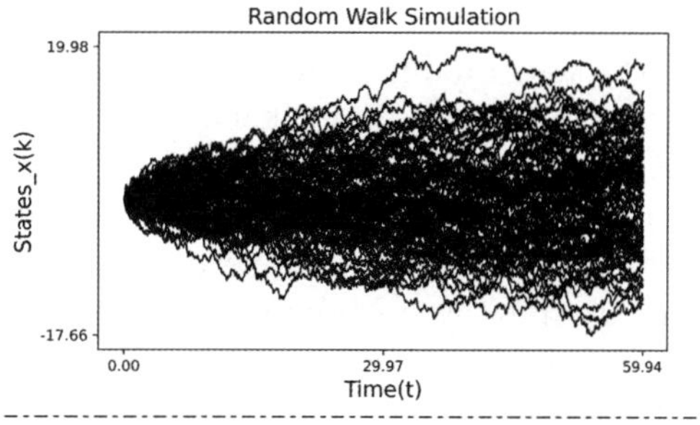

Figure 2.5.10. Representation of entropy output

We also display the frequency distribution, probability distribution, and entropy.

[In]

```
np.random.seed(0)  # For reproducibility

for k in range(N):
    x[k, 0] = 0
    for n in range(999):
        e = np.random.normal(0, sigma)
        xnew = x[k, n] + e
        if xnew > 20:
            diff = xnew - 20
            xnew = 20 - diff
        elif xnew < -20:
            diff = xnew + 20
            xnew = -20 - diff
        x[k, n+1] = xnew
```

We generate a 2D array x in which each row represents a random walk limited between −20 and 20, with each step influenced by a Gaussian noise. Then, we calculate and stores the entropy at each time step, utilizing bins between −20 and 20; total of 1000 time steps.

[In]

```
bins = np.arange(-20, 21, 1)
entropy_values = np.zeros(1000)
for n in range(1000):  # Loop over time
    freq, _ = np.histogram(x[:, n], bins)
    prob = freq / N
    entropy_values[n] = entropy(prob, base=np.e)
```

The variable bins is a numpy array containing the values from −20 to 20, inclusive, with a step size of 1. The numpy function histogram is used to count the frequency of each value in bins appearing in the nth column of x. This produces a frequency count, freq, which is then divided by N to get the probabilities of each value in bins. These probabilities are stored in the variable prob.

The code then computes the entropy of prob using the entropy function from the scipy.stats module, which is added to the nth element of the entropy values array. This process is repeated for each time step (i.e., for each column of x), resulting in a time series of entropy values.

The final part of the code creates a plot of the entropy values over time, with time represented by the timepoints variable. The x-axis and y-axis of the plot are simplified to show only the minimum, mean, and maximum values, and the line in the plot represents the entropy values. The function is used to ensure that the x and y tick labels are formatted to two decimal places.

Finally, we can see the output of entropy production (Figure 2.5.11).

```
[In]
fig, ax = plt.subplots()

ax.plot(timepoints,     entropy_values,
color='k', linewidth=1)
```

→

[Out]

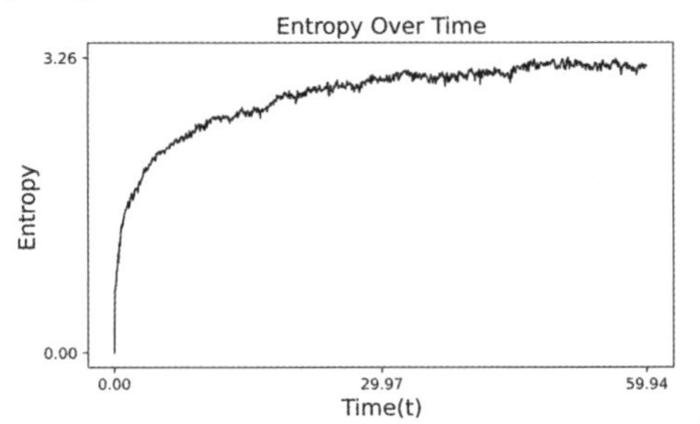

Figure 2.5.11. Representation of entropy production

This is a measure of the amount of uncertainty in a system. In general, data result calculated by the above procedure with higher entropy is to be more random and more difficult to predict, while data result with lower entropy is to be more structured and more predictable.

3. Maximum Likelihood Estimation

Now, with a grasp on entropy and its importance in characterizing the randomness or predictability in a system, we turn our attention to another significant statistical tool: Maximum Likelihood Estimation.

Introduced by R. A. Fisher, an eminent English mathematical statistician, in 1912, this technique allows us to determine the optimal parameters for a given model (Myung, 2003). Not only does it provide a simple method for estimating an unknown parameter, but it also offers broad applicability and an intuitive appeal. As such, maximum likelihood estimation often furnishes a reasonable estimator for the unknown parameter, playing a critical role in diverse system modeling tasks.

+ **Define Processes:** The parameter values are such that they maximize the likelihood that the product review process described by the model produces the rating suggested by the table. Let us build a model and estimate the parameters such that they maximize the likelihood of obtaining the values observed in the reviews.

[Table 2.5.2] Artificially Created Data (Not Real Values)

Item A			Item B		
Rating	Frequency	Observed ratings	Rating	Frequency	Observed ratings
1	10	6%	1	57	14%
2	6	4%	2	33	8%
3	10	6%	3	29	7%
4	27	17%	4	45	11%
5	109	67%	5	246	60%

In Table 2.5.2, we apply the estimation technique by comparing two products using hypothetical observed ratings derived from public opinions about the products. In particular, we are looking at the number of times a product has received a given rating (between 1 and 5) reflecting how people feel about it. We are interested in estimating π_1, or the probability that people rate the product with i number of levels. Because each new rating has a value between 1 and 5, it follows a categorical distribution π.

$$\pi = [\pi_1, \dots, \pi_5]$$

For a given product, we observe a vector of ratings $\vec{v} = [\vec{v}_1, \vec{v}_1, \vec{v}_1, \vec{v}_1, \vec{v}_1]$. Each rating $\vec{v}_1$ is the total number of i-level reviews received by that product. Each i-level review is made with probability π_1. Because each review $\vec{v}$ is independent of the others, we can model the probability by a categorical distribution $h(\pi)$:

$$h(\vec{v} = i|\pi) = \pi_i, where \sum_{n=1}^{5} \pi_i = 1$$

The likelihood of observing a vector of occurrences $\vec{v}$ out of n independent reviews is given by the multinomial distribution parametrized by π:

$$\hbar(\vec{v}|\pi, n) = \hbar(\vec{v}_1, \dots, \vec{v}_5|\pi_1, \dots, \pi_5, n) = \frac{ni}{\vec{v}_1 i, \dots, \vec{v}_5 i} \prod_{i=1}^{5} \pi_i^{\vec{v}_i}$$

We can calculate its log-likelihood as follows:

$$l(\pi_1, \dots, \pi_5) = \log n\, i - \sum_{i=1}^{5} \log(\vec{v}_i i) + \sum_{i=1}^{5} \vec{v}_i * \log(\pi_i i)$$

+ **Maximum Likelihood Estimator:** This allows us to determine the maximum likelihood estimator of π for each of the two products. For this purpose, we are looking for the π that maximizes the likelihood l. This is typically done by differentiating the likelihood function with respect to π using the so-called Lagrange multipliers.

$$l(\pi_1, \ldots, \pi_5) = l(\pi_1, \ldots, \pi_5) + \lambda(1 - \sum_{i=1}^{5} \pi_i)$$

By constraining all the derivatives to 0, we get the most natural estimate:

$$\hat{\pi}_i = \vec{v}_i / n$$

The maximum likelihood estimators are listed below.

```
[In]
Produced_A = [10,6,10,27,109]
Produced_B = [57,33,29,45,246]
```

Calculate the maximum likelihood estimators and print them out.

```
[In]
theta_A = Produced_A / np.sum(R_A)
theta_B = Produced_B / np.sum(R_B)
print("MLE for Product A: ")
print("MLE for Product B: ")
```

→

```
[Out]
MLE for Product A:  [0.062 0.037
0.062 0.167 0.673]

MLE for Product B:  [0.139 0.08
0.071 0.11  0.6  ]
```

Based on the maximum likelihood estimators of $\boldsymbol{\pi}$ for both models, we can compute the maximum likelihood.

```
[In]
import math
import numpy as np
def f(n):
  return math.factorial(n)
def mle(v, theta):
  return (f(np.sum(v)) / (np.prod([f(r) for r in v]))) * \ np.prod([theta[i]**v[i]
for i in range(len(v))])
```

The function takes a single argument n and returns the factorial of n which is the product of all positive integers less than or equal to n. The function takes two argument of v and 'theta'. The 'mle' is computed by first calculating the numerator of the likelihood function which is the factorial of the sum of the elements in the data vector v, divided by the product of the factorial's elements in v. The denominator is then computed as the product of 'theta' raised to the power of the corresponding element in v for each element in v.

The maximum likelihood

```
[In]
ml_A = mle(Produced_A,theta_A)
ml_B = mle(Produced_B,theta_B)
print("ML for Product A
{:.7f}".format(ml_A))
print("ML for Product B
{:.7f}".format(ml_B))
```

→

```
[Out]
ML for Product A 0.0002345
ML for Product B 0.0000207
```

We have some evidence that product *B* seems to be preferred to product *A* by people who gave reviews. We have just seen how to use maximum likelihood estimation in the simple context of a product review.

+ **Maximum Likelihood with Regression:** Let us now discuss how to use the maximum likelihood estimation with a regression on normally distributed data. Here we perform simple linear regression on synthetic data. We ensure that the data are normally distributed by incorporating some random Gaussian noises. Data can be said to be normally distributed if their residual follows the normal distribution (in programming). Import the necessary libraries.

```
[In]
import numpy as np
import pandas as pd
import seaborn as sns
from statsmodels import api
from scipy import stats
from scipy.optimize import minimize
```

Create an independent variable and a normally distributed residual.

```
[In]
x = np.linspace(-10, 30, 100)
e = np.random.normal(10, 5, 100)
```

Next, generate ground truth with a data distribution.

```
[In]
y = 10 + 4*x + e
df =
pd.DataFrame({'x':x
, 'y':y})
df.head()
sns.regplot(x='x',
y='y', data = df)
plt.show()
```

→

[Out]

	x	y
0	-10.000000	-17.220807
1	-9.595960	-17.773363
2	-9.191919	-23.880360
3	-8.787879	-15.207193
4	-8.383838	-12.147515

→

[Out]

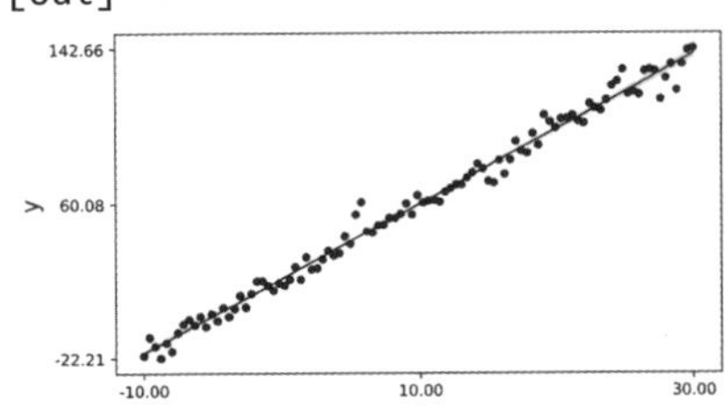

Figure 2.5.12. Generated distribution (right) by the ground truth (left)

A DataFrame is constructed using these variables, and a regression plot is generated to visualize the relationship between x and y. Figure 2.5.12 illustrates the generated distribution by the ground truth, showcasing the linear relationship between the two variables. Hence, we can use maximum likelihood estimation on the synthetic data to find the model parameters. Solution by maximum likelihood estimation.

```
[In]
def MLE_Norm(parameters):
  const, beta, std_dev = parameters
  pred = const + beta*x
  LL = np.sum(stats.norm.logpdf(y, pred, std_dev))
  neg_LL = -1*LL
  return neg_LL
mle_model = minimize(MLE_Norm, np.array([2,2,2]),
method='L-BFGS-B')
mle_model
```

→

```
[Out]
Intercept: 19.999
Regression coefficient
(Beta): 3.957
Standard deviation: 4.645
```

The maximum likelihood approach here is a general template for any kind of problem. With expertise in maximum likelihood estimation, users can formulate and solve their own problems with raw data in hand. In this section, we discussed the concept behind maximum likelihood estimation and how it can be applied to any kind of machine-learning problem with structural data. We went through a hands-on programing implementation to solve a linear regression problem with normally distributed data. You can practice more by solving your data problems with maximum likelihood estimation.

Remarks: In Part 4 of our Probability Theory series, we not only ventured into the realm of the exponential probability density function, which is often used to describe intervals between occurrences in a process, but we also delved into key concepts in information theory such as entropy, mutual information, and divergence. These concepts provide a solid framework for quantifying the information contained within a message. We also discussed the application of Maximum Likelihood Estimation, a statistical method introduced by R. A. Fisher, which allows us to estimate the optimal parameters of a model. These concepts collectively form the backbone of effective information processing and transmission in communication systems. Understanding these principles enhances our ability to navigate and make use of these powerful systematical tools in various real-world scenarios.

Supplemental Information 2.1: Theoretical Functions of Various Distributions

We provide a comprehensive review of commonly used probability distributions in time series analysis in this resource, explaining their key concepts in a clear and concise manner. Our aim is to enhance your understanding of these essential tools by presenting these standard distributions in an accessible and straightforward way. Whether you are a novice or an experienced practitioner, this resource will serve as a valuable reference for your needs.

+ **The Ultimate (Probability) Distribution:** The most typical probability distribution for continuous variables is the normal distribution, commonly referred to as the Gaussian distribution or bell curve. It is defined by the variance and the mean, which also corresponds to the median and mode. The central limit theorem, which asserts that the sum of n random variables (independent of their mean, variance, and distribution) approximates a normal distribution as n rises, is what makes the normal distribution relevant. The mean, which represents the typical value of the random variable, serves as a gauge of the distribution's central tendency. The variance quantifies the spread of the distribution and reveals how widely distributed the random variable is from the mean.

This flexible and important probability distribution is crucial in many different domains.

Parameters: μ = mean, σ = standard deviation

```
[In]
# Define Gaussian distribution parameters
mu = 0
sigma = 1

# Create data points for the distribution
x = np.linspace(mu - 3 * sigma, mu + 3 * sigma,
100)
y = (1 / (sigma * np.sqrt(2 * np.pi))) *
np.exp(-0.5 * ((x - mu) / sigma) ** 2)
```

→ [Out]

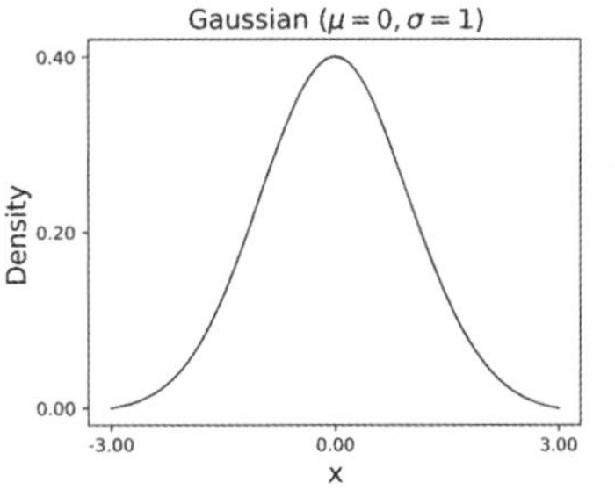

Figure S2.1.1. Representation of the Gaussian distribution

+ **Binomial Distribution:** The discrete likelihood of receiving a certain number of successes (n) in a certain number of independent trials (N) in a binary experiment is modelled using the binomial distribution, a fundamental probability distribution used in statistics. A binary experiment consists of N Bernoulli trials, each having a probability of success of p and a probability of failure of $q = p$, which is the outcome for failure in the experiment. Two essential factors, size (N) and success probability, determine the binomial distribution (p). The number of independent trials in the experiment is denoted by the size, and each Bernoulli trial's chance of success is denoted by the probability of success (p). Many disciplines, such as biology, economics, and engineering, utilize the binomial distribution to model and evaluate.

Parameters: $size$ = number of independent trials, $prob$ = probability of success

[In]

```
# Define Binomial distribution parameters
size = 10
prob = 0.5

# Create data points for the distribution
x = np.arange(0, size + 1)
y = binom.pmf(x, size, prob)
```

→

[Out]

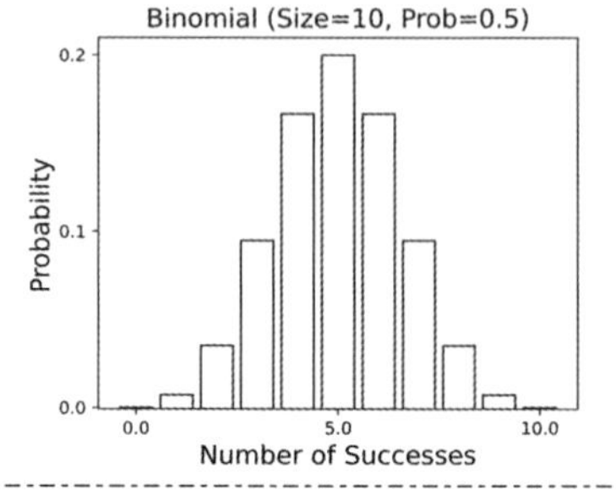

Figure S2.1.2. Representation of the Binomial distribution

+ **Negative Binomial Distribution:** The number of Bernoulli trials required to attain a specific number of r successes is modeled by the negative binomial distribution, which is a probability distribution. The size and the likelihood of success (p) in each trial serve as the two factors that define this distribution. The dispersion parameter, the gamma mixing distribution's shape parameter used to calculate the degree of variance in the distribution, may also be used to build the negative binomial distribution, which is an extension of the geometric distribution. The negative binomial distribution becomes the geometric distribution with $r = 1$. Numerous disciplines, including economics, biology, and physics, employ the negative binomial distribution to model and examine a wide range of events.

Parameters: $size$ = a target for numbers of successful trials or dispersion parameter (the shape parameter of the gamma mixing distribution), $prob$ = the probability of success in each trial

[In]

```
# Define Negative Binomial distribution
parameters
size = 5
prob = 0.5

# Create data points for the distribution
x = np.arange(0, 21)
y = nbinom.pmf(x, size, prob)
```

→

[Out]

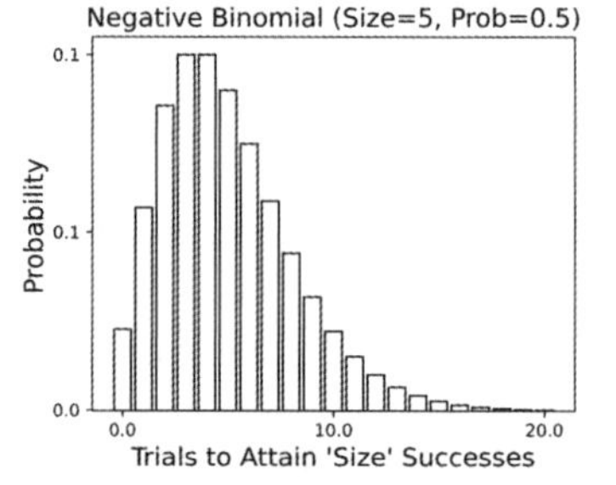

Figure S2.1.3. Representation of the Negative Binomial distribution

+ **Hypergeometric Distribution:** The hypergeometric distribution is a probability distribution which represents the number of elements of class A in a sample of k items taken without replacement from a finite population of $m + n$ elements, where m elements are classed as successes and n elements are not.

The hypergeometric distribution is used when the population size is finite, which is a typical situation in practical applications. It is a counterpart to the binomial distribution, which is used when the population size is big enough to make sampling with replacement a workable approximation. The three parameters that define this distribution are: m, the proportion of the population's items that are classed as successes; n, the proportion of the population's items that are not classified as successes; and k, the number of items in the sample that are classified as success. This distribution gives the probability of obtaining a specific number of successes in the sample.

Parameters: m = number of items in the population that are classified as successes, n = number of items in the population that are not classified as successes, k = number of items in the sample that are classified as successes

[In]

```
# Define Hypergeometric distribution parameters
m = 15
n = 10
k = 8

# Create data points for the distribution
N = m + n
x = np.arange(0, k + 1)
y = hypergeom.pmf(x, N, m, k)
```

→ [Out]

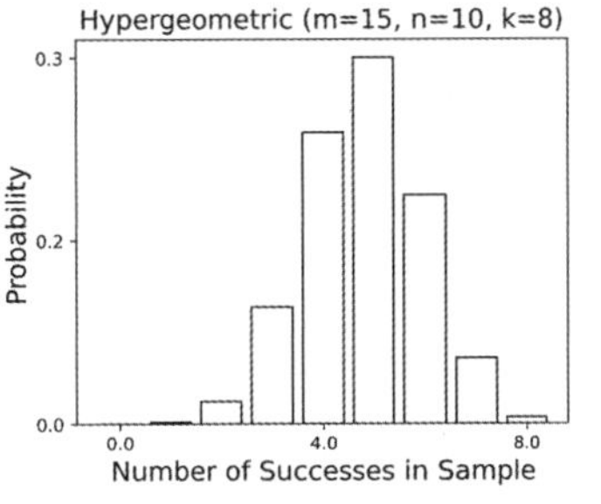

Figure S2.1.4. Representation of the Hypergeometric distribution

+ **Lognormal Distribution:** A probability distribution called the lognormal distribution occurs when a variable's logarithm follows a normal distribution. It is frequently used to simulate skewed distributions with low means, high variances, and only positive values. This distribution is also known as the Galton distribution. The two parameters log, which is the mean of the logarithm of the variable, and log, which is the standard deviation of the logarithm of the variable, define the lognormal distribution.

The lognormal distribution is frequently used to describe and study a wide range of phenomena, including stock prices, particle size distributions, and response times. It is also used extensively in physics, engineering, and finance.

Parameters: μ_log = mean, σ_log = standard deviation

[In]

```
# Define Lognormal distribution parameters
mu_log = 0
sigma_log = 1

# Create data points for the distribution
x = np.linspace(0.001, 10, 1000)
y = lognorm.pdf(x, sigma_log,
scale=np.exp(mu_log))
```

→ [Out]

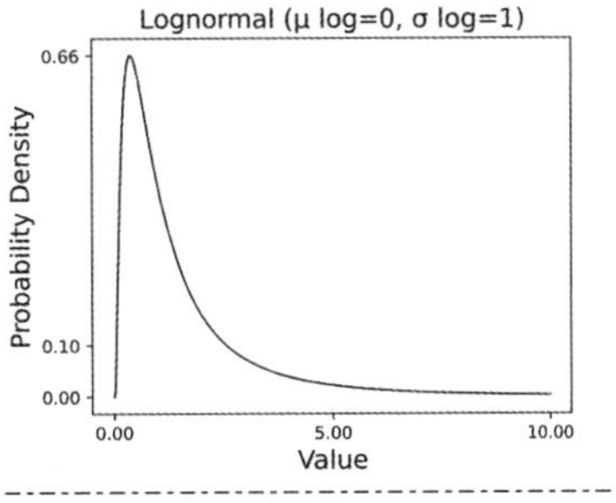

Figure S2.1.5. Representation of the Lognormal distribution

+ **Uniform Distribution:** The uniform distribution, commonly known as the rectangle distribution, is a continuous probability distribution with a constant probability for the range $[a, b]$. It has uses in many areas, including computer simulations, game creation, and statistical modeling, and is mostly used to produce random numbers.

Two parameters, a and b, which stand for the bottom and upper bounds of the interval, respectively, define the uniform distribution. It is an easy distribution to deal with because of its simple cumulative distribution function and probability density function. The uniform distribution may be used to generate random numbers inside a particular range since every value between $[a, b]$ has an equal probability of being chosen.

Parameters: min = lower limit, max = upper limit

[In]

```
# Define the Uniform Distribution parameters
min_value = 0
max_value = 10

# Create data points for the Distribution
x = np.linspace(min_value-1, max_value+1, 1000)
y = uniform.pdf(x, loc=min_value,
scale=max_value-min_value)
```

→ [Out]

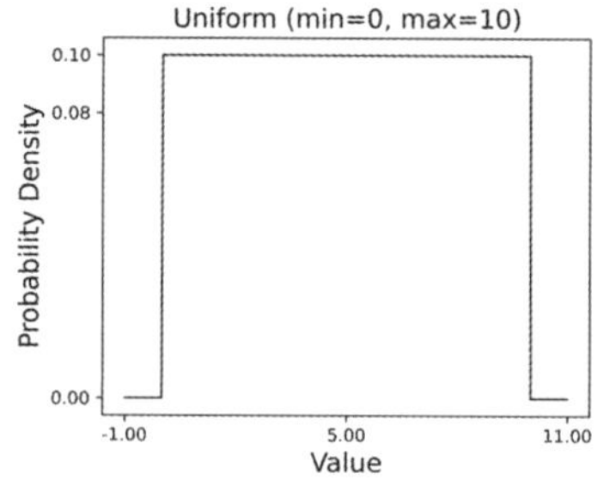

Figure S2.1.6. Representation of the Uniform distribution

+ **Beta Distribution:** The beta distribution is a continuous probability distribution that is commonly used in Bayesian inference as it represents the a priori distribution of other distributions. It is defined for values between 0 and 1, and its shape is determined by two positive parameters called shape1 and shape2. The beta distribution is flexible and can model a wide range of distributions with different shapes, such as U-shaped, bell-shaped, or J-shaped distributions. It is widely used in many fields, including biology, engineering, and finance, to model and analyze various phenomena.

Parameters: $shape1$: positive shape 1 parameter, $shapce2$: positive shape 2 parameter

[In]

```
# Define the Beta distribution parameters
shape1 = 2
shape2 = 5

# Create data points for the distribution
x = np.linspace(0, 1, 1000)
y = beta.pdf(x, shape1, shape2)
```

→ [Out]

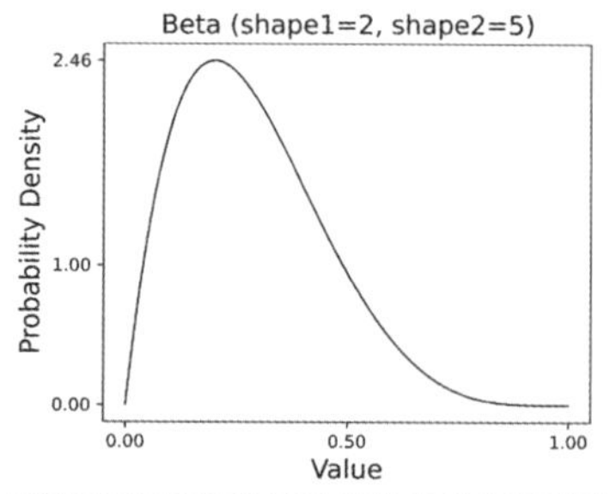

Figure S2.1.7. Representation of the Beta distribution

+ Student's *t* Distribution: The Student's t distribution is a probability distribution that is frequently employed in statistical inference to determine quantiles for a specific degree of confidence or significance. When the sample size is small—generally less than 30 or 40—it is a sampling distribution and is used in place of the normal distribution in such circumstances. The t distribution's degrees of freedom parameter, which is equal to the sample size minus 1, is what gives the distribution its characteristics. There are several statistical techniques that employ the t distribution, including regression analysis, confidence interval calculation, and hypothesis testing, confidence interval estimation, and regression, among other methods.

Parameters: df = degrees of freedom (sample size minus 1)

[In]

```
# Define the Student's t-distribution parameters
df = 5

# Create data points for the distribution
x = np.linspace(-10, 10, 1000)
y = t.pdf(x, df)
```

→ [Out]

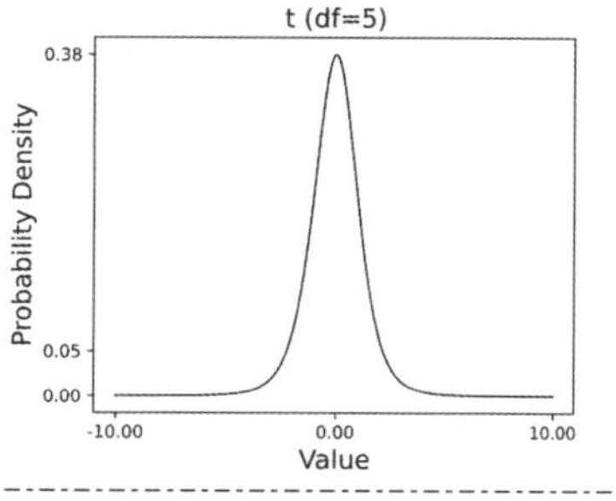

Figure S2.1.8. Representation of the Student's t distribution

+ Chi-Square Distribution: The chi-square distribution is a probability distribution that is used as a sampling distribution to determine quantiles at a specific degree of confidence or significance. It is a particular instance of the gamma distribution, where a chi-square distribution with n degrees of freedom is identical to a gamma distribution with $n/2$ for the shape parameter and 0.5 (or 2) for the scale parameter. The sample size minus one is the degrees of freedom parameter, which is a crucial component of the chi-square distribution. Applications in statistics include tests for homogeneity in multinomial populations, independence in contingency tables, and goodness-of-fit tests. Additionally, while constructing confidence intervals, the chi-square distribution is utilized.

Parameters: df = degrees of freedom (sample size minus 1)

[In]

```
# Define the Chi-square distribution parameters
df = 5

# Create data points for the distribution
x = np.linspace(0, 20, 1000)
y = chi2.pdf(x, df)
```

→ [Out]

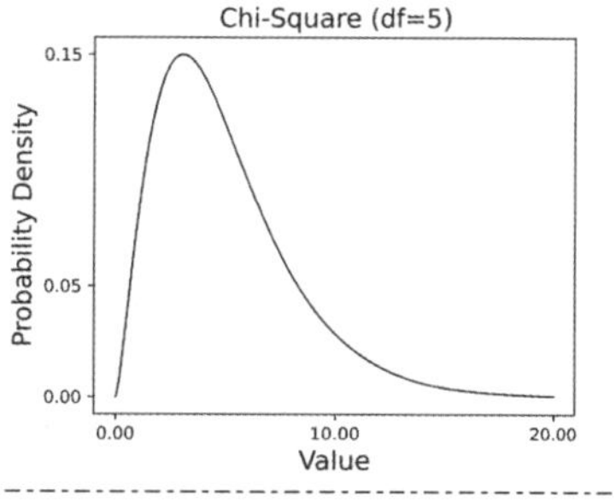

Figure S2.1.9. Representation of the Chi-Square distribution

+ ***F* Distribution:** The *F* distribution is a probability distribution that is frequently used in statistical analysis to check if the variances of two samples are identical. It is connected to the *F* statistic, which measures how much the variances of the two samples differ from one another. The degrees of freedom connected to the two samples are represented by the two parameters of the F distribution, $df\ 1$ and $df\ 2$; first sample are represented by $df\ 1$ (sample size 1 minus 1), while the degrees of freedom for the second sample are represented by $df\ 2$ (sample size 2 minus 1). The *F* distribution is frequently used in analysis of variance (ANOVA) to determine if the sample variances are equal, as well as in regression analysis to test the significance of the overall regression model.

Parameters: $df\ 1$= degrees of freedom from the first sample (sample size 1 minus 1), $df\ 2$ = degrees of freedom from the second sample (sample size 2 minus 1)

```
[In]
# Define the F distribution parameters
df1 = 10
df2 = 20

# Create data points for the distribution
x = np.linspace(0, 5, 1000)
y = f.pdf(x, df1, df2)
```

[Out]

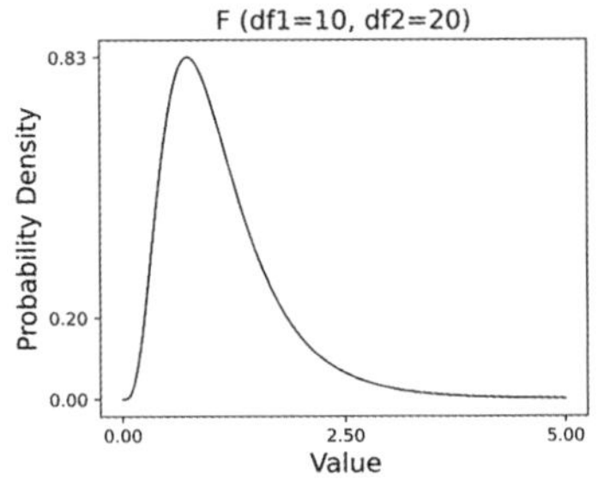

Figure S2.1.10. Representation of the F-distribution

+ **Cauchy Distribution:** The Cauchy distribution is a continuous probability distribution that resembles the normal distribution in general but differs in that it has a higher peak and fatter, slower-decaying tails. Another name for it is the Lorentz distribution. The *scale* parameter represents half the breadth of the probability density function at half the maximum height, while the *location* parameter specifies the distribution's apex. The Cauchy distribution is significant in physics and engineering because it commonly appears as the Cauchy-Lorentz distribution in scattering and dispersion phenomena and is the solution to a number of linear differential equations.

Parameters: *location* = location parameter (indicates where the peak is), *scale* = half the width of the probability density function at half the maximum height

```
[In]
# Define the Cauchy distribution parameters
location = 0
scale = 1

# Create data points for the distribution
x = np.linspace(-10, 10, 1000)
y = cauchy.pdf(x, location, scale)
```

[Out]

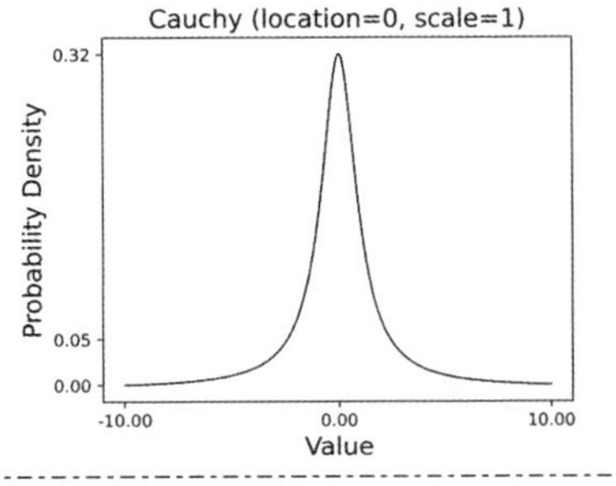

Figure S2.1.11. Representation of the Cauchy distribution

Supplemental Information 2.2: Preliminary Requisites for Stochastic Processes

There is inherent stochasticity in a system being modeled. We somehow fail to model the situation completely, which is almost always true. In most cases, it is better to use an uncertain but simple model than a complicated system that is more accurate. Navigating through the realm of stochastic processes necessitates a firm grasp of several foundational concepts. There include different aspects of process and perspectives.

+ **Frequentist Probability:** A frequentist probability explains situations in which we know that an event can be simulated or triggered again (e.g., drawing a card).

+ **Bayesian Probability:** A Bayesian probability explains situations that are hard to simulate and trigger, yet because of its uncertain nature, we treat it as a frequentist probability.

+ **Random Variables:** Random variables represent an unknown quantity. By itself, a random variable just represents a bunch of possible states. It must be coupled with a probability distribution describing how likely each of these states is.

+ **Probability Distributions:** The probability distribution is a description of how likely a random variable or set of random variables is to take on each of its possible states.

+ **Discrete Variables:** Probability over discrete values is described using a probability mass function. You can have Px, Py, Pxy. This indicates the probability distribution for single random values. Or joint probability distribution functions. In a probability mass function, P must explain all possible states of x. For all x, $p(x)$ must be between 0 and 1 inclusive. The sum of all $p(x)$s must be 1. A uniform probability distribution is $1/k$.

+ **Continuous Variables:** Probability over continuous variables is described using a probability density function. In a probability density function, p must explain all possible states of x. For all x, p must be greater than or equal to 0. The integral of $p(x) = 1$. The probability that x lies in the interval $p(x)[\mathrm{a, b}]$ is given by the integral over $p(x)dx$.

+ **Marginal Probability:** The probability of a subset of variables is called *marginal probability*. For example, if we know Pxy, and we want to know Px, we sum over the y of Pxy. For a continuous distribution, we integrate over the y of Pxy.

+ **Conditional Probability:** The probability of an event happening given another event happening is called *conditional probability*. $Py|x = Pyx|Px$, where Px must be greater than 0.

+ **The Chain Rule of Conditional Probabilities:** Any joint probability distribution over many random variables can be decomposed into conditional distributions over only one variable.

$$P(\vec{x}_1, \ldots, \vec{x}_n) = P(\vec{x}_1) \prod_2^n P(\vec{x}|\vec{x}_1, \ldots, \vec{x}_{i-1})$$

This is also called the *chain rule* or *product rule* of probability. The formula states that the probability of the entire sequence is equal to the probability of the first variable in the sequence multiplied by the probability of each subsequent variable in the sequence give the previous variables.

+ **Independence and Conditional Independence:** Two random variables x and y are independent if their probability distribution can be expressed as a product of two factors, one exclusive x and one exclusive y.

$$p(x = X, y = Y) = p(x = X)p(y = Y)$$

Two random variables x and y are conditionally independent given a random variable z if the conditional probability over x and y factorizes in this way for every value of z.

$$p(x = X, y = Y|z = Z) = p(x = X|z = Z)p(y = Y|z + Z)$$

+ **Expectation, Variance, Covariance:** Expected value, or the expectation of a function $f(x)$ with respect to a probability distribution $p(x)$, is the average or mean that f takes on when x is drawn from p.

$$E_x p[f(x)] = \sum_x p(x)f(x), \qquad E_x p[f(x)] = \int p(x)f(x)dx$$

There are some algebraic properties of $\mathrm{E}(f(x))$ that are convenient to study.

$$E_x[\alpha f(x) + \beta g(x)] = \alpha E_x[f(x)] + \beta E_x[g(x)]$$

Variance gives a measure of how much the values of a function of a random variable x vary as we sample different values of x from its probability distribution.

$$Var[f(x)] = E_x[f(x) - E_x[f(x)]^2]$$

When the variance is low, all values are around the $\mathrm{E}(f(x))$. When the variance is high, the distribution is more spread out. The square root of the variance is called the *standard deviation.*

Covariance gives some sense of how linearly related two variables are as well as the scale of these variables.

$$Cov[f(x), g(x)] = E_x[(f(x) - E_x[f(x)])(g(x) - E_x[g(x)])]$$

When absolute values are high, they change a lot and are far from the expected values. When values are positive, both values take on high values together. When values are negative, one takes on high values, whereas the other takes on low values. Independent variables must have no covariance. However, no covariance does not necessarily mean independence. The covariance of a random vector x element of R^n is an $n * n$ matrix.

$$Cov(\vec{x})_{i,j} = Cov(x_i, x_j)$$

The equation reflects the strength of the relationship between the two variables. It measures the degree to which they vary together. A positive covariance indicates that the variables tend to increase or decrease together, while a negative covariance indicates that one variable tends to increase as the other decrease.

+ **Deterministic and Stochastic:** Deterministic models, by definition, follow a structured and predictable path. When a specific set of initial conditions is provided, these models invariably produce a singular trajectory that the system will follow. A common example of deterministic models is differential equation models, which provide deterministic solutions based on the defined equations and the initial conditions.

For example, consider a simple linear differential equation, which is a common type of deterministic model. A classic example is a first-order linear differential equation used to model the cooling or heating of an object:

$$\frac{dT}{dt} = -k(T - T_{env})$$

Here, T is the temperature of the object, T_{env} is the constant temperature of the environment, and k is a positive constant that describes how quickly the object cools or heats up. For a given set of initial conditions (like the initial temperature of the object), this equation will yield one unique solution trajectory.

Contrarily, stochastic models introduce an element of randomness, representing the uncertainty and variability often encountered in real-world systems. Even with the same initial conditions, stochastic models can yield varying outcomes upon each run. This quality makes stochastic models particularly valuable in systems analysis, where inherent randomness or chance can significantly influence the system's behavior.

For example, consider a simple form of a stochastic model, like a basic random walk, which can be represented as:

$$X(n+1) = X(n) + Z(n)$$

Here, $X(n)$ is the state at time n, and $Z(n)$ is a random variable representing a step that may be positive, negative, or zero, representing the inherent randomness in the system. Every time you run this model, even with the same starting condition $X(0)$, you will get a different trajectory due to the randomness $Z(n)$.

Please note that these are basic, illustrative examples and real-world systems analysis often involves more complex deterministic and stochastic models. Now, let us consider an example for each of deterministic and stochastic models within the context of systems analysis. First, we import library and create initial conditions for both models

```
[In]
X0 = 0
T0 = 0
N_steps = 500      # Number of steps
t = np.linspace(0, 100, N_steps + 1)     # Time grid
```

We increase or decrease by 1 at each step for deterministic model while generate random walk for stochastic model.

```
[In]
# Deterministic model: increase or decrease by 1 at each step
T = [T0]
for i in range(1, N_steps + 1):
    if i % 2 == 0:
        T.append(T[-1] + 1)
    else:
        T.append(T[-1] - 1)

# Stochastic model: random walk
Z = np.random.choice([-1, 0, 1], size=N_steps)
X = np.concatenate(([X0], Z)).cumsum()
```

Let's see comparing what happens in both model

```
[In]
plt.figure()
plt.plot(t, T, label='Deterministic
Model: T(n+1) = T(n) + (-1)^n')
plt.plot(t, X, label='Stochastic Model:
X(n+1) = X(n) + Z(n)')
```

[Out]

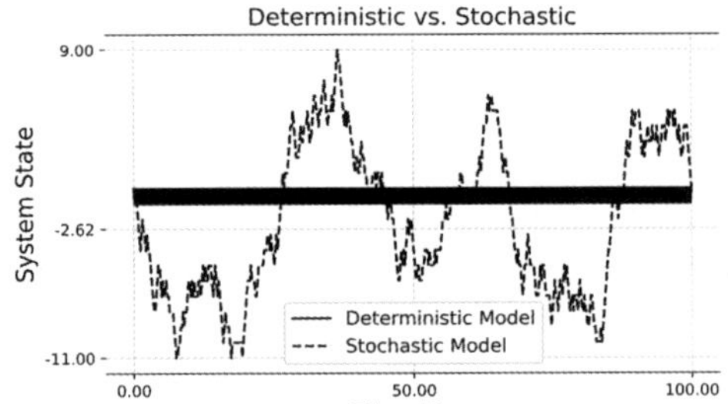

Figure S2.2.1. Deterministic and Stochastic Models

In essence, deterministic models offer a degree of precision and consistency, ideal for systems with predictable behavior. Stochastic models, on the other hand, while not providing a single, fixed outcome, excel in representing complex systems where variability and unpredictability are integral components.

[2.6]

Stochastic Processes, Part 1

Definition and Trajectory, Time-Dependent Moments, Joint Probability

Stochastic processes pervade our daily experiences and are characterized by their dependence on the initial conditions of a given model (Cox & Miller, 2017). This captivating feature has attracted the attention of numerous mathematicians over the past century. Today, this phenomenon is referred to as deterministic chaos, which is distinct from genuine randomness.

In 1963, the study of chaotic systems underwent a significant advancement, largely due to the work of Edward Norton Lorenz, who was researching methods to enhance weather forecasting at the time (Palmer, 2009). Through his analysis, Lorenz observed that even minor atmospheric perturbations could lead to substantial climatic alterations. This observation is famously encapsulated in his assertion that 'a butterfly flapping its wings in Brazil can produce a tornado in Texas' (Lorenz, 2000). Consequently, chaos theory is often colloquially known as the butterfly effect.

1. Definition and Trajectory

A family of random variable $[X(t), t \in T\ ]$ is a stochastic process for each $t \in T$, where T is the index set of the process and $X(t)$ is a random variable. An element of T is usually referred to as a time parameter. The state space of the process is the state of all possible values that random variables $X(t)$ can assume. Each of these values is called the *state* of the process.

+ **Process:** In a broader systemic analysis perspective, the term 'process' denotes the progression or development of a system or a variable's state. It encompasses the transformation of a system's state over time or the alteration of a variable within a given context. This understanding of 'process' transcends the confines of physics and mathematics, allowing for a more comprehensive exploration of change and development in various domains

- Evolution of the state of a system (physics)
- Evolution of a variable (math)

+ **Stochastic Process:** Incorporating the systemic analysis perspective, a stochastic process represents the development of a system's state or a variable when influenced by random factors. Also referred to as a random process, it comprises a sequence of random variables that evolve over time.

Stochastic processes are widely employed across various fields, including physics, engineering, finance, biology, and computer science, often necessitating the application of advanced mathematical techniques.

- Evolution of the state of a system subjected to random impacts
- Evolution of a random variable; $\{X(t_1), X(t_2), X(t_3), \dots\},\ t_1 < t_2 < t_3 < \dots$

As defined earlier (see Figure S2.2.1), a stochastic process can be understood as a collection of random variables indexed by time, t, and a set of deterministic functions of time indexed by outcome, s. This approach is frequently employed to represent the progression of a random value or system over time. In contrast to deterministic processes, which evolve in a single, predictable manner, stochastic or random processes encompass elements of uncertainty and indeterminacy.

+ **Defining Probability Based on the State**: The state of a system can be employed to ascertain probability, and various combinations of time and state can be utilized to construct stochastic models. These models provide a conceptual framework for understanding the evolution of a system or variable while accommodating random influences and uncertainties.

- Discrete time and discrete state

$$P_j(t_n) = \frac{1}{N} \sum_{k=1,\dots,N;X^{(k)}(t_n)=j} 1$$

This equation describes a probability distribution over discrete states at a given time step, t_n. The probability of being in state j at time t_n is given by the sum of $\frac{1}{N}$ for all k such that $X^{(k)}(t_n) = j$, where N is the total number of states and $X^{(k)}$ represents the state at time.

- Discrete time and continuous state

$$P(X \in [a,b]; t_n) = \frac{1}{N} \sum_{k=1,\dots,N;X^{(k)}(t_n)\in[a,b]} 1, \qquad P(x, t_n) = \lim_{\Delta x \to 0} \frac{p(X \in [x, x+\Delta x]; t_n)}{\Delta x}$$

The equation represents the probability that the random variable X lies within the interval $[a,b]$ at a specific time t_n, calculated as an average over N realizations, with $P(x, t_n)$ denoting the limiting probability density function as the interval width Δx approaches zero.

- Continuous time and discrete state

$$P_j(t) = \frac{1}{N} \sum_{k=1,\ldots,N; X^{(k)}(t)=j} 1$$

This equation describes the probability of being in state j at time t. The probability is given by the sum of $\frac{1}{N}$ for all k such that $X^{(k)}(t) = j$, where N is the total number of states and $X^{(k)}$ represents the state at time.

- Continuous time and continuous state

$$P(X \in [a,b]; t) = \frac{1}{N} \sum_{k=1,\ldots,N; X^{(k)}(t) \in [a,b]} 1, \qquad P(x,t) = \lim_{\Delta x \to 0} \frac{p(X \in [x, x+\Delta x]; t)}{\Delta x}$$

This equation describes the probability that the value of the random variable X falls within the range $[a, b]$ at time t. Note that the continuous-time stochastic model (i.e., the Wiener process) is the fundamental mathematical stochastic process for diffusion (see 2.14 for the detail), which is defined as

$$P(x,t) = \frac{1}{\sqrt{2\pi\sigma^2(t)}} exp\left\{-\frac{x^2}{2\sigma^2(t)}\right\}, \qquad \sigma^2(t) = Qt$$

Here, the x represents the value of the random variable, t represents time, σ^2 represents the variance at time t. and Q is a constant. The term 'exp' represents the exponential function. This is the continuous probability distribution that is often used to model real-valued random variables. The distribution is a continuous probability distribution that is frequently used to represent real-valued random variables, as the phrase illustrates. It also gives the probability density formula, where the variance rises proportionately to $\sigma^2(t) = Qt$, $Q > 0$, where Q is a positive constant, and the mean is always zero for all t (Figure 2.6.1).

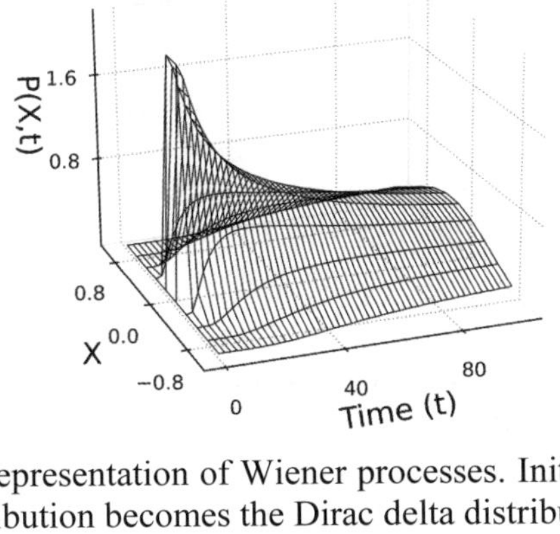

Figure 2.6.1. Schematic representation of Wiener processes. Initially ($t = 0$), all realizations are at $x = 0$. The normal distribution becomes the Dirac delta distribution (peak ∞).

On the notion of change, in a stochastic process, the idea of change is represented by a series of state changes (ΔX_n) that are prone to fluctuations across successive time periods ($\Delta t_n > 0$).

$$\{\Delta X_1, \Delta X_2, \Delta X_3, \dots\}, \{\Delta t_1, \Delta t_2, \Delta t_3, \dots\}, \qquad X(t_M) = X(t_0) + \sum_{n=1}^{M} \Delta X_n, \qquad t_M = t_0 + \sum_{n=1}^{M} \Delta t_n$$

Here, the equation describes a sum of discrete increments of a random variable X at times $t_0, \dots, t_M$. The sum is defined as the initial value of X at time t_0 plus the sum of the increments of X at each time step from $t_0, \dots, t_M$. The ΔX_n represents the increment of X at time t_n, Δt_n represents the time interval between t_{n-1} and t_n, and t_M represents the final time step. These equations describe various stochastic processes, capturing the evolution of a system or variable over time while accounting for random influences and uncertainties in discrete or continuous states and time domains. Building upon this foundation, we can explore the concept of trajectories.

+ **Trajectories:** In the context of a stochastic process, a 'trajectory' refers to a sequence representing the progression of a state variable or vector over time. Comprising state variables or state vectors at different time intervals expressed as functions of time, an example of a trajectory is

$$\{\vec{X}(t_1), \vec{X}(t_2), \vec{X}(t_3), \dots\}$$

Here, the $\vec{X}$ is the state vector and t_1, t_2, t_3,... are the corresponding time points. By starting with a simple parameterization, such rolling a dice, we can simulate the trajectory using a Monte Carlo method. For initial parameterization (rolling a die)

```
[In]
total_time_steps = 10
dievalues = np.zeros(total_time_steps)
```

The variable total_time_steps is a scalar that is set to the value 10. And the dievalues is an array of zeros with size 1-by-total_time_steps. This means that dievalues is a row vector with total_time_steps number of elements, all of which are initialized to 0.

Create the for loop

```
[In]
for n in range(total_time_steps):
    epsilon = np.ceil(np.random.rand() * 6)
    dievalues[n] = epsilon
```

The code is a loop that executes a set of instructions a specific number of times. This will run from n=1 to n=total_time_steps. The 'epsilon' is generated using the ceil and rand function and it will be a random integer between 1 and 6. The value of epsilon is stored in the n-th element of the 'dievalues' array.

Create the output

```
[In]
import numpy as np
import matplotlib.pyplot as plt

plt.plot(range(1, total_time_steps
+ 1), dievalues, marker='o')
```

→

[Out]

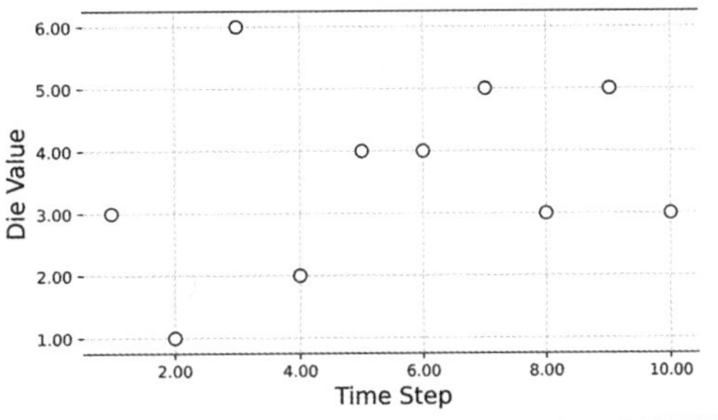

Figure 2.6.2. A trajectory: Three cards are shuffled into a sequence of states. Vertical axis: 1=ABC, 2=ACB, … 6=CBA. The time points are $t_1 = 1, t_2 = 2, t_3 = 3, \ldots$

In Figure 2.6.2, a trajectory is shown, with each sequence representing a state at each time point. Three cards are jumbled into various sequences. The cards' potential permutations are represented along the vertical axis, and the time points are denoted by the letters $t_1 = 1, t_2 = 2, t_3 = 3, \ldots.$ etc. A discrete-time stochastic process, represented by this trajectory, is one in which the system's state changes at regular intervals of time. These types of trajectories are frequently employed in stochastic process analysis and modeling, and they can aid researchers in better understanding how a system changes over time under particular circumstances.

+ **Ensemble Description of a Stochastic System:** Imagine a system is studied in an experiment. The experiment is repeated *N* times. Each repetition yields a trajectory (i.e., a realization) → *N* trajectories (realizations). A stochastic system can be described by an ensemble of trajectories. Consider an *N*-times repeated experiment where a system is being researched. There are *N* trajectories or realizations since each repeat produces a trajectory, also known as a realization. An ensemble of trajectories is created as *N* gets closer to infinity from an endlessly large set of realizations ($N \to \infty$). The ensemble of trajectories can be denoted as

$$\{\vec{X}^{(1)}(t_1), \vec{X}^{(1)}(t_2), \vec{X}^{(1)}(t_3), \ldots\} \ldots \{\vec{X}^{(k)}(t_1), \vec{X}^{(k)}(t_2), \vec{X}^{(k)}(t_3), \ldots\} \ldots$$

in the particular case of a continuous time scale, where each $\vec{X}^{(1)}(t)$ represents a realization of the system's state at time t ($\vec{X}^{(1)}(t) \ldots \vec{X}^{(k)}(t)$).

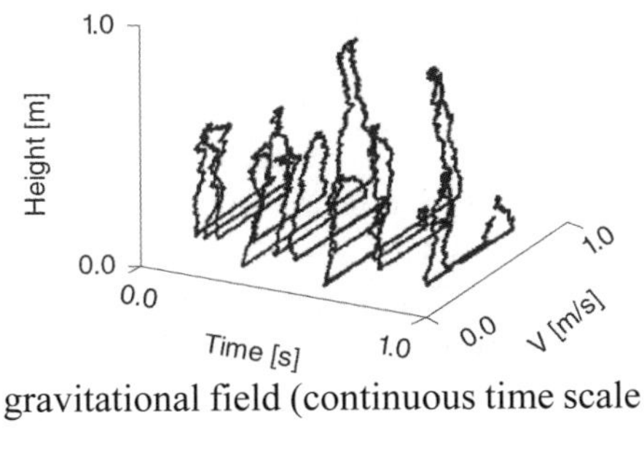

Figure 2.6.3. A particle in a gravitational field (continuous time scale in three-dimensional space)

A complete description of the stochastic process (stochastic system) contains all possible pieces of information. From the ensemble (Figure 2.6.3), all quantities of interest can be calculated (e.g., moments as functions of time, time-dependent expectation values [energy, entropy], probabilities, density functions, correlations, the power spectrum). To further elaborate on these concepts, let us consider time-dependent moments.

2. Time-dependent Moments

Imagine that a system is investigated in a series of repeated experiments. Each repetition produces a trajectory or realization of the system. As the number of repetitions approaches infinity, we may consider the entire collection of realizations as an ensemble of trajectories. This approach proves particularly useful when studying systems across continuous time spans.

+ **Moments Depend on Time**: We may establish time-dependent probabilities and probability densities to examine the system's behavior across time. These explain the probability of the system existing in a specific condition at a specific moment. The time-dependent probabilities and probability densities can be defined as

$$m(t) = \langle x(t) \rangle = \lim_{n\to\infty} \frac{1}{N} \sum_{k=1}^{N} x^{(k)}{}_{(t)}, \qquad E_{kin}(t) = \langle \frac{m}{2} v^2(t) \rangle = \frac{m}{2} \lim_{n\to\infty} \frac{1}{N} \sum_{k=1}^{N} \left[v^{(k)}{}_{(t)} \right]^2$$

Here, $m(t)$ denotes mean and $E_{kin}(t)$ represents energy (i.e., kinetic) as functions of time t. The notation $\langle \frac{m}{2} v^2(t) \rangle$ represents the expected value of the kinetic energy at time t, and the limit expression as N approaches infinity indicates that the average is taken over an infinite number of time steps. It is a measure of the energy of motion of the components of the system and is typically defined as the product of the mass of the system m and the squre of its velocity v^2 (Figure 2.6.4).

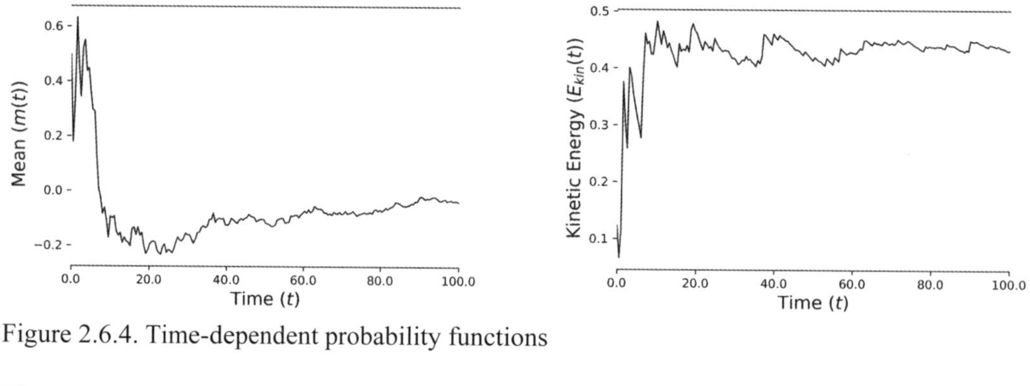

Figure 2.6.4. Time-dependent probability functions

Figure 2.6.4: Illustration of time-dependent moments, showcasing the calculation of mean and kinetic energy as functions of time, which provide insights into the system's behavior and energy of motion across different time steps. For more detail, let us consider the example of student attendance (N = 28 students). The link between the students' attendance and time is depicted on a graph in Figure 2.6.5.

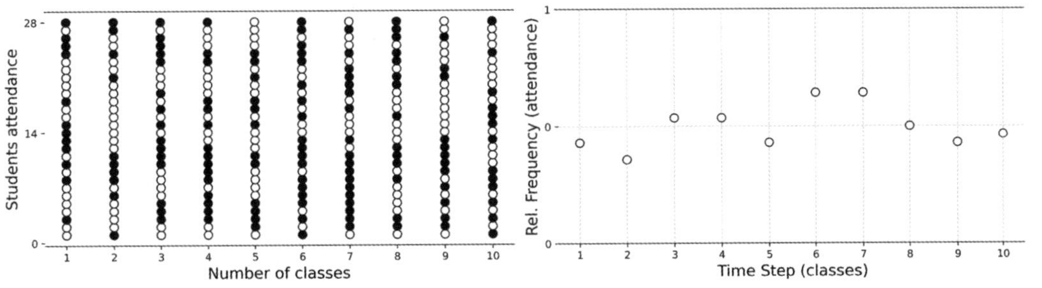

Figure 2.6.5. Time-dependent probability functions with discrete time and states. Rel. Attendance Freq. = relative attendance frequency

Discrete-state and discrete-time variables are shown in the graph on the x-axis, respectively (on the y-axis). The y-axis shows attendance or absence, while the x-axis shows various courses or moments in time. According to the time-dependent probability function, the graph depicts the relative attendance frequency of students as a function of time.

The stochastic modeling of student attendance is a random walk model in that future attendance is independent of present attendance.

- Model is described by probabilities p(A), p(NoA)

By starting with a simple parameterization using a Markov chain model, programing can simulate the student attendance model. For initial condition

```
[In]
import numpy as np
T = 7
pA = np.zeros(T)
```

The code initializes a variable T to the value 7 and creates a '1 $*$ T' array of zeros called pA. It then sets the first element of pA(1) = 0.5. Create the for loop

```
[In]
for n in range(T - 1):
    pA[n + 1] = 0.85 * pA[n] + 0.7 * (1 - pA[n])

pA[0] = 0.5
```

The code iterates over the range of values from 1 to T−1. On each iteration, it updates the value of pA(n+1) based on the value of pA(n); the rule is pA(n +1) = 0.85 $*$ pA(n) + 0.7 $*$ (1 – pA(n)). This means that the value of pA(n+1) is a weighted sum of the value of pA(n) and the value of 1 – pA(n).

Create the output

```
[In]
import matplotlib.pyplot as plt
import matplotlib.ticker as ticker

t = np.arange(1, T + 1)

ax.plot(t, pA, 'o')
```

→

[Out]

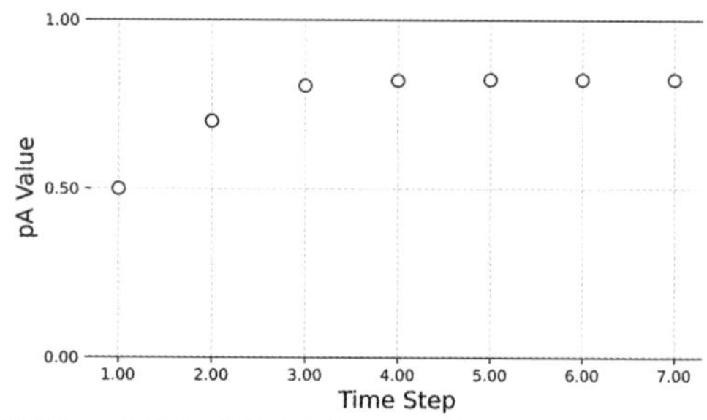

Figure 2.6.6. Markov chain model defined by transition probabilities. For example, $p(A \rightarrow A) = .85$, $p(A \rightarrow NoA) = .15$, $p(NoA \rightarrow A) = .70$; $p(NoA \rightarrow NoA) = .30$

Now, for the Monte Carlo simulation of student-attendance trajectories, let us create the random walk model for student attendance.

```
[In]
import numpy as np
nstud = 10
total_classes = 7
attendance_traj = np.zeros(total_classes)
attdata_sim = np.zeros((nstud, total_classes))
```

The code initializes the variables 'nstud' and 'total_classes' to the values 10 and 7, respectively. It also creates two arrays; attendance_traj, and attdata_sim. The attendance_traj array has total_classes elements, while the attdata_sim array has nstud rows and total_classes columns. Create the for loop of the students including trajectory for an individual student

```
[In]
for irun in range(nstud):
    for n in range(total_classes):
        u = np.random.rand()
        if u < 0.8:  # try 0.8 and 0.5
            epsilon = 1  # attendance
        else:
            epsilon = 0  # does not attend
        attendance_traj[n] = epsilon

    attdata_sim[irun, :] = attendance_traj
```

The code iterates over the range of values from 1 to 'nstud' for the outer loop and the range of values from 1 to 'total_classes' for the inner loop. On each iteration of the inner loop, the code generates a random number between 0 and 1 and stores it in the variable 'u'. Then, it checks if 'u' is less than 0.8. If it is, it sets the variable epsilon to 1. If 'u' is not less than 0.8, it sets epsilon to 0. And the 'attendance_traj(n)' updates the value of the nth element of the array based on the value of epsilon. Store the information from the loop function, then, see what is going on with the simulation

```
[In]
import matplotlib.pyplot as plt
import matplotlib.ticker as ticker

fig, ax = plt.subplots()

# Plotting
for k in range(nstud):
    for n in range(total_classes):
        x_ofk_atn = attdata_sim[k, n]
        if x_ofk_atn == 0:
            ax.plot(n+1, k+1, 'o')
        else:
            ax.plot(n+1, k+1, 'o')
```

→

[Out]

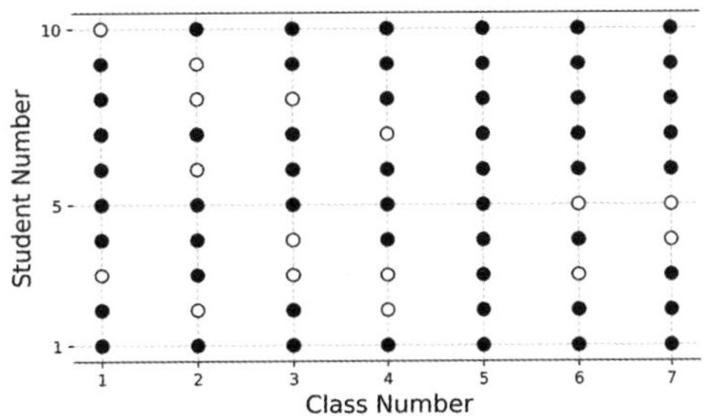

Figure 2.6.7. Monte Carlo simulation of student-attendance simulation trajectories

The scatter plot visualizes a simulation of student attendance (Figure 2.6.7). The y coordinate indicates the student, and the x coordinate denotes the class number. The 'attdata_sim' array records attendance, with '1' representing presence and '0' indicating absence. The color of the marker denotes attendance status: a white marker for absence and a black marker for presence. The visual representation simplifies the analysis of attendance patterns among students.

Now that we have explored these concepts, let us delve into a practical application of stochastic methods.

+ **Practice with Stochastic Methods:** To reinforce the fundamental mechanisms behind the stochastic process, let us practice with the same programing platform. The probability function $F(x) = p(X < x)$ of an exponentially distributed random variable X defined on $[0, \infty]$ is given by

$$F(x) = 1 - \exp(-ax), \qquad a > 0$$

The function $F(x)$ is a decreasing function of x, with values ranging from 1 to 0. The rate at which the function decrease is controlled by the parameter a.

Problem 1: Let us calculate the function values (i.e., probabilities) $F(x)$ for $a = 0.3$ at the values $x = 0$, 1, 2, 3, and 6 using a hand calculator. Round the probabilities to two decimal places. Sketch $F(x)$. How does $F(x)$ behave for $x \to \infty$?

Differentiate $F(x)$ with respect to $P(x)$. That is, calculate the probability density $P(x) = dF/dx$. Sketch the function $P(x)$ qualitatively (Figure 2.6.8).

$F(X) = 1 - \text{EXP}(-aX)$	
$X = 0$	0
$X = 1$	0.259182
$X = 2$	0.451188
$X = 3$	0.59343
$X = 4$	0.698806
$X = 5$	0.77687
$X = 6$	0.834701

→

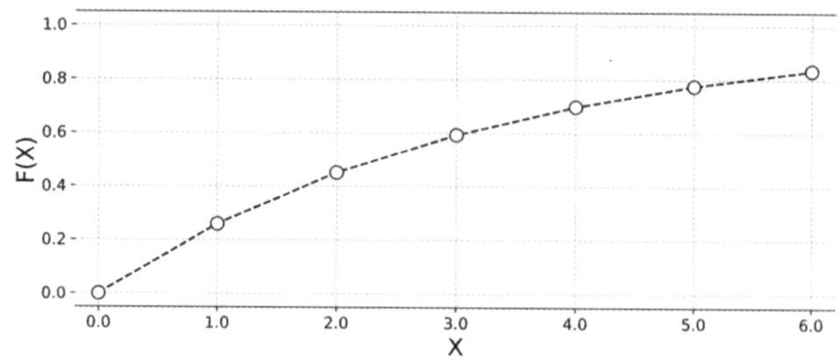

$P(X) = a\ \text{Exp}(-1aX)$	
$X = 0$	0.3
$X = 1$	0.2222454
$X = 2$	0.1646434
$X = 3$	0.1219708
$X = 4$	0.0903582
$X = 5$	0.0669390
$X = 6$	0.0495896

→

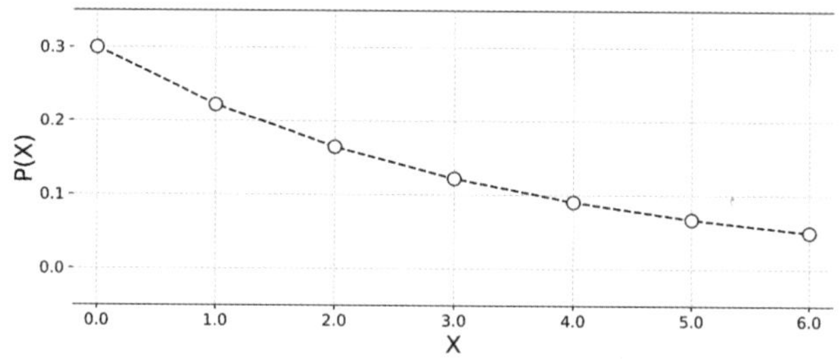

Figure 2.6.8. Calculated solution to Problem 1. As $F(x)$ approaches ∞, values approach 1. As $P(x)$ approaches ∞, values approach 0

Problem 2 (white noise [trajectories]): One of the most fundamental processes in statistics is white noise. (Note that white noise is not the Wiener process; it is even more fundamental than the Wiener process.) White noise involves a continuous-state variable X. It can be defined on a discrete time scale or a continuous time scale. For a discrete time scale $t_1, t_2, t_3, \ldots$, white noise is given by

$$X(t_n) = \varepsilon(t_n)$$

At every time point t_n the variable ε is a normally distributed random variable with mean 0 and standard deviation (*SD*) $\sigma > 0$. The *SD* is the same for all time points. Random variables ε from two different time points are selected independently of each other. That is, there are no correlations in time. A trajectory is shown above for a sequence of five time points. We choose an *SD* σ between 2 and 10 and then use this value for all simulations in Problem 2 and Problem 3. Report the *SD* (i.e., write it down explicitly in the solutions). Generate the trajectory $X^k(t_n)$ of a realization X^k of white noise in two different ways. We can use the command to generate one sequence of five independent normally distributed random variables and plot the trajectory.

```
[In]
import numpy as np
X = np.random.normal(0, sigma=2, 5)
tgrid = np.arange(1, 6)
```

Now use the command random.normal() five times in the row. The code will select five independent random variables as long as you do not manipulate the seed of program's random generator.

[In]

```
import matplotlib.pyplot as plt
Xkis1_t    =    np.random.normal(0,
sigma=2, 5)

fig, ax = plt.subplots()
ax.plot(tgrid, X, 'ko')
ax.plot(tgrid, Xkis1_t, 'ko')
```

→ [Out]

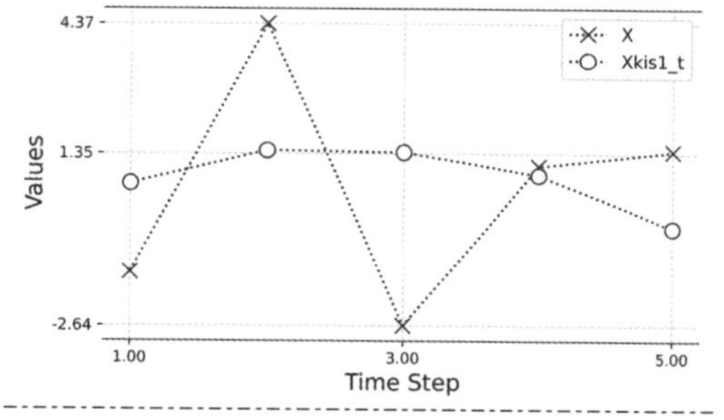

Figure 2.6.9. Created representation with SD = 2

The Figure 2.6.9 shows two trajectories of white noise generated using different methods with a standard deviation of 2, and provides a visual representation of how these different methods result in different trajectories of white noise.

Problem 3 (white noise [time-dependent moments and expectation values]): The first moment (mean) of a discrete-time process is defined by

$$m(t_n) = \lim_{n\to\infty} m_e\,(N, t_n), \qquad m_e(N, t_n) = \frac{1}{N}\sum_{k=1\ldots,N} X^k\,(t_n)$$

where $X^k(t_n)$ are realizations (i.e., statistically independent scores) from a statistical ensemble at a given time step. Expectation values are defined in a similar way.

Let us determine the mean (first moment) numerically and the *SD* of white noise for the first five-time steps 1, 2, 3, 4, and 5 using the limiting case procedure suggested by the equation. To this end, we determine m_e and the *SD* for a sample of *N* trajectories at the five time points 1, 2, 3, 4, and 5 and increase *N* gradually, like *N* = 10, 100, 1,000, 10,000. Fill out Tables 2.6.1 below.

[Table 2.6.1] Example Table for m_e (left) and Example Table for *SD* (right)

	t1=1	t2=2	t3=3	t4=4	t5=5
N=10					
N=100					
N=1,000					
N=10,000					

	t1=1	t2=2	t3=3	t4=4	t5=5
N=10					
N=100					
N=1,000					
N=10,000					

For comparison purposes, plot the four different graphs for m_e in four different panels that show the same range on the *y*-axis. Use the range [- σ, σ], where σ is the value between 2 and 10 that you chose in Problem 2. (Alternatively, you may plot all four graphs in the same diagram, but then you need to use color coding or different styles of lines to make the four figures distinguishable).

Plot the four different graphs for *SD* in four different panels that show the same range on the *y*-axis. Use the range [0, 2σ], where σ is the value between 2 and 10 that you chose in Problem 2. (Alternatively, you may plot all four graphs in the same diagram with color coding or different styles of line.) You are free to solve Problem 3 any way you choose by writing a code script. You are required to report the two tables and the graphs of m_e and SD.

Below is a step-by-step guide on how to do this exercise with code using the command window only (i.e., without writing a script). Generate trajectories for a sample of N realizations.

Use the following:

```
[In]
N = 10      # Define N and sigma
sigma = 2

# Generate normally distributed random numbers for each time step
Xkis1toN_t1 = np.random.normal(0, sigma, N)
Xkis1toN_t2 = np.random.normal(0, sigma, N)
Xkis1toN_t3 = np.random.normal(0, sigma, N)
Xkis1toN_t4 = np.random.normal(0, sigma, N)
Xkis1toN_t5 = np.random.normal(0, sigma, N)
```

This produces N realizations of trajectories. For example, realization $k = 1$ has the values Xkis1toN_t1(1), Xkis1toN_t2(1), …, Xkis1toN_t5(1) at the five time steps 1, 2, 3, 4, and 5.

Next compute the mean at every given time step.

```
[In]
# Calculate sample mean for each time step
SampleMean_t1 = np.mean(Xkis1toN_t1)
SampleMean_t2 = np.mean(Xkis1toN_t2)
SampleMean_t3 = np.mean(Xkis1toN_t3)
SampleMean_t4 = np.mean(Xkis1toN_t4)
SampleMean_t5 = np.mean(Xkis1toN_t5)
```

Finally, generate a vector that contains all five values.

```
[In]
SampleMean_t = [SampleMean_t1, SampleMean_t2, SampleMean_t3, SampleMean_t4,
SampleMean_t5]
# Calculate standard deviation for each time step
SD_t1 = np.std(Xkis1toN_t1)
SD_t2 = np.std(Xkis1toN_t2)
SD_t3 = np.std(Xkis1toN_t3)
SD_t4 = np.std(Xkis1toN_t4)
SD_t5 = np.std(Xkis1toN_t5)
```

At this stage you may write the numbers in Table. Then the graph for $m_e(t_n)$ can be plotted. Make sure that the y-axis is in the range $[-\sigma, \sigma]$.

Next compute the *SD* at the five different time points from the sample of *N* trajectories. Write them down in Table and plot them as a graph. Combine *SD* into one array, define time grid, and create the results as follows;

```
[In]
SD_t = [SD_t1, SD_t2, SD_t3, SD_t4, SD_t5]  # Combine SD into one array
tgrid = np.arange(1, 6) # Define time grid

fig, ax = plt.subplots()  # Create a figure and a set of subplots
ax[0].plot(tgrid, SampleMean_t, 'k-o')
ax[1].plot(tgrid, SD_t, 'k-o')
```

Repeat these steps for N = 100, 1,000, and 10,000.

[Table 2.6.2] Results for m_e

	$t1 = 1$	$t1 = 2$	$t1 = 3$	$t1 = 4$	$t1 = 5$
N = 10	1.257438	–0.82332	0.018184	–0.09709	0.833391
N = 100	0.011618	–0.2098	0.066145	0.246605	–0.10757
N = 1,000	0.014868	0.04575	0.086931	0.038028	–0.01471
N = 10,000	–0.01701	–0.00424	0.005582	–0.01997	–0.01555

[Table 2.6.3] Results for *SD*

	$t1 = 1$	$t1 = 2$	$t1 = 3$	$t1 = 4$	$t1 = 5$
N = 10	2.055413	1.719543	0.95992	1.730017	1.833077
N = 100	2.09178	1.950527	1.994863	1.918195	1.950661
N = 1,000	1.986302	2.055075	1.976095	2.017465	2.006194
N = 10,000	2.002772	1.967761	1.990293	2.01704	2.011262

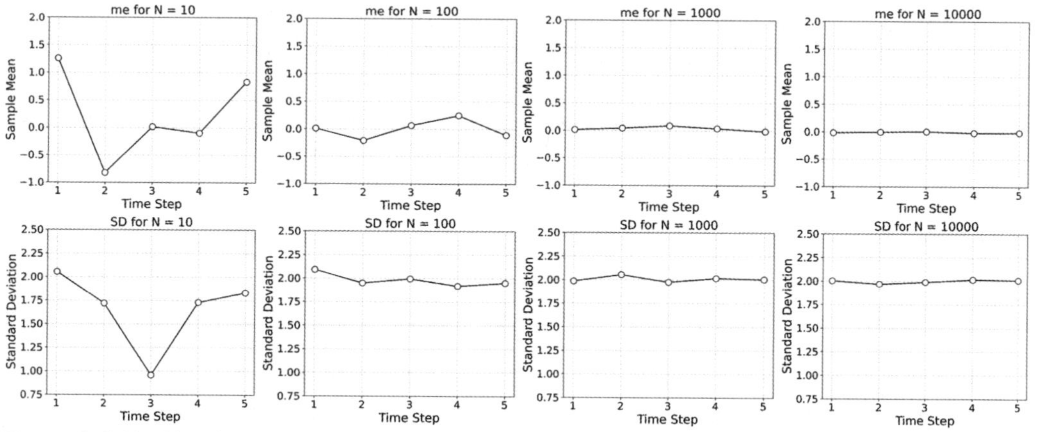

Figure 2.6.10. Simulation results. As the sample size increases, m_e approaches 0 (upper set), as the sample size increases, *SD* approaches 2 (bottom set).

This problem examines the simulation of a white noise stochastic process. In such a process, the system's future states involve an element of randomness.

We simulate white noise, a simple stochastic process with constant mean and variance over time, and statistically independent values at each step. To generate the sequence, we use the normrnd() function, producing *N* realizations at each time step. We then compute the mean and standard deviation (*SD*) of these *N* realizations, recording the results in tables and plotting them in figures (Figure 2.6.10). The plots show the key takeaway: as sample size *N* increases, both the sample mean and *SD* approach their theoretical values, which for white noise is a mean of 0 and an *SD* equal to the standard deviation of the noise distribution. This is a consequence of the Law of Large Numbers, indicating that the larger the sample size, the closer the sample average gets to the population average.

Understanding these fundamental behaviors of stochastic processes is an important step towards studying more complex aspects, such as joint probabilities.

3. Joint Probability

Let us move on to the concept of joint probability, including single time point probabilities and probability densities as special cases. Joint probability is a measure of two events occurring together and at the same point in time. Simply, it is the probability of event *Y* occurring at the same time as event *X*. The important point here is that we will come to understand a basic description of the stochastic process as an alternative to a description via an ensemble of trajectories.

In this section, we will explore the fundamental classification of variables, with further details to be provided in the next chapter. The types of variables we will consider are as follows

- Type A (general cases): involve multiple random variables
- Type B: involve random variables from multiple time points

One specific type of Type A is a special case, which is specifically related to stochastic processes.

[Table 2.6.4] Classification of Stochastic Processes

	Discrete (D) state variable	Continuous (C) state variable
Type A: Multiple random variables	D state, D time	C state, D time
Type B: Multiple time points	D state, C time	C state, C time

Table 2.6.4 displays the classification of stochastic processes based on discrete or continuous state variables, and whether they involve multiple random variables or random variables from multiple time points. For instance, imagine a game where we

shake a box containing three dice of different colors and randomly choose one (Figure 2.6.11), recording the color and the quantity of dots on the die as two random variables.

X = die value {1, 2, 3, 4, 5, 6}
Y = die color (white, grey, dark), Y in {1, 2, 3}
Number of dices: 4 white, 5 grey, 6 dark

Joint probability: $p_{jk} = p(X = k, Y = j)$
for example, $p(X = 1, Y = white)$ = ?

(Three dice of different colors)

Figure 2.6.11. Joint probability: Dice of different colors

The formula for the combined probability of the three random variables is

$$p_{jk} = p(X = k, Y = j)$$

Note that X and Y are statistically independent and the solution of the joint probability implies that

$$p(X, Y) = p(X)P(Y), [p(X = 1, Y = white) = \frac{1}{6} * \frac{4}{15} * \frac{4}{90} = 0.044]$$

+ **Joint Probability Distributions (Discrete State Space):** To display the characteristics of discrete state space, let us define the state vector as w components and define its domains as follows:

$$\vec{X} = (X_1, \dots, X_w), \qquad X_n \in \Omega_n$$

For example, $x = x_1$ in {1, 2, 3, 4, 5, 6}, $y = x_2$ in {1, 2, 3}. Thus, the joint probability would be

$$p_{i_1, i_2, \dots, i_w} = p(X_1 = i_j \wedge X_2 = i_2 \wedge \dots \wedge X_w = i_w)$$

And the probability that the state of the system is $(X_1 = i_1, \dots, X_w = i_w)$

$$\begin{aligned} p_{i_1, i_2, \dots, i_w} &= p(X_1 = i_j \wedge X_2 = i_2 \wedge \dots \wedge X_w = i_w) \\ &= \lim_{n \to \infty} \frac{1}{N} \sum_{k=1,\dots,N: X_1{}^{(k)} = i_1 \wedge X_2{}^{(k)} = i_2 \wedge \dots \wedge X_w{}^{(k)} = i_w} 1 \end{aligned}$$

The equation says that the probability of the sequence $(i_1, \dots, i_w)$ is equal to the limit as N approaches infinity of the sum of 1 over N, where N is a large number, over all values of k from 1 to N such that the k-th value of X_1 is equal to i_1, the k-th value of X_2 is equal to i_2, and so on up to the k-th value of X_w being equal to i_w. It is calculating the probability of a sequence of values for the random variables by taking the average of the number of times that occurs over a large number of iterations.

+ **Joint Probability Distributions (Continuous State Space):** For continuous state space, we can define the state vector as *w* components and define its domains as follows:

$$\vec{X} = (X_1, \dots, X_w), \qquad X_n \in \Omega_n = [A_n, B_n]$$

For example, $x = x_1$ in $\{x_{\min}, x_{\max}\}$, $y = x_2$ in $\{y_{\min}, y_{\max}\}$. Thus, the joint probability would be

$$p(X_1 \in I_j \wedge X_2 \in I_2 \wedge \dots \wedge X_w \in I_w)$$

And the probability that the intervals in the *w*-dimensional cell define the state of the system $(I_1, \dots, I_w)$

$$p(X_1 \in I_j \wedge X_2 \in I_2 \wedge \dots \wedge X_w \in I_w) = \lim_{n\to\infty} \frac{1}{N} \sum_{k=1,\dots,N: X_1{}^{(k)} \in I_1 \wedge X_2{}^{(k)} \in I_2 \wedge \dots \wedge X_w{}^{(k)} \in I_w} 1$$

For example, the probability that a ball is in cell *jk*

$$\begin{matrix} X \ in\ \Omega_x = [X_{min}, X_{max}] \\ Y \ in\ \Omega_y = [Y_{min}, Y_{max}] \end{matrix} \xrightarrow{2D\ lattice} \begin{matrix} \mathrm{M}_x\Delta_x = X_{max}, X_{min} \\ \mathrm{M}_y\Delta_y = Y_{max}, Y_{min} \end{matrix}$$

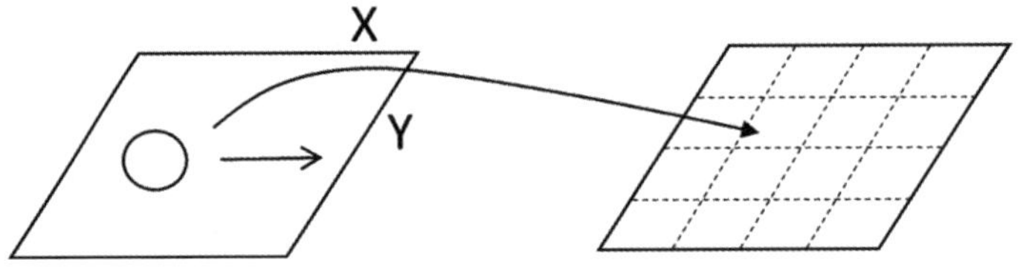

Figure 2.6.12. A ball on a two-dimensional surface and two-dimensional lattice. The lattice grid points are (x_j, y_k), $j = 1, \dots, \mathrm{M}_x$ and $k = 1, \dots, \mathrm{M}_y$.

In Figure 2.6.12, the joint probability distributions for continuous state space involve defining a state vector with w components and their respective domains. The probability of a state is calculated as the limit of the sum of indicators of events that satisfy the state, as the number of events approaches infinity.

An example of this is the probability of a ball being in a particular cell of a two-dimensional lattice grid.

$$p_{jk} = p(X \in [x_j, x_j + \Delta x] \wedge Y \in [y_k, y_k + \Delta y])$$

Here, the probability p_{jk} is the probability that X is between x_j and $x_j + \Delta x$, and Y is between y_k and $y_k + \Delta y$, at the same time. This means that both of these events must occur for the probability p_{jk} to be non-zero.

+ **Joint Probability Density:** The density of the final points of the joint probability density is

$$P(x,v) = \lim_{\Delta x \to 0, \Delta v \to 0} \frac{p(X \in [x, x+\Delta x] \wedge V \in [v, v+\Delta v])}{\Delta x \Delta v} = \lim_{\Delta x \to 0, \Delta v \to 0} \frac{p(x+\Delta x, v+\Delta v)}{\Delta x \Delta v}$$

In general, $\vec{X} = (X_1, \dots, X_w)$

$$P(\vec{x}) = \lim_{\Delta x \to 0, \dots, \Delta v \to 0} \frac{p(X_1 + \Delta X_1, \dots, X_w + \Delta X_w)}{\Delta X_1, \dots, \Delta X_w}$$

Here, the probability density $P(\vec{x})$ is calculated as the limit as Δx approaches 0, up to ΔX_w approaching 0, of the probability $p(X_1 + \Delta X_1, \dots, X_w + \Delta X_w)$ divided by the product of $\Delta X_1, \dots, \Delta X_w$ which represents the volume in the w-dimensional space defined by the x variables.

An example of joint probability density (N trajectories, wait for a long time [i.e., consider $t \to \infty$, stationary], and its initial condition is forgotten) is as follows:

$$Density: P(x,v) = p(x+\Delta x, v+\Delta v)/\Delta x \Delta v$$

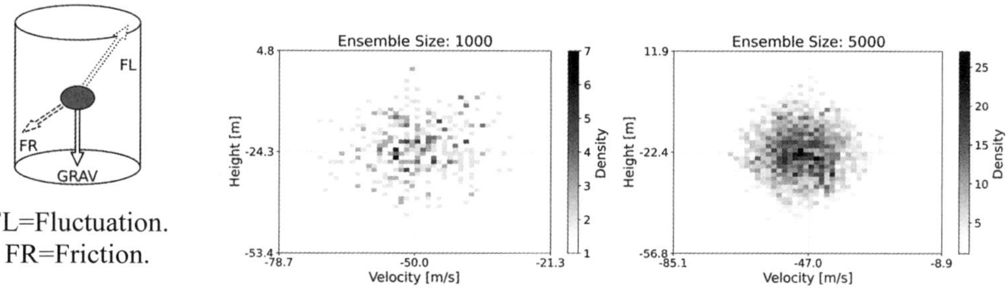

Figure 2.6.13. Fluctuating force and friction. Plot x, v final points $N \to \infty$. FL = fluctuating force, FR = friction, GRAV = gravity

We show the situation where there are N paths and we wait for a very long period before assuming that the system has achieved a stationary state and that the starting condition is no longer significant. The final positions of N trajectories, where N is getting close to infinity, are shown in Figure 2.6.13. The graphic illustrates the effects of several forces, including gravity, friction, and the fluctuation force (FL) and gravity (GRAV).

Bayes' Theorem (by hand)

Bayes' theorem (Thomas Bayes, 1763) is one of the most prominent implementations of conditional probability. It is used to calculate the probability of a particular event more accurately based on prior knowledge of conditions related to the event.

$$P(A|B) = \frac{P(B|A)P(A)}{P(B)}$$

The equation says that we can calculate *P*(*A*|*B*) using *P*(*B*|*A*), *P*(*A*), and *P*(*B*). For example, if we want to know the probability of a person suffering from cancer, we will initially output the percentage of people suffering from cancer. However, if the person is an avid smoker, we need to update our previous calculation because the probability will be much higher in this case. The definitions of each term in the equation (in this smoking example) are as follows:

- $P(A|B)$ is a posterior, and it represents the probability of a patient having cancer given that he is a smoker.
- $P(B|A)$ is a likelihood, and it represents the probability of a patient being a smoker if he is suffering from cancer.

- $P(A)$ is a prior, and it represents the probability of a patient suffering from cancer.
- $P(B)$ is a marginal likelihood, and it represents the probability of a patient being a smoker.

To use the Bayes algorithm, we find all the probabilities required for the calculation of posterior probability

$$P(smokers|X) = \frac{P(X|smokers)P(smokers)}{P(X)},$$

$$P(X|smokers) = \frac{3}{10}, P(smokers) = \frac{10}{30}, P(X) = \frac{4}{30}$$

Calculate all the obtained probabilities and find that posterior probability values belong to the category

$$P(smokers|X) = \frac{\frac{3}{10} * \frac{10}{30}}{\frac{4}{30}} = 0.75$$

Thus, Bayes' theorem utilizes prior knowledge to refine probability calculations. It is one of the most prominent implementations of conditional probability.

Remarks: The definitions of stochastic processes, their trajectories, time-dependent moments, which describe how the processes change over time, and joint probabilities and joint probability densities, which can be used to examine how different random variables change over time, were all covered in Part 1 along with other important ideas. These ideas, which have applications in many disciplines including as engineering, economics, and the physical sciences, can help us understand how complex systems behave. It will be possible to comprehend stochastic processes and their applications better after further consideration of these ideas in the portions that follow this one.

[2.7]

Stochastic Processes, Part 2

Empirical Detail of Joint Probability, Stationarity, Markov Property

In an analysis, we are likely to work with many random variables. For example, in a table of data, such as an Excel spreadsheet, each row represents a separate observation or event, and each column represents a separate random variable. Variables may be either discrete, which means that they take on a finite set of values, or continuous, which means that they take on real or numerical value (Papoulis & Saunders, 1989). Thus, we are interested in probability across two or more random variables.

The many ways in which random variables can interact affects their probabilities. We can simplify things by reducing the discussion to just two random variables (X and Y), although the principles generalize to multiple variables. We can also simplify things by discussing the probability of just two events, one for each variable ($X = A, Y = B$), although we could just as easily be discussing groups of events for each variable. Therefore, here we discuss the probability of multiple random variables as the probability of event A and event B, which in shorthand is $X = A$ and $Y = B$.

Let us assume that the two variables are related or dependent in some way. There are three main types of probability we might want to consider:

- Joint probability: the probability of events A and B
- Marginal probability: the probability of event $X = A$ given variable Y
- Conditional probability: the probability of event A given event B

These types of probability form the basis of much predictive modeling with problems such as classification and regression. For example, the probability of a row of data is the joint probability across each input variable. The probability of a specific value of one input variable is the marginal probability across the values of the other input variables.

The predictive model itself is an estimate of the conditional probability of output given an input example. Joint, marginal, and conditional probability are foundational in time series analysis. Let us take a closer look at each in turn.

1. Empirical Detail of the Joint Probability

We may be interested in the probability of two simultaneous events (e.g., the outcomes of two different random variables). The probability of two (or more) events is called the joint probability (Fine, 1982).

The joint probability of two or more random variables is referred to as the joint probability distribution. For example, the joint probability of event A and event B is written formally as $P(A \text{ } and \text{ } B)$. The conjunction is denoted using a caret (^) or sometimes a comma (,)

$$P(A \text{ } and \text{ } B) = P(A \hat{} B) = P(A, B)$$

The joint probability of events A and B is calculated as the probability of event A given event B multiplied by the probability of event B. This can be stated formally as $P(A \text{ } and \text{ } B) = P(A \text{ } given \text{ } B) \times P(B)$. The calculation of the joint probability is sometimes called the fundamental rule, the *product rule*, or the chain rule of probability. Here $P(A \text{ } given \text{ } B)$, the probability of event A given that event B has occurred, is called the conditional probability.

Joint probability is symmetrical, which means that $P(A \text{ } and \text{ } B)$ is the same as $P(B \text{ } and \text{ } A)$. The calculation of conditional probability is also symmetrical:

$$P(A \text{ } and \text{ } B) = P(A \text{ } given \text{ } B)P(B) = P(B \text{ } given \text{ } A)P(A)$$

+ **Joint Probability and Joint Probability Density:** In general cases as defined, joint probability involves multiple variables from multiple time points (Type A), and involves random variables from multiple time points (Type B).

- Special case of type A
- Specific for stochastic processes

[Table. 2.7.1] Classification of Stochastic Process

	Discrete (D) state variable	Continuous (C) state variable
Type A: Multiple random variables	D state, D time	C state, D time
Type B: Multiple time points	D state, C time	C state, C time

Table 2.7.1 displays the classification of stochastic processes based on their state variables and time points. This process involves random variables from multiple time points and can have either discrete or continuous state variables.

Joint probability that the system is in state ($X_i = i_1, \ldots, X_w = i_w$)

$$p(X_1 = i_j \hat{} X_2 = i_2 \hat{} \ldots \hat{} X_w = i_w) = \lim_{n \to \infty} \frac{1}{N} \sum_{k=1,\ldots,N: X_1{}^{(k)} = i_1 \hat{} X_2{}^{(k)} = i_2 \hat{} \ldots \hat{} X_w{}^{(k)} = i_w} 1$$

Joint probability that the state of the system is in the w-dimensional cell defined by the intervals $I_i, \ldots, I_w$

$$p(X_1 \in I_j \wedge X_2 \in I_2 \wedge \ldots \wedge X_w \in I_w) = \lim_{n\to\infty} \frac{1}{N} \sum_{k=1,\ldots,N: X_1{}^{(k)} \in I_1 \wedge X_2{}^{(k)} \in I_2 \wedge \ldots \wedge X_w{}^{(k)} \in I_w} 1$$

Joint probability density for a system with a w-dimensional state vector

$$P(\vec{x}) = \lim_{\Delta x\to 0,\ldots,\Delta v\to 0} \frac{p(X_1 + \Delta X_1, \ldots, X_w + \Delta X_w)}{\Delta X_1, \ldots, \Delta X_w}$$

Here, the equation represents the probability that the system will be in a particular state, given by the values of the variables in the state vector. It is defined as the limit of a discrete probability distribution as the size of the discrete intervals represented by Δx and Δv approaches zero. The distribution is given by the function $p(X_1 + \Delta X_1, \ldots, X_w + \Delta X_w)$, which denotes the probability that the system will be in a particular state by the values of the variables in the state vector plus the corresponding the increments $\Delta X_1, \ldots, \Delta X_w$. Having discussed this aspect of joint probability, let us now consider the case of the same random variable at multiple, different time points

+ **With the Same Random Variable at Multiple, Different Time Points:** Alternative notation for time (time step): Replace t_n with n = 1, 2, 3, … The probability that the system is in state j at time step n and in state j' at a previous time step n'

$$p(X(n) = j, X(n') = j'), \qquad n > n'$$

In previous, $p_j(n) = \lim_{n\to\infty} \frac{1}{N} \sum_{k=1,\ldots,N; X^{(k)}(n)=j} 1$, generalize the formula for $p_j(n)$ to calculate joint probabilities from ensembles

$$p(X(n) = j, X(n') = j') = \lim_{n\to\infty} \frac{1}{N} \sum_{k=1,\ldots,N; X^{(k)}(n)=j \wedge X^{(k)}(n')=j'} 1, \qquad n > n'$$

Here, the equation represents the joint probability of the two variables $X(n)$ and $X(n')$ in the limit as n approaches infinity.

$$p(X(n) = j, X(n') = j', X(n'') = j'') = \lim_{n\to\infty} \frac{1}{N} \sum_{k=1,\ldots,N; X^{(k)}(n)=j \wedge X^{(k)}(n')=j' \wedge X^{(k)}(n'')=j''} 1,$$
$$n > n' > n''$$

The equation represents the joint probability that a system with a discrete state space in three specific sates at three different time points by the variables $X(n), X(n'), X(n'')$. The probability of the states at each time point is represented by the sum of the probabilities of all samples k for which the system is in the corresponding state at each time point. To be specific, we consider the example of a three-card shuffle (Table 2.7.2):

[Table 2.7.2] Three-Card Shuffle

#	1	2	3	4	5	6
States	ABC	ACB	BAC	BCA	CAB	CBA

$$p(X(n+1) = BAC, X(n) = ABC = ?$$

Note that their notation is about $\sum_{k=1,\dots,N; X^{(k)}(n+1)=BAC} {}_{X^{(k)}(n)=ABC} 1$ because there are no one-step transitions from state 1 to state 3 $[p(X(n+1) = BAC, X(n) = ABC = 0)$. Thus, a hierarchy from the joint probability is

$$p(X(n_1) = j_1 \wedge X(n_2) = j_2 \wedge \dots \wedge X(n_w) = j_w, \qquad w\ time\ points\ n_1 > n_2 > \dots > n_w$$

The equation represents the joint probability that the system will be in states at time points, represented by the variables $(X(n_1) = j_1, X(n_2) = j_2, \dots, X(n_w) = j_w$. The time points are listed in decreasing order with $n_1 > n_2 > \dots > n_w$.

$$p(X(n) = j, X(n') = j'), \qquad two\ time\ points\ n > n'$$

The equation is represented by the variables $X(n) = j$, and $X(n') = j'$. The time points are listed in decresing order with $n > n'$. After discussing single time point probability, let us explore joint probability for multiple time points

$$p(X(n) = j), \qquad single\ time\ point$$

The equation is represented by the variable $X(n) = j$.

+ **Joint Probability for Multiple Time Points:** Type B cases involve random variables from multiple time points, providing a more comprehensive understanding of the system's behavior across time.

[Table 2.7.3] Variables from Multiple Time Points (a)

	Discrete (D) state variable	Continuous (C) state variable
Type A: Multiple random variables	D state, D time	C state, D time
Type B: Multiple time points	**D state, C time**	**C state, C time**

Table 2.7.3 represents the classification of variables from multiple time points, including both discrete and continuous state variables. Mechanisms of each ensemble are as follows:

- Discrete time, discrete state: $p(X(n) = j, X(n') = j') \rightarrow p(X(n) = j, X(n') = j', X(n'') = j'')$

The equation represents the joint probability that a system with a discrete state where the time points are discrete $X(n) = j, X(n') = j'$.

- Discrete time, continuous state: $p(X(n) \in I, X(n') \in I') \rightarrow p(X(n) \in I, X(n') \in I', X(n'') \in I'')$

The equation represents that a system with a continuous state where the time points are discrete $X(n) \in I, X(n') \in I'$.

- Continuous time, discrete state: $p(X(t) = j, X(t') = j') \rightarrow p(X(t) = j, X(t') = j', X(t'') = j'')$

The equation represents that a system with a discrete state where the time points are continuous $X(t) = j, X(t') = j'$.

- Continuous time, continuous state: $p(X(t) \in I, X(t') \in I') \rightarrow p(X(t) \in I, X(t') \in I', X(t'') \in I'')$

The equation represents that a system with a continuous state where the time points are continuous $X(t) \in I, X(t') \in I'$. The four equations represent the joint probability in a specific state at different time points, under different assumptions about the time and state spaces of the system.

+ **Joint Probability Density for Multiple Time Points:** Also, in cases of Type B, density of the random variables from multiple time points is as shown in the Table below.

[Table 2.7.4] Variables from Multiple Time Points (b)

	Discrete (D) state variable	**Continuous (C) state variable**
Type A: Multiple random variables	D state, D time	**C state, D time**
Type B: Multiple time points	D state, C time	**C state, C time**

Table 2.7.4 shows the joint probability density for Type B cases, which involve random variables from multiple time points. Mechanisms of each ensemble are as follows: Discrete time, continuous state:

$$p(X(n) \in [x, x + \Delta x], X(n') \in [x, x + \Delta x])$$

$$p(x, n; x', n') = \lim_{\Delta x \to 0, \Delta \to x'} \frac{p(x(n) \in [x, x + \Delta x], x(n') \in [x', x' + \Delta x'])}{\Delta x \Delta x'}$$

The equation is defined as the limit as the size of the intervals Δx and $\Delta x'$ approach zero of the probability of the system being in the intervals $[x, x + \Delta x]$ and $[x', x' + \Delta x']$ at discrete tiems n and n', respectively, divided by the product of the intervals $\Delta x \Delta x'$. Continuous time, continuous state:

$$p(X(t) \in [x, x + \Delta x], X(t') \in [x, x + \Delta x])$$

$$p(x, t; x', t') = \lim_{\Delta x \to 0, \Delta \to x'} \frac{p(x(t) \in [x, x + \Delta x], x(t') \in [x', x' + \Delta x'])}{\Delta x \Delta x'}$$

The equation is defined as the limit as the size of the intervals Δx and $\Delta x'$ approach zero of the probability of the system being in the intervals $[x, x + \Delta x]$ and $[x', x' + \Delta x']$ at continuous times t and t', respectively, divdied by the product of the intervals $\Delta x \Delta x'$. Now, let us discuss the concept of equivalent stochastic systems with discrete state spaces.

+ Two Equivalent of Stochastic Systems with Discrete State Spaces: Equivalent means that both descriptions provide the same information.

- Ensemble of trajectories

$\{X^{(1)}{}_{(n_1)}, X^{(1)}{}_{(n_2)}, X^{(1)}{}_{(n_3)}, \dots\}$

...

$\{X^{(k)}{}_{(n_1)}, X^{(k)}{}_{(n_2)}, X^{(k)}{}_{(n_3)}, \dots\}$

...

$X^{(1)}{}_{(t)}$

...

$X^{(k)}{}_{(t)}$

- Hierarchy of joint distributions

$p(X(n \; or \; t) = j)$

$p(X(n \; or \; t) = j, X(n' \; or \; t') = j')$

$p\big(X(n \; or \; t) = j, X(n' \; or \; t') = j', X(n'' \; or \; t'')\big)$

...

These hierarchy can be used to characterize the statistical properties of the system and to make predictions about its future behavior, for example: rolling a fair die 1, 2, …, k times

$$p(X = j) = \frac{1}{6} \quad \rightarrow \quad p(X(1) = j, X(2) = j') = \frac{1}{36}$$

The left equation is represented by the variable $X = j$, at a single time point which probability is given by 1/6. The right equation is represented by the variables $X(1) = j$ and $X(2) = j'$, where j and j' are specific states which probability is give by 1/36.

$$p(X(1) = j_1 \wedge X(2) = j_2 \wedge \dots X(k) = j_k) \left(\tfrac{1}{6}\right)^k \rightarrow \text{e.g., } p(365) = \left(\tfrac{1}{6}\right)^3 \approx 0.005$$

This equation is represented by the variables $X(1) = j_1 \wedge X(2) = j_2 \wedge \dots X(k) = j_k$ and the probability is given by '$\left(\frac{1}{6}\right)^k$' where '$k$' is the number of time points. As it shows, if the system is in the state '365' at three time points, the probability is given by $\left(\frac{1}{6}\right)^3$ which is approximately 0.005.

+ Two Equivalent of Stochastic Systems with Continuous State Spaces: Let us delve deeper into the idea of equivalent stochastic systems when dealing with continuous state spaces. There are two ways to represent these systems. In the ensemble of trajectories, we have a collection of k trajectories, each represented by a set of state variables at different time points n or continuous times t. On the other hand, the hierarchy of joint distributions describes the probability distributions associated with these state variables across time.

- Ensemble of trajectories

$$\{X^{(1)}{}_{(n_1)}, X^{(1)}{}_{(n_2)}, X^{(1)}{}_{(n_3)}, \dots\}$$

...

$$\{X^{(k)}{}_{(n_1)}, X^{(k)}{}_{(n_2)}, X^{(k)}{}_{(n_3)}, \dots\}$$

...

$$X^{(1)}{}_{(t)}$$

...

$$X^{(k)}{}_{(t)}$$

- Hierarchy of joint distributions

$$p(x, n\ or\ t)$$

$$p(x, n\ or\ t; x', n'\ or\ t')$$

$$p(x, n\ or\ t; x', n'\ or\ t'; x'', n''\ or\ t'')$$

...

Complete descriptions from hierarchies: We recall the previous stochastic process as a sequence of state changes ΔX_n subject to fluctuations between consecutive time intervals $\Delta t_n > 0$. For example:

- Hierarchy of joint probability densities

$$p(x,t) \rightarrow p(x,t;x',t') \rightarrow p(x,t;x',t';x'',t'') \rightarrow \cdots$$

- Joint distributions describe how state changes are distributed

$$\{\Delta X_1, \Delta X_2, \Delta X_3, \dots\} \rightarrow X(t_M) = X(t_0) + \sum_{n=1}^{M} \Delta X_n$$

This equation can be used to compute the state of the system at a particular time point, given the initial state of the system and the state changes that have occurred over time.

The states changes are represented by the set $\{\Delta X_1, \Delta X_2, \Delta X_3, \dots\}$, which denotes the change is the stae of the system from one time point to the next by the variable $X(t)$.

The state of the system at the final time point t_M is given by the initial stae of the system at time t_0 plus the sum of all of the state changes that have occurred between t_0 and t_M. This relationship is expressed as $X(t_0) + \sum_{n=1}^{M} \Delta X_n$.

2. Stationarity

When analyzing time series data, it is often useful to treat the observed time series as part of a realization of a stochastic process. To fully grasp the concept of stationarity in stochastic processes, we first need to review two preliminary definitions.

Probability space: A probability space is a triple space (Ω, F, P), where Ω is a non-empty set called the sample space, F is a σ-algebra of subsets of Ω (i.e., a family of subsets closed with respect to countable union and complementary with respect to Ω), and P is a probability measure defined for all members of F.

Random variable: A real random variable or real stochastic variable on (Ω, F, P) is a function $x: \Omega \rightarrow R$, such that the inverse image of any interval $(-\infty, a)$ belongs to F (i.e., a measurable function).

Stochastic process: A real stochastic process is a family of real random variables $X = \{x_i(\omega); i \in T$, all defined on the same probability space (Ω, F, P). The set T is called the index set of the process. If $T \subset Z$, then the process is a discrete stochastic process. If T is an interval of R, then the process is a continuous stochastic process, which for a stochastic process X is also commonly denoted

$$F_x(x_{t1}, \dots, x_{tn})$$

Intuitively, this represents a projection of the process onto a finite-dimensional vector space (in this case, a finite set of time points).

+ **Definitions of Stationarity:** We can now consider the concept of stationarity. Stationarity means that the statistical properties of the process do not change over time (Witt et al., 1998). Most methods are designed to work on a stationary time series.

Why is it important? A stationary time series is necessary for some time series forecasting models (such as stochastics), as they are simpler to model due to their consistent statistical characteristics. Therefore, if a time series is not stationary, we should make it stationary in order to see what more we can accomplish.

A nonstationary series is often transformed into a stationary one as the first stage in the procedure. Before delving into these definitions, it is crucial to note that, despite the variety of stationarity concepts that have been proposed, stationarity of any sort is a stochastic process property, not a property of any finite or infinite realization of it (i.e., a time series of values).

A stochastic process is considered stationary if all of its properties remain unchanged regardless of a specific reference time point.

- Implications of stationarity (unaffected by time shifts): Stationarity implies that when time is shifted by an arbitrary finite interval, the properties of the process remain constant.

- Consequences for single time point probabilities (Independence from Time): As a result of stationarity, single time point probabilities become independent of time, which can be expressed as follows:

$$p(X(n\ or\ t) = j) = p(X = j) \quad \rightarrow \quad p(x, n\ or\ t) = p(x)$$

This suggests that the probability of a specific state is not influenced by the time it is observed.

- Time-invariant joint probabilities (dependence on time differences): In a stationary process, all joint probabilities depend solely on time differences, meaning they do not explicitly rely on time.

For stationarity example when rolling a fair die, the process is stationary because the single time point distribution remains independent of time. Additionally, all joint distributions are time-independent, further emphasizing the stationary nature of this process.

$$p(X(1) = j_1 \wedge X(2) = j_2 \wedge \ldots \wedge X(k) = j_1) = \left(\frac{1}{6}\right)^k$$

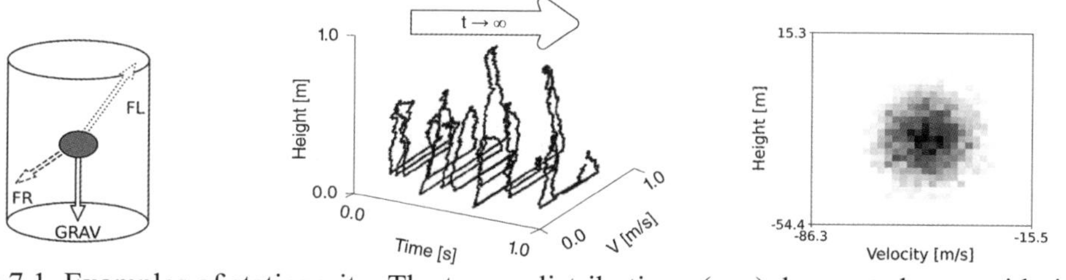

Figure 2.7.1. Examples of stationarity. The t → ∞ distribution $p(x, v)$ does not change with time. FL = fluctuating force, FR = friction, GRAV = gravity.

This indicates that the product of the probabilities of each event, which is $\left(\frac{1}{6}\right)^k$, equals the likelihood of having a certain combination of outcomes for the k rolls of a fair die. A few instances of stationary processes are shown in Figure 2.7.1. Even if the variables themselves may fluctuate arbitrarily, when a system enters a stationary state, the probability distribution of its variables no longer varies over time. The distribution of position (x) and velocity (v) is depicted in the image along with the impacts of other forces, such as the fluctuation force (FL), friction (FR), and gravity (GRAV). A steady distribution is one that does not vary over time, such as the t distribution, $p(x, v)$.

+ Practice with Stationary Processes: To understand in more detail, let us describe and conduct a stochastic calculation and its stationary process.

There are two equivalent ways to describe a stochastic process completely. Again, the stochastic process is a collection of random variables. It is often used to represent the evolution of a random value, system, or time. A stochastic process is a sequence of state changes ΔX_n subject to fluctuations between the consecutive time interval $\Delta t_n > 0$, $(X(t_n) = X(t_0) + \sum_{n=1}^{m} \Delta X_n)$. The trajectory of a stochastic process describes the evolution of a state variable or vector as a function of time: $(\underset{X}{\rightarrow}(t_1), \underset{X}{\rightarrow}(t_2), \underset{X}{\rightarrow}(t_3) \ldots)$.

There are two equivalent and complete descriptions of stochastic systems with discrete state spaces.

- Ensemble of trajectories:

$$\{X^1(n_1), X^1(n_2), X^1(n_3) \ldots\ldots\} \ldots\ldots \{X^k(n_1), X^k(n_2), X^k(n_3) \ldots\ldots\}$$

- Hierarchy of joint distributions:

$$P(X(n\ or\ t) = j) \rightarrow P(X(n\ or\ t) = j = X(n'\ or\ t') = j')$$
$$\rightarrow P(X(n\ or\ t) = j = X(n'\ or\ t') = j', X(n''\ or\ t'') = j'') \rightarrow \ldots$$

Both descriptions provide the same information.

Stationary processes: Consider continuous-time stochastic processes with continuous random variables $X(t)$, where t is time. In general, the processes exhibit time-dependent probability densities $P(X,t)$. Which of the two probability densities shown below describes a stationary process? Which describes a nonstationary process? Explain your answers. X defined on a real line

$$P = \frac{1}{\sqrt{2\pi\varrho t}} exp\left(-\frac{X^2}{2\varrho t}\right), \qquad \varrho > 0$$

- X defined on the interval from 0 to 1

$$P = 2(1 - X)$$

Nonstationary process: When t changes, the process is affected because t depends on time.

$$P = \frac{1}{\sqrt{2\pi\varrho t}} exp\left(-\frac{X^2}{2\varrho t}\right)$$

$$P = \frac{e - \dfrac{X^2}{2\varrho t}}{\sqrt{2\pi}\sqrt{\varrho t}}$$

$$P = \lim_{t\to\pm\infty} \frac{1}{\dfrac{X^2}{e2t\varrho}\sqrt{2\pi}\sqrt{t\varrho}} = 0$$

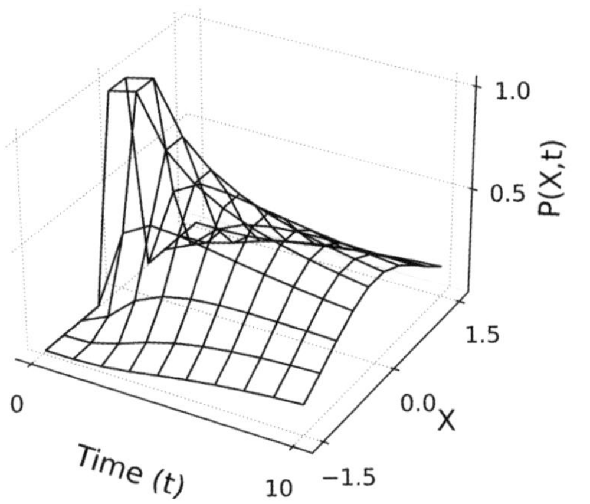

Figure 2.7.2. A nonstationary process with a three-dimensional distribution

The probability density has a normal distribution (Figure 2.7.2). Mean = 0 for all t, the variance increases like $\sigma^2(t) = \varrho * t, \varrho > 0$, initially ($t = 0$), all realizations are at $X =$ 0. The normal distribution becomes the delta distribution (peak ∞).

The probability distribution of its variables does not rely on time in a stationary process. This indicates that even as time passes, the process behaves in the same way. For instance, the stationary process described by the equations $P = 2(1 - X)$ and $P = -2(X - 1)$ has a probability distribution of the variable P that is independent of time.

$$P = 2(1 - X), \qquad P = -2(X - 1)$$

These equations may be rewritten as

$$P + 2X = 2, \qquad P + 2X - 2 = 0$$

respectively, demonstrating that the variable P is both a function of X and a constant number. In a stationary process, P and X's joint probability distribution does not vary over time, and the process' behavior is constant regardless of when it is observed. This problem concerns the joint probability density and joint probabilities of two continuous random variables calculated from a linear regression model of body weight and body height. The aim of this exercise is threefold. First, we want to illustrate joint probabilities as shown in the Figure 2.7.3 below.

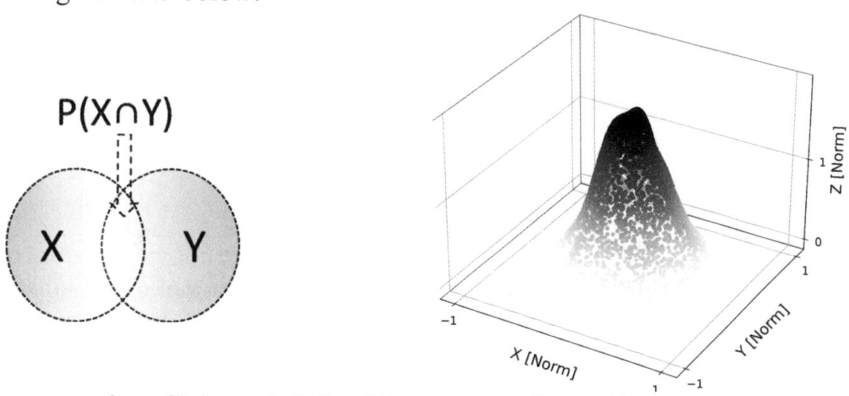

Figure 2.7.3. Representation of joint probability (Norm = normalized arbitrary unit)

Second, we would like to understand that a linear regression model defines a joint probability density. Third, we want to determine numerically a particular joint probability (not probability density) from the linear regression model. The following model was created for the purpose of this exercise and does not reflect reality. Let X_1 and X_2 denote two random variables. X_1 describes body height. X_2 describes body weight. We assume that a linear model holds between weight and height with a slope parameter a and an intercept parameter b. Accordingly, we have

$$X_2 = a \cdot X_1 + b + Error$$

The error term represents normally distributed random numbers with 0 mean and a particular standard deviation, which is given below. The error term accounts for individuality. That is, individuals can deviate from the linear relationship $X_2 = aX_2 = b$. The regression model allows us to compute a number N of realization k as follows:

$$X_2{}^{(k)} = a \cdot X_1{}^{(k)} + b + \varepsilon^{(k)}$$

where $\varepsilon^{(k)}$ are the realizations of the normally distributed population of errors. In what follows, we use a = 0.9 kg/cm and b = –80 kg (note the minus sign). The standard deviation of the error term is sigma(error) = 1.0.

Solving a problem: Now, let us consider a couple of problems and solve them via stochastic calculation and its stationary process.

- **Problem a:** Generate height and weight data for a sample of N = 10 individuals. Plot the probability densities *P*(Height) and *P*(Weight) and illustrate the joint probability density *P*(Height, Weight) by means of a scatter plot (similar to the scatter plot in the figure above). Consider a population of individuals with a mean height of 160 cm and standard deviation 1 that follows a normal distribution.

Here is a step-by-step guide. Using programing, first determine the height of the individuals.

```
[In]
import numpy as np
N = 10
mu = 160
sigma = 1
np.random.seed(0)  # For reproducibility
Height = np.random.normal(mu, sigma, N)
```

We generated a sample of random variables from a normal distribution with mean 160 and standard deviation 1. The sample size is determined by the variable N which is set to 10 in this case. Next generate 10 normally distributed errors.

```
[In]
Error = np.random.normal(mu, sigma, N)
```

The code takes three arguments; the mean and standard deviation value, and the size of the sample 'N,1' which sample is stored in the variable Error. Finally, compute the weights of the individuals from the regression model.

```
[In]
Weight = 0.9 * Height - 80 + Error
```

The variable 'Weight' is calculated by multiplying the values in the 'Height' sample by 0.9, subtracting 80 from the result, and adding the values in the 'Error' sample. The 'Error' sample was previously generated by the code above which generates a sample of random variables from a normal distribution with mean 0 and standard deviation 1. Next use density to estimate and plot the functions *P*(Height) and *P*(Weight). Make a scatter plot with height on the x-axis and weight on the y-axis. To make the scatter plot, use the following:

```
[In]
plot(Height,Weight,'o')
```

Note that in the option field of the plot command there is a lowercase letter *o* (i.e., this is not a zero). In the end, you should produce three panels like the panels shown below. Note that you will have different graphs and a different scatter plot because you will have different numbers.

Solution to Problem a

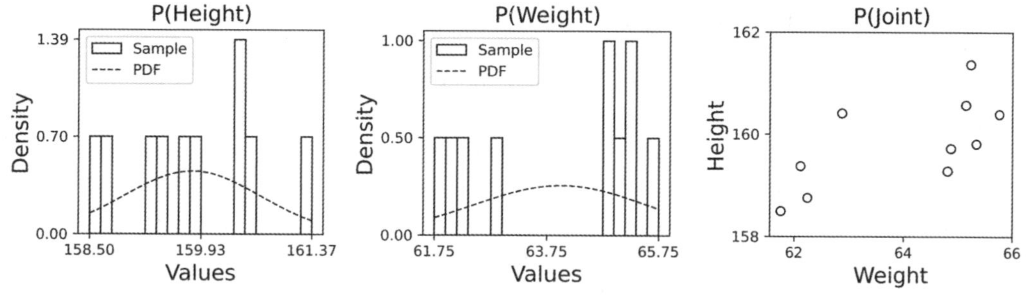

Figure 2.7.4. Output of the simulation (solution to problem a)

- **Problem b**

From the sample of $N = 10$ individuals, determine the following four relative frequencies:

Relative frequency of individuals with height ≥ 160 AND weight ≥ 64
Relative frequency of individuals with height < 160 AND weight ≥ 64
Relative frequency of individuals with height ≥ 160 AND weight < 64
Relative frequency of individuals with height < 160 AND weight < 64

Note that these four relative frequencies must add up to 1. To determine them, ask program to display the height and weight values of your 10 individuals (Height, Weight). Then assign each participant to one of the four groups. Make a table and enter the participant number k in the corresponding cell. For illustration purposes, we have put the participant numbers k in the corresponding four cells. In our data set (see the rightmost panel above with the scatter plot) participants 2 and 6 have height ≥ 160 and weight ≥ 64. Note that you will have different numbers. Because we know N, we can calculate the four relative frequencies. In summary, for Problem b you should report the table and the four relative frequencies. Report the relative frequencies.

Solution to Problem b

[Table 2.7.5] Number k (left) and Relative Frequency (right)

	Height < 160	Height ≥ 160
Weight ≥ 64	0	5 (1, 2, 7, 8, 10)
Weight < 64	3 (4, 5, 9)	2 (3, 6)

	Height < 160	Height ≥ 160
Weight ≥ 64	0	0.5
Weight < 64	0.3	0.2

Table 2.7.5 shows the number of participants (k) height and weight measurements, calculated from a sample of N=10 individuals in Problem b. The four groups are: height ≥ 160 AND weight ≥ 64, height < 160 AND weight ≥ 64, height ≥ 160 AND weight <

64, and height < 160 AND weight < 64. The relative frequencies are reported to one decimal place and add up to 1.

- **Problem c**: Let us consider next the limiting case when N goes to infinity.

First, we would like to get a better idea of what the probability densities P(Height), P(Weight), and P(Height, Weight) look like. Set N = 10,000. Repeat the steps in Problem 3a and make three graphs again that show P(Height), P(Weight), and a scatter plot for P(Height, Weight). Set N = 100,000. Show P(Height), P(Weight), and a scatter plot for P(Height, Weight). State whether you think the graphs and scatter plot change much from N = 10,000 to N = 100,000. If they do not, then they are good approximations for N as it goes to infinity. If they do differ substantially, then you should try even larger sample sizes. Second, we want to determine the joint probability P(Height ≥ 160, Weight ≥ 64) using the limiting case procedure. Consider samples of size N = 10, 50, 100, 500, and 1,000, and determine the frequency and the relative frequency of individuals with height ≥ 160 and weight ≥ 64. You cannot use the method in Problem 2b to solve this problem. Instead, you can either write a script or follow the step-by-step instructions below.

Step-by-step instructions: This part requires some patience on your part because you need to write a for-loop in the command window (alternatively, you may try to write a script to solve the problem). As in Problem 3a, generate height and weight data for a sample of N individuals using the linear regression model.

```
[In]
N = 50
mu_height = 160
sigma_height = 1
mu_error = 0
sigma_error = 1
```

Generate height, error, and weight data

```
[In]
Height = np.random.normal(mu_height, sigma_height, N)
Error = np.random.normal(mu_error, sigma_error, N)
Weight = 0.9 * Height - 80 + Error
```

Next define a variable sum that counts the number of individuals with height ≥ 160 and weight ≥ 64. That is, the variable sum gives the frequency. Write the following in the command window:

```
[In]
sum = 0
for k in range(N):
    if Height[k] >= 160 and Weight[k] >= 64:
        sum += 1
```

This is the 'sum' sign in the code that goes over all realization. If the condition you are interested in is satisfied, then increase the variable sum by 1. After that, you begin a for-loop. It is not closed yet. Next write the condition to increase the sum if the conditions you are interested in are satisfied. Finally, close the for-loop.

To find out the number that is stored in the variable sum, enter 'sum' in the command window. Read off the number and report it in the table. Then calculate the relative frequency. Repeat the procedure with $N = 100$, 500, and 1,000.

Solution to Problem c: The frequency and relative frequency of individuals with height ≥ 160 and weight ≥ 64 are calculated for $N = 10$, 50, 100, 500, and 1,000, and the results are reported in Table 2.7.6.

[Table 2.7.6] Solution of the Relative Frequency Problem

	$N = 10$	$N = 50$	$N = 100$	$N = 500$	$N = 1{,}000$
Frequency of individuals with height $\geq$ 160 and weight $\geq$ 64	3	19	41	184	383
Relative frequency of individuals with height $\geq$ 160 and weight $\geq$ 64	0.3	0.38	0.41	0.368	0.383

In Problem c, the goal is to estimate the joint probability of Height and Weight in the limiting case as N goes to infinity. To achieve this, P(Height) and P(Weight) are plotted for $N = 10{,}000$ and $N = 100{,}000$, and it is concluded that they are good approximations for N going to infinity.

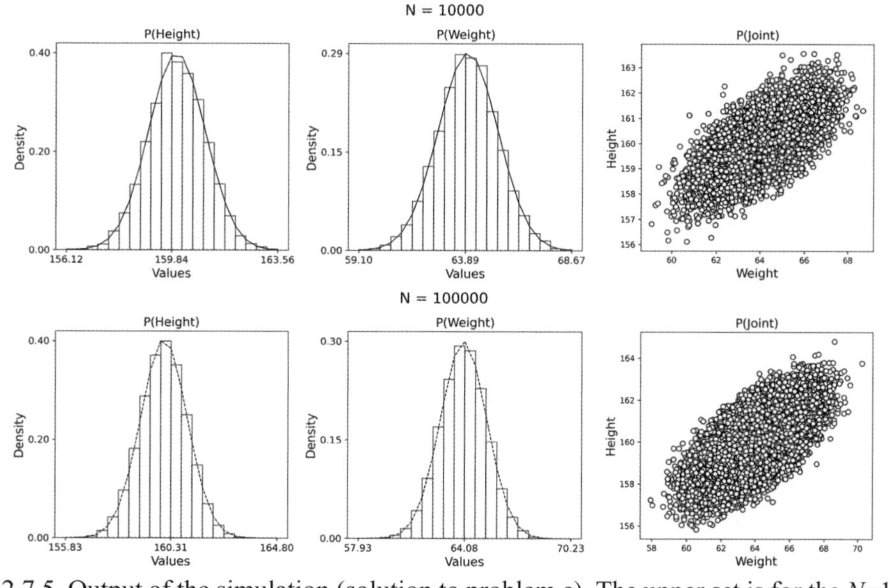

Figure 2.7.5. Output of the simulation (solution to problem c). The upper set is for the N=10,000, and the bottom set is for the N=100,000.

The results show that the distributions of height and weight, and their joint relationship, stabilize as the sample size increases, indicating that the estimates converge as N goes to infinity. This suggests that the distributions at larger sample sizes (N = 10000 and N = 100000) are good approximations of the true, unknown distributions.

In short, the simulation results suggest that larger sample sizes give more accurate estimates of probability densities and joint probabilities in this context. Having explored joint probability densities and joint probabilities in a linear regression model of body weight and height, we will now delve into the concept of Markov Property, a powerful method for determining parameter values in various models.

3. Markov Property

Before introducing Markov chains, let us briefly review some basic but essential notions of probability theory.

In nonmathematical terms, a random variable X is a variable whose value is the outcome of a random phenomenon. This outcome can be a number (or number-like, including a vector) or not. For example, we can define a random variable as the outcome of rolling a die (number) or the outcome of flipping a coin (not a number, unless we assign, e.g., 0 to heads and 1 to tails).

Notice that the space of possible outcomes of a random variable can be discrete or continuous.

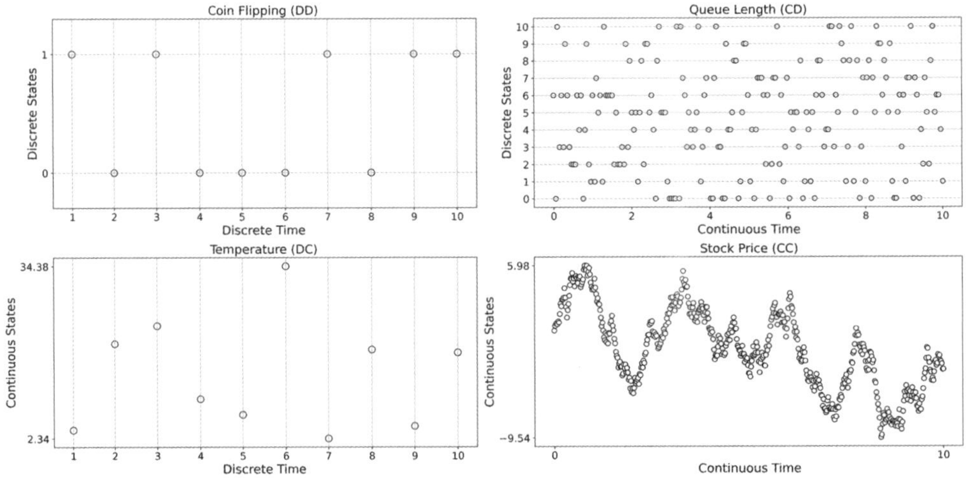

Figure 2.7.6. Schematic representation of the different random processes (discrete or continuous in space and time) associated with a real-world example

We can define a random process (also called a stochastic process) as a collection of random variables indexed by a set T that often represent different instants of time (we assume this in the following). The two most common cases are either that T is the set of natural numbers (discrete-time random process) or that T is the set of real numbers (continuous-time random process). For example, flipping a coin every day defines a discrete-time random process, whereas the price of a stock market option varying continuously defines a continuous-time random process.

Random variables at different times can be independent of one other (e.g., coin flips) or dependent in some way (e.g., stock price) and can have a continuous or discrete state space (Figure 2.7.6) the space of possible outcomes at each instant of time).

+ **Property:** There exist some well-known families of random processes: Gaussian processes, Poisson processes, autoregressive models, moving average models, Markov chains, and others. These particular cases each have specific properties that allow us to better study and understand them. One feature that makes the study of random processes much easier is the Markov property.

The Markov property says that for a random process, if we know the value taken by the process at a given time, we will not obtain any additional information about the future behavior of the process by gathering more knowledge of the past. Stated in slightly more mathematical terms, for any given time, the conditional distribution of future states of a process given present and past states depends only on the present state and not at all on the past states (the process is memory-less). A random process with the Markov property is called a *Markov process* (Howard, 1960).

$$P(future|\, present, pase) = P(future \mid present, Markov\ property)$$

The Markov property expresses the fact that at a given time step and knowing the current state, we will not obtain any additional information about the future by gathering information about the past (Figure 2.7.7).

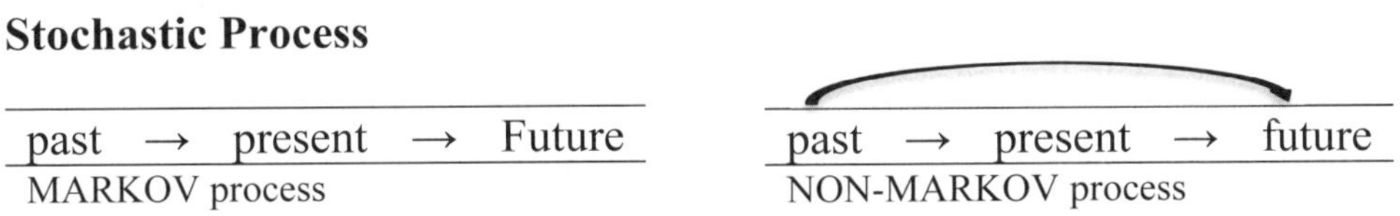

Figure 2.7.7. Classification of the Markov property

+ **Definition:** The future behavior of a Markov process depends only on the current state and not on past states. Transition probabilities of Markov processes rely solely on current and future states (i.e., link only two time points; math). We can iterate a Markov process from an initial state (numeric). For example, discrete state, discrete time ($n > n' > n''$)

$$p(X(n) = \leftarrow j, X(n') = j', X(n'') = j'') = p(X(n) = \leftarrow j, X(n') = j')$$

If and if analogous relations for all possible transition probabilities hold, then the process is a Markov process.

+ **Important Note:** All processes in nature are Markovian provided we study them in sufficiently large state spaces. Consider a harmonic oscillator:

- State space = real line: The initial value $x(0)$ is not sufficient to solve this equation → non-Markov process

$$\frac{d^2}{dt^2}x(t) = -k\,x(t)$$

- State space = (x, v), that is, R^2: A problem with the initial value $x(0)$, $v(0)$ can be solved → Markov process

$$\frac{d}{dt}x(t) = v(t), \qquad \frac{d}{dt}v(t) = -k\,x(t)$$

The two equations describe the time evolution of the variables. The left side of the equation represents the time derivative of the variable $x(t)$ which is denoted by $\frac{d}{dt}x(t)$. It is equal to the variable $v(t)$. It shows how the variable $x(t)$ changes over time in terms of the variable $v(t)$. The right side of the equation is denoted by $\frac{d}{dt}v(t)$ which is equal to the negative of the variable $x(t)$ multiplied by a constant k. It shows how the variable $v(t)$ changes over time in terms of the variable $x(t)$ and the constant k.

Consider the case when time is discrete. The time steps are $n = 1, 2, 3, \ldots$ or $n = t_0, t_1, t_2, \ldots$ (alternative notation: $t_1, t_2, t_3, \ldots$ or $t_0, t_1, t_2, \ldots$).

This dynamic evolves on a set of discrete states. Usually the number of states is finite (final set). Discrete-time process with a discrete state space. Thus, according to the Markov property, the probability of finding the process in a particular state in the future depends only on the present state, not on states preceding this present state. The future of the process is conditioned on the present process but not the past. The future of the process is independent of all past states except for the one the process currently occupies.

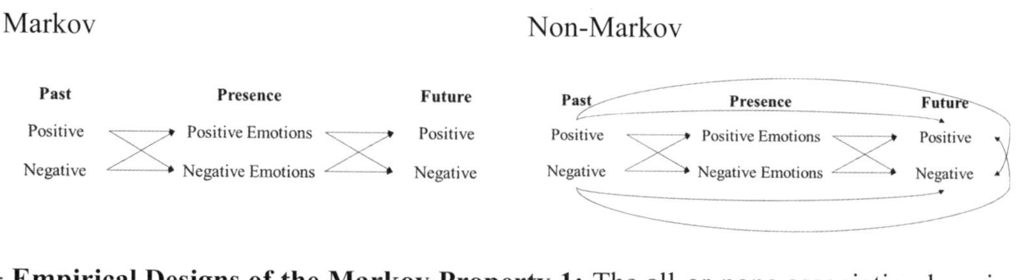

+ **Empirical Designs of the Markov Property 1:** The all-or-none association learning approach (Wickens, 1982, for instance) has participants learn about pairings of objects through a learning activity. In this particular example, the nonwords CIM, IEH, NAD, and EOJ are specifically related to the integers 1, 2, 3, and 4 in the following manner: CIM is connected to 1, IEH to 2, NAD to 3, and EOJ to 4.

- Condition: Nonwords {CIM, IEH, NAD, EOJ} are related to integers {1, 2, 3, 4} in a particular way: CIM = 1, IEH = 2, NAD = 3, EOJ = 4.

Participants receive feedback throughout the assignment, enabling them to understand how nonwords and integers relate to one another. The all-or-none association model may be used in this activity to assist researchers better understand how people learn and recall pairings of objects. In this learning task, each item can be in one of two states: G for guessing or L for learned. If an item is in the guessing state, it means that the participant does not know the relationship between the item and its associated value. On the other hand, if an item is in the learned state, it means that the participant has successfully learned the association between the item and its value.

- Item is the guessing state: Participant does not know the relationship of the item.
- Item is in the learned state: Participant has learned the association of the item.

Researchers can better understand learning and find elements that influence it by differentiating between guessing and taught states. This can provide insight into how the brain retains and processes information, assisting in the development of more efficient learning interventions and tactics.

Homogeneous process: A homogeneous process is one in which all items can be treated separately, without any interaction. In the context of the learning task, this means that the learning process for each item is independent of the others. If there is no forgetting, meaning that participants do not transition from the learned state (L) back to the guessing state (G), then the learning process can be modeled as a homogeneous process.

This simplifies the analysis of the learning task and allows researchers to focus on the learning process for individual items

- No forgetting → no transitions from L to G.

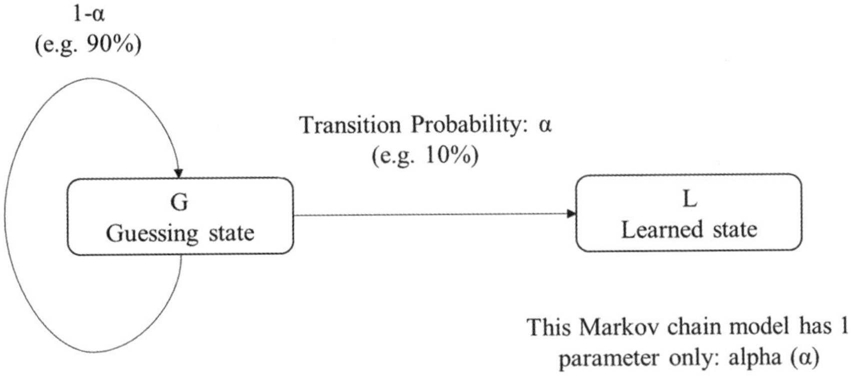

Figure 2.7.8. Modeling perspective of the model

Figure 2.7.8 provides a modeling perspective of the learning task, which can help researchers to better understand and analyze the learning process. An alpha close to 1 means that learning is good, whereas an alpha close to 0 means that learning is bad. This Markov chain model has one parameter only: alpha.

Response from the guessing state: When the item is in the guessing state, there is a change G that by accident a participant gives the correct answer.

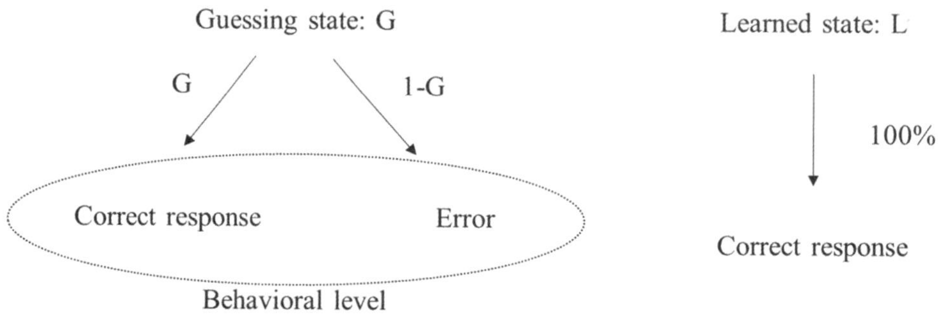

Figure 2.7.9. Item in the guessing state (left = G in [0, 1]) and item in the learned state (right)

In the Figure 2.7.9, the knowledge state space is G (guessing) or L (learned), and the G and L states cannot be observed directly. The behavioral level of the response must be C (correct) or E (error).

Recorded responses of the experimental observations will follow the two-level model:

- Markov chain for {G, L} linked to a behavioral level.
- All-or-none model for learning

Let us assume the data represent inferences or interpretations, not observations.

- C = correct response
- E = incorrect response (error)
- G = guessing
- L = learned

[Table 2.7.7] All-or-none Model for Learning

N=1	N=2	N=3
E	E	C
G	G	G or L ?

Table 2.7.7 shows the All-or-none Model for Learning, where the behavioral level of response must be either correct (C) or incorrect (E), and the knowledge state space can be either guessing (G) or learned (L).

The table provides possible responses for $N = 1$, 2, and 3 trials. Simply put the math results (see Wickens, 1982):

Learning curve and mean errors

$$p_E(n) = (1 - g) * (1 - a)^{n-1}$$

$$p_E(n = 1) = 1 - g$$

$$p_E(n = 2) = (1 - g) * 0.9$$

$$p_E(n = 3) = (1 - g/0.9^2)$$

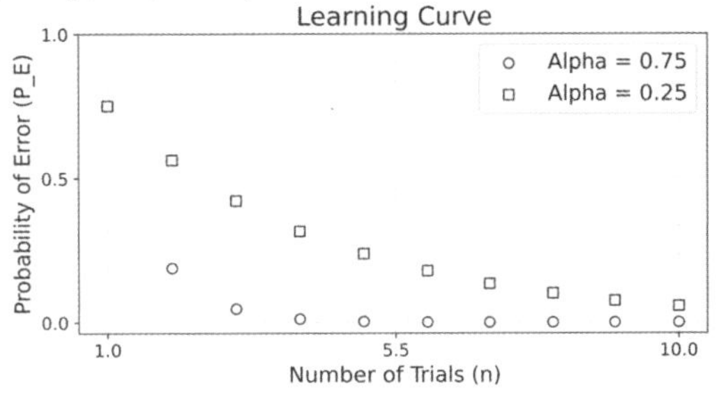

Figure 2.7.10. Representation of the learning curve and mean errors

- Alpha close to 1: fast decay, fast learning
- Alpha close to 0: slow decay, bad learning

We consider items separately. n is the nth presentation of an item, and the probability of making an error at step n is $p_E(n)$. The error probability decays exponentially and is a function of $n \rightarrow$ interpretation as the learning curve (Figure 2.7.10). Thus, mean number of errors is

$$\mu_E = (1 - g)/a$$

Parameter estimation (moment method): Let us compute two expectation values (moments) from the model. The moments depend on parameters a and g, and they have an inverse relationship $\rightarrow$ formula for a and g given experimental estimates for moments. Moment method assuming that g is known

- Solve $\mu_E = (1 - g)/a$ for alpha
- Result: $a = (1 - g)/\mu = a(estim) = (1 - g)/mena\ number\ of\ errors$

Example

- Mean number of errors: $M = (0 + 0 + 2 + 2)/4 = 1$
- Assume that $g = 25\% = 0.25$
- Estimate alpha: $(estim) = (1 - g)/M = 0.75 = 75\%$

+ **Empirical Designs of the Markov Property 2:** Emotional dynamics (Isenhower et al., 2012). Participants were asked to report experiences of positive and negative emotions: 'When you feel something positive or negative during the day, report it.' This is called *event-based recording*.

Time denotes events when something was experienced, n = 1, 2, 3, 4, …, and their discrete time and discrete state is given by

- Time is discrete: events n = 1, 2, 3, …
- The state is discrete: positive or negative feelings.

Markov chain modeling: Consider the only qualitative aspect of a 'positive' or 'negative' response → two states $\{p, n\}$.

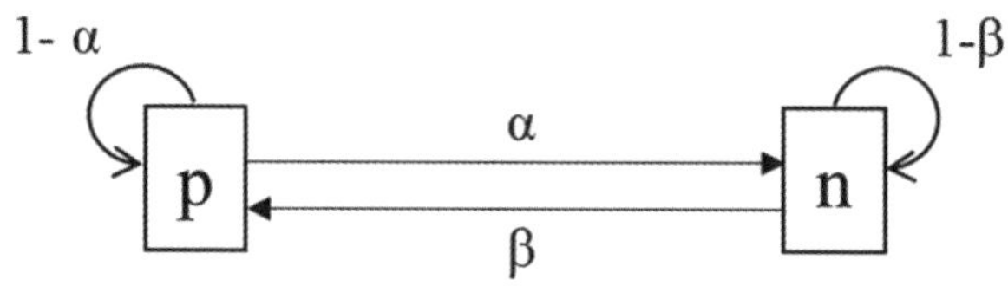

Figure 2.7.11. Schematic representation of the model with two independent parameters: α and β and their results

- Transition probability $p \rightarrow n$: α = 26%
- Transition probability $n \rightarrow p$: β = 53%

Both probabilities are different from the chance level of 50% (Figure 2.7.11; the difference is statistically significant). And the difference between probabilities α and β is significant. Attractor strength for $p(Y_p = 0.63)$ is significant; it is different from the chance level of 0.5 and greater than the chance level. Thus, we can conclude that emotions do not just change by accident; dynamics underlying. There is a greater chance that you will become happy when you are sad than sad when you are happy.

Remarks: In sum, the empirical detail of joint probability, stationarity, the Markov property, and designs based on the Markov property are only a few of the significant ideas that were presented in Part 2 of this topic. These ideas are crucial for comprehending the patterns and behavior of stochastic processes, estimating model parameters from observable data, and creating models for experimental or decision-making purposes. By using these techniques, we may improve stochastic systems for desired outcomes and make better informed.

[2.8]

Markov Chain Modeling, Part 1

Markov Chains, Marginal vs. Conditional Probability, Practical Practice

Now that we have learned about the Markov property, we can define homogenous discrete-time Markov chains. A Markov chain is a Markov process with discrete time and a discrete state space (Tierney, 1996). It is a discrete sequence of states, each drawn from a discrete state space (finite or not), and it follows the Markov property. We can denote a Markov chain by

$$X = (X_n)_{n \in N} = (X_0, X_1, X_2, \dots)$$

The notation means that X is a sequence, where each element of the sequences is denoted by X_n, representing the index or position of the element in the sequence. Where at each instant of time the process takes its values in a discrete set E such that

$$X_n \in E \qquad \forall_n \in N$$

Here, the equation states that for every natural number n, the variable X_n belongs to the set E. In notation, the symbol '∀' is used to represent the universal quantifier 'for all' and the symbol '∈' denotes membership in a set. The Markov property implies that we have

$$P(X_{n+1} = s_{n+1} | X_n = s_n, X_{n-1} = s_{n-1}, X_{n-2} = s_{n-2}, \dots) = P(X_{n+1} = s_{n+1} | X_n = s_n)$$

This is the probability that the value of X at time $n+1$ is equal to s_{n+1} given the value of X at time n is equal to s_n. The vertical bar '|' separates the condition $X_n = s_n$ from the event $X_{n+1} = s_{n+1}$ or which we are calculating the probability.

Notice that this latter formula expresses the fact that for a given historic (where we are now and where we were before), the probability distribution for the next state (where we go next) depends only on the current state and not on past states.

past → present → Future
MARKOV

1. Markov Chains

System analysis comes across the terms Markov chain and Markov process when they start working on models. Markov chains are used nearly everywhere in everyday life (throughout the finance, automobile, food, and gaming industries, etc.). Hence, all analysts must know about them.

+ **Markov Process:** Let us imagine an object moving in a random manner. This object could be a football, a chess player making his or her next move, exchange rates, the price of a stock, the movement of a car, the position of a customer in a queue, a person moving on the road, players on a football field, and so on. These things move in a random manner. Therefore, we can conclude that they are stochastic in nature. They essentially represent an entire system. It is interesting to note that all of these systems can have a Markov property.

As we defined in the previous section (Chapter 2.7' Markov Property), when an object moves in a random fashion and its movements are memoryless, then that object has a Markov property. For instance, imagine that our target object is a football being kicked around by a group of football players. The football can be in any state next (it can move right or left). If our target object is the price of a share of a company, then it can have any numerical value next, such as 0.999, 1, 2, 3, and so on. Again, we can state that the Markov property is memoryless. It implies that the future movement of an object depends only on its present state. This is the most important concept to understand: The Markov property is memoryless.

+ **Markov Chain:** This brings us to the concept of Markov chains (Hermanns, 2002). Imagine an object moving in a random manner; the state of the object (or system) can change. This change is known as a transition, and each transition has a probability associated with other transitions. Simply put, this mathematical system can transition from one state to another, and that transition is based on a probability. Let us assume that the state can be in A or B. When the object is in state A, there is a 40% chance it will remain in state A and a 60% chance it will transition into state B as demonstrated below (Figure 2.8.1):

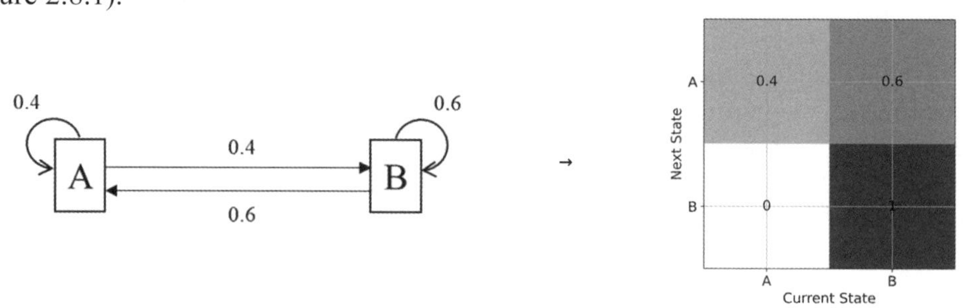

Figure 2.8.1. Schematic representation of the state machine and its matrix

The key to note is that the process has a Markov property, which implies it is memoryless. As a result, the probability of future transitions is not dependent on its past states. Future transitions only depend on the current state. This is the reason why we consider it to be memoryless. This property highlights the two main components of a Markov chain:

- The first component = the state space

The state space is the set of all possible states (places) that a random system can be in. It is denoted as S. For instance, S can be {up, down} or {top, bottom} or {1.1, 2.2, ...} or {positive, negative, neutral} or {AAA, AA, A, B, D, E, ...} and so on. The point to note is that S is the state space that is the set of all possible states the object can be in.

- The second component = transition probabilities

Transition probabilities are a table of probabilities. Each entry i, j in the table informs us about the probability of an object transitioning from state i to state j. There is a probability associated with all the states that needs to be equal to or greater than 0. Plus, the sum of the probability values needs to be 1.

+ **Markov Chain Assumptions:** Let us provide an easy-to-understand explanation of Markov chains, using a practical example. Time is represented as a sequence of discrete time steps, starting from $n = 1$ and extending indefinitely. Dynamics evolve on a set of discrete states.

- State variable = categorical variable (coding by integer) or discrete numerical variable
- Discrete-time process with a discrete state space

The probability of finding the process in a particular state in the future depends only on the present state; it does not depend on states preceding this present state. The future of the process is conditioned on the present of the process but not the past. The future of the process is independent of all past states except for the one the process currently occupies.

Transition probabilities and transition probabilities matrix: States $j = 1, \ldots, M$. Transition probability $p(X(n+1) = j' \leftarrow X(n) = j)$ is the probability that a transition to state j' at time step $n + 1$ occurs if the system is in state j at time step n

[Table 2.8.1] Matrix of a Three-State Example with States A, B, and C

	$X(n) = \text{A}$	$X(n) = \text{B}$	$X(n) = \text{C}$
$X(n+1) = \text{A}$	$P(\text{A} \leftarrow \text{A})$	$P(\text{A} \leftarrow \text{B})$	$P(\text{A} \leftarrow \text{C})$
$X(n+1) = \text{B}$	$P(\text{B} \leftarrow \text{A})$	$P(\text{B} \leftarrow \text{B})$	$P(\text{B} \leftarrow \text{C})$
$X(n+1) = \text{C}$	$P(\text{C} \leftarrow \text{A})$	$P(\text{C} \leftarrow \text{B})$	$P(\text{C} \leftarrow \text{C})$

Table 2.8.1 shows the transition probabilities matrix for a three-state system with states A, B, and C, where each entry represents the probability of transitioning from one state to another at the next time step.

In time-homogeneous processes, transition probabilities depend on states but not on time step n: $p(j \leftarrow j')$. In non-homogeneous processes, the transition probabilities depend on states and depend explicitly on time step n: $p(j \leftarrow j', n)$.

In nonlinear Markov processes, the transition probabilities depend on states and depend implicitly on time via simple probabilities $p_k(n)$: $p(j \leftarrow j', \{p_1, \dots, p_m\})$.

\+ **Markov Chain Model:** The Markov chain model is heuristically derived in this section, and the resulting two-state Markov chain is illustrated in Figure 2.8.2

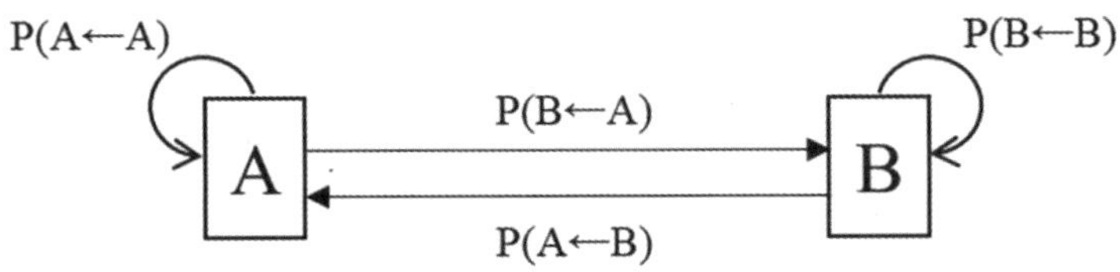

Figure 2.8.2. Representation of the two-state Markov chain

Let us think about what the probability is that the system is in state A at step $n + 1$? Consider state A

- Probability of realizations in A at step n: $p(X(n) = A)$
- Probability that those realizations make the transition to A: $p(A \leftarrow A)$
- Probability that realizations from A go to A: $p(A \leftarrow A) \times p(X(n) = A)$

Consider state B

- Probability of realizations in B at n: $p(X(n) = B)$
- Probability that those realizations make the transition to A: $p(\mathrm{A} \leftarrow B)$
- Probability that realizations from B go to A: $p(\mathrm{A} \leftarrow \mathrm{B}) \times p(X(n) = B)$

Here, the probability that the state of a Markov chain at time n is A, denoted as $p(X(n) = \mathrm{A})$, can be calculated using the following formula

$$p(X(n) = A) = p(A \leftarrow A) \times p(X(n) = A) + p(A \leftarrow B) \times p(X(n) = B)$$

where, the probability is calculated as that A is the result of a transition from A, multiplied by probability that X at time n is equal to A, plus the probability that A is the result of a transition from B, multiplied by the probability that X at time is equal to B.

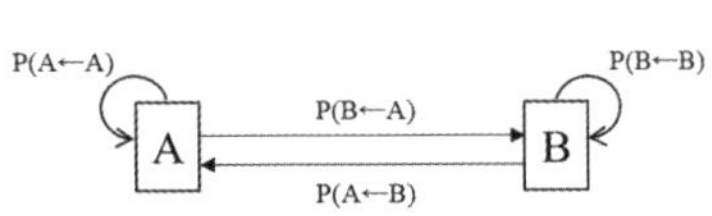

$$\begin{aligned} A: p(X(n+1) = A) &= p(X(n+1) = A \leftarrow X(n) = A)p(X(n) = A) \\ &+ p(X(n+1) = A \leftarrow X(n) = B)p(X(n) = B) \end{aligned}$$

$$\begin{aligned} B: p(X(n+1) = B) &= p(X(n+1) = B \leftarrow X(n) = A)p(X(n) = A) \\ &+ p(X(n+1) = B \leftarrow X(n) = B)p(X(n) = B) \end{aligned}$$

For a simplified notation:

$$p_j(n) = p(X(n) = j) and\ p(j \leftarrow k) = p(X(n+1) = j \leftarrow X(n) = k)$$

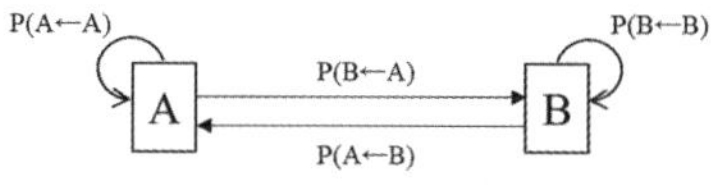

$$p_A(n+1) = p(A \leftarrow A)p_A(n) + p(A \leftarrow B)p_B(n)$$
$$p_B(n+1) = p(B \leftarrow A)p_A(n) + p(B \leftarrow B)p_B(n)$$

For a probability distribution vector: Vector and matrix notation

$$\begin{pmatrix} p_A(n+1) \\ p_B(n+1) \end{pmatrix} = \begin{pmatrix} p(A \leftarrow A) & p(A \leftarrow B) \\ p(B \leftarrow A) & p(B \leftarrow B) \end{pmatrix} * \begin{pmatrix} p_A(n) \\ p_B(n) \end{pmatrix}$$

Component notation

$$p_j(n+1) = \sum_{k \in \{A,B\}} p(j \leftarrow k)p_k(n), \qquad j \in \{A, B\}$$

The equation represents a recursive relationship between the probability of different events at different points in time. The probability $p_j(n+1)$ at time $n+1$ depends on the probabilities $p_k(n)$ at time n. The symbol of sum is taken over all possible values of k which are specified by the set $\{A, B\}$.

We discussed how a system evolves over time in which the chance of changing states solely depends on the present state. The transition probabilities can be expressed as matrices for two, three, or M states.

- For two states

$$\begin{pmatrix} p_A(n+1) \\ p_B(n+1) \end{pmatrix} = \begin{pmatrix} p(A \leftarrow A) & p(A \leftarrow B) \\ p(B \leftarrow A) & p(B \leftarrow B) \end{pmatrix} * \begin{pmatrix} p_A(n) \\ p_B(n) \end{pmatrix}$$

- For three states

$$\begin{pmatrix} p_A(n+1) \\ p_B(n+1) \\ p_C(n+1) \end{pmatrix} = \begin{pmatrix} p(A \leftarrow A) & p(A \leftarrow B) & p(A \leftarrow C) \\ p(B \leftarrow A) & p(B \leftarrow b) & p(B \leftarrow C) \\ p(C \leftarrow A) & p(C \leftarrow B) & p(C \leftarrow C) \end{pmatrix} * \begin{pmatrix} p_A(n) \\ p_B(n) \\ p_C(n) \end{pmatrix}$$

- For M states

$$\vec{p}(n+1) = T * \vec{p}(n), \qquad \vec{p}(n) = \left(p_1(n), \dots, p_M(n)\right)^T,$$
$$T \leftarrow Transition\ probability\ (M * M) matrix$$

- Thus, the M states component notation is

$$p_j(n+1) = \sum_{k=1}^{M} p(j \leftarrow k)p_k(n), \qquad j = 1, \dots, M$$

This equation is similar to the previous, but here the sum is taken over a range of values for k, from 1 to M. This means that the probability $p_j(n+1)$ at time $n+1$ is calculated as the sum of the probabilities of transitioning from each possible event k to event j, multiplied by the probability of event k at time n.

2. Marginal vs. Conditional Probability

To get into more detail about the Markov chain model through actual practice, we may be interested in the probability of an event for one random variable, irrespective of the outcome of another random variable (e.g., the probability of $X = A$ for all results of Y). The probability of one event in the presence of all (or a subset of) outcomes of another random variable is called the marginal probability or the marginal distribution (Gelfand & Smith, 1990) (Figure 2.8.3).

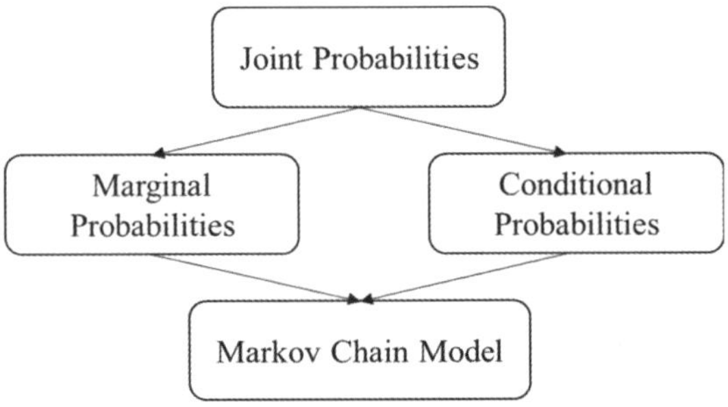

Figure 2.8.3. The Markov chain model: Rigorous derivation

Figure 2.8.3 shows a visual representation of the Markov chain model, demonstrating the marginal probability or the marginal distribution of one event in the presence of all or a subset of outcomes of another random variable.

+ **Marginal Probability:** It is called the marginal probability because if all outcomes and probabilities for the two variables were laid out together in a table (X as columns, Y as rows), then the marginal probability of one variable (X) would be the sum of the probabilities of the other variable (Y rows) on the margin of the table. There is no special notation for the marginal probability; it is just the sum or union over all of the probabilities of all events for the second variable for a given fixed event for the first variable.

$$P(X = A) = sum\ P\ (X = A, Y = Yi) for\ all\ Y$$

This critical foundational rule in probability is referred to as the sum rule. The marginal probability is different from the conditional probability (described next) because it considers the union of all events for the second variable rather than the probability of a single event.

+ **Conditional Probability:** We may be interested in the probability of an event given the occurrence of another event. The probability of one event given the occurrence of another event is called the conditional probability (Uttley, 1959).

The conditional probability of one variable to one or more random variables is referred to as the conditional probability distribution. For example, the conditional probability of event A given event B is written formally as follows:

$$P(A \; given \; B)$$

The 'given' is denoted using the pipe operator (|): *P*(A|B). The conditional probability of event A given event B is calculated as follows:

$$P(A \; given \; B) = P(A \; and \; B)/P(B)$$

This calculation assumes that the probability of event B is not 0 (i.e., is not impossible). The notion of event A given event B does not mean that event B has occurred (i.e., is certain); instead, it is the probability of event A occurring after or in the presence of event B for a given trial.

Let us return to multiple random variables. Consider the following:

- Two random variables *X* and *Y*

For example, gender (female [F], male [M]) and academic degree (has a master's degree [MSc], does not have a master's degree [NoMSc])

+ **Relation to Joint Probability:** Using conditional probability, we may determine a connection between the probability of a single variable, $p(X = i)$ or $p(y = j)$, and the probability of a pair of variables, $p(x = i, y = j)$. To do this, we must take into account the conditional probability $p(x = i| \; y = j)$.

$$p(X = i, Y = j) = p(X = i \;|Y = j) \cdot p(y = j)$$

This equation shows how the two variables, $p(x = i, y = j)$, can be related to the conditional probability of one variable given the other, $p(x = i| \; y = j)$, and the probability of the second variable, $p(y = j)$.

$$p(X = i, Y = j) = \lim_{N \to \infty} \frac{1}{f(y = j)} \sum_{k=1,\ldots,N: X^{(k)}=i \,\wedge\, Y^{(k)}=j} 1, \qquad f(y = j) = \sum_{k=1,\ldots,N: Y^{(k)}=j} 1$$

Here, the equation defines the probability of two events occurring simultaneously, $X = i$ and $Y = j$ which is calculated as the limit of a series as N approaches infinity. The $f(y = j)$ is a normalizing factor that ensures that the probability is a valid value between 0 and 1.

Calculation Example

The mathematics above might not be intuitive to everyone. Visualizations, however, can be a great tool for reinforcing complex topics. A tree diagram is one example that can help you break down a general problem into smaller components, which is perfect for probability problems that involve multiple events that lead to a variety of outcomes.

For example, let us create a tool that helps answer the following question: If you have a bag of 23 marbles (5 green, 8 blue, and 10 red), what is the probability that you will randomly pull out a blue marble and a green marble?

The probability of grabbing a blue marble is 35%, because there are 8 ways you can get a blue marble and 23 total potential outcomes.

Now, if you pull out a blue marble, the probability of grabbing a green marble from the bag is 23%: 5 green marbles divided by 22 potential outcomes (notice how the total number of outcomes changes the second time, hence the change in probability).

Finally, calculating the probability of both these events happening involves multiplying the probability of both events ($.35 \times .23 = 8\%$).

Even though conditional probability may seem elementary compared to the more advanced concepts in data analysis, having a solid understanding of the foundation on which data science is built is extremely important.

+ Probability Practice Model (Effect of Gender on Academic Degree): To analyze the effect of gender on obtaining a degree in the academic field, a probability model may be developed. We will take into account the following variables when building this model: X, which stands for gender (Female [F], Male [M]), and Y, which stands for academic standing (Tables 2.8.2 and 2.8.3; Master's Degree in Science [MSc], No Master's Degree in Science [NoMSc]).

[Table 2.8.2] Variable representation 1

Simple probability	Among males, how likely is MSc?	What is the probability of a male with MSc and without MSc?
$p(X = \text{M}) = 40\% = 0.4$	$p(\text{MSc}\|\text{M}) = 10\% = 0.1$	$p(\text{MSc} \wedge \text{M}) = 0.1 \times 0.4 = 0.04 = 4\%$ $\rightarrow p(\text{MSc} \wedge \text{M}) = p(\text{MSc}\|\text{M})p(\text{M})$
	$p(\text{NoMSc}\|\text{M}) = 90\% = 0.9$	$p(\text{NoMSc} \wedge \text{M}) = 0.9 \times 0.4 = 0.36 = 36\%$ $\rightarrow p(\text{NoMSc} \wedge \text{M}) = p(\text{NoMSc}\|\text{M})p(\text{M})$

[Table 2.8.3] Variable representation 2

Simple probability	Among females, how likely is MSc?	What is the probability of a female with MSc and without MSc?
$p(X = \text{F}) = 60\% = 0.6$	$p(\text{MSc}\|\text{F}) = 5\% = 0.5$	$p(\text{MSc} \wedge \text{F}) = 0.05 \times 0.6 = 0.03 = 3\%$ $\rightarrow p(\text{MSc} \wedge \text{F}) = p(\text{MSc}\|\text{F})p(\text{F})$
	$p(\text{NoMSc}\|\text{M}) = 95\% = 0.95$	$p(\text{NoMSc} \wedge \text{F}) = 0.95 \times 0.6 = 0.57 = 57\%$ $\rightarrow p(\text{NoMSc} \wedge \text{F}) = p(\text{NoMSc}\|\text{F})p(\text{F})$

For marginal probability, how can we compute simple probabilities from joint probabilities? For example, the joint probability is given as

$$\begin{array}{ll} p(\mathrm{MSc} \wedge \mathrm{M}) = 0.04 = 4\% & p(X = \mathrm{M}) = p(\mathrm{MSc} \wedge \mathrm{M}) + p(\mathrm{NoMSc} \wedge \mathrm{M}) = 0.4 \\ p(\mathrm{NoMSc} \wedge \mathrm{M}) = 0.36 = 36\% & p(X = \mathrm{F}) = p(\mathrm{MSc} \wedge \mathrm{F}) + p(\mathrm{NoMSc} \wedge \mathrm{F}) = 0.6 \\ p(\mathrm{MSc} \wedge \mathrm{F}) = 0.03 = 3\% \quad \rightarrow & p(Y = \mathrm{MSc}) = p(\mathrm{MSc} \wedge \mathrm{M}) + p(\mathrm{MSc} \wedge \mathrm{F}) = 0.07 \\ p(\mathrm{NoMSc} \wedge \mathrm{F}) = 0.57 = 57\% & p(Y = \mathrm{NoMSc}) = p(\mathrm{NoMSc} \wedge \mathrm{M}) + p(\mathrm{NoMSc} \wedge \mathrm{F}) = 0.93 \end{array}$$

In general: X in $\{1, \ldots, M\}$, Y in $\{1, \ldots, M'\}$

$$p(X = i) = \sum_{j=1}^{M\prime} p(X = i \wedge Y = j), \qquad p(X = j) = \sum_{i=1}^{M} p(X = i \wedge Y = j)$$

This equation means that marginal probabilities (reduced probabilities) are simple probabilities calculated from joint probabilities.

+ **M-state Markov Chain Model:** Then, For an M-state Markov chain model, we may utilize the marginal probability that is generated by adding up all potential prior states k to predict the likelihood of being in state j at time $n + 1$, or $p_j(n + 1)$.

Calculate $p(X(n + 1) = j)$ and add the results over k.

$$p(X(n + 1) = j, (n) = k) \quad \rightarrow \quad p_j(n + 1) = \sum_{j=1}^{M} p(X(n + 1) = j, (n) = k)$$

How can we relate this to $p_j(n)$? To relate this to the probability of being in state j at time n, $p_j(n)$, we can use conditional probability, specifically the transition probability $p(j \leftarrow k)$, which is the probability of transitioning from state k to state j.

$$p(X(n + 1) = j, X(n) = k) = p(j \leftarrow k)p_k(n)$$

$$p(j \leftarrow k) = p(X(n + 1) = j \mid X(n) = k)$$

This can be calculated using the conditional probability $p(X(n + 1) = j \mid X(n) = k)$, and multiplied by $p_k(n)$ to obtain the probability of being in state k at time n and transitioning to state j at time $n + 1$.Taken together, this gives us the following:

$$p_j(n + 1) = \sum_{j=1}^{M} p(j \leftarrow k){p_k}^{(n)}$$

In order to update the Markov chain from time n to time $n + 1$, the equation employs the summation symbol to add up all feasible prior states. Given the transition probabilities $p(jk)$ and the probabilities of being in each state at time n, ${p_k}^{(n)}$, for k = 1, 2,..., M, the equation determines the likelihood of being in state j at time $n + 1$.

Empirical Understanding of the Joint, Marginal, and Conditional Probability (by hand)

Now we may need to understand random variables and their probability distributions to work on many different problems. We are going to start this discussion with the following scenario.

Scenario: A survey was carried out among 500 strangers in a city to determine people's favorite sports. The options were football and rugby; all other sports were grouped together under 'other.' The results of the survey are displayed in Table 2.8.4.

[Table 2.8.4] Results of the Survey

	Male	Female	Total
Football	120	75	195
Rugby	100	25	125
Other	50	130	180
Total	270	230	500

Table 2.8.4 is not quite a probability distribution. If we want to get the probability distribution we can simply divide each number in Table 1 by 500 (the number of observations), and the results will be as given in Table 2.8.5.

[Table 2.8.5] Probability Distribution

	Male	Female	Total
Football	0.24	0.15	0.39
Rugby	0.2	0.05	0.25
Other	0.1	0.26	0.36
Total	0.54	0.46	1

Joint Probability: As previously noted, the joint probability is a measure that is used to calculate the probability of two events occurring together at the same time: P(A and B) or P(A, B). For example, using Table 2.8.5, we can see that the joint probability of someone being a male and liking football is 0.24.

[Table 2.8.6] Joint Probability Distribution

	Male	Female	Total
Football	0.24	0.15	0.39
Rugby	0.2	0.05	0.25
Other	0.1	0.26	0.36
Total	0.54	0.46	1

Notice that the cells highlighted in Table 2.8.6 (the joint probability distribution) must sum to 1 because everyone in the distribution must be in one of the cells. The joint probability is symmetrical, which means that P(Male and Football) = P(Football and Male). We can also use it to find other types of distributions, such as the marginal distribution and the conditional distribution.

Marginal Distribution: In probability theory and statistics, the marginal distribution of a subset of a collection of random variables is the probability distribution of the variables contained in the subset. It gives the probabilities of various values of the variables in the subset without reference to the values of the other variables. Simply put, the marginal probability is the probability of an event irrespective of the outcome of another variable: P(A) or P(B).

[Table 2.8.7] Marginal Distribution

	Male	Female	Total
Football	0.24	0.15	0.39
Rugby	0.2	0.05	0.25
Other	0.1	0.26	0.36
Total	0.54	0.46	1

Notice whether we ignore gender or sport our marginal distributions must sum to 1 (Table 2.8.7). An interesting fact about marginal probability is that all the marginal probabilities appear in the margins. Hence, P(Female) = 0.46, which completely ignores the sport the female prefers, and P(Rugby) = 0.25 completely ignores gender.

Conditional Probability: Conditional probability is one of the most fundamental concepts in probability theory and in general opinion. It defines the probability of an event occurring given that another event has occurred (by assumption, presumption, assertion, or evidence).

To make sense of this, let us again use equation $[P(A|B) = P(A,B)/(P(B)]$. If we want to calculate the probability that a person will like rugby given that they are a female, we must divide the joint probability that the person is female and likes rugby (P[Female and Rugby]) by the probability of the condition. In this case, the probability is that the person is female (P[Female]), which we can work out from the margin of Table 2 to be 0.46. Hence, we get 0.11 (to two decimal places). Let us write that up more neatly: P(Female, Rugby) = 0.05, P(Female) = 0.46, P(Rugby | Female) = 0.05 / 0.46 = 0.11 (to two decimal places). If we continued to fill in the probability of preferring a sport given that the person being observed is a female, then we would have a conditional probability distribution.

+ **Hierarchy:** It is possible to determine the joint probabilities for the complete Markov chain using the probabilities at a single time point and the transition probabilities between several time points.

We can denote the Markov chain hierarchy simply by

- One time point

$$p_j(n)$$

- Two time point

$$p(X(n+1) = j', X(n) = j) = p(j' \leftarrow j)p_j(n)$$

The equation represents the probability of an event j' occurring at time $n+1$ given that event j occurred at time n. It is calculated as the probabtility of transitioning from event j to event j', multiplied by the probability of event j occurring at time n.

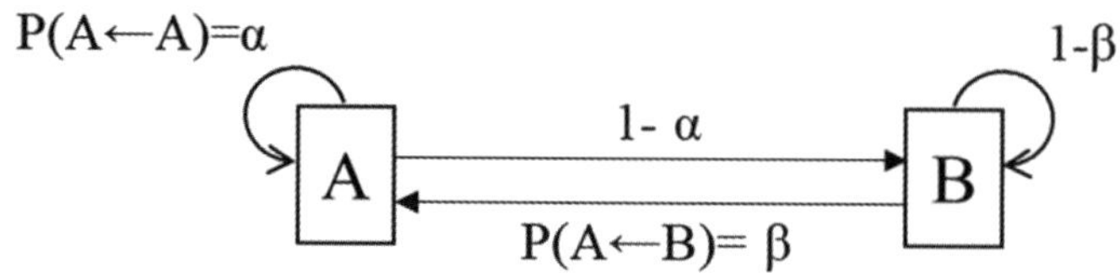

Figure 2.8.4. Schematic representation of the parameters (only two independent case)

Again, according to Figure 2.8.4, the two-state model only contains two independent parameters, denoted by two independent parameters: α and β.

- Three time point:

$$p(X(n+2) = j'', X(n+1) = j, X(n) = j)) = p(j'' \leftarrow j)(j' \leftarrow j)p_j(n)$$

The equation then represents the probability of an event j'' occurring at time $n+2$ given that event j' and j occurred at times $n+1$ and n, respectively.

+ **Stationarity:** We can use time series to extrapolate interesting information by analyzing the behavior of a given variable over time. If relevant findings emerge, it is possible to predict the trend of the value in the future. However, before we can analyze a time series, we need to make some assumptions, as our object of interest is stochastic.

To understand the component of time series, we want this quantity to converge (or diverge) toward the true mean of the time series, and we can achieve that under the logics that follow.

Two-state model: general (nonstationary and stationary)

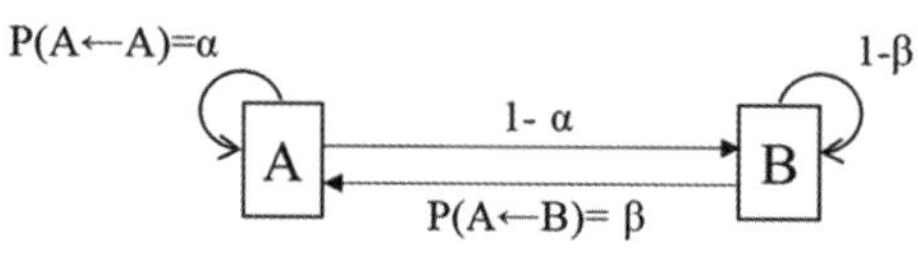

$$\begin{pmatrix} p_A(n+1) \\ p_B(n+1) \end{pmatrix} = \begin{pmatrix} \alpha & \beta \\ 1-\alpha & 1-\beta \end{pmatrix} * \begin{pmatrix} p_A(n) \\ p_B(n) \end{pmatrix}$$

Stationary distribution is independent of time step n and is determine the distribution (p_A, p_B) such that

$$\begin{pmatrix} p_A \\ p_B \end{pmatrix} = \begin{pmatrix} \alpha & \beta \\ 1-\alpha & 1-\beta \end{pmatrix} * \begin{pmatrix} p_A \\ p_B \end{pmatrix}$$

In case of three-state model: stationary

$$\begin{pmatrix} p_A \\ p_B \\ p_C \end{pmatrix} = \begin{pmatrix} p(A \leftarrow A) & p(A \leftarrow B) & p(A \leftarrow C) \\ p(B \leftarrow A) & p(B \leftarrow B) & p(B \leftarrow C) \\ p(C \leftarrow A) & p(C \leftarrow B) & p(C \leftarrow C) \end{pmatrix} * \begin{pmatrix} p_A \\ p_B \\ p_C \end{pmatrix}$$

M-state model: Stationary

$$\vec{p} = T \cdot \vec{p} \quad \rightarrow$$

Stationary distribution = eigenvector of matrix T with eigenvalue 1

M-state model in components: Stationary

$$p_j = \sum_{j=1}^{M} p(j \leftarrow k) p_k \quad \rightarrow$$

The stationary distribution satisfies M algebraic relations

Stationary distribution for a two-state model

$$\begin{pmatrix} p_A \\ p_B \end{pmatrix} = \begin{pmatrix} \alpha & \beta \\ 1-\alpha & 1-\beta \end{pmatrix} * \begin{pmatrix} p_A \\ p_B \end{pmatrix} \quad \rightarrow \quad \begin{aligned} p_A &= \alpha p_A + \beta p_B \\ p_B &= (1-\alpha) p_A + (1-\beta) p_B \end{aligned}$$

Because $1 = p_A + p_B$ holds, there is only one independent equation and one independent variable, so we use the first equation.

$$p_A = \alpha p_A + \beta p_B \rightarrow p_A = \alpha p_A + \beta(1 - p_A), \quad p_A = \frac{\beta}{1-\alpha+\beta}, \quad p_B = 1 - p_A$$

Here, the equation represents a relationship between the probabilities of two events A and B. The event A is calculated as the sum of two terms; A occurring given that it occurred in the previous time step, multiplied by the probability of A occurring at the previous time step, plus the probability of A occurring given that B occurred in the previous time step, multiplied by the probability of B occurring at the previous time step.

The probability of event B is then calculated as 1 minus the probability of event A. For example: Student attendance model

$p(\mathrm{A} \rightarrow \mathrm{A}) = 0.85 = \alpha$
$p(\mathrm{NoA} \rightarrow \mathrm{A}) = 0.70 = \beta$ $\rightarrow$

$$p_A = \frac{\beta}{1-\alpha+\beta} = \frac{0.7}{1-0.85+0.7} \approx 0.82$$

$$p_B = 1 - p_A = 1 - 0.82 = 0.18$$

+ **Why Do We Need It?** Imagine we have a stochastic process, that is, an infinite sequence of random variables that are defined over Ω (the event space) and return a number within R.

$$\{X_t\}_{t=-\infty}^{+\infty} = \{..X_{-k}, .., X_0, ..., X_{+k} ...\}$$

Each random variable is indexed at a fixed point in time and has a probability distribution. Nevertheless, at each point in time t only one realization of our stochastic process in t will be shown: We have no information about the other possible realizations. Thus, we cannot compute the cross-sectional (for each t) sample mean of our process, that is,

$$E(X_t) = \frac{1}{N}\sum_{i=1}^{N} X(\omega_i)$$

Here, the $E(X_t)$ is the expected value of X at time t, and N is the number of possible outomes that the random variable can take on. The sum is taken over all possible values of i from 1 to N, and $X(\omega_i)$ represents the value of X at the i-th outcome. The impossibility of computing this quantity relies on the impossibility of observing the realization of X for more than one ω at a time.

We can visualize this restriction as follows (Figure 2.8.5):

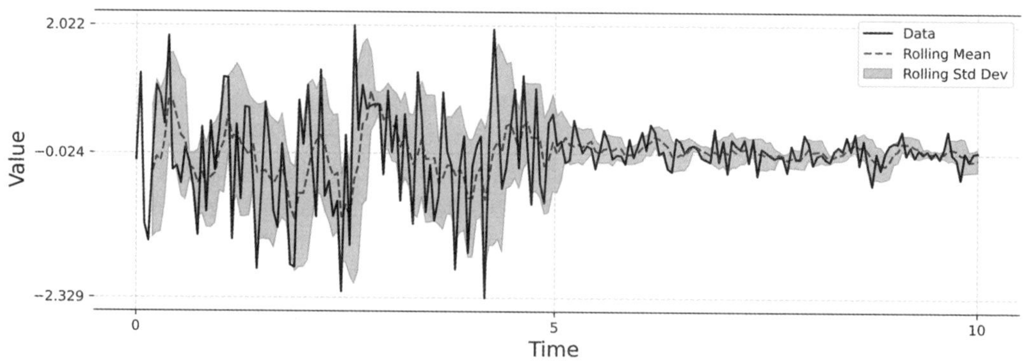

Figure 2.8.5. Output of the data set: The x-axis denotes time, and the y-axis denotes the value of the data. Notice the Rolling Mean denotes the average of the previous n data points. And the Rolling Std Dev is the standard deviation of the previous n data points.

At each moment t in time, we have a probability distribution of our random variable X_t, but we can observe only one realization. The only way to compute the cross-sectional sample mean is by drawing, for the same phenomenon, multiple time series, but this is impossible because it requires the ability to go back in time.

As an alternative, we might change our framework and compute the sample average of our process across time.

$$E(X_t) = \frac{1}{T}\sum_{i=1}^{T} X_t(\omega)$$

We need the quantity to converge toward the true mean of the time series, and we can achieve that under two conditions. Stationarity is somewhat sufficient for the sample to be a good approximation of the real mean (trend), but it is not all the information needed to show the data's convergence or divergence characteristics, because it can be manipulated by other further conditions to reach goals.

3. Practical Practice with Markov Chain Models

In this exercise we run Markov chain models and the probability process.

Problem 1 (Markov chain models): The trajectories of Markov chain models for a deterministic three-card shuffle show that the three-card shuffle game exhibits periodic solutions under certain conditions. Let us specify with an example of three-card shuffle as below (Table 2.8.8):

[Table 2.8.8] Three-card Shuffle

#	1	2	3	4	5	6
State	ABC	ACB	BAC	BCA	CAB	CBA

Moving the first card: Period of 3	Moving the second card: Period of 2	Moving the last card: Period of 1
If the initial state is ABC, ABC → BCA → CAB → ABC 1 → 4 → 5 → 1 If the initial state is BAC, BAC → ACB → CBA → BAC 3 → 2 → 6 → 3	If the initial state is ABC, ABC → ACB → ABC 1 → 2 → 1 If the initial state is BAC, BAC → BCA → BAC 3 → 4 → 3	If the initial state is ABC, ABC → ABC 1 → 1 If the initial state is BAC, BAC → BAC 3 → 3

(a) Consider the deterministic case in which a participant always takes the card in the first position (i.e., the card to the left). What do solutions in this case look like when the participant starts with ABC? What do they look like when he or she starts with BAC? Describe the trajectories using both three-letter arrangements and integer-state labels (1, 2, …, 6). How long is the period? (b) Describe the periodic solutions for a participant who always picks the middle card. Give two examples of periodic trajectories. How long is the period? (c) What happens when a participant always takes the last card (i.e., the card to the right)?

Problem 2 (joint, conditional, and marginal probability): In a box are marbles of two sizes (small and large) and three colors (red, blue, and yellow). There are 4 small red marbles, 12 small blue marbles, 8 small yellow marbles, 8 large red marbles, 3 large blue marbles, and 1 large yellow marble. We shake the box and pick out one marble. Assume that every marble in the box has the same chance of being picked.

(a) Calculate all six possible joint probabilities: p(small, red), p(small, blue), and so on (round the results to two decimal places). Because of rounding errors, the six probabilities will not add up to 1. Take one of the probabilities (it does not matter which one) and add 0.01 to make sure that the sum of all joint probabilities equals 1.

[Table 2.8.9] Probability of p(small|red) = p(small/red) × p(red/marble)

p(small, red) = 4/36 = .33 × .33 = .11	p(large, red) = 8/36 = .67 × .33 = .22
p(small, blue) = 12/36 = .8 × .42 = .34	p(large, blue) = 3/36 = .2 × .42 = .08
p(small, yellow) = 8/36 = .89 × .25 = .22	p(large, yellow) = 1/36 = .11 × .25 = .03

Table 2.8.9 shows the six possible joint probabilities for small and large marbles of red, blue, and yellow colors, and calculates the probabilities for *p*(small|red), *p*(large|red), *p*(small|blue), *p*(large|blue), *p*(small|yellow), and *p*(large|yellow), with rounding errors corrected by adding 0.01 to one of the probabilities.

(b) Determine the two conditional probabilities that you will find a marble of a particular size in your hand given that you have picked out a red marble; that is, determine *p* (small|red) and *p* (large|red). Likewise, determine *p* (small|blue) and *p* (large|blue). Finally, determine *p*(small|yellow) and *p*(large|yellow).

[Table 2.8.10] Probability of *p*(small/red) = *p*(*p*[small]/red)

p(small, red) = 4/(4 + 8) = 1/3 = .33	*p*(large, red) = 8/(4 + 8) = 2/3 = .67
p(small, blue) = 12/(12 + 3) = 4/5 = .8	*p*(large, blue) = 3/(12 + 3) = 1/5 = .2
p(small, yellow) = 8/(8 + 1) = .89	*p*(large, yellow) = 1/(8 + 1) = .11

Table 2.8.10 shows the conditional probabilities of finding a small or large marble given the color of the marble drawn from the bag, where *p*(small|red) = 1/3, (large|red) = 2/3, *p*(small|blue) = 0.8, *p*(large|blue) = 0.2, *p*(small|yellow) = 0.89, and *p*(large|yellow) = 0.11.

(c) Calculate from the joint probabilities determined in Problem 2a the marginal probability *p*(red) of picking out a red marble. In addition, calculate *p*(red), that is, the probability of picking out a red marble, directly from the numbers of marbles given in the text above and compare the results. Then calculate the probabilities *p*(blue) and *p*(yellow) using the two methods (round the results for the second, direct method to two decimal places).

[Table 2.8.11] Probability of *p*(red) = *p*(red/small) + *p*(not red/small)

p(red) = (.33 × .33 = .11) + (.67 × .33 = .22) = .33	*p*(red) = 12/36 = 1/3 = .33
p(blue) = (.8 × .42 = .34) + (.2 × .42 = .08) = .42	*p*(blue) = 15/36 = 5/12 = .42
p(yellow) = (.89 × .25 = .22) + (.11 × .25 = .03) = .25	*p*(yellow) = 9/36 = 1/4 = .25

Table 2.8.11 shows the probability calculations for picking out marbles of different colors. The table includes the calculated marginal probability of picking out a red marble *p*(red), and the probabilities of picking out blue and yellow marbles calculated using both the joint probabilities determined in Problem 2a and the direct method. The results from both methods are rounded to two decimal places.

(d) Calculate the joint probability of picking out a marble that has a particular size and is red, that is, *p* (size, red), for size = small and size = large using the conditional probability *p*(size|red) determined in Problem 2b and the marginal probability *p*(red) determined in Problem 2c—use *p* (red) obtained from the first method. Likewise, calculate in the same way the joint probabilities *p*(size, blue) and *p*(size, yellow) for size = small and size = large (round the results to two decimal places). Compare your results to those for Problem 2a.

[Table 2.8.12] Probability of p(red) = p(small/red) × p(red)

p(S_size|red) = p(small/red) × p(red) = .33 × .33 = .11	p(L_size|red) = p(large/red) × p(red) = .67 × .33 = .22
p(S_size|blue) = p(small/blue) × p(blue) = .8 × .42 = .34	p(L_size|blue) = p(large/blue) × p(blue) = .2 × .42 = .08
p(S_size|yellow) = p(small/yellow) × p(yellow) = .89 × 25 = .22	p(L_size|yellow) = p(large/yellow) × p(yellow) = .11 × 25 = .03

Table 2.8.12 shows the joint probabilities of picking out a marble that has a particular size and is red, blue, or yellow. The probabilities are calculated using the conditional probability p(size|red) determined in Problem 2b and the marginal probability p(red), p(blue), and p(yellow) determined in Problem 2c. The joint probabilities for size = small and size = large are presented for each color.

(e) Determine the three conditional probabilities that you will find a marble of a particular color in your hand given that you have picked out a small marble; that is, determine p(red|small), p(blue|small), and p(yellow|small). Likewise, determine the conditional probabilities p(red|large), p(blue|large), and p(yellow|large).

[Table 2.8.13] Probability of p(red) = p(red/small)

p(red|small) = 4/(4 + 12 + 8) = 1/6 = 0.17	p(red|large) = 8/(8 + 3 + 1) = 2/3 = 0.67
p(blue|small) = 12/(4 + 12 + 8) = 1/2 = 0.5	p(blue|large) = 3/(8 + 3 + 1) = 1/4 = 0.25
p(yellow|small) = 8/(4 + 12 + 8) = 1/3 = 0.33	p(yellow|large) = 1/(8 + 3 + 1) = 1/12 = 0.08

Tables 2.8.13 represent various calculations of joint probabilities and conditional probabilities for a set of marbles with different sizes and colors. These calculations involve determining the probabilities of picking out a marble of a certain size and/or color given certain conditions, such as having picked out a red marble or a small marble. The probabilities are rounded to two decimal places in some cases, and the results are compared to those obtained from previous calculations.

Remarks: For the analysis and forecasting of probabilistic systems with time dependency, Markov chain modeling is a potent and popular method. A hierarchy of joint probability distributions that explain the behavior of the system over time are produced using the notions of marginal and conditional probability, which are crucial to the construction of Markov chain models. Furthermore, Markov chains have the potential to stay stationary throughout time, which denotes that the system's behavior does not change over time. Markov chain models may help practitioners and scholars better understand complex systems and make defensible judgments in a range of industries, including banking, engineering, and biology.

[2.9]

Markov Chain Modeling, Part 2

Markov Components, Random Walks and Monte Carlo Simulation, Practical Practice

Time series is the use of stochastic methods to predict future behavior based on data (i.e., historical data). It is similar to other statistical learning approaches, such as supervised or unsupervised learning. However, stochastic forecasting has many nuances that make it different from regular analysis. From data processing all the way to model validation, time series analysis is a different beast.

Venturing into the realm of systems analysis, we aim to bolster decision-making, recognizing that confident forecasts of future events can be advantageous across various sectors and challenges, as suggested by Twiss (1992). For instance, consider a major sporting event like the World Cup or Olympics; with accurate anticipations of visitor numbers, the host country can strategically plan for aspects such as staff deployment, budget allocation, or potential infrastructure expansion. Integrating this, we then traverse the intricacies of the Markov chain, an indispensable component of time series. We shed light on the operation of basic forecasting techniques, pinpoint commonly observed patterns in time series data, and gradually transition towards more sophisticated concepts like Random Walks and Monte Carlo Simulation. The goal remains constant to refine our understanding of past events in order to make informed projections about the future.

1. Markov Components (In Time Series)

A stochastic process describes the evolution of data over time (as we learned). Thus, when we are dealing with stochastic methods, order matters. Specifically, the values in a stochastic process express a dependency on time. Consequently, if we change the order of a time series, we may change the meaning of the data.

+ **Time Series Data**: In general, time series data are measured sequentially, and they are equally spaced in time. Each time unit has at most one data measurement. In addition, when doing time series analysis, we usually have two goals:

- First, we want to identify patterns that explain the behavior of the time series.
- Second, we want to use these patterns to forecast (predict) new values.

+ **Forecasting:** Time series forecasting has a rich family of models. Some of the most basic ones include the following:

- Average method: The average method assumes that a future event is best described by the average of all past events.
- Moving average method: The moving average method builds on simple average methods. Instead of using the average of all past events, it predicts a new event as the average of a predefined number of recent values.
- Naïve method: The naïve method assumes that the next event will be equal to the most recent one.

These models are simple to understand. Each uses a different assumption to predict new values. However, with such simple methods, we certainly will not get good forecasting. Hence, our decisions might turn out badly more frequently than we expect.

These methods do not account for the many fluctuations that are usually present in time series data. But how can we do better? To answer this question, we first need to understand the main patterns we usually find in time series.

+ **Patterns:** A pattern is a particular way in which something is done, is organized, or happens. Most time series data usually have at least one of three patterns: trend, seasonality, and/or cycles (Bandt & Shiha, 2007). Let us briefly consider each one.

Trend: The trend describes the general behavior of a time series. If a time series manifests a positive long-term slope over time, it has an upward trend. If it describes a general negative slope, it has a downtrend. The overall trend may change direction. There can be an up-to-down trend or a down-to-up trend. Finally, a stationary or horizontal trend defines a time series with neither positive nor negative long-term patterns.

Seasonality: A seasonal pattern is any kind of fluctuation (change) in a time series that is caused by calendar-related events. These events can be the time of year (like winter or summer) or the time of day or week. Seasonality always has fixed frequencies. That is, a seasonal pattern always starts and ends in the same period of the day, week, year, and so on. Take a data center as an example. If we consider the cooling system as the primary source of energy consumption, it is easy to imagine that in the summer energy costs probably go up, whereas winter might show a decrease in energy consumption. Also, a clothing store that sells heavy coats might have more sales in the winter than the summer.

Cycles: A cyclical pattern in a time series is a change that is not related to seasonal factors. These are rises and falls with nonfixed magnitudes that can last for more than a calendar year. Cyclical patterns are not repetitive. They usually result from external factors that make them much harder to predict. Forecasting methods usually take advantage of these patterns to produce reliable predictions.

Below we can see time series data for sales of new single-family houses (Figure 2.9.1).

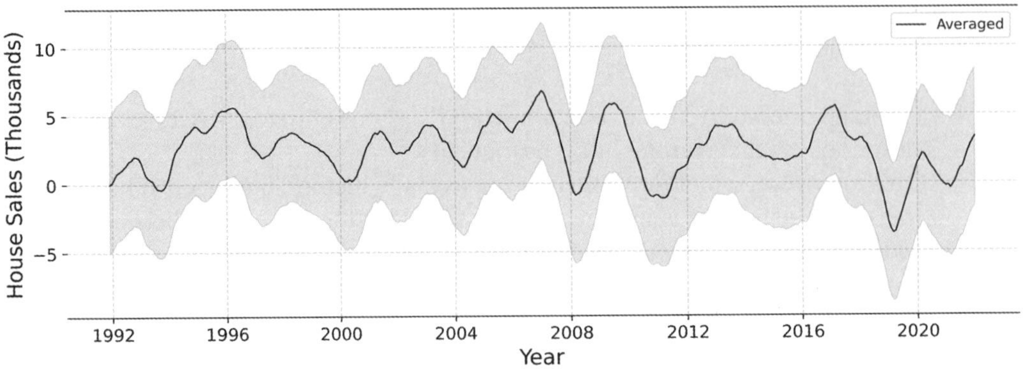

Figure 2.9.1. Cyclical pattern data for home sales. Notice the strong seasonality. House sales are normally strong cycles that might range from 6 to 10 years. Note, cycles do not have fixed periods.

It is important to note that not all time series are predictable. Some have no predictable patterns over the long term. Such time series are difficult, if not impossible, to forecast because future movements are equally likely to be up or down. To forecast these data, we usually use random walk models. These models assume that the next event is completely uncorrelated with the previous one. Hence, forecasts from random walk models are equal to the last observation plus some noise. Random walk models are used with many actual data.

2. Random Walks and Monte Carlo Simulation

Building upon the understanding of time series data and Markov chain models, we delve into the concept of random walks. A random walk is characterized as a sequence of equal-length, discrete steps that move in arbitrary directions. While we explore random walks within a one-dimensional context, such as a number line, it is important to note that these principles equally apply to multi-dimensional spaces.

+ **State Space (Discrete Time Steps over a Finite State)**: Let us imagine that we are in a park and can see a dog looking for food. He is currently in position 0 on the number line below, and he has an equal probability of moving left or right to find food.

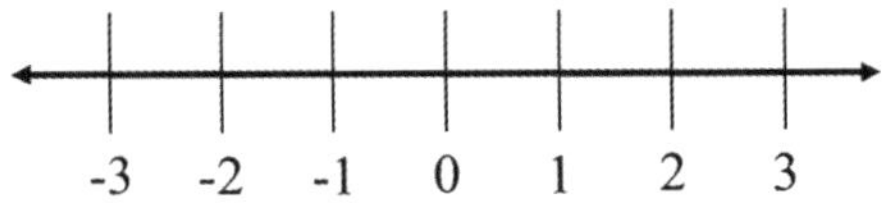

Figure 2.9.2. Sequence of discrete steps

If we want to determine the position of the dog after a certain number of N steps, we can use the law of large numbers (Figure 2.9.2). Using this law, we find that as N goes to infinity, our dog will probably be back at its starting point. Random walks are used to describe a discrete-time process (Brownian motion can be used to describe a continuous-time random walk). Thus, a random walk is the process by which randomly moving objects wander away from where they start. A stochastic process whose temporal structure exhibits no correlations with any objects will start in one place and randomly walk away (i.e., a die roll).

+ **Random Walk:** A stochastic process with a time-dependent state variable $X(n)$ defined on $\{1, \ldots, M\}$ for $n = 1, 2, \ldots$ such that $X(n)$ and $X(n')$ are statistically independent for any $n > n'$. Note that in the previous section, all joint probabilities factorize. All conditional probabilities become independent of any constraints and correspond to single time point probabilities.

- A random walk is a special case of a Markov process.

$$p(X(n) = j, X(n') = j') = p(X(n) = j) \cdot p(X(n') = j')$$

The equation represents the probability of two events occurring simultaneously; X at time n being equal to j, and X at time n' being equal to j'.

$$p(X(n) = j \mid X(n') = j'), X(n'') = j'', \ldots) = p(X(n) = j)$$

This equation represents the probability of an event occurring at time n given that another event occurred at a previous time n' which is calculated as the probability of X at time n being equal to j.

+ **Property:** In a random walk, the future is independent of the present, whereas in a Markov process, the future is dependent only on the present and not the past, and in non-Markov processes, the future is dependent on both the present and the past.

Random walk: Future does not depend on the present, No memory	past present future
Markov process: Future depends on the present but not the past, Short-term memory	past → Present → Future
Non-Markov process: Future depends on the present and past, Long-term memory	past → Present → Future

+ **Time Homogeneous (M-state Random Walk):** This is a property of Markov chains, which stipulates that the probability of transitioning from one state to another is independent of time. The M-state random walk is time homogeneous, stationary, has a probability distribution vector at time step n, and its probability distribution is time independent.

- Also features conditional probabilities and a hierarchy of joint probabilities

Time homogeneous, probability distribution vector at time step n

- General: $\vec{p}(n) = (p_1(n), \dots, p_M(n))$
- Time homogeneous: Probability distribution is independent of time (i.e., stationary: $\vec{p}(n) = (p_1, \dots, p_M) \quad \forall n \geq 1$)

Hierarchy of joint probabilities: $p(X(n) = j, X(n') = j') = p_j \cdot p_{j'}$ $p(X(n) = j, X(n') = j', X(n'') = j'') = p_j \cdot p_{j'} \cdot p_{j''}$

Conditional probabilities: $p(X(n) = j \mid X(n') = j'), X(n'') = j'', \dots) = p_j)$

+ **Transition Matrix (Two-State Random Walk):** This is a key component of a Markov chain. It represents the probability of moving from one state to another within one time step. In a two-state random walk, where the initial probability distribution is stationary, consists of probability distributions for states A and B, and transition probabilities for A and B to stay in their respective states.

- States A, B: Probability distribution

$$\begin{matrix} p_A(1) = & p_A = a \\ p_B(1) = & p_B = b = 1 - a' \end{matrix} \qquad a \in [0,1]$$

This can be modeled as a two-state Markov chain where the probability vector at time step $n + 1$ is dependent on the transition probabilities and the probability vector at time step n. In a time-homogeneous two-state random walk, the transition probabilities are fixed and the probability distribution vector is stationary, with values for A and B determined, respectively.

- Transition probability

$$\begin{matrix} P(A \leftarrow A) = & p(A \leftarrow B) = p_A = a \\ P(B \leftarrow A) = & P(B \leftarrow B) = p_B = b = 1 - a \end{matrix}$$

- Two-state Markov chain model …

$$\begin{pmatrix} p_A(n+1) \\ p_B(n+1) \end{pmatrix} = \begin{pmatrix} p(A \leftarrow A) & p(A \leftarrow B) \\ p(B \leftarrow A) & p(B \leftarrow B) \end{pmatrix} * \begin{pmatrix} p_A(n) \\ p_B(n) \end{pmatrix}$$

Therefore, the probability distribution vector at any time step n can be calculated from the probability distribution vector at time step 1, where the probability of being in state

- … becomes

$$\begin{pmatrix} p_A(n+1) \\ p_B(n+1) \end{pmatrix} = \begin{pmatrix} a & a \\ b & b \end{pmatrix} * \begin{pmatrix} p_A(n) \\ p_B(n) \end{pmatrix} = \begin{pmatrix} p_A(n) \\ p_B(n) \end{pmatrix} = \begin{pmatrix} a \\ b \end{pmatrix} \forall n > 1$$

For examples, the two provided examples show how to apply Markov chain modeling to two different situations: modeling student attendance in a random walk model (with states of attendance and no attendance) and modeling coin tossing (with states of heads and tails), with the corresponding probabilities determining the transition matrix and the stationary distribution.

Tossing coins → states: heads (A) and tails (B) (probability of heads = 0.5, probability of tails = 0.5)

$$\begin{pmatrix} p_A(n+1) \\ p_B(n+1) \end{pmatrix} = \begin{pmatrix} 0.5 & 0.5 \\ 0.5 & 0.5 \end{pmatrix} * \begin{pmatrix} p_A(n) \\ p_B(n) \end{pmatrix} = \begin{pmatrix} p_A(n) \\ p_B(n) \end{pmatrix} = \begin{pmatrix} 0.5 \\ 0.5 \end{pmatrix}$$

$$\begin{pmatrix} p_A(n+1) \\ p_{NoA}(n+1) \end{pmatrix} = \begin{pmatrix} 0.8 & 0.8 \\ 0.2 & 0.2 \end{pmatrix} * \begin{pmatrix} p_A(n) \\ p_{NoA}(n) \end{pmatrix} = \begin{pmatrix} p_A(n) \\ p_{NoA}(n) \end{pmatrix} = \begin{pmatrix} 0.8 \\ 0.2 \end{pmatrix}$$

Random walk model of student attendance → states: attendance (A) and no attendance (NoA) (probability of attendance = 0.8, probability of no attendance = 0.2)

+ **Monte Carlo Simulation (Random Walk Transitions)**: Monte Carlo simulations, which use random sampling to yield numerical outcomes, fosters a powerful, versatile tool for navigating uncertainties and making informed decisions in various fields. How can we simulate trajectories of random walks? With Monte Carlo simulation. M-state random walk on n = 1, …, T. What are the Monte Carlo simulation methods?

They are solution methods that involve a probabilistic element. The simulations are based on a repetitive drawing of random variables. Let us generate random variables $\mathcal{E}(n)$ for n = 1, 2, …, T. The variables are statistically independent for n and n'. For every n, variables must be distributed on {1, …, M} according to the given distribution → $p(\mathcal{E}(n)$ $= j)\ p_i$, Put $X(n) = \mathcal{E}(n)$

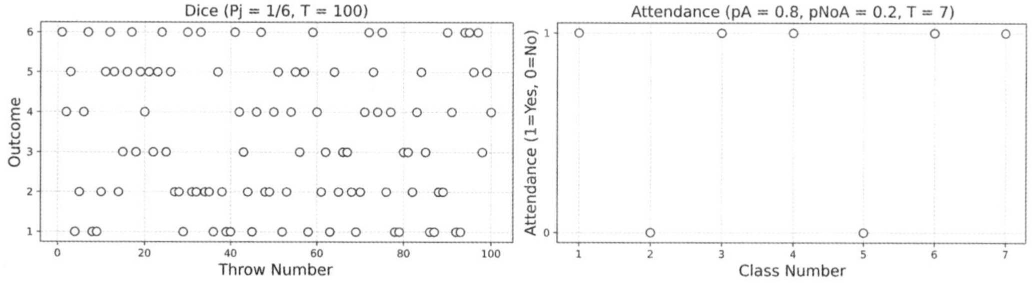

Figure 2.9.3. Random walk simulations: (left) a random walk for dice (p_j = 1/6, T = 100), (right) a random walk for student attendance (p_A = 0.8, p_{NoA} = 0.2, T = 7).

We simulate independent random variables (n) with distributions using the probability distribution $p(n = j)p_j$ as a guide. As shown in Figure 2.9.3, simulation shows the setting $X(n) = n$ to replicate random walks for the dice game (p_j = 1/6, T = 100) and the student attendance model (p_A = 0.8, p_{NoA} = 0.2, T = 7).

+ **Two-State Markov Chain:** Trajectories for a two-state Markov chain may be created using the Monte Carlo technique. To achieve this, draw a random number $\mathcal{E}$ uniformly distributed in [0, 1]. If $X(n) = A$

- Make a transition $A \rightarrow A$ with probability $p(A \leftarrow A)$
- Make a transition $A \rightarrow B$ with probability $p(B \leftarrow A)$

How? If $\mathcal{E} < p(\leftarrow A)$ then make $A \rightarrow A$, that is, put $X(n+1) = \text{A}$; otherwise, make $A \rightarrow B$, that is, put $X(n+1) = B$. If $X(n) = B$

- If $\mathcal{E} < p(B \leftarrow B)$ then make $B \rightarrow B$, that is, put $X(n+1) = B$; otherwise, make $B \rightarrow A$, that is, put $X(n+1) = A$

The process is then repeated to produce new trajectories by going back to Step 1. This method's two-state Markov chain trajectories are displayed in the left panel of Figure 2.9.4 while Markov chain attendance trajectories are displayed in the right panel.

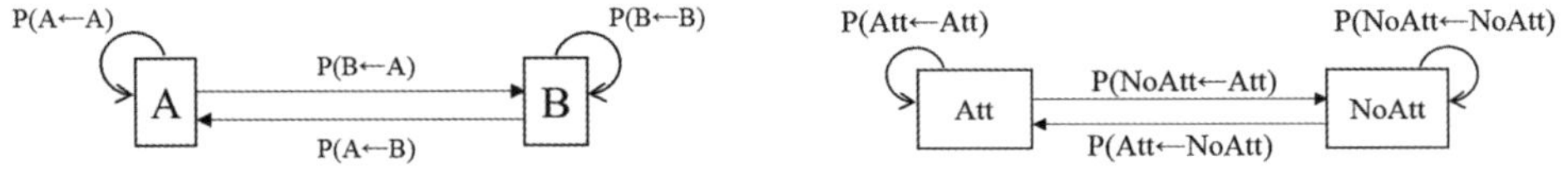

Figure 2.9.4. Two-state Markov chain method: (left) two-state Markov chain trajectories, (right) Markov chain attendance trajectories

Calculation with Student Attendance

- Two states: A = attendance, N = no attendance
- Given trajectories of two students (Kate and Gerry for $n = 1, \ldots, 6$)

[Table 2.9.1] Trajectories of the Two Students

	$A \rightarrow A$	$A \rightarrow N$	$N \rightarrow A$	$N \rightarrow N$	$N\ (A \rightarrow \cdot)$	$N\ (N \rightarrow \cdot)$
2=Kate	3	1	1	0	4	1
1=Gerry	0	2	3	0	2	3
Totals	3	3	4	0	6	4

In a Table 2.9.1, that details the number of transitions between the stages for each student, the trajectories of two students, Kate and Gerry, are shown. After then, the passage determines estimates for the probability of the observed transitions.

$$p(A \leftarrow A, estim) = \frac{N(A \leftarrow A)}{N(A \rightarrow \cdot)} = \frac{3}{6} = \frac{1}{2}, \quad p(N \leftarrow A, estim) = \frac{N(N \leftarrow A)}{N(A \rightarrow \cdot)} = \frac{1}{2}$$

$$p(A \leftarrow N, estim) = \frac{4}{4} = 1, \quad p(N \leftarrow N, estim) = \frac{0}{4}$$

The solution uses these observations to estimate the probability of the observed transitions. Specifically, the estimates for the probability of a student attending class given they attended the previous day and the probability of a student attending class given they did not attend the previous day are calculated.

+ **Example:** Then, let us produce a simulation using the two-state Markov chain model by programing. First, we need to create students and their attendance (Coding: 0=No Attendance, 1=Attendance)

```
[In]
import numpy as np
nstud = 2

# number of students (try 2 and 10)
total_classes = 7
attendance_traj = np.zeros(total_classes)
attdata_sim = np.zeros((nstud, total_classes))
```

We set up variables to store attendance data for a group of students. The 'nstud' variable represents the number of students, the 'total_calsses' represents the total number of classes, the 'attendance_traj' is a row vector with 1 row and 'total_classes' columns which all initialized to 0, the 'attdata_sim' is a matrix with 'nstud' rows' and 'total_classes' columns.

For (in)dependent parameters: we try with

```
[In]
pAtoA = 0.85
pNoAtoA = 0.70
pAtoNoA = 1 - pAtoA
pNoAtoNoA = 1 – pNoAtoA
```

To give the initial condition (initial probability [1st class] of attendance)

```
[In]
pAinitial = 0.5
np.random.seed(1)

for k in range(nstud):
    e = np.random.rand()
    if e < pAinitial:
        attdata_sim[k, 0] = 1
    else:
        attdata_sim[k, 0] = 0
```

The code uses a for loop to iterate over all students. For each student, it generates a random number between 0 and 1 using the 'rand' function. If the random number is less that 'pAinitial', it indicates that the student attended the first class, and the value 1 is assigned to the corresponding element in the 'attdata_sim' matrix. Otherwise, it indicates that the student did not attend the first class, and the value 0 is assigned to the corresponding element in the matrix. To loop copy the initial condition including the trajectory for an individual student and his or her transition, we implement

```
[In]
for irun in range(nstud):
    attendance_traj[0] = attdata_sim[irun, 0]  # copy initial condition
    for n in range(total_classes-1):  # trajectory for an individual student
        x = attendance_traj[n]
        e = np.random.rand()
        if x == 1:  # state Att
            if e < pAtoA:
                xnew = 1  # transition Att to Att
            else:
                xnew = 0  # transition Att to NoAtt
        if x == 0:  # state NoAtt
            if e < pNoAtoNoA:
                xnew = 0  # transition NoAtt to NoAtt
            else:
                xnew = 1  # transition NoAtt to Att
        attendance_traj[n+1] = xnew
    attdata_sim[irun, :] = attendance_traj  # store
```

This code simulated attendance data for a group of students for multiple classes. If the student attended the previous class 'x==1', the code checks if the random number is less than 'pAtoA'. If it is, the student attended the next class 'xnew=1', otherwise the student did not attend the next class 'xnew=0. If the student did not attend the previous class 'x=0', the code checks if the random number is less than 'pNoAtoNoA'. If it is, the student did not attend the next class 'xnew=0', otherwise the student attended the next class 'xnew=1'. After generating the attendance data for all classes, the code stores the attendance data for each student in the 'attdata_sim' matrix. Then, create its plot

```
[In]
fig, ax = plt.subplots(figsize=(7, 4), dpi=100)
ax.tick_params(axis='both',         which='major',
labelsize=20)
for k in range(nstud):
    for n in range(total_classes):
        x_ofk_atn = attdata_sim[k, n]
        if x_ofk_atn == 0:
            ax.plot(n+1, k+1, 'o', color='black')
        else:
            ax.plot(n+1, k+1, 'o', color='black')
# set axis limits
ax.set_xlim([0, total_classes+1])
ax.set_ylim([0, nstud+1])
plt.show()
```

→

[Out]

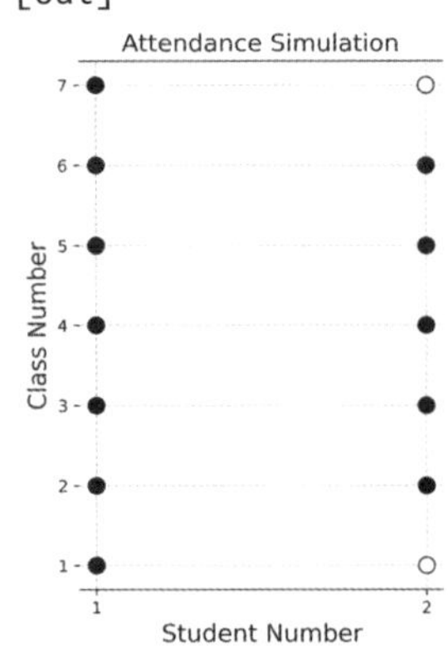

Figure 2.9.5. Markov chain model using a two-state model

Each student's attendance data, retrieved from the 'attdata_sim' matrix, is represented as a point on the plot (Figure 2.9.5). If the attendance data is 0 (non-attendance), the point is colored red, whereas if the attendance data is 1 (attendance), the point is colored green. This visual representation offers us insights into individual patterns and overall trends in attendance. With this understanding, we are well-prepared to delve deeper into the world of Markov chain probability and its diverse applications.

+ **Evolution of Markov Chain:** As we shift focus to the Evolution of Markov Chain, an interesting question arises: what insights can we glean from the evolution of a probability distribution? Many of the stochastic processes utilized in systems modeling are Markovian in nature. This attribute makes it relatively straightforward to simulate these processes, revealing their evolutionary dynamics and predictive potential in various application domains. Here we discuss a brief simulation of Markov chains in a general state space.

$$\begin{pmatrix} p_A(n+1) \\ p_B(n+1) \end{pmatrix} = \begin{pmatrix} p(A \leftarrow A) & p(A \leftarrow B) \\ p(B \leftarrow A) & p(B \leftarrow B) \end{pmatrix} * \begin{pmatrix} p_A(n) \\ p_B(n) \end{pmatrix}, \qquad 2 - state$$

Interpretation as an iterative map → solve the map numerically

$$\begin{pmatrix} x_{(n+1)} \\ y_{(n+1)} \end{pmatrix} = \begin{pmatrix} a_{11} & a_{12} \\ a_{21} & a_{22} \end{pmatrix} * \begin{pmatrix} x_n \\ y_n \end{pmatrix}$$

Interpretation as an M-dimensional iterative map (involving $M-1$ independent variables) → solve by numerical algorithm; solution for two-state Markov Chain models

$$\vec{p}(n+1) = T \cdot \vec{p}(n)$$

$$\begin{pmatrix} p_A(n+1) \\ p_B(n+1) \end{pmatrix} = \begin{pmatrix} \alpha & \beta \\ 1-\alpha & 1-\beta \end{pmatrix} * \begin{pmatrix} p_A(n) \\ p_B(n) \end{pmatrix},$$

$$p_A(n+1) = \alpha p_A(n) + \beta[1 - p_A(n)] = (\alpha - \beta)p_A(n) + \beta$$

```
[In]
T = 7
pA = np.zeros(T)
for n in range(T - 1):
    pA[n + 1] = 0.85 * pA[n] + 0.7 * (1 - pA[n])
pA[0] = 0.5

t = np.arange(1, T + 1)
ax.plot(t, pA, 'o')
```

→ [Out]

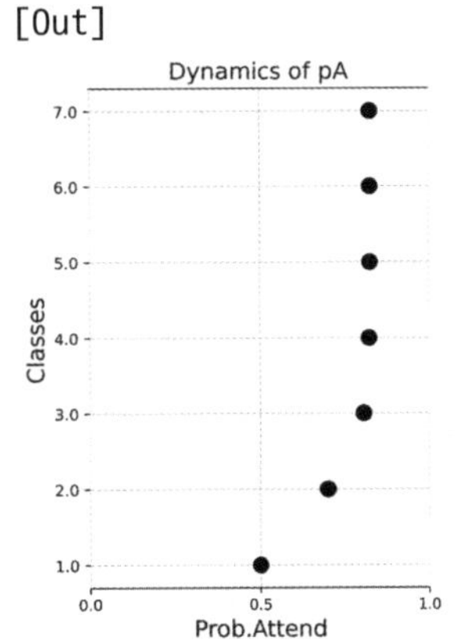

Figure 2.9.6. Markov Chain representation: Evolution of probability p_A

$$x_{n+1} = (\alpha - \beta)x_n + \beta$$

The Figure 2.9.6, and the equation represents the value of x at time $n+1$, the x_n represesnt the value of x at time n, and the α, β are constatnts that determine the relationship between x at different points in time.

The value of x at time $n+1$ is calculated as the sum of two terms $\alpha - \beta$ times the value of x at time n, and β. It is crucial to understand that many stochastic processes used for modeling in systems are Markovian (Markov chain models), making simulations relatively straight-forward.

By examining the numerical solution for two-state Markov chain models, we can interpret the iterative map and comprehend how Figure 2.9.6 represents the evolution of probability p_A in terms of variables x, α, and β.

Now, let us delve into a practical example using Markov Chain models to analyze math grades collected from children ages 7 through 10. We will walk through constructing a time-homogeneous Markov chain model that describes learning over a 4-year period and apply this model to two children, Ursula and Carlos, to observe the evolution of their math grades

3. Practical Practice with Markov Chain Models

In this practice we run Markov Chain models and describe their calculation processes.

We collect math grades from children ages 7, 8, 9, and 10. We only consider grades in the range A, B, C. You collect one grade every year. We want to construct a time-homogeneous Markov chain model that describes learning over the 4-year period.

The model has three states: grades A, B, and C. That is, trajectories $X(n)$ describe how the math grades evolve over the 4-year observation period. We have data from two children: Ursula and Carlos.

The math grade trajectories are B, B, A, A for Ursula and C, C, B, A for Carlos.

[Table 2.9.2] Data from Students

Ursula	B	B	A	A
Carlos	C	C	B	A

Table 2.9.2 shows the math grade trajectories for two children Ursula and Carlos. After organizing the data in Table 2.9.3, the estimated transition probabilities for the three states A, B, and C are calculated based on the number of transitions between the states for each child.

[Table 2.9.3] Organized Data for Students

	A ← A	A ← B	A ← C	B ← A	B ← B	B ← C	C ← A	C ← B	C ← C	N(A → ·)	N(B → ·)	N(C → ·)
Ursula	1	1	0	0	1	0	0	0	0	1	2	0
Carlos	0	1	0	0	0	1	0	0	1	0	1	2
Frequency (N)	1	2	0	0	1	1	0	0	1	1	3	2
$P(j \leftarrow k$, estim) = N(A ← A)/N(· ← A) =	1/1 = 1	2/3 = .66	0/2 = 0	0/1 = 0	1/3 = .33	1/2 = .5	0/1 = 0	0/3 = 0	1/2 = .5	1/6 = .17	3/6 = .5	2/6 = .33

Problem 1a: The Markov chain model has nine parameters, which are the nine transition probabilities $P(j \leftarrow k)$ for j = A, B, C and k = A, B, C. Determine from the data given above the nine estimated transition probabilities $p(A \leftarrow \text{A, estim})$ for j = A, B, C and k = A, B, C. Use the data from both children for the estimates.

First determine the frequencies N(A to …), N(B to …), N(C to …) to observe changes (including no change) from grade A, grade B, and grade C, respectively. Then determine all non-zero frequencies $N(j \leftarrow k)$ of finding particular changes in grade from k to j (including the case j to j) in the data sets. Finally, calculate the estimated transition probabilities. Report all transition probabilities as ratios (e.g., 7/8).

$$p(A \leftarrow \text{A, estim}) = N(A \leftarrow \text{A})/\text{N}(\cdot \leftarrow \text{A}) = 1/1$$

$$p(A \leftarrow \text{B, estim}) = N(A \leftarrow \text{B})/\text{N}(\cdot \leftarrow \text{B}) = 2/3$$

$$p(B \leftarrow \text{B, estim}) = N(B \leftarrow \text{B})/\text{N}(\cdot \leftarrow \text{B}) = 1/3$$

$$p(B \leftarrow \text{C, estim}) = N(B \leftarrow \text{C})/\text{N}(\cdot \leftarrow \text{C}) = 1/2$$

$$p(C \leftarrow \text{C, estim}) = N(C \leftarrow \text{C})/\text{N}(\cdot \leftarrow \text{C}) = 1/2$$

Problem 1b: Determine from the two trajectories given above the relative frequency f(improve) that the children exhibit year-to-year changes in grade that reflect an improvement (e.g., for Ursula we have three observations of change from year to year; in one case her math grade improved).

In addition, calculate f(neutral), which is the relative frequency that grades do not change, and f(worsen), which is the relative frequency that grades become worse.

[Table 2.9.4] Solution to Problem 1b

Ursula rating			Ursula_improvement		
Math rating j	Frequency hj	Percent (Relative frequency)	Improve rating j	Frequency hj	Percent (Relative frequency)
A	2	2/4 = 0. 5 =50%	Improve	1	1/3 = 0.33 = 33%
B	2	2/4 = 0.5 = 50%	Neutral	2	2/3 = 0.66 = 66%
C	0	0/4 = 0 = 0%	Worsen	0	0/3 = 0 = 0%

Carlos rating		
Math rating *j*	Frequency *hj*	Percent (Relative frequency)
A	1	1/4 = 0.25 = 25%
B	1	1/4 = 0.25 = 25%
C	2	2/4 = 0.5 = 50%

Carlos_improvement		
Improve rating *j*	Frequency *hj*	Percent (Relative frequency)
Improve	2	2/3 = 0.66 = 66%
Neutral	1	1/3 = 0.33 = 33%
Worsen	0	0/3 = 0 = 0%

Rating Total = *N*(rating)/*N*(total)		
Math rating *j*	Frequency *hj*	Percent (Relative frequency)
A	3	3/8 = 0.375 = 37.5%
B	3	3/8 = 0.375 = 37.5%
C	2	2/8 = 0.25 = 25%

Improvement Total = *N*(improve)/*N*(total)		
Improve rating *j*	Frequency *hj*	Percent (Relative frequency)
Improve	3	3/6 = 0.5 = 50%
Neutral	3	3/6 = 0.5 = 50%
Worsen	0	0/6 = 0 = 0%

Relative frequencies: $f(\text{improve}) = N(\text{improve})/N(\text{total}) = (2 + 1)/(1 + 3 + 2) = 3/6 = 1/2$
$f(\text{neutral}) = N(\text{neutral})/N(\text{total}) = (1 + 1 + 1)/(1 + 3 + 2) = 3/6 = 1/2$
$f(\text{worsen}) = N(\text{worsen})/N(\text{total}) = 0/6 = 0$

Problem 1c: How are the relative frequencies from Problem 1b related to the Problem 1a and the relative frequencies? The relative frequencies $f(A)$, $f(B)$, and $f(C)$ are the relative frequencies of observing grades A, B, and C if we ignore the data points at the ends of the trajectories. That is, $f(A)$, $f(B)$, and $f(C)$ can be computed by normalizing the frequencies N(A to …), N(B to …), and N(C to …). Calculate the relative frequencies $f(A)$, $f(B)$, and $f(C)$ using either N(A to …), N(B to …), and N(C to …) or the data from the two trajectories. In the latter case, use only data for which transitions can be observed (i.e., B, B, A for Ursula and C, C, B for Carlos). Then estimate the joint probabilities $p(j[n+1], k[n], estim)$ that grades improve using the estimates for the conditional probabilities from Problem 1b and $f(k)$. The $f(k)$ from Problem 1b are then certain superpositions (i.e., sums) of the estimated joint probabilities.

[Table 2.9.5] Solution to Problem 1c

	Frequency *hj*	Percent (Relative frequency)
$N(A \rightarrow)$	1	1/6 = 0.17
$N(B \rightarrow)$	3	3/6 = 0.5
$N(C \rightarrow)$	2	2/6 = 0.33

[Table 2.9.6] Reorganized Data for Students

	A ← A	A ← B	A ← C	B ← A	B ← B	B ← C	C ← A	C ← B	C ← C
Ursula	1	1	0	0	1	0	0	0	0
Carlos	0	1	0	0	0	1	0	0	1
Frequency (*N*)	1	2	0	0	1	1	0	0	1
$P(j \leftarrow k$, estim) $= N(A \leftarrow A)/N(\cdot \leftarrow A)$	1/1 = 1	2/3 = .66	0/2 = 0	0/1 = 0	1/3 = .33	1/2 = .5	0/1 = 0	0/3 = 0	1/2 = .5
p(A, A, estim) = p(A ← A, estim) × f(A)	1/1 × 1/6 = .166	2/3 × 3/6 = .33	0/2 × 2/6 = 0	0/1 × 1/6 = 0	1/3 × 3/6 = .165	1/2 × 2/6 = .165	0/1 × 1/6 = 0	0/3 × 3/6 = 0	1/2 × 2/6 = .165

Table 2.9.5 shows the frequency and relative frequency for the different grades in a Markov chain model. Table 2.9.6 reorganizes the data for two students, Ursula and Carlos, and shows the estimated probability of transitioning from one grade to another. The last row in the table shows the estimated probability of a specific grade for Ursula and Carlos.

$$p(A \leftarrow \text{A}, \text{estim}) = P(A \leftarrow \text{A}, \text{estim}) \times f(\text{A})1 \ \times \ 1/6 \ = \ 1/6$$

$$p(A \leftarrow \text{B}, \text{estim}) = P(A \leftarrow \text{B}, \text{estim}) \times f(\text{B})2/3 \ \times \ 1/2 \ = \ 1/3$$

$$p(B \leftarrow \text{B}, \text{estim}) = P(B \leftarrow \text{B}, \text{estim}) \times f(B)1/3 \ \times \ 1/2 \ = \ 1/6$$

$$p(B \leftarrow \text{C}, \text{estim}) = P(B \leftarrow \text{C}, \text{estim}) \times f(\text{C})1/2 \ \times \ 1/3 \ = \ 1/6$$

$$p(C \leftarrow \text{C}, \text{estim}) = P(C \leftarrow \text{C}, \text{estim}) \times f(C)1/2 \ \times \ 1/3 \ = \ 1/6$$

Then

f(improve) = p(A ← B, estim) + p(B ← C, estim) = (2/3 × 3/6) + (1/2 × 2/6) = .495 = .5

f(neutral) = p(A ← A, estim) + p(B ← B, estim) + p(C ← C, estim) = (1/1 × 1/6) + (1/3 × 3/6) + (1/2 × 2/6) = .496 = . 5

f(worsen) = p(B ← A, estim) + p(C ← B, estim) = (0/1 × 1/6) + (0/3 × 3/6) = 0

Problem 2: Two-state Markov chain models with Boltzmann distributions. The two-state Markov chain model can alternatively be written as

$$\begin{pmatrix} p_A(n+1) \\ p_B(n+1) \end{pmatrix} = \begin{pmatrix} \alpha & \beta \\ 1-\alpha & 1-\beta \end{pmatrix} * \begin{pmatrix} p_A(n) \\ p_B(n) \end{pmatrix} \ \rightarrow \ \begin{pmatrix} p_A(n+1) \\ p_B(n+1) \end{pmatrix} = \begin{pmatrix} 1-\gamma & \beta \\ \gamma & 1-\beta \end{pmatrix} * \begin{pmatrix} p_A(n) \\ p_B(n) \end{pmatrix}$$

with $\gamma = 1 - \alpha$. In this case, the stationary probability distribution is given by

$$p_A = \frac{1}{1 + \gamma/\beta}, \qquad p_B = \frac{1}{1 + \beta/\gamma}$$

Consider a system with potential energy that exhibits two levels, A and B, defined by the energies EA and EB. Assume that B has a higher energy than A. Assume that the transitions from A to B occur with transition probability γ = p(B ← A) = exp(–ΔE) < 1 with ΔE = EB – EA > 0. Moreover, assume that β = p(A ← B) = 1. Note that in this case the equation describes a Boltzmann distribution with pk = exp(–Ek)/Z and Z = exp(–EA) + exp(–EB). In what follows, set ΔE = 1.

Problem 2a (Calculating probabilities): Calculate γ = p(B ← A) = exp(–ΔE) and p(A ← A) = 1 – γ, and calculate the stationary probabilities pA and pB. Round your results to three decimal places.

Solution to Problem 2a: $\gamma = p(\mathrm{B} \leftarrow \mathrm{A})\, exp(-\Delta E) < 1\; with\; \Delta\mathrm{E} = \mathrm{E_B} - \mathrm{E_A} > 0,$
$\beta = p(\mathrm{A} \leftarrow \mathrm{B}) = 1$

Calculate: $\gamma = p(\mathrm{B} \leftarrow \mathrm{A})\, exp(-\Delta\mathrm{E}) = 0.367879, P(\mathrm{A} \leftarrow \mathrm{A}) = 1 - \gamma = 0.632121$

Calculate the stationary probabilities: γ =0.367879, $\beta = 1$

$$pA = \frac{1}{1+\gamma/\beta} = \frac{1}{1+\ 0.368/1} = \frac{1}{1.368} = 0.731$$
$$pB = \frac{1}{1+\beta/\gamma} = \frac{1}{1+\ 1/0.368} = \frac{1}{3.717} = 0.269$$

Problem 2b (Showing stationarity): Show explicitly that the values for $p\mathrm{A}$ and $p\mathrm{B}$ obtained in Problem 2a are stationary values of the Markov chain model. To this end, substitute the numerical values $p\mathrm{A}(n) = p\mathrm{A}$ and $p\mathrm{B}(n) = p\mathrm{B}$ obtained in Problem 2a into the equation. Then substitute the values of $p\mathrm{A}$ and $p\mathrm{B}$ in the right sides and show explicitly that the left and right sides are equal; that is, show that $p\mathrm{A}(n+1) = p\mathrm{A}$ and $p\mathrm{B}(n+1) = p\mathrm{B}$. Round your work to three digits.

Solution to Problem 2b: Given $p\mathrm{A}(n = 0.731$ and $p\mathrm{B}(n = 0.269,$

$$p\mathrm{A}(\mathrm{X(n+1) = A}) = p(\mathrm{A} \leftarrow \mathrm{A}) * p(\mathrm{X(n) = A}) + p(\mathrm{A} \leftarrow \mathrm{B}) * p(\mathrm{X(n) = B})$$
$$= 0.632 * 0.731 + 1 * 0.269 = 0.462 + 0.269 = 0.731$$

$$\mathrm{pB(X(n+1) = B) = p(B \leftarrow A)*p(X(n) = A) + p(B \leftarrow A)*p(X(n) = B) = 0.368*0.731 +}$$
$$\mathrm{0*0.269 = 0.269 + 0 = 0.269}$$

Thus,

$$p\mathrm{A(n+1)} = p\mathrm{A(n)} = 0.731\; and\; p\mathrm{B(n+1)} = p\mathrm{B(n)} = 0.296$$

→ Stationarity supported

Remarks: We explored deeper into the stochastic processing of Markov chains in this section of Markov Chain Modeling, utilizing ideas like random walks and Monte Carlo simulation for random walk transitions. We also examined how the transition matrix, a potent tool for simulating intricate real-world systems, changed the Markov Chain probability with time. As demonstrated in the student attendance study example, we may estimate the transition probabilities in actual applications using observed data. Numerous fields, including physics, biology, economics, and others, use the Markov Chain Modeling framework to forecast future occurrences based on observations of the past and to create models for a variety of processes.

[2.10]

Stochastic Iterative Maps, Part 1

Moving Average Model, Autoregressive Model, Practical Practice

The moving average (MA) model, a commonly utilized method in time series analysis, posits a linear dependency of the output variable on the present and various prior values of a stochastic term - a term that is not perfectly predictable (Durbin, 1959). In essence, the MA model can be viewed as a linear regression of the series' current value against both the present and previous white noise error terms, or random shocks. It is premised on the assumption that these random shocks at each point are mutually independent and derived from an identical distribution, typically normal, centered at 0 and with a consistent scale.

The autoregressive (AR) model, on the other hand, forecasts future trends based on past behaviors, and is typically employed when a time series' values display correlation with their preceding and succeeding values (Akaike, 1998). In essence, the AR process can be likened to a linear regression of the data in the current series against one or more of its past values. Similar to traditional linear regression, the AR model directly links the outcome variable (Y) at a certain time point t, with the predictor variable (X). However, the key differentiator between simple linear regression and AR models lies in the fact that, in the latter, Y is dependent on not just X, but also on its own previous values.

1. Moving Average Model

The MA(q) time series model means that the value Xt of random variable X is a linear combination of one or more stochastic values lagged at times 0. The average of X can be added to the model if it is significantly different from 0. Together with the autoregressive (AR) model, the MA model is a special case and key component of the more general autoregressive moving average (ARMA) models of time series, which have a more complicated stochastic structure.

The representation of MA processes defined by Stochastic Iterative Maps is shown in Table 2.10.1 and is divided into categories for discrete-state variables and continuous-state variables. The mixed process known as the Autoregressive Moving Average (ARMA) combines AR and MA processes of orders k and q, where MA(0), MA(1),..., and AR(0), AR(1),..., respectively, reflect the order of attributes for MA and AR processes. Discrete-time stochastic processes with continuous, numerical state variables

[Table 2.10.1] Representation of MA, Processes Described by Stochastic Iterative Maps

	Discrete-state variable	Continuous-state variable	
Discrete time		**Moving Average**	→
Continuous time			

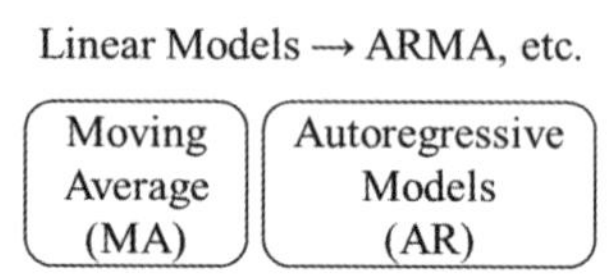

* ARMA = autoregressive moving average

To represent the combined AR and MA process of order k and q, we use the notation in general;

- MA, AR, ARMA have property order
- MA(0), MA(1), …
- MA(q): MA process of order q
- AR(0), AR(1), …
- AR(k): AR process of order k
- ARMA(k, q) = mixed AR and MA process of order k and q

+ **MA Model of Order 0:** MAs are used and discussed quite commonly by systems analysts and scientists alike. They can help an analyst filter noise and create a smooth curve from an otherwise noisy curve. It is important to note that MAs lag because they are based on historical data, not current data. MA(0) is a discrete-time process, with $X(n)$ defined on a real line

$$X(n) = \varepsilon_n$$

The equation represents a random variable X at a particular point in time n which value of the random variable is equal to ε_n as a random error term characterized by random variables. Statistically independent → $X(n)$ has no autocorrelation, normal distribution, 0 mean, variance $\sigma^2{}_\varepsilon$ (the properties do not change with time), or $X(n)$ normal distribution → $X(n)$ stationary process.

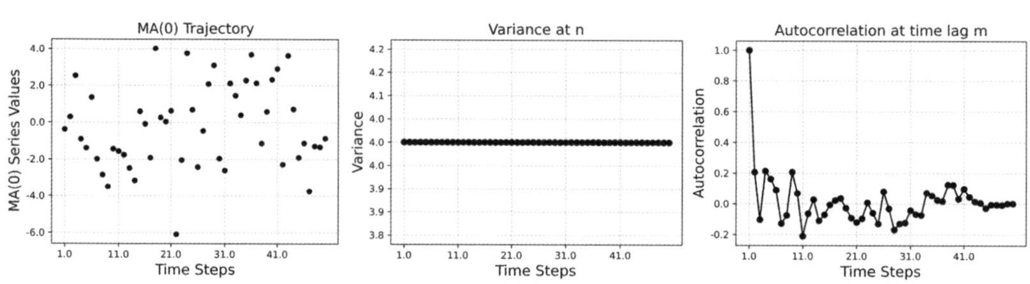

Figure 2.10.1. Moving average model of order 0: (left) MA(0) trajectory, (middle) variance at n (here $\sigma^2{}_\varepsilon$ = 4), (right) autocorrelation at time lag $m = n - n \geq 0$

The MA model of order 0 is shown in Figure 2.10.1, where the left plot displays the MA(0) trajectory, and the middle plot provides the variance at time index n (with a value of $\sigma^2{}_\varepsilon = 4$ in this example). And the right plot displays the autocorrelation at time lag $m = n - n \geq 0$, (See Supplement 2.3 for further information on the autocorrelation function.)

We can use the above description to generate a simulation of the MA(0) model in programing. For initial parameters with white noise

```
[In]
import numpy as np
sigma = 2
T = 10    # use 10 and 10000
X = np.zeros(T)
```

We defined a variable 'sigma' and assigns it a value of 2, assigns T as a value of 10, and defined a X as $1 * 10$ initialized it with all zeros. For loop

```
[In]
for n in range(T):
    epsilon = np.random.normal(0, sigma)
    X[n] = epsilon
```

We iterate from $n = 1$ to $n = T$, on each iteration, it generates a random value for a normal distribution with mean 0 and 'sigma' which assigned this value to 'epsilon'. Then the code assigned the value of 'epsilon' to the n-th element of the matrix X.

Create the output

```
[In]
fig, ax = plt.subplots()

# Plotting
t = np.arange(1, T + 1)
ax.plot(t, X, 'o',
markerfacecolor='black')

# Save
out = X.T
np.savetxt('timeseries_ma0.txt', out)
```

→

[Out]

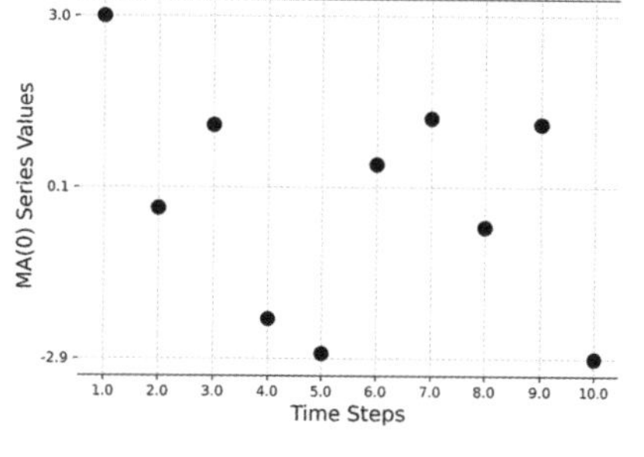

Figure 2.10.2. Visualization of model of the MA(0) trajectory

We generate the values in the matrix X against a sequence of indices from 1 to T (Figure 2.10.2). The 'plot' function is being passaged several optional arguments including 'o' which specifies that the data points would be plotted as circles. After the plot is created, the code transposes the matrix X and stores the result in a new variable called 'out' and saves it as 'timesereis_ma0.txt' format.

+ **MA Model of Order 1:** The MA(1) is characterized by the following parameters: b_1 (positive or negative) and variance ${\sigma^2}_{\varepsilon}$. ε has normal distribution $\rightarrow X$ is the superposition of two normally distributed variables. X has a normal distribution.

$$X(n) = \varepsilon_n + b_1\varepsilon_{n-1}$$

The equation states that the value of the n-th element in the matrix X is equal to the value of ε_n plus b_1 times the value of the $n-1$-th element in the matrix. For example, when we have ($\varepsilon = \{2,3,-1,4,\dots\},\ b_1 = 7$), the calculation is as follows:

$$X(n) = \varepsilon_n + b_1\varepsilon_{n-1}$$

$$\varepsilon = \{2,3,-1,4,\dots\}, \qquad b_1 = 7$$

$$X(1) = \varepsilon_n + b_1\varepsilon_0 = 3 + 7 * 2 = 17$$

$$X(2) = \varepsilon_2 + b_2\varepsilon_1 = -1 + 7 * 3 = 20$$

$$X(3) = \varepsilon_3 + b_3\varepsilon_2 = 4 + 7 * (-1) = -3$$

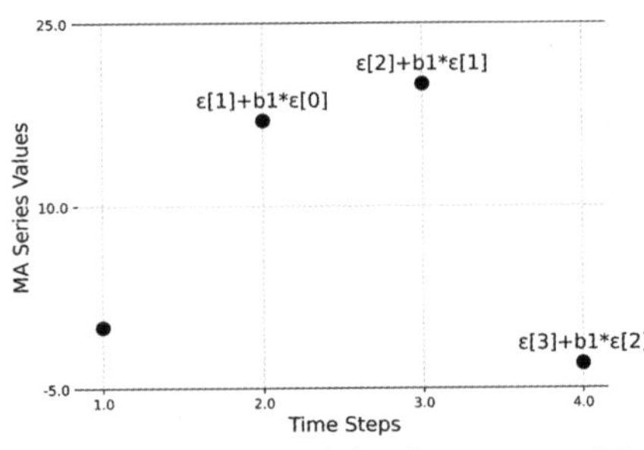

MA(1) → moving weighted average of 2
MA(2) → moving weighted average of 3

Figure 2.10.3. Moving weighted average of MA(1) and MA(2). $X(t)$ is the time series, ε is the white noise at time t, and the b_1 is the weight for the noise at the previous time step.

In Figure 2.10.3, this is the computation of the elements in a matrix X given a sequence of values ε and a coefficient b_1. To compute the value of $X(2)$ it calculates $\varepsilon_2 + b_2\varepsilon_1$ and it continuous in this way for the $X(3)$ and so on. Again, let us produce a simulation of the MA(1) model by programing.

For initial parameters

```
[In]
import numpy as np
sigma = 2
b1 = 0.3
T = 10    # use 10 and 10000
X = np.zeros(10)
epsilon = np.random.normal(0, sigma, T)
```

We defined the variables and the ‘epsilon’ is defined as a matrix of random values drawn from a normal distribution with mean 0 and standard deviation ‘sigma’. For loop

```
[In]
for n in range(1, T):  # Python uses 0-based indexing
    X[n] = epsilon[n] + b1 * epsilon[n - 1]
```

The loop then iterates from $n = 2$ to $n = T$, on each iteration, it sues the equation of MA(1) to compute the value of the n-th element in the matrix 'X' which is equal to the value of the n-th element in the matrix 'epsilon' plus 'b1' times the value of the $n-1$-th element in the matrix 'epsilon'.

Create the output

```
[In]
fig, ax = plt.subplots()

# Plotting
t = np.arange(1, T + 1)
ax.plot(t, X, 'o',
markerfacecolor='black')
```

→

[Out]

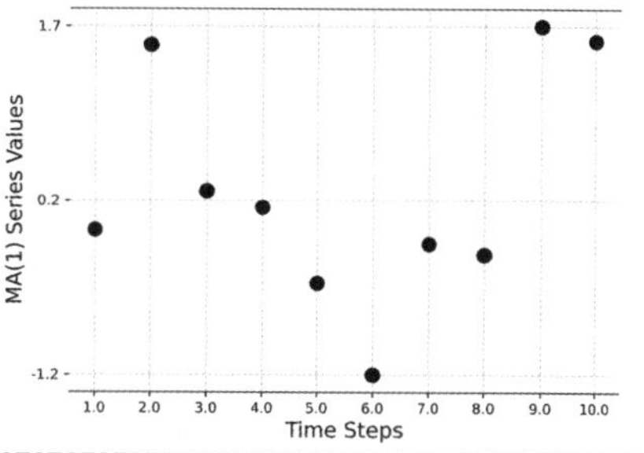

Figure 2.10.4. Visualization of model of the MA(1) trajectory

The code creates plot which the first input is the x-values of the data points, and the second input is the y-values of the data points (Figure 2.10.4). The 'o' specifies that the data points should be plotted as circles wish a line width of 4.

+ **MA Model of Order q**: The Moving Average (MA) model of order q with mean is denoted by the $Y(n)$ as below are independent and identically distributed random variables with a mean of zero and variance of 2.

- MA(q) with mean μ:

$$Y(n) = \mu + \varepsilon_n + b_1\varepsilon_{n-1} + \cdots + b_q\varepsilon_{n-q}$$

This equation describes how the value of the n-th element in a matrix Y depends on the value of the previous q elements represented by

$$\varepsilon_{n-1}, \quad \varepsilon_{n-2}$$

and so on which set of coefficients b_1, b_2 and so on.

- MA(q) with zero mean (i.e., $X(n) = Y(n) - \mu$): $X(n) = \varepsilon_n + b_1\varepsilon_{n-1} + \cdots + b_q\varepsilon_{n-q}$

This equation states that the value of the n-th element in the matrix X can be expressed as a sum of the value of ε_n and b_1 times the value of the $n-1$-th element in the matrix ε and so on up to b_q times the value of the $n-q$-th element in the matrix ε. Therefore, it relates that the value of the n-th in the matrix X is equal to the value of the n-th element in a matrix Y minus a constant μ, or equivalently $Y(n) - \mu$.

What Is a Noise Source?

A random, unpredicted signal is referred to as a noise source. Noise sources can be divided into two categories: Gaussian noise and non-Gaussian noise, depending on how the signal is distributed. The term 'Gaussian noise' refers to noise that is regularly distributed, whereas the term 'non-Gaussian noise' refers to noise that is not normally distributed. Distribution:

- Normally distributed → Gaussian noise
- Not normally distributed → non-Gaussian noise

Moving Average (MA) models, where $X(n)$ represents the noise source at time n, can be thought of as noise sources. The MA(0) process, where $X(n) = n$, represents Gaussian white noise, and $X(n)$ is a source of Gaussian noise in an MA(q) process.

MA(q) as noise sources:

- MA(q): $X(n)$ are Gaussian noise sources
- MA(0): $X(n)$ Gaussian white noise

$$X(n) = \varepsilon_n$$

However, X(n) represents Gaussian colored noise for MA(1), MA(2), and higher orders, respectively (Table 2.10.2).

- MA(1), MA(2), ...: $X(n)$ Gaussian colored noise

$$X(n) = \varepsilon_n + b_1\varepsilon_{n-1}, \qquad X(n) = \varepsilon_n + b_1\varepsilon_{n-1} + b_2\varepsilon_{n-2}$$

[Table 2.10.2] Random Walks and Noise Sources

	Random walks	Noise sources
Stationarity	Stationary or not	Stationary
Autocorrelation	Zero	Zero (white) or Not zero (colored)

* Stationary random walk = white noise source

The Table 2.10.2 classifies noise sources into Gaussian and non-Gaussian, with Gaussian noise further divided into white and colored based on their properties in moving average models, while emphasizing that stationary random walks are associated with white noise sources.

Understanding the distinctions between Gaussian and non-Gaussian noise, as well as white and colored noise within Gaussian noise, is crucial for properly analyzing and interpreting different noise sources in various applications.

We now know that MA processes may be seen as noise sources, with the autocorrelation function being able to quantify the degree of dependency between observations made at various times. Furthermore, we have shown that the distribution of a noise source establishes whether it is a Gaussian noise or a non-Gaussian noise. MA models have a variety of applications in real-world settings.

2. Autoregressive Model

The AR process is an example of a stochastic process, which has a degree of uncertainty or randomness built in. The randomness means that you might be able to predict future trends well with past data, but you are never going to get 100% accuracy. Usually, the process gets close enough for it to be useful in most scenarios.

+ **Notation:** The parameters are the AR coefficients a_k and the variance of the error component, denoted by $Var(\varepsilon)$, and the notation for state $X(n)$ is specified on a real line.

The following is a representation of the AR procedures for orders one through three:

$$\text{AR(1)}: X(n) = a_1X(n-1) + \varepsilon_n$$

$$\text{AR(2)}: X(n) = a_1X(n-1) + a_2X(n-2) + \varepsilon_n$$

$$\text{AR(3)}: X(n) = a_1X(n-1) + a_2X(n-2) + a_3X(n-3) + \varepsilon_n$$

These notational rules give a basis for comprehending and interpreting autoregressive models, which are crucial for time series analysis and have applications in a number of disciplines, including engineering, economics, and finance.

+ **AR Model of Order 0:** The formula for the AR(0) model is a function of n and ε_n stands for the error term.

$$X(n) = \varepsilon_n$$

The AR(0) model, represented by the equation $X(n) = \varepsilon_n$, is unique because it does not include any autoregressive coefficients. It is characterized by Gaussian white noise and is equivalent to both a moving average of order 0 (MA(0)) and a mixed autoregressive moving average of orders 0 and 0 (ARMA(0,0)), all representing Gaussian white noise processes.

+ **AR Model of Order 1:** The formula for the AR(1) model is

$$X(n) = a_1X(n-1) + \varepsilon_n$$

This equation defines $X(n)$ as a function of n, with the value of $X(n)$ being a function fo the previous value of $X(n-1)$ and ε_n, and its constant coefficient a_1. Initial value: $X(1)$ or $X(n=0)$. First AR parameter: a_1

. ε : stationary, uncorrelated, normal distribution, mean = 0, variance $\sigma^2{}_\varepsilon$

- ε = Gaussian white noise
- Two parameters: a_1 and $\sigma^2{}_\varepsilon$
- Trajectory: $X(1)$, $X(2)$, …

Let us examine the deterministic part of the computation in practice, when $a_1 > 0$ below.

$$X(n) = a_1 X(n-1)$$

		0.1	2
$X(1) = a_1 X(0)$	N=1	$a_1 = 0.1$	$a_1 = 2$
$X(2) = a_1 X(0) = a_1(a_1 X(0)) = a_1{}^2 X(0)$	N=2	$a_1{}^2 = 0.01$	$a_1{}^2 = 4$
$X(3) = a_1{}^3 X(0)$	N=3	$a_1{}^3 = 0.001$	$a_1{}^3 = 8$
...			
$X(n) = a_1{}^n X(0)$			

The AR(1) model is defined by the equation $X(n) = a_1 X(n-1) + \varepsilon_n$, where ε represents Gaussian white noise, and the model has two parameters, a1 and σ²_ε. The deterministic part of the computation, $X(n) = a_1 X(0)$, reveals the trajectory of the model, with different values of a_1 leading to distinct behaviors. We demonstrate that for $a_1 < 1$, there is a stable fixed point at $X = 0$ using the numbers in the following table when a_1 is equal to 0.1 or 2. However, there is an unstable fixed point when $a_1 < 1$.

- $a_1 < 1$ stable fixed point at $X = 0$, $a_1 < 1$ unstable fixed point

[Table 2.10.3] Deterministic Part: Illustration for $a_1 > 1$

Calculation results (interest calculated yearly)			
Year	Annual interest	Total interest	Total sum
1	$2.00	$2.00	$102.00
2	$2.04	$4.04	$104.04
3	$2.08	$6.12	$106.12
4	$2.12	$8.24	$108.24
5	$2.16	$10.41	$110.41

*Data are from www.thecalculatorsite.com. Base amount: $100.00, interest rate: 2%, years: 5

Table 2.10.3 shows that when $a_1 < 1$, there is a stable fixed point at $X = 0$, while when $a_1 > 1$, an unstable fixed point occurs. The table illustrates these concepts using a_1 values of 0.1 and 2, highlighting the distinct behaviors depending on the a_1 value.

[Table 2.10.4] Compound Interest in a Savings Account

$FV = PV * (1 * r)^n$ $X(1) = X(0) * a_1{}^n$ $a_1 = 1 * r$	PV: present value FV: present value after *n* years *n*: years *r*: interest rate (%)

In Table 2.10.4, the first line of the equation defines the future value FV of an investment as a function of the present value PV, a constant interest rate r, and a number of periods n. The second line of the equation defiens the value of $X(1)$ as a function of the initial value of $X(0)$, a constant coefficient a_1 and a number of periods n. And the third line of the equation defines the constant coefficient a_1 as the product fo 1 and the interest rate

r. All they are being used to predict the behavior of some financial or mathematical system over time.

We illustrate the distinct behaviors of stable and unstable fixed points based on the a_1 value, highlighting the importance of understanding the relationship between a_1 and the stability of the system. Then, we visually represent these concepts, demonstrating the differences between stationary (a_1 = 0.8) and nonstationary (a_1 = 1.02) processes, as well as a deterministic process.

Let us provide a clear visual representation of the various processes and their relationship to the a_1 value, complementing the information provided in the tables.

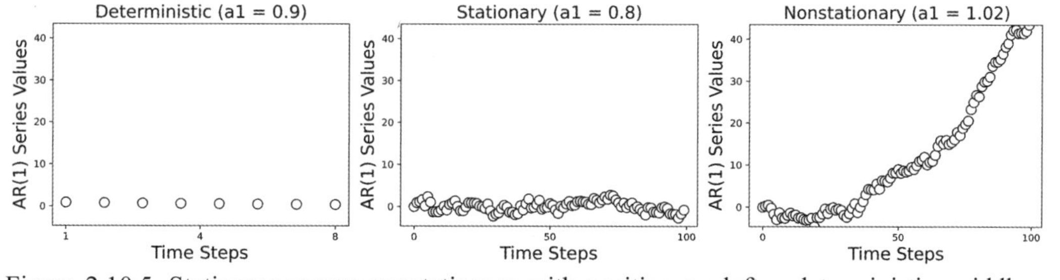

Figure 2.10.5. Stationary versus nonstationary with positive a_1: left = deterministic, middle = stationary (a_1 = 0.8), right = nonstationary (a_1 = 1.02)

The comparison between processes is shown in the Figure 2.10.5, with a positive value for a_1. The middle and right plots, meanwhile, show stationary (a_1 = 0.8) and nonstationary (a_1 = 1.02) processes, respectively. The left graph illustrates the deterministic process.

With an a1 value of 1.02, the nonstationary process is depicted in the center graph. While the right graph displays a stationary process with an a1 value of 0.8, the opposite is true.

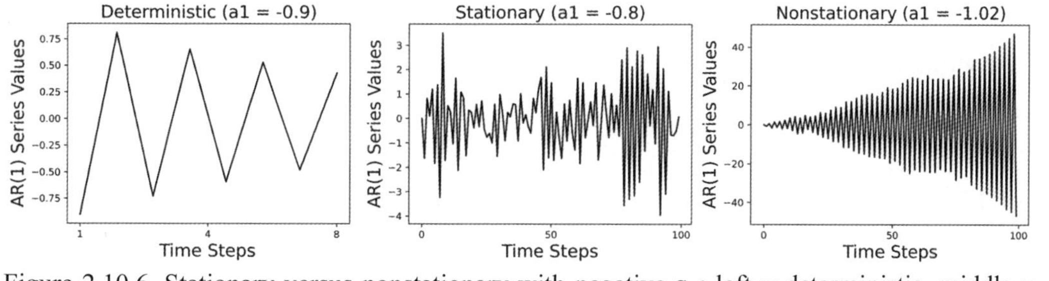

Figure 2.10.6. Stationary versus nonstationary with negative a_1: left = deterministic, middle = stationary (a_1 = 0.8), right = nonstationary (a_1 = 1.02)

The contrast between stationary and nonstationary processes is shown in Figure 2.10.6 with a negative value for a_1. The middle and right graphs, however, show stationary and nonstationary processes, respectively, while the left graph represents the deterministic process.

A nonstationary process with an a1 value of –1.02 is depicted in the center graph. The right graph, on the other hand, displays a stationary process with an a1 value of –0.8.

+ **Stationary AR(1) Processes:** A stationary AR(1) process is represented by the process value at time n, a_1 controls the process' persistence, with a normally distributed random variable with a mean of zero and a constant variance.

$$X(n) = a_1X(n-1) + \varepsilon_n$$

As we defined, the $X(n)$ is a function of n, with the value of $X(n)$ being a function of the previous value of $X(n-1)$ and ε_n, and its constant coefficient a_1. AR parameter a_1 satisfies $|a_1| < 1$: Mean, Correlation function and variance.

Mean (first moment): We may use the expected value operator E on both sides of the equation to get the mean (first moment) of the stationary AR(1) process $X(n)$.

- Apply $E(\cdot)$ to the left-hand side and right-hand side of the AR(1) equation:

$$X(n) = a_1X(n-1) + \varepsilon_n \gg E\big(X(n)\big) = a_1E\big(X(n-1)\big) + E(\varepsilon_n)$$

$$\gg E\big(X(n)\big) = a_1E\big(X(n-1)\big) \gg m(n) = a_1m(n-1)$$

- The mean behaves like the deterministic part of the AR(1) model.
- $|a_1|$ implies $m \rightarrow 0$ (i.e., stationary AR(1) has mean = 0).

Variance $X(n)$: To calculate the variance of the stationary AR(1) process, we can define the variance

$$Var(x) = Var(\varepsilon)/1 - {a_1}^2$$

This equation calculated as the average of the squared differences between the values of the variable and the mean of the values. It denotes the variance of a variable x as a function of the variance of another variable ε and a constant coefficient a_1.

$Var(x)$ increases when the variance of noise ε increases. Minimal for the special case of MA(0) is white noise (i.e., $a_1 = 0$). $Var(x)$ also increases when a_1 approaches instability thresholds ± 1.

In the case of visual comparisons using the other function (i.e., power spectrum function, $P(f) = \frac{Var(\varepsilon)}{1-2a_1\cos(2\pi f)+{a_1}^2}$; see Supplements 2.4 for more details). Plot in Figure 2.10.7 displays two distinct values for the parameter a_1.

The plot on the left side of the picture depicts the non-oscillatory behavior of the signal in the scenario when $0 < a_1 < 1$. An oscillatory behavior of the signal is shown by the figure on the right side, which corresponds to the scenario when $-1 < a_1 < 0$.

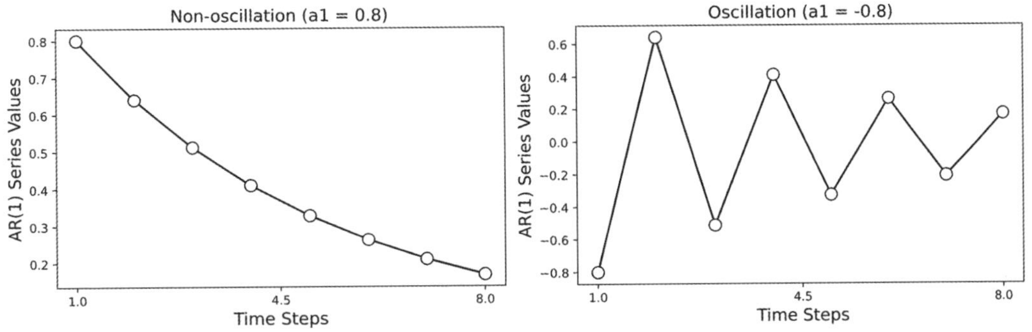

Figure 2.10.7. Non-oscillation. $0 < a_1 < 1$ (left), oscillation. $-1 < a_1 < 0$ (right). Both plots show a deterministic case.

Both figures display a deterministic scenario, which implies that the signal's characteristics and beginning circumstances govern it entirely. They illustrate how the signal's behavior can change based on the value of a parameter, and the caption offers additional precise information about the plots and the signal's behavior. Higher-order AR processes: Higher-order AR processes only display stationarity if their AR coefficients meet specific requirements, as in the case of AR-2 models with a triangular parameter domain.

- Stationarity: Only if AR coefficients satisfy certain conditions, AR-2, triangular parameter domain

Higher-order AR processes degrade their autocorrelation function exponentially. Please notice the Supplements 2.3 and 2.4 for further information on the autocorrelation function and the power spectrum function.

- Autocorrelation function: Decays exponentially

+ **Calculate the AR Process:** Let us delve deeper into the calculation of the autoregressive model of order 1, abbreviated as AR(1), focusing on the deterministic case (i.e., no noise) for two special cases: stable, non-alternating, and stable, alternating.

$$X(n) = a_1 X(n-1) + \varepsilon_n$$

The first scenario, (a), pertains to the stable, non-alternating case, where the coefficient a1 lies between 0 and 1. In this case, the AR(1) process follows a simple model equation: $X(t) = a_1 X(t-1)$. Here, the value at time t depends on the value at the previous time $(t-1)$ multiplied by the coefficient a_1. Given that a1 is between 0 and 1, this model will gradually converge towards zero, provided the initial condition $X(t) = 1$.

The second scenario, (b), covers the stable, alternating case, where the coefficient a_1 is between −1 and 0. In this case, the AR(1) process remains the same: $X(t) = a_1 X(t-1)$. But due to the negative value of a_1, the model will oscillate around zero, gradually decreasing in amplitude. That is because every subsequent value alternate in sign from the previous one, but the absolute value decreases since $|a_1| < 1$. For part (c), the task is to sketch the trajectories $X(n)$ for $n = 1, \ldots, 8$ in two different panels, given the initial condition $X(1) = 1$.

[Table 2.10.5] Solution to Problem 1

Problem 1a: Nonalternating ($a_1 = 0.9$)		Problem 1b: Alternating ($a_1 = -0.9$)	
$X(1)$	1	$X(1)$	–1
$X(2)$	0.9	$X(2)$	–0.9
$X(3)$	0.81	$X(3)$	0.81
$X(4)$	0.729	$X(4)$	–0.729
$X(5)$	0.6561	$X(5)$	0.6561
$X(6)$	0.59049	$X(6)$	–0.59049
$X(7)$	0.531441	$X(7)$	0.531441
$X(8)$	0.478297	$X(8)$	–0.4783

↓ ↓

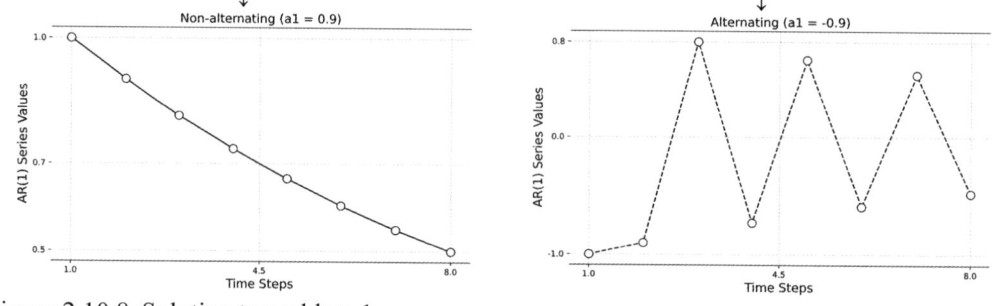

Figure 2.10.8. Solution to problem 1

In the first panel (for case (a)), you would start at 1 ($X(1) = 1$) and successively multiply by a_1 for each new n. The trajectory would decrease gradually towards zero, given that a_1 is between 0 and 1. In the second panel (for case (b)), you would also start at 1 ($X(1) = 1$). But here, each successive value would alternate in sign and gradually get closer to zero, given that a_1 is between -1 and 0.

This exercise helps visualize the behavior of the AR(1) process under different conditions, providing insight into its stability and response to initial conditions. Next, for the AR-1 case, calculate the five state values $X(2), \ldots, X(6)$ of the AR(1) process defined by

$$X(n) = a_1 X(n-1) + \varepsilon_n$$

For this computation, we also need the value of a1 and the initial condition $X(1)$. As these values are not provided in your question, we will have to make an assumption for the purpose of this example. Let us use the following sequence of random numbers taken from a normally distributed population with a standard deviation of 0.1:

$$e(2) = -0.12, \quad e(3) = 0.30, \quad e(4) = 0.07,$$

$$e(5) = -0.01, \quad e(6) = 0.07$$

Report $X(1), \ldots, X(6)$ to two decimal places.

[Table 2.10.6] Solution to Problem 2

solution: $a_1 = 0.9$, initial condition $X(1) = 1$
$X(2) = 0.9 \times 1 + (-0.14) = 0.76$
$X(3) = 0.9 \times 0.76 + (0.30) = 0.984$
$X(4) = 0.9 \times 0.984 + (0.07) = 0.9556$
$X(5) = 0.9 \times 0.9556 + (-0.01) = 0.85004$
$X(6) = 0.9 \times 0.85004 + (0.07) = 0.835036$

→

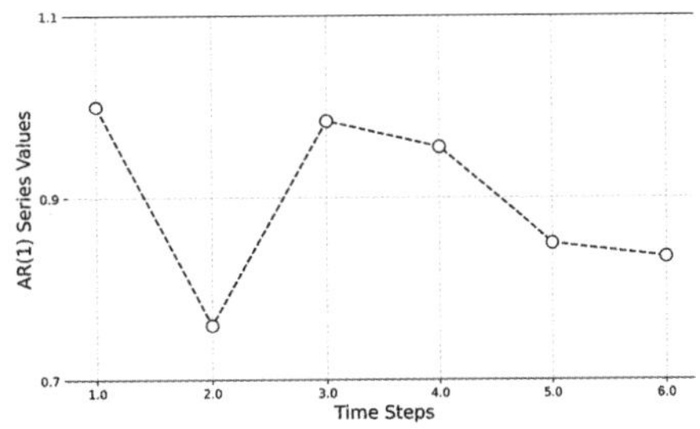

Figure 2.10.9. Solution to problem 2

Problem 2's solution is shown in this table, where a1 is the parameter and $X(1)$ is the starting condition, both of which are equal to 1. Based on a linear equation that includes the previous value of X and a constant, the following rows display how the value of X evolves over time. There is probably a graphic representation of the answer to Problem 2 in the Figure 2.10.9.

3. Practice with Time Series AR Models

Take the equation ($X(n) = a_1 X(n-1) + \epsilon(t)$), where a_1, ..., a_p are the model parameters, and $\epsilon(t)$ is white noise with mean 0 and standard deviation σ. Each lagged element is multiplied by a parameter. Let us simulate this with a random function.

```
[In]
from collections import deque
from numpy.random import randn
```

Make a model to fill self. X.

```
[In]
class AR:
    def __init__(self, a, x_init=None, sigma=1):
        self.p = len(a)
        self.a = a
        self.sigma = sigma
        self.x = deque(x_init or [0], maxlen=self.p)
        while len(self.x) < self.p:
            self.x.append(self.x[-1])
    def __iter__(self):
        return self
    def __next__(self):
        noise = randn() * self.sigma
        deterministic = sum(a*x for a, x in zip(self.a, reversed(self.x)))
        self.x.append(deterministic + noise)  # Record the history.
        return self.x[-1]
```

We defined a class called 'AR' which stands for autoregressive which is a type of statistical model that describes the evolution of a time series as a function of its past values. The 'p' represents the order of the model, 'a' represents the coefficient of the model, and 'sigma' represents the standard deviation of the noise term in the model. The '__iter__' method initializes the instance variables of the class and returns the instance of the 'AP' class itself, and the 'next' method generates the next value in the time series.

Let us apply methods of iteration to the value initialized using the existing history.

```
[In]
import itertools
import matplotlib.pyplot as plt


def simulate_AR(n, *args, **kwargs):
    model = AR(*args, **kwargs)
    x = list(model.x)  # Initialize with existing history.
    while len(x) < n:
        x.append(next(model))
    return x
```

The code defined a function called 'simulate_AR' that function takes as input the number of values to generates 'n' and any additional arguments and keyword arguments that should be passed to the 'AR' class when it is initialized. And then uses the 'next' function to generate successive values of the time series appending them to a list 'x'.

This continues until the length of the list 'x' is equal to the desired number of values, and finally returns the list of generated values.

See what is going on (Figure 2.10.10).

```
[In]
plt.plot(simulate_AR(100, [0.5]))
```

→

[Out]

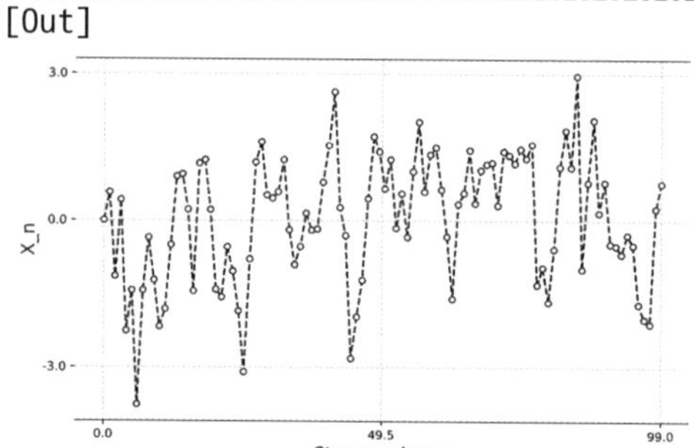

Figure 2.10.10. Output of the AR simulation 1

In short, each successive element of a gets multiplied by a lagged value of X: a_1 is multiplied by $Xt - 1$, a_2 is multiplied by $Xt - 2$, and so on. Its current value is only affected by the history at lag – 1.

For more practice, let us see if we can figure out what is going on.

You may need to consult the Python documentation *deque* and *islice*.

```
[In]
from collections import deque
from itertools import islice
from numpy.random import randn
```

Create the AR model with parameters a: list, Autoregressive parameters. len(a) is the order p of the AR model. x_init: list, optional, History at $t = 0$. sigma, optional: float, Standard deviation of white noise in model.

```
[In]
class AR:
    def __init__(self, a, x_init=None, sigma=1):
        self.p = len(a)
        self.a = a
        self.sigma = sigma
        self.x = deque(x_init or [0], maxlen=self.p)
        # An AR model must begin with a history.
        while len(self.x) < self.p:
            self.x.append(self.x[-1])
    def __iter__(self):
        return self
    def __next__(self):
        deterministic = sum(a*x for a, x in zip(self.a, reversed(self.x)))
        noise = randn() * self.sigma
        self.x.append(deterministic + noise)
        return self.x[-1]
```

Simulate the model with the number of steps.

```
[In]
    def simulate(self, n):
        return list(islice(self, n))
```

To test it, we need to plot it (Figure 2.10.11).

```
[In]
import matplotlib.pyplot as plt
%matplotlib inline

plt.plot(AR([0.4, -
0.7]).simulate(100))
```

→

[Out]

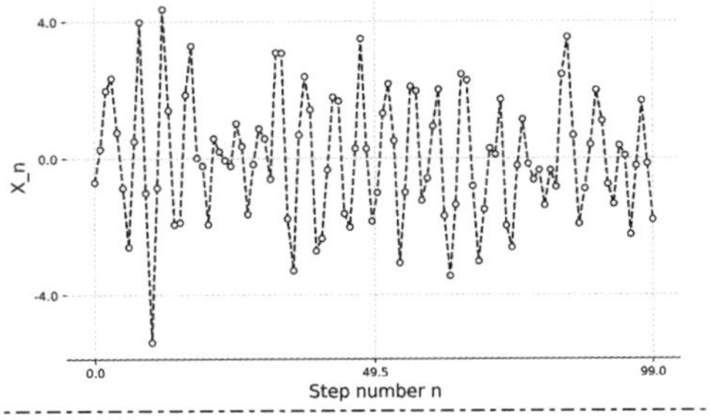

Figure 2.10.11. Output of AR simulation 2

See it with different values (Figure 2.10.12).

[Out]

[In]

```
plt.plot(AR([-0.4,
0.7]).simulate(100))
```

→

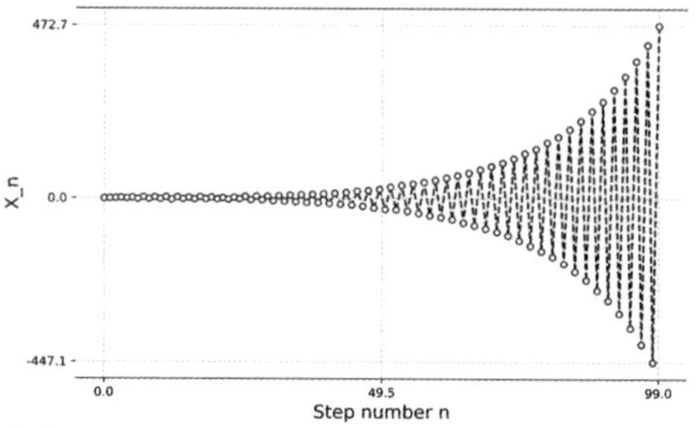

Figure 2.10.12. Output of AR simulation 3

We create two plots of the time series data. Each plot is generated by calling 'simulate()' method of the AR model with an argument of 100. The 'AR()' function takes coefficients [0.4, – 0.7] for the first plot and the second plot takes coefficients [– 0.4, 0.7].

Remarks: Moving average (MA) models, frequently utilized in the systems analysis, have proven instrumental in forecasting future event (i.e., stock prices, volatility, and other metrics) by effectively modeling time series data. These models function akin to signal processors, extracting pertinent information from a signal by eliminating noise, which is crucial for anyone keen on studying and modeling time series data. The knowledge derived from these MA models can be employed across various domains to tackle real-world problems and formulate robust decisions. In the same vein, autoregressive (AR) models represent time series data by postulating that each observation is a linear function of prior observations, augmented by a random error component. These models echo the dynamics of stochastic iterative mappings, which necessitate the iterative application of a nonlinear function to a random variable. AR models offer practical solutions in real-world contexts, facilitating predictions, parameter estimations, and signal processing operations such as filtering and noise reduction. Stochastic iterative mappings, meanwhile, enable the simulation of complex dynamic systems and the investigation of their behavior under diverse conditions. In essence, both MA and AR models, along with stochastic iterative mappings, provide powerful tools for data analysis and interpretation. They possess wide-ranging applicability in a broad spectrum of scientific contexts, offering significant insights and solutions.

[2.11]

Stochastic Iterative Maps, Part 2

Autoregressive Moving Average Model, Function and Simulation, Practical Practice

Building on the concepts of autoregressive (AR) and moving average (MA) models, autoregressive moving average (ARMA) processes represent a key class of stochastic processes for systems analysis. Distinguished from Gaussian processes, ARMA models possess autocovariance functions that rely on a finite number of unknown parameters, enabling a more efficient and focused analysis of time series data (Fan & Yao, 2008).

1. Autoregressive Moving Average Model

An ARMA is a model that uses time series data to either better understand the data set or to predict future trends. ARMA processes assume that a time series fluctuates uniformly around a time-invariant mean. If we are trying to analyze a time series that does not follow this pattern, then the series will need to be differenced until it achieves stationarity.

The acronym ARMA can be broken down in two main parts:

- Autoregressive (AR) = the model takes advantage of the connection between a predefined number of lagged observations and the current one.
- Moving average (MA) = the model takes advantage of the relationship between the residual error and the observations.

ARMA models make use of two main parameters $(p,\ q)$. p is the number of lag observations, and q is the size of the moving average window.

+ **Notation:** The two main parameters of ARMA models, p and q, define them. The parameters p and q specify the size of the moving average window and the number of lag observations, respectively.

- Moving average model of order q: MA(q)

$$X(n) = \varepsilon_n + b_1\varepsilon_{n-1} + b_2\varepsilon_{n-2} + \cdots + b_q\varepsilon_{-q}$$

The equation defines $X(n)$ as a function of ε_n, ε_{n-1}, ε_{n-2}, and so on up to ε_{-q}. $X(n)$ is given as the sum of these values with each value multiplied by a corresponding coefficient b_1, b_2, etc.

- Autoregressive model of order p: AR(p)

$$X(n) = a_1X(n-1) + a_2X(n-2) + \cdots a_pX(n-p) + \varepsilon_n$$

This equation defines $X(n)$ as a function fo $X(n-1)$, $X(n-2)$, and so on up to $X(n-p)$, as well as ε_n. The $X(n)$ is given as the sum of these values with each value multiplied by a corresponding coefficient a_1, a_2, etc.

- Mixed ARMA model of order q, p: ARMA(q, p)

$$X(n) = \underbrace{a_1X(n-1) + \cdots a_pX(n-p)}_{p\text{ terms}} + \underbrace{\varepsilon_n + b_1\varepsilon_{n-1} + \cdots + b_q\varepsilon_{-q}}_{q+1\text{ terms}}$$

Here, the equation has the form of an autoregressive-moving average which is a type of statistical model that describes the evolution of a time series as a function of both its past values and the past values of an external. The present state $X(n)$ is affected by the weighted average of p previous states. And the present state $X(n)$ is affected by the weighted average of present and q previous perturbations.

Let us assume why do we set $b_0 = 1$ and not weight ε_n in ARMA models? This is because the variance of the error term (ε) is used to account for the weighting in ARMA models.

+ **Process:** Consider Table 2.11.1, which shows an example dataset for a linear regression model with two predictors, age (X_1) and years in school (X_2), and one response variable, income (Y).

The relationship between income and the predictors can be modeled as

[Table 2.11.1] Data Set Example for a_1

Person	Age (X_1)	Years in school (X_2)	Income (Y)
Walter	33	13	$$$
Susan	28	10	$$$
Mohamed	41	18	$$$

Income = a_1 Age
+ a_1 Education
+ Random effects

$$Y = a_1X_1 + a_1X_1 + \varepsilon$$

where a_1 and a_2 are coefficients and ε is the random error term. To prediction Y, as well as the values of X_1 and X_2, the model can be used:

$$X(n) = a_1X(n-1) + a_2X(n-2) + \varepsilon_n$$

Here, $X(n)$ represents the predicted variable, while $X(n-1)$ and $X(n-2)$ are the predicting variables.

- Predicted variable = predicting variables (autoregressive model)

In ARMA models, we set b_0 to 1 and do not weight ε_n because the variance of the error term, var(ε), is already used to account for the weighting of the model

+ **Function and Simulation:** Here we combine these models to produce more sophisticated examples called ARMA and autoregressive integrated models. As you can see, ARMA models are simply a combination of AR(p) and MA(q) models.

- AR(p) models are used to explain the momentum and mean reversion effects often observed in trading markets. These effects can be attributed to market participant behavior.
- On the other hand, MA(q) models aim to capture the shock effects seen in white noise terms. These unexpected events, such as surprise earnings or wars, can affect the observation process.

To ensure stationarity in ARMA models, we can refer to Table 2.11.2. For MA(q) models, stationarity is guaranteed for all parameters. However, for AR(p) models, stationarity depends on the value of a_k. For instance, an AR(1) model is stationary if $|a_1| < 1$.

[Table 2.11.2] ARMA Stationarity

	Stationarity	
MA(q)	Is stationary for all parameters	
AR(p)	Depends on a_k	AR(1): stationary if $|a_1| < 1$
ARMA(p, q)	Depends on a_k	ARMA(1, q): stationary if $|a_1| < 1$

[Table 2.11.3] ARMA Function

	Corr(m)	
MA(q)	Cutoff after lag q	Can be used for model selection and model identification
AR(p)	Exponential decay or damped oscillation	
ARMA(p, q)	Exponential decay or damped oscillation After lag q	

Table 2.11.3 demonstrates the function of MA(q) models, which feature a cutoff after lag q that may be utilized for model identification and selection. On the other hand, AR(p) models show exponential decay or damped oscillation. After latency q, the same function is applicable to ARMA(p, q) models.

+ **Simulation (ARMA):** We can take the following actions to simulate ARMA models. Let us outline the process of defining initial conditions, creating sequences of perturbations and states, and calculating subsequent values of X using updated values and new perturbations for a given model.

- Define the initial state at time point $n = 1$, including the initial conditions for the model
- Create an initial sequence of $q + 1$ perturbations: $\varepsilon_{1-q}, \dots, \varepsilon_0, \varepsilon_1$
- Create an initial sequence of p states: $X(1 - p)$, …, $x(-1)$, $X(0)$
- Calculate $X(1)$ using the initial states and perturbations
- Draw ε_2 and use it to compute $X(2)$ using the updated values of X and the new perturbation

This procedure can be performed as many times as necessary to replicate the ARMA model's behavior throughout time (Figure 2.11.1).

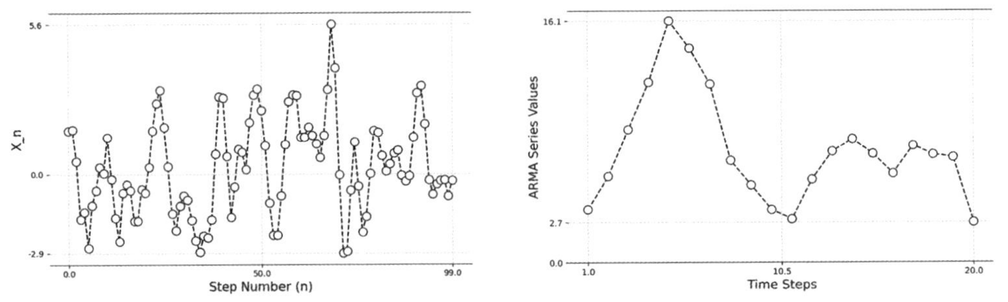

Figure 2.11.1. Autoregressive moving average simulation. Trajectory for $n = 1, \dots, 20$. $a_1 = 0.6$, $b_1 = 0.8$, $b_2 = 0.8$, var(ε) = 4, $X(1) = 1$, $T = 50{,}000$ data points

Figure 2.11.1, left side displays a time series of 100 data points, which is directly linked to the right side of the ARMA series value; depicts a trajectory for $n = 1, \dots, 20$. Using programing, we can simulate autoregressive moving average (ARMA) models.

To implement this, we must first define these parameter values in order to build the simulation.

```
[In]
T = 50000
X = [0]*T
```

The 'T' is assigned the value 50,000 which represents limit, and the 'X' is assigned the value of an array of zeros with shape '1, T'. this creates an array with one row and 'T' columns, all of which are initialized to 0.

For parameters including Var(epsilon)=sigma-squared

```
[In]
a1 = 0.6
sigma = 2

b1 = -0.8 * 0  # put b1=0 and b2=0 to simulate AR(1)
b2 = -0.8 * 0
```

The code above sets up an AR(1) model with an a1 value of 0.6, a variance of 2 for the noise term, and both b1 and b2 coefficients equal to 0. For initial time ($n = 1$) and the sequence of T+2 perturbation and shift

```
[In]
import numpy as np
X = np.zeros(T)
X[0] = 1

# sequence of T+2 perturbations: e(n=-1), e(0), e(1),…,(T)
e = np.random.normal(0, sigma, T+1)
shift = 1       # use shift=1 to avoid negative indices
for n in range(1, T):  # Python uses 0-based indexing
    X[n] = a1*X[n-1] + e[n+shift] + b1*e[n-1+shift] + b2*e[n-2+shift]
```

We defined a variable X as an array with one element, initialized to 1. And the variable 'e' is an array of normally distributed random values with mean 0 and standard deviation 'sigma' with shape '1, T+1' which is used to generate the noise term in this ARMA process. The 'shift' as 1 is used to avoid negative indices when accessing elements of the array 'e', and the conde enters a loop that runs from 2 to 'T'. On each iteration of the loop, it updates the value of $X(n)$ using the defined equation 'a1 $*$ $X(n-1)$+e(n+shift)+b$*$e($n-1$+shift)+b2$*$e($n-2$+shift)' which includes previous s and current values with constants. Create the plot

```
[In]

fig, ax = plt.subplots()

# Plot the series
ax.plot(range(1, T+1), X, 'ko--')
```

→

[Out]

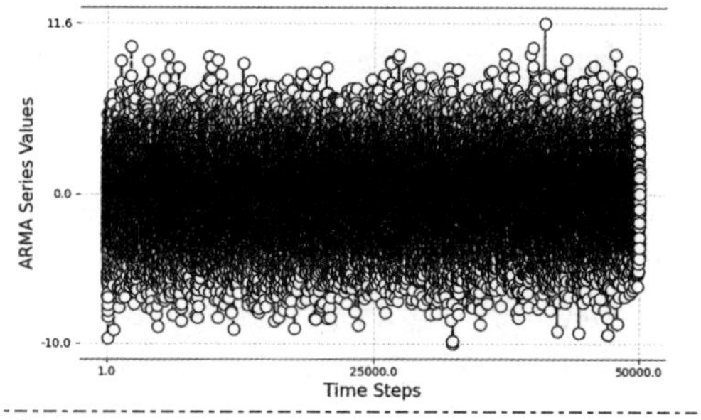

Figure 2.11.2. Output of the ARMA(1, 2) simulation

The output is seen in Figure 2.11.2. We start by invoking the 'figure' function. Then, we create the plot using the 'plot' function, where the y-axis represents the values in the X array and the x-axis represents the number of data points, which runs from 1 to T.

2. Practice with the Time Series ARMA Models

Moving forward with our practice of time series, we will conduct simulations and calculations using the ARMA model. The goal of this models is to integrate both autoregressive and moving average components when analyzing systemically time series data. However, it is crucial to acknowledge that these models do not consider volatility clustering, a significant empirical feature commonly observed in numerous time series.

The ARMA(1, 1) model is

$$X(t) = a_1 X(t-1) + b_1 \varepsilon(t-1) +, \varepsilon(t), \qquad \varepsilon(t) = white\ noise\ with\ E[\varepsilon(t)] = 0$$

An ARMA model often requires fewer parameters than an AR(p) or MA(q) model alone. That is, it is redundant in its parameters.

We simulate an ARMA(2, 2) process with specific parameters. These parameters are set as alphas to [0.5, - 0.25] and betas to [0.5, - 0.3]. Using the 'statsmodels' library in Python, we generate a sample from this ARMA process and visualize it.

```
[In]
import pandas as pd
import numpy as np
import statsmodels.tsa.api as smt
import statsmodels.api as sm
import matplotlib.pyplot as plt
```

Define the ARMA model parameters, for time series:

```
[In]
alphas = np.array([0.5, -0.25])
betas = np.array([0.5, -0.3])
ar = np.r_[1, -alphas]
ma = np.r_[1, betas]
```

Generate a sample from the ARMA process

```
[In]
n = int(5000) # lots of samples to help estimates
burn = int(n/10) # number of samples to discard before fit
arma22 = smt.arma_generate_sample(ar=ar, ma=ma, nsample=n, burnin=burn)
```

And plot the time of the ARMA process

```
[In]
plt.figure(figsize=(10, 5))
plt.plot(arma22)
plt.title('ARMA(2,2) Time Series')
plt.show()
```

→ [Out]

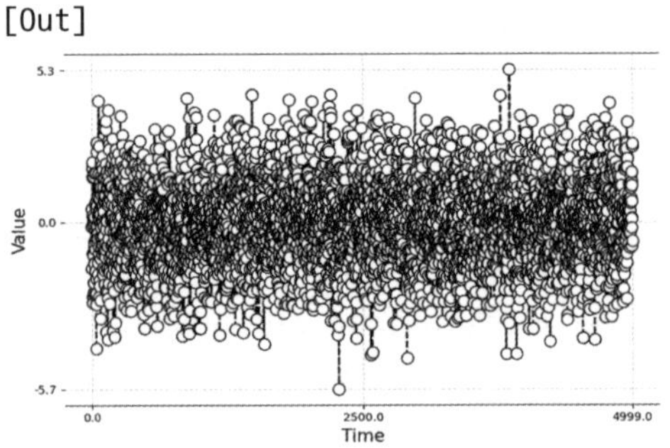

Figure 2.11.3. Output of the ARMA (2, 2) process

Figure 2.11.3 displays the results of the simulation. We can also check the normality of the simulated data with QQ (Quantile-Quantile) and PP (Probability-Probability) plots. These plots can help verify if the data follow a particular theoretical distribution. The QQ plot compares the quantiles of our data against the quantiles of a normal distribution. The PP plot compares the cumulative distribution function (CDF) of our data against the CDF of a normal distribution. If our data follow a normal distribution, points in these plots should lie on a straight line.

```
[In]
import scipy.stats as scs
def tsplot(y, figsize=(10, 8), style='bmh'):
    if not isinstance(y, pd.Series):
        y = pd.Series(y)
    with plt.style.context(style):
        fig, axes = plt.subplots(nrows=1, ncols=2, figsize=figsize)
        sm.qqplot(y, line='s', color='b', ax=axes[0])  # QQ plot
        axes[0].set_title('QQ Plot')
        scs.probplot(y, sparams=(y.mean(), y.std()), plot=axes[1])  # PP plot
        axes[1].set_title('PP Plot')
        plt.tight_layout()
    return

# Generate the ARMA(2,2) time series
alphas = np.array([0.5, -0.25])
betas = np.array([0.5, -0.3])
ar = np.r_[1, -alphas]
ma = np.r_[1, betas]
n = int(5000)
burn = int(n/10)
arma22 = smt.arma_generate_sample(ar=ar, ma=ma, nsample=n, burnin=burn)
tsplot(arma22)
```

```
[Out]
```

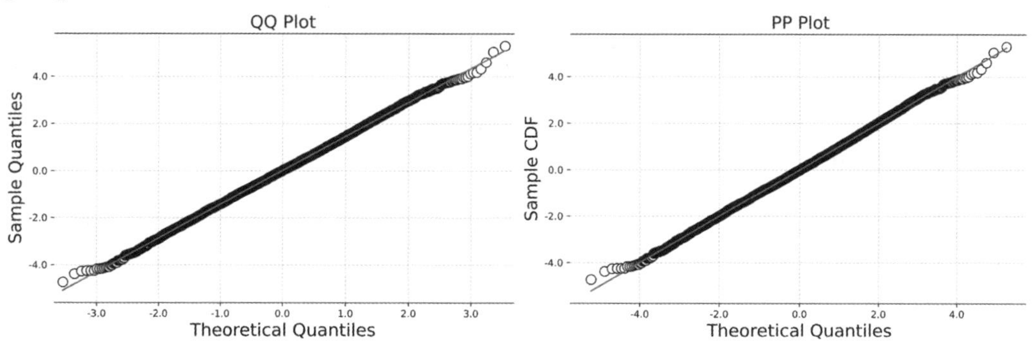

Figure 2.11.4. Output of the QQ and PP plots

To assess the normality of this data (illustrated in the Figure 2.11.3. ARMA(2,2)), we employ two powerful statistical visualizations, the QQ and PP plots, as shown in Figure 2.11.4.

The QQ plot matches the quantiles of our data with those of a normal distribution. If the points form a straight line, this suggests our data might follow a normal distribution. Any deviation from the line provides insights into the ways our data differs from this assumption. The PP plot, on the other hand, compares the cumulative probabilities of our data to a normal distribution. Here too, points lining up along the diagonal suggest a normal distribution.

In essence, these plots serve as an essential check, helping us understand if our simulated data aligns with the theoretical normal distribution, a common assumption in many statistical models.

In statistics, these ARMA(2, 2) process are frequently used to evaluate the normality of a dataset.

```
[In]
from statsmodels.tsa.arima.model import ARIMA

# Fit an ARMA(2,2) model (equivalent to ARIMA(2,0,2)) and print the summary
mdl = ARIMA(arma22, order=(2, 0, 2), trend='n').fit()

print(mdl.summary())
```

[Table 2.11.4] ARMA Model Results

```
==============================================================================
Dep. Variable:                    y   No. Observations:                 5000
Model:                   ARMA(2, 2)   Log Likelihood               -7115.319
Sample:                           0   HQIC                         14252.059
==============================================================================
                 coef    std err          z      P>|z|      [0.025      0.975]
------------------------------------------------------------------------------
ar.L1.y        0.4933      0.059      8.400      0.000       0.378       0.608
ar.L2.y       -0.2374      0.015    -15.387      0.000      -0.268      -0.207
ma.L1.y        0.5009      0.060      8.408      0.000       0.384       0.618
ma.L2.y       -0.3120      0.055     -5.668      0.000      -0.420      -0.204
sigma2         1.0079      0.020     50.585      0.000      -0.969      1.1047
==============================================================================

Ljung-Box (L1) (Q):                   0.02   Jarque-Bera (JB):                0.57
Prob(Q):                              0.90   Prob(JB):                        0.75
Heteroskedasticity (H):               1.07   Skew:                           -0.00
Prob(H) (two-sided):                  0.15   Kurtosis:                        3.05
==============================================================================
```

Table 2.11.4 presents results for an ARMA model fitting an ARMA(2,2) model to our simulated time series data.

Here is a simplified explanation:

- The 'coef' column shows the weights that the model has learned for each term, which are pretty close to the values we used to simulate the data, showing the model's effectiveness.
- The 'p>|z|' column is the p-value associated with the hypothesis test for each coefficient. If this value is less than 0.05, we can reject the hypothesis that the corresponding coefficient is zero. Here, all p-values are below 0.05, indicating all terms are statistically significant.
- The Ljung-Box (Q) test checks whether the residuals from our model have any remaining autocorrelation. The high p-value (0.90) shows that there is no significant autocorrelation left unexplained by our model.
- The Jarque-Bera test assesses whether the residuals from our model are normally distributed. The high p-value (0.75) suggests they are.
- The 'sigma2' term indicates the variance of the residuals or the noise, which is close to 1 as in our simulation settings.

Diagnostic tests indicate that the residuals of the model (the differences between the observed and predicted values) are not significantly different from a random noise process. In short, the ARMA(2,2) model fits our simulated data well, indicating that the model is good at capturing the patterns in the data. .

Remarks: We have now covered two key ideas in time series analysis: stochastic iterative maps and autoregressive moving average (ARMA) models. The value of a variable at each time point in a stochastic iterative map depends on the value of the variable at the previous time point as well as a random shock term. We have shown that stochastic iterative maps may display intricate and fascinating behavior, such as chaotic dynamics. The ARMA models, which combine moving average (MA) and autoregressive (AR) models, may capture both the momentum and mean reversion effects frequently seen in real-world data as well as the impacts that can be viewed as unexpected occurrences altering the observation process. We have learned how to create output such time series plots, as well as how to parameterize and simulate ARMA models using simulation. In general, comprehension of these ideas is crucial for modeling and evaluating systems analysis in a variety of domains.

Supplemental Information 2.3: Autocorrelation

A time series is a series of data points indexed in time. The fact that time series data are ordered makes them unique in the data space because they often display serial dependence. Serial dependence occurs when the value of a data point at one time is statistically dependent on the value of another data point in another time.

Often one of the first steps in any data analysis is performing regression analysis. However, one of the assumptions of regression analysis is that the data are not autocorrelated. Perhaps the most compelling aspect of autocorrelation analysis is how it can help you uncover hidden patterns in your data and help you select the correct forecasting methods (Kandananond, 2012). Specifically, you can use it to help identify seasonality and trends in time series data. In addition, analyzing the autocorrelation function in conjunction is necessary to select the appropriate model for your time series prediction.

1. Definition

Autocorrelation, as a concept, is known as serial correlation, which provides information about temporal structure. It refers to the degree of correlation among the same variables between two successive time intervals and measures how the lagged version of the value of a variable is related to its original version in a time series.

[Table S2.3.1] Schematic Representation of Autocorrelation

	Discrete-state variable	Continuous-state variable
Discrete time Continuous time		**Autocorrelation**

* State (*X*)= continuous, numerical variable

Table S2.3.1 displays a graphical illustration of autocorrelation in discrete and continuous time for a continuous state variable and a discrete state variable. In this sense, the state X is an ongoing, numerical variable. A crucial idea in time series analysis is autocorrelation, which quantifies the relationship between a variable and a lagged version of itself.

Autocorrelation is a linear relationship between states X_1 and X_2 occupied at two different times t_1 and t_2 (also known as *lags*). To what extent are states X at times t_1 and t_2 related to each other?

- Strong linear relationship: Autocorrelation is large.
- Weak linear relationship: Autocorrelation is small.

It shows whether there is a sign change between states at t_1 and t_2 or whether $X(t_2)$ has on average the same sign as $X(t_1)$ or the opposite sign.

- Same sign: Autocorrelation is positive.
- Opposite sign: Autocorrelation is negative.

Robert Brown (1827) studied the motion of pollen grains in water. Another example of Brownian motion is the motion of macroparticles in a fluid layer (Uhlenbeck & Ornstein, 1930).

2. Processes

Autocorrelation is a type of serial dependence. A time series is autocorrelated when it is linearly related to a lagged version of itself. By contrast, two independent variables are correlated if they are linearly related (Taylor, 1990).

+ **Notation (Corr):** The notation 'Corr' is used to define the autocorrelation of a stochastic process $X(t)$ in terms of expectation values. Autocorrelation depends on two time points, denoted as Corr(t_1, t_2), and is represented as an expectation value involving $X(t_1)$ and $X(t_2)$, such that Corr(t_1, t_2) = $E[X]$.

- Stochastic process $X(t)$
- Autocorrelation depends on two time points → Corr(t_1, t_2)
- Autocorrelation is an expectation value involving $X(t_1)$ and $X(t_2)$ → Corr(t_1, t_2) = $E[X]$

We can represent the expected values (means) of two random variables $X(t_1)$ and $X(t_2)$, respectively.

$$\mu_1 = E[X(t_1)], \qquad \mu_2 = E[X(t_2)]$$

It shows a discrete random variable with a probability mass function $f(x)$ which is given by $\mu_1 = Ex * f(x)$, if if is a continuous with a probability density function $f(x)$, the expected value of t_1 will be given by

$$\mu_1 = \int x * f(x)dx$$

Standard deviation at t_1 and t_2: σ_1, σ_2

$$Corr(t_1, t_2) = \frac{1}{\sigma_1 , \sigma_2} E[X(t_1) - \mu_1) \cdot (X(t_2) - \mu_2)]$$

This equation defines the correlation between two random variables $X(t_1)$ and $X(t_2)$ at time t_1 and t_2, respectively. It is calculated as the expected value of the product fo the deviations of $X(t_1)$ and $X(t_2)$ from their respective means μ_1 and μ_2 normalized by the product of their standard deviations σ_1 and σ_2.

In general, a high positive correlation indicates that the two variables tend to increase or decrease together, while a high negative correlation indicates that one variable tends to increase as the other decrease. Definition in terms of realizations: Realizations $X^{(k)}(t), k = 1, \dots, N$, and $N \rightarrow$ infinity

$$Corr(t_1, t_2) = \frac{1}{\sigma_1 , \sigma_2} \sum_{k=1}^{N} \left(X^{(k)}(t_1) - \mu_1 \right) \cdot \left(X^{(k)}(t_2) - \mu_2 \right) \Big)$$

This equation is similar to the previous one, but instead of calculating the correlation between two random variables $X(t_1)$ and $X(t_2)$, it calculates the correlation between two samples of those variables. The samples are denoted as $X^{(k)}(t_1)$ and $X^{(k)}(t_2)$, where k is an index running from 1 to N, and N is the number of samples in each set. Again, the equation does not measure the correlation between two random variables (see Part 1, section 'Statistical Model, Part 3' about correlation). Instead, it measures the correlation between a variable against itself; this is why it is called auto-correlation.

+ **Range:** The correlation coefficient, which indicates the degree and direction of a linear relationship between two variables, can have values between −1 and 1, inclusive ($1 \leq$ correlation ≤ 1).

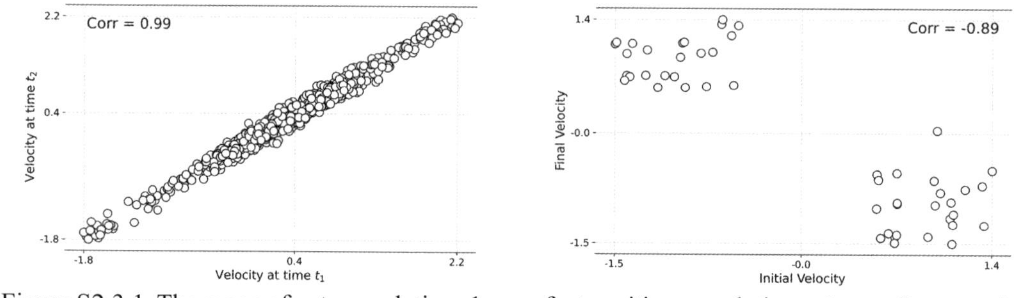

Figure S2.3.1. The range of autocorrelation: 1 = perfect positive correlation, –1 = perfect negative correlation. Corr = correlation

In Figure S2.3.1, the value indicates how strongly the variables are related to each other. If the value is 1, the variables are perfectly positively correlated; if it is −1, they are perfectly negatively correlated; and if it is 0, there is no correlation.

Thus, autocorrelation allows us to gain insight into the characteristics of Seasonality with Trend. Therefore, we would need to carry out some differencing to render the time series stationary.

+ **Stationary Case:** When the system is stationary, autocorrelation is determined using time averaging, where all joint probabilities depend only on time differences.

- Autocorrelation via time averaging → all joint probabilities depend only on time differences.

This has implications for the autocorrelation function, which is a function of time differences $t_2 - t_1$.

- Implication for the autocorrelation function → the function depends on the time differences $\tau = t_2 - t_1$.

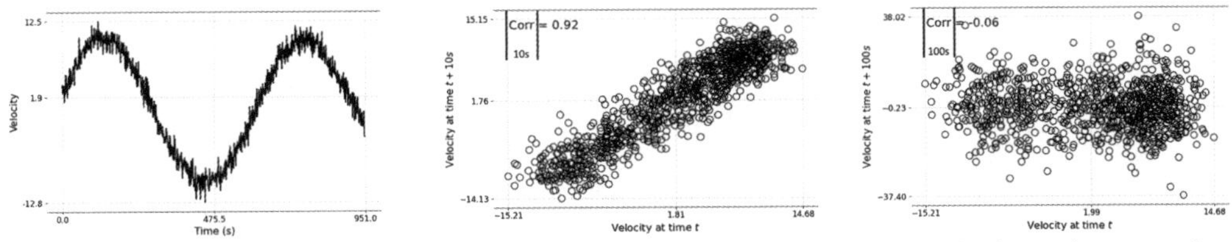

Figure S2.3.2. Autocorrelation via time averaging (stationary case). *V*(*t*) is the trajectory of one Brownian particle

The stationary instance of autocorrelation by time averaging is shown in Figure S2.3.2. V stands for the trajectory of a single Brownian particle (t). In the time-averaged autocorrelation, we consider a stationary process with a single trajectory, $X(1)$, $X(2)$,..., $X(N)$, and we use time averaging to get the sample mean M. Utilizing time averaging under constant, we may calculate the correlation at a specified time lag.

- Stationary process: One trajectory $[X(1), X(2),..., X(N)]$
- Sample mean M (time averaged)

How can you estimate Corr at a time lag? Solution: Time averaging under constant

$$t_{n+m} = t_n + \tau, \qquad Corr(\tau, estim) = \frac{1}{(N-m)\sigma^2} \sum_{n=1}^{N-m} (X(t_n) - M) \cdot X(t_{n+m}) - M)$$

This equation defines the correlation between two variables $X(t_n)$ and $X(t_{n+m})$, where n and m are indices running from 1 to N and m, respectively. The correlation is calculated as the sum of the products of the deviations of $X(t_n)$ and $X(t_{n+m})$ from their respective means M normalized by the product of the number of samples N and the variance σ^2. This is useful tool for understanding the relationship between two variables, and for making prediction about one variable based on the values of the other. Example;

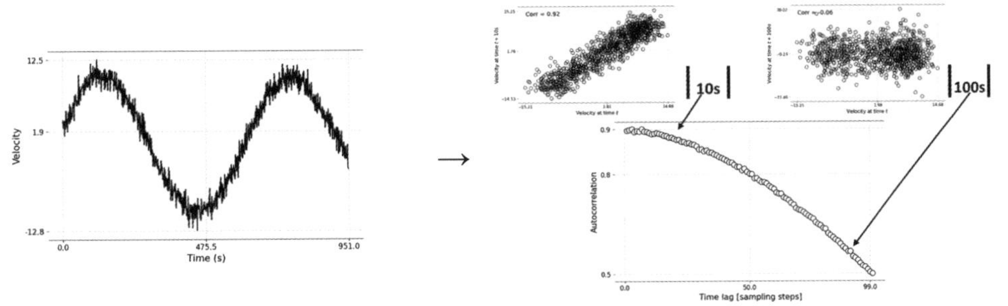

Figure S2.3.3. The dynamic velocity of a Brownian particle

Figure S2.3.3 depicts the dynamic velocity of a Brownian particle. For discrete-time case

$$t_1 \to n', t_2 \to n\,, t_2 - t_1 \to m = n - n'$$

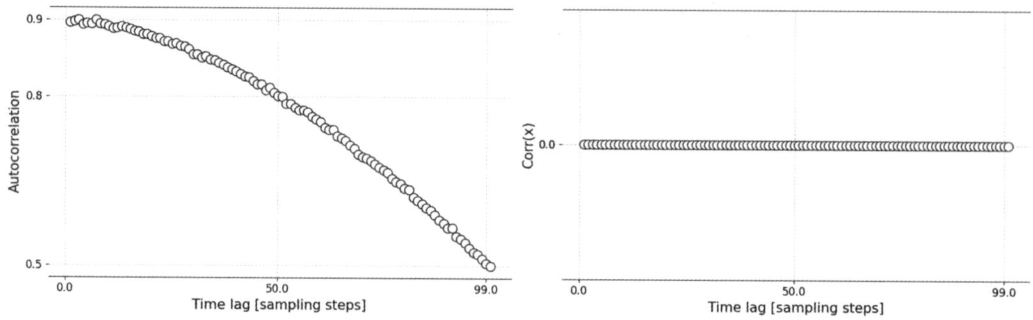

Figure S2.3.4. Two functions of the (left) stationary (time averaging, function of time lag), (right) and the discrete time

Figure S2.3.4 illustrates two functions that may be seen in the discrete-time case: Left (Stationary Function) plot illustrates the autocorrelation function, which depends on time lag and diminishes as the time lag increases. This represents a stationary function that depends on time lag. Right (Discrete-Time Function) plot illustrates a function that is independent of time lag. This function does not change with time lag, representing a discrete-time function that is independent of time lag.

3. Simulation

To simulate computing autocorrelations from a single trajectory via time-averaging in programing, first load the data using the provided Supplement data set or a noisy function. Use the following command to load the data using the Supplement data set.

```
[In]
import matplotlib.ticker as ticker
from scipy import signal
data = np.loadtxt('timeseries_ac.txt') # Load data
```

Alternately, you may create the data using function below.

```
[In]
if 1==2:
    dt=100/1000  # sampling 100 Hz
    timepoints=np.arange(0,10,dt)  # duration 10 sec

    T=2.5  #oscillation period 2.5  sec
    sigma=0.01  # try sigma 0.01  and sigma 1
    noise=np.random.normal(0,sigma,len(timepoints))
    data=np.sin(2*np.pi/T*timepoints)+noise
```

We defined several variables above code. The 'dt' is the sampling rate of the loaded data, in Hz. The 'timepoints' is an array of time points at which the data is sampled range from 0 to 10. The 'T' is the period of the sinusoidal wave and the wave is generated using the 'sin' function. The 'sigma' is the Gaussian noise and the 'noise' is an array of noise values generated using the 'normal' function from the 'stats' module. The code also contains an 'if' statement that checks whether 1 is equal to 2. For a one-sided autocorrelation function and sampling steps on the horizontal axis

```
[In]
maxlags=600  # Autocorrelation from the loaded data set case
csigned = signal.correlate(data, data, mode='full', method='auto') / np.var(data)
c = csigned[len(data)-1 : len(data)-1+maxlags]
lags = np.arange(len(c))
```

The function calculates the degree to which data at different lags are correlated with each other. The 'numpy.correlate()' function is used to compute the autocorrelation of the data, which is the cross-correlation of the data series with itself. The autocorrelation is essentially normalized autocovariance, so the same effect is achieved. The result is stored in the array c, which contains autocorrelation values for different lag values, ranging from zero to 'maxlags'. The corresponding lag values are stored in the lags array. The autocorrelation values c are then plotted against these lag values, providing a visual representation of how the autocorrelation in the data series changes with different lags. Then, see output of the function

```
[In]
# Output autocorrelation plot
fig1, ax1 = plt.subplots()
ax1.plot(lags, c, 'ko--')
```

→ [Out]

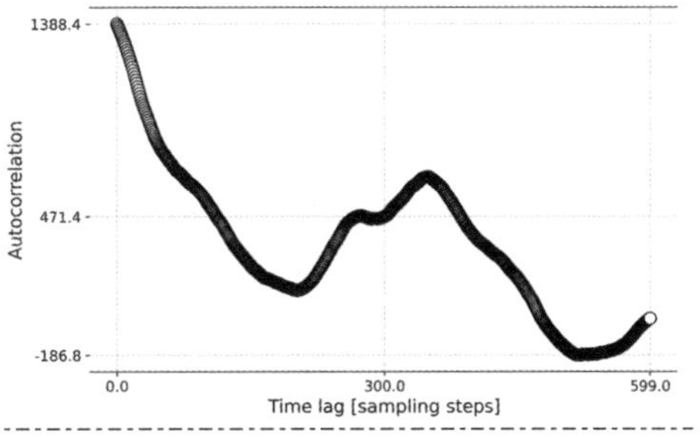

Figure S2.3.5. Output of autocorrelation with time lag

Then, create its output trajectory

```
[In]
# Output trajectory
fig2, ax2 = plt.subplots()
ax2.plot(np.arange(1,     len(data)+1),
data, 'ko--')
```

→ [Out]

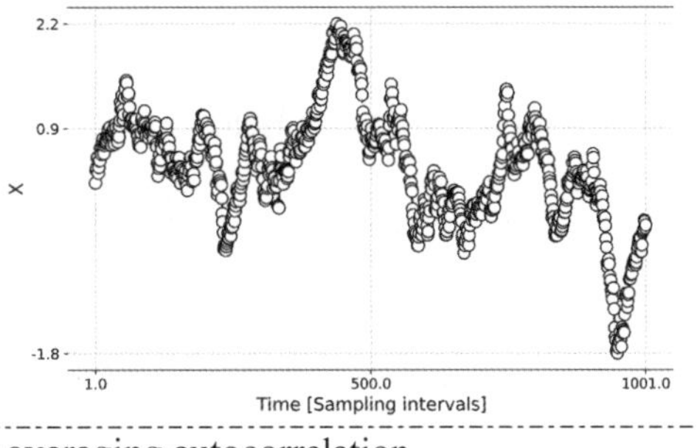

Figure S2.3.6. Output trajectory defined by the time averaging autocorrelation

The code supplied in the earlier messages demonstrates how to compute the autocorrelation function in programing and present the generated data as a figure. The subsequent plots (Figures S2.3.5 and S2.3.6) illustrate the output of this autocorrelation function, with the former showing the correlation values over time delays and the latter showing the trajectory of the time series itself.

These charts can be used to learn more about the time series' underlying structure and to provide guidance.

Let us practice one more through an example in actual data set to make this function more concrete. For this walkthrough, we will use our data set called (see the Supplement data set; airPassengers.csv).

First step: Import packages as follows:

```
[In]
import plotly.express as px
import pandas as pd
```

Second step: read the time series data as follows:

```
[In]
data = pd.read_csv('AirPassengers.csv')
```

Third: Plot the data for your own edification (Figure S2.3.7):

```
[In]
fig = px.line(data, x='Month',
y='#Passengers',
              labels=({'#Passengers':
'Abundance', 'Month': 'Time'}))

fig.update_layout(template="simple_white
", font=dict(size=15))
```

→

[Out]

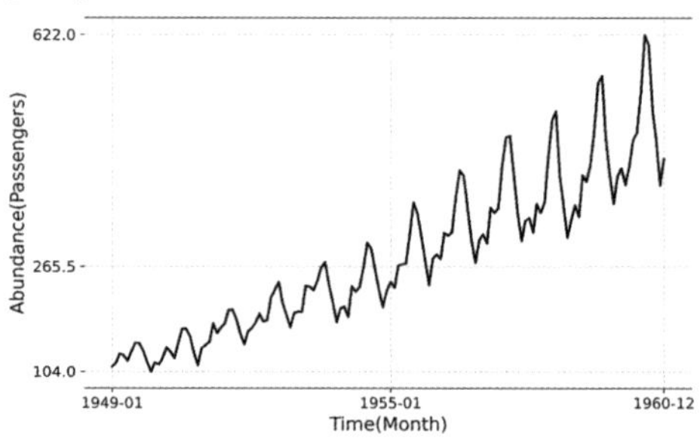

Figure S2.3.7. Output generated by actual data: Seasonality

The data indexed by the time series show us a clear trend and seasonality.

Now, we use the autocorrelation function from the package (statsmodels) to represent the autocorrela-tion as below:

Third step: Import packages as follows:

```
[In]
from statsmodels.graphics.tsaplots import plot_acf
import matplotlib.pyplot as plt
```

Take a look at the plot as follows:

[In]

```
plt.rc("figure",
figsize=(11,5))
plot_acf(data['#Passengers'],
lags=48)
plt.tight_layout()
plt.show()
```

[Out]

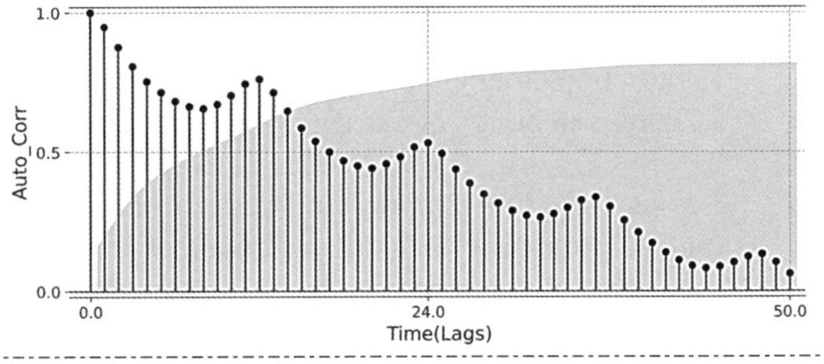

Figure S2.3.8. Output generated by actual data: Cyclical trend

We can see that there is a cyclical trend in the lags corresponding to the time index (Figure S2.3.8). Also, the plot shows that the correlation decreases as the lags increase. The shaded area denotes which lags are statistically significant.

We have practiced its application on an actual data set, namely the AirPassen-gers.csv data. By plotting the data (Figure S2.3.7), we were able to observe a clear trend and seasonality within the dataset. Following the preliminary visual analysis, we applied the autocorrelation function to our data. The resulting plot (Figure S2.3.8) shows the autocorrelation as a function of the time lags.

An autocorrelation plot, provides a visualization of the correlation between the observations at the current point in time and the observations at previous points in time. This can be useful in identifying seasonality in the data, as well as any cyclical trends that may be present. Observing Figure S2.3.8, a cyclical trend is clearly visible. The shaded area in the plot represents the confidence intervals, giving an indication of the statistical signify-cance of the correlations. If an autocorrelation value is beyond this area, it can be considered statistically significant. The autocorrelation is a valuable tool in time series analysis, providing insights into the temporal dependence structure of a dataset.

The use of this tool allows for the identification of trends, seasonality, and cyclical patterns, which can help guide the selection of appropriate models for forecasting.

Supplemental Information 2.4: Power Spectrum

A time series is a sequence of measurements of the same variable(s) made over time. Usually, the measurements are made at evenly spaced intervals (e.g., monthly or yearly). In this supplement, we discuss power in the frequency domain models using very simple examples in hopes of clarifying this explanation (Neumaier & Schneider, 2001).

Let us describe temporal process structures. In general, joint probability functions (and/or probability densities). Processes with continuous-state variables: two additional measures (autocorrelation, the power spectrum).

1. Definition

The power spectrum indicates the power of each frequency component of a source time domain. It shows at which frequencies variations are strong and at which frequencies variations are weak (Bingham et al., 1967).

[Table S2.4.1] Stochastic Iterative Maps of the Power Spectrum

	Discrete-state variable	Continuous-state variable
Discrete time		**Autocorrelation**
Continuous time		**Power Spectrum**

Table S2.4.1 provides a summary of the types of stochastic iterative maps that can be used to describe various types of processes in terms of their power spectrum, auto-correlation, and whether they are discrete or continuous in nature. The power spectrum is a very useful tool if you want to know frequencies and amplitudes of oscillatory signals in your time series data. For example, let us assume you are operating a factory with many machines and some of them have motors inside.

You detect unwanted vibrations from somewhere. You might be able to get a clue about which machines are the offending machines by looking at the power spectral density function, which would give you the frequencies of vibrations. The power spectral density is useful even if data do not contain any purely oscillatory signals. For example, if you have sales data from an ice cream parlor, you can get a rough estimate of the summer sales peak by looking at the power spectral density of your data.

$$P(f), \qquad f = Fourier\ analysis$$

We quite often compute and plot the power spectral density to get a feel for data in the early stage of time series analysis. Looking at the power spectral density is like looking at a simple time series plot except that we look at a time series as a function of frequency instead of a function of time. Frequency is a transformation of time and looking at variations in the frequency domain is just another way of looking at variations in time series data. The power spectral density tells us at which frequency ranges variations are strong, and this might be quite useful for further analysis.

+ **Stationarity**: The process or single trajectory $X(t)$ is a function defined on time t. The power spectrum P of $X(t)$ does not refer to probability density.

- Function defined on the frequency domain f or defined on angular frequencies $\omega = 2\pi f$
- Used to determine periodic subprocesses underlying $X(t)$
- Used to distinguish deterministic cycles from randomness

Decomposition (representation) of $X(t)$ as a superposition of oscillatory processes with various frequencies f($\rightarrow$ Fourier analysis). In power spectrum $P(f)$, the function $P(f)$ tells you which oscillation frequencies f are important and which are not.

Alternative name and symbol: Spectral density is $S(f)$ or $S(\omega)$.

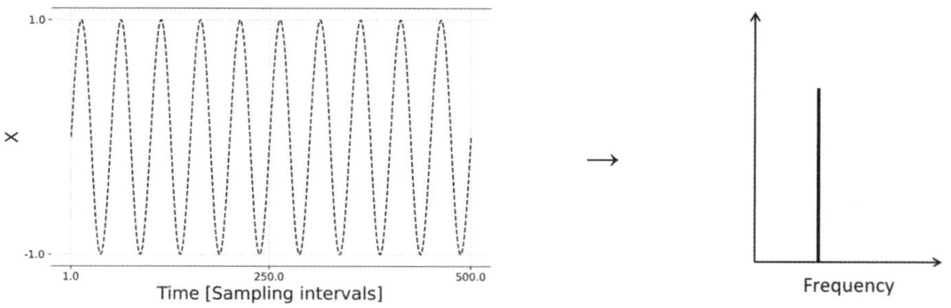

Figure S2.4.1. Representation of the harmonic oscillation: (left) a pure sine-wave process, (right) density of the power spectrum

A density function of the power spectrum (Figure S2.4.1) serves as a representation for the spectral density, often known as S(f). A pure sine-wave process is depicted in the figure on the left, and the matching density of the power spectrum is shown in the figure on the right.

+ **White Noise:** White noise is stationary, and the power spectrum is constant.

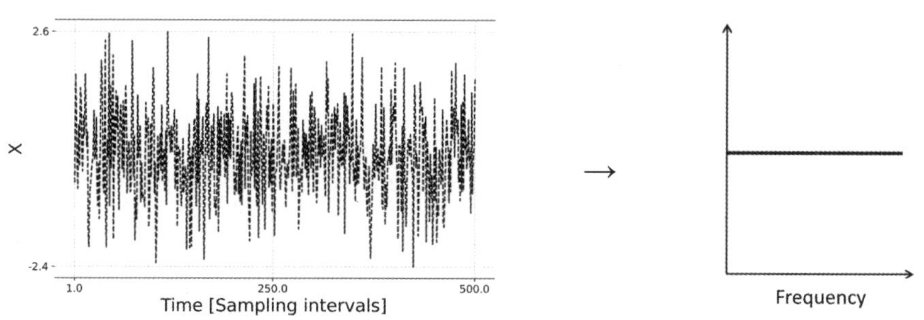

Figure S2.4.2. Representation of the white noise: (left) purely random white noise, (right) density of the power spectrum

The Figure S2.4.2 displays a pure white noise process together with the power spectrum density.

White noise, according to the assertion, is a stationary process with a consistent power spectrum. The harmonic oscillation with white noise is;

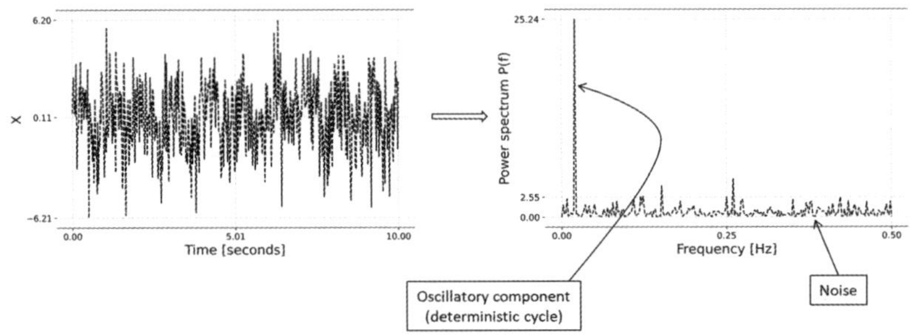

Figure S2.4.3. Harmonic oscillation with white noise: (left) purely random white noise with harmonic oscillation, (right) density of the power spectrum

The harmonic oscillation signal in this image (Figure S2.4.3) has been tainted by white noise. The signal's time series is displayed in the left panel, while the power spectrum's density is displayed in the right panel.

2. Process

The function in the stationary case and with temporal averaging are described by the supplied formulae. The average relationship between deviations from the mean at a given time lag is shown by $Corr(m)$, while $Cov(m)$is identical to Corr but not normalized with regard to variance.

$$Corr(m) = \frac{1}{\sigma^2}\frac{1}{(N-m)}\sum_{n=1}^{N-m}(X(t_n) - M)\cdot X(t_{n+m}) - M)$$

$$Cov(m) = \frac{1}{(N-m)}\sum_{n=1}^{N-m}(X(t_n) - M)\cdot X(t_{n+m}) - M)$$

The time lag is denoted by the symbol $\tau \rightarrow m$, and the expression: $\tau \rightarrow m\ (t_{n+m} = t_n + \tau)$

- $Corr$: describes how deviations from the mean at a fixed time lag τ are related to one another on average
- Autocovariance (Cov): same as Corr but not normalized with respect to variance

+ **The Power Spectrum and Autocorrelation Function:** According to the Wiener-Khintchine theorem, power spectrum $p(f)$ is a Fourier transformation of the autocovariance function $Cov(m)$. Normalized power spectrum $p \times (f) = p(f)/\sigma^2$ is also a Fourier transformation of the autocorrelation function $Corr(m)$ (Lampard, 1954).

$$p(f) = Cov(0) + 2\sum_{m\geq 1} Cov(m)\cos(2\pi\, f\, m)$$

The equation defines a data as a function of frequency f which can be used to identify periodic patterns in the data and to estimate the amount of noise present. It is the sum of the autocovariance at different lags, each weighted by a cosine function.

The autocovariances are denoted as $Cov(m)$ where m is the lag. The term $Cov(0)$ is the autocovariance at zero lag, and represents the variance of the data. The other terms in the sum represent the contributions to the variance at different frequencies. The cosine function weights the autocovariances by the frequency, with higher frequencies receiving higher weights.

$$p* = \frac{p(f)}{\sigma^2} = 1 + 2\sum_{m\geq 1} Corr(m)\cos(2\pi\, f\, m)$$

This equation is similar to the previous one, but is normalized by the variance σ^2 of the data which is used to compare the power of the data at different frequencies, relative to the total variance of the data. The term is the sum '1' is the autocorrelation at zero lag, and represents the variance of the data.

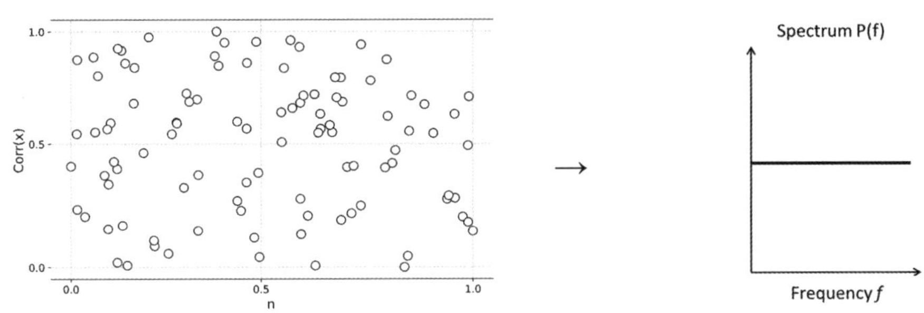

Figure S2.4.4. Output of white noise and the power spectrum

White noise, which is a random signal with equal strength across all frequencies, is depicted in Figure S2.4.4 as its output. The picture also exhibits the white noise's power spectrum, which is a graph that illustrates the strength of the signal at various frequencies.

$$p* = \frac{p(f)}{\sigma^2} = 1 + 2\sum_{m\geq 1} Corr(m)\cos(2\pi\, f\, m) = 1$$

The equation connects the white noise's power spectrum to its autocorrelation function. The power spectrum and autocorrelation function of white noise are shown in Table S2.4.2. It makes a distinction between stationary and nonstationary processes and offers instructions for figuring out the autocorrelation and power spectrum for each kind of process.

[Table S2.4.2] Power Spectrum and Autocorrelation Function

	Nonstationary	Stationary
Autocorrelation	Ensemble averaging	Ensemble averaging or time averaging
Power spectrum	Tricky and difficult	Fourier spectrum Via *Cov* or *Corr*

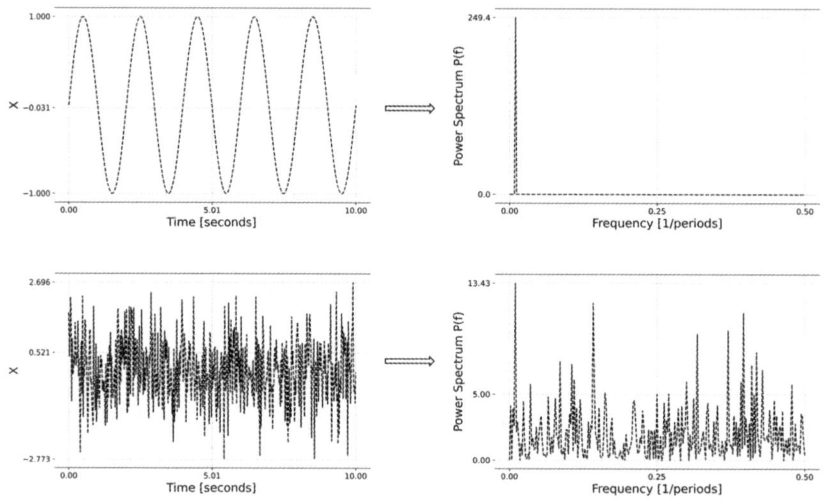

Figure S2.4.5. Schematic representation of oscillation with a period of 2 (upper set). On the bottom is MA(0) with white noise.

Two 2-period oscillations, one on the left and an MA(0) process with white noise, the other, are schematically depicted in Figure S2.4.5. White noise may be modeled using a moving average process called MA(0), which has a mean of zero.

3. Simulation

The function in programming will be used to model the power spectrum for time-discrete systems. To achieve this, the data set from the Supplementary materials (use the Supplement data set) should be imported first in order to create a simulation.

```
[In]
import pandas as pd  # import the necessary library
data = pd.read_csv('timeseries_periodtwo.txt')
```

For the power spectrum P(f) via periodogram

```
[In]
import scipy.signal as signal
fs = 1  # Set the sampling frequency
freqs, pxx = signal.periodogram(data, fs)  # Use the periodogram function
pxxn = pxx * fs / (4*np.pi)  # Normalize the power spectral density
```

We calculate the power spectral density of a time series data set 'data' using the periodogram method which takes the arguments above. The 'periodogram' function is from the 'signal' module which is used to calculate the density. The '[]' is an optional window to apply to the data.

The 'fs' is the sampling rate of the data in Hz. The function returns the density 'pxx' as a function of frequency 'freqs' which is normalized by dividing by the length of the data set. The code then scales the density by the sampling rate 'fs' ad the factor '4∗pi', and stores the result in 'pxxn'.

Create its states according to the time steps

[In]

```
plt.figure()  # Create a figure
time_steps = range(1, len(data) + 1)

# Plot data as points and lines
plt.plot(time_steps, data, 'o')
plt.plot(time_steps, data)
```

→ [Out]

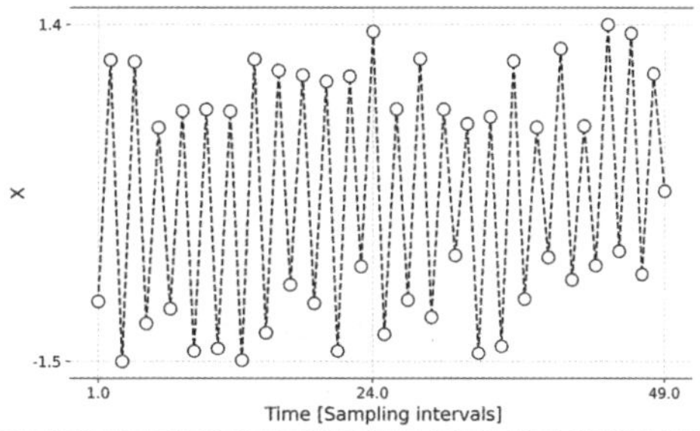

Figure S2.4.6. Visualizing the $P(f)$ via periodogram

In Figure S2.4.6, the x-axis of the plot represents the time steps of the data, and the *y*-axis represents the values of the data at those time steps. It plots the data twice which are used to allow multiple plots to be overlaid as circles connected with a line.

Then, we can create the Power spectrum

[In]

```
plt.figure() # Create a figure

# Plot data
plt.plot(freqs, pxxn)
plt.show()
```

→ [Out]

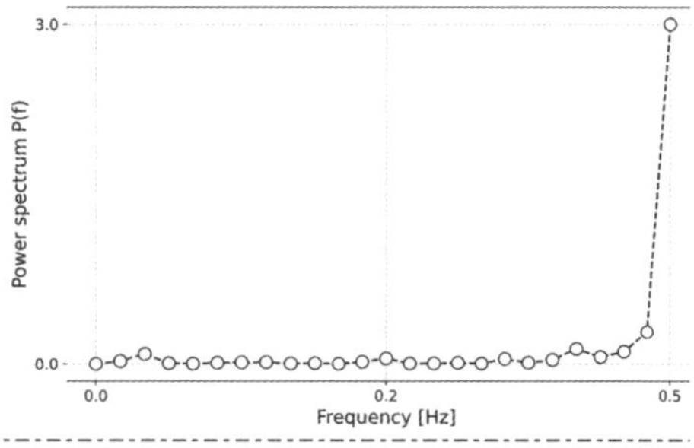

Figure S2.4.7. Visualizing the Power spectrum

Finally, we generate a plot of the power spectral density (Figure S2.4.7). The x-axis of the plot represents the frequencies of the data, and the y-axis represents the normalized power at those frequencies. The x-axis data is given by the array 'freqs' and the y-axis data is given by the array 'pxxn'. It reflects the continuous state, discrete time: discrete-time stochastic processes with continuous, numerical state variables. Processes described by (nonlinear) stochastic iterative maps.

The Fourier Transforms

The Fourier transform is a way of splitting something up into a bunch of sine waves (Smith, 2011). As usual, the name comes from a person who lived a long time ago called Fourier. In mathematical terms, the Fourier transform is a technique that transforms a signal into its constituent components and frequencies. The simplest possible implementation of the fast Fourier transform can be done using programming.

First step: Create a time vector as follows:

```
[In]
import numpy as np
t = np.linspace(1, 1000, 1000)[:, np.newaxis] / 100
```

This creates a column vector of 1000 rows, starting from .01 and going up to 10, in steps of .01 (as determined by the denominator, 100). We will interpret t's units as seconds in the following. Second step:

Create a bunch of time series as follows:

```
[In]
from scipy.signal import square
s1 = np.sin(2 * np.pi * 3 * t)
s2 = s1 + np.sin(2 * np.pi * 6 * t)
s3 = 10 * (np.random.rand(s1.size, 1) - 0.5)
s4 = s1 + s3
s5 = np.sin(2 * np.pi * 3.05 * t)
s6 = np.sin(2 * np.pi * 95 * t)
s7 = square(2 * np.pi * 3 * t)
s8 = np.sin(2 * np.pi * (3 + 0.2 * t) * t)
```

Plot each one for your own edification (Figure S2.4.8):

```
[In]
import matplotlib.pyplot as plt
plt.plot(t, s1)
plt.show()
```

→

[Out]

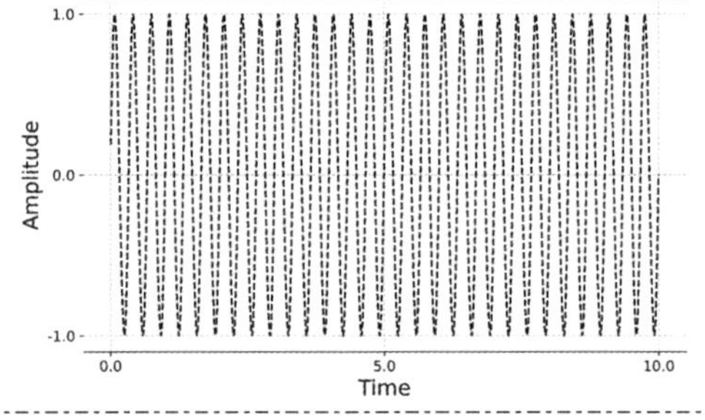

Figure S2.4.8. Visualizing the time series

Third step: Now do the Fourier transform, computing the amplitude and phase spectrum for each of the above time series. Our little routine should be in your computer path or the current directory. The variable f is frequency in Hz and is the same for all of the Fourier transforms.

Here .01 is the sampling interval, also constant:

```
[In]
from scipy.fft import fft
def magf(signal, dt):
    N = len(signal)
    fft_signal = fft(signal)
    freq = np.fft.fftfreq(N, dt)
    amplitude = np.abs(fft_signal) / N
    phase = np.angle(fft_signal)
    return freq, amplitude, phase
dt = 0.01
f, a1, p1 = magf(s1, dt)
f, a2, p2 = magf(s2, dt)
```

Take a look at the amplitude (left side) and phase (right side) plots as follows (Figure 2.4.9):

```
[In]
plt.plot(f, a1)
plt.show()
plt.plot(f, p1)
plt.show()
```

→ [Out]

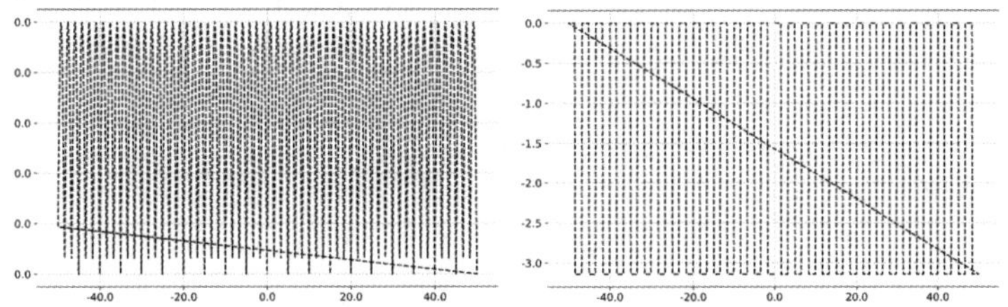

Figure S2.4.9. Visualizing the amplitude and the phase

Fourth step: Now use, and repeat steps 2 and 3 (the same time step, but with only 100 rows).

```
[In]
t = np.linspace(1, 100, 100)[:, np.newaxis] / 100
t = np.linspace(1, 100, 100)[:, np.newaxis] / 10
```

and repeat steps 2 and 3. Except in step 3, generate a plot that shows the amplitude spectrum.

```
[In]
dt_new = 0.1
f, as1, ps1 = magf(s1, dt_new)
plt.plot(f, as1)
plt.show()
```

→ [Out]

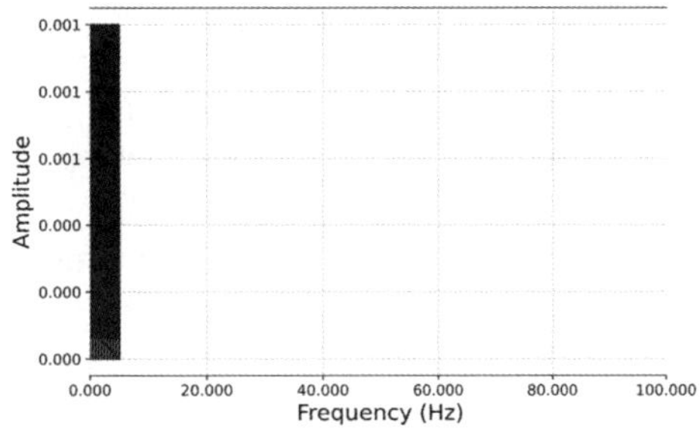

Figure S2.4.10. Visualizing the amplitude spectrum

The example shows how to produce time series data with different features (plot the data; Figure 2.4.10), do a Fourier transform on the data to extract its amplitude and phase spectrum, and plot those spectra. The illustration also demonstrates how altering the time step alters the spectrum.

[2.12]

Master Equations

Identification, Numerical Simulation for Stochastic Systems, Practical Practice

Building upon the foundational concepts of autoregressive (AR), moving average (MA), and autoregressive moving average (ARMA) models in time series analysis, master equations emerge as differential equations that depict the evolution of probabilities for continuous-time Markov processes in systems transitioning between discrete states.

These master equations serve as the continuous-time counterparts of recurrence relations for Markov chains (Norris & Norris, 1998) and find applications in various systems analysis, such as chemical reactions, radioactive decay, epidemic spread, and numerous other areas of interest.

[Table 2.12.1] Stochastic Processes for Discrete State Spaces in Continuous Time

	Discrete states	Continuous-state variable
Discrete time	Monte Carlo simulations Markov chains	Stochastic iterative maps Special case of the Chapman-Kolmogorov equation (integral evolution equation for probability densities)
Continuous time	**Master equations**	

Table 2.12.1 gives an overview of several stochastic modeling techniques for continuous time discrete state space systems. The methods given include stochastic iterative mapping for Markov chains, Monte Carlo simulations for systems with discrete states and discrete time, and the particular case of the Chapman-Kolmogorov equation for integral evolution equations with probability densities. The master equation is a suitable modeling technique for continuous time systems with discrete states.

1. Identification

Here we consider the generating function method of solving master equations and some approximate solution methods. We consider all of the possible ways particles can move in a system at very small-time steps and keeping track of either the number of particles around or the probability of being at that position at a specific time.

+ **Discrete State Spaces:** Let us be more explicit and draw some images. We assume that we live in a one-dimensional world. Along this dimension, we have a string of spaces (or position) of a width (constant = 1) which can either be empty or contain a particle. At any point in time, a given particle can move to either the left or right at some rate k, which we will call the jump frequency, or it can stay put.

Image of these trajectories and their weights are shown below.

Random jumping on a grid: Calculating the odds of going from one point to another using random hopping on a grid is possible.

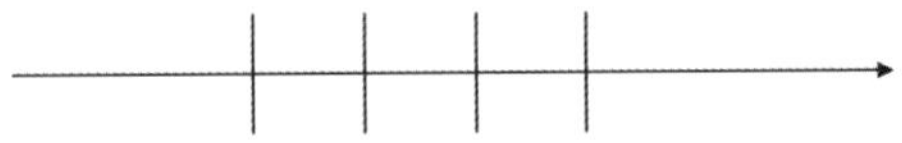

Quantum systems with energy levels E_0, E_1, E_2, ...: Systems of quantum particles with energies E0, E1, E2,... explains how the wave function has changed throughout time.

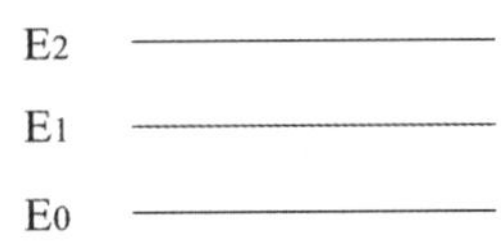

The number of molecules produced in a biochemical reaction: The chemical can be used to model how many molecules are created during a biological process.

Categorical states: Depending on the unique features of the system, many modeling strategies may be used to represent categorical states.

Healthy	Pre-clinical	Clinical
k-1	k-2	k-3

These weights can be used to formulate our master equation. We simply have to ask, given the probability of a particle being at position x at time t, what is the probability at time $t + \Delta t$ knowing the jump frequency k? We have to keep track of the movement of particles from our space of interest as well as the movement of particles from the spaces on the left and right.

+ **Transition Rates of Continuous-Time Processes:** The transition rate is the rate of transition from a state k to a state j (e.g., the number of transitions from k to j per second).

Notation: $w(j \leftarrow k)$, $w(j,k)$, w_{jk}, semi-positive: $w \geq 0$. This is similar to but different from transition probability.

- Time-homogeneous processes: Rates $w()$ independent of time
- Non-homogeneous processes: Rates depend on time: $w(j,k;k,t')$ with $t > t'$

Master equations are used in many applications. In physics: Pauli master equation → quantum mechanical systems in contact with their environments (damping, pumping, and irreversibility), quantum optics (photon fluctuations). In biochemistry: chemical master equation → stochastic reaction kinetics. And in life sciences: birth and death processes → stochastic population dynamics, stochastic disease state models.

+ **Define the Processes:** A master equation is a phenomenological set of first-order differential equations describing the time evolution of (usually) the probability of a system occupying each one of a discrete set of states with regard to a continuous-time variable *t*.

Let us consider a situation in which something can switch between two states that we call 1 and 2. There are many examples of this situation:

$M = 2$ states

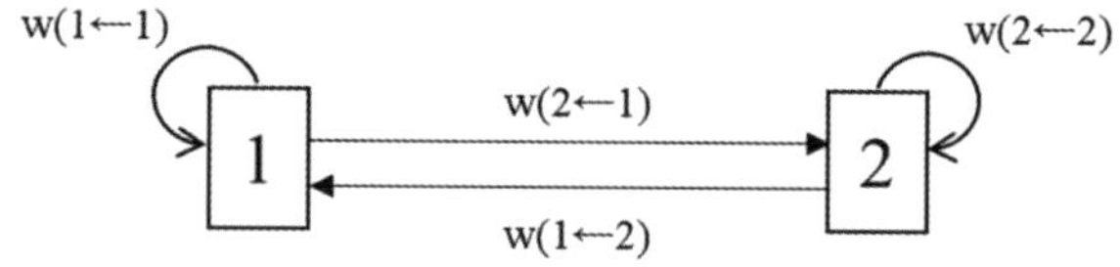

Figure 2.12.1. Representation of 2 states. p(1, t) and p(2, t) are the probabilities that the process is in state (w) 1 or 2 at time t.

The scenario is a two-state Markov chain representation of a system having two potential states. The probabilities of being in each state at a certain time t are represented by the symbols $p(1,t)$ and $p(2,t)$, where $p(1,t)$ denotes the likelihood of being in state 1 and $p(2,t)$, the likelihood of being in state 2. Figure 2.12.1 illustrates this. The goal of the master equation is to determine the change in probability $p(1,t)$. Transitions 1 → 1 do not result in a change.

- $p(1,t)$ increases with 2 → 1 jumps. Rate in: $w(1 \leftarrow 2) \times p(2,t)$.
- $p(1,t)$ decreases with 1 → 2 jumps. Rate out: $w(2 \leftarrow 1) \times p(1,t)$.

$$\frac{d}{dt}p(1,t) =' rate\ in' - 'rate\ out' = w(1 \leftarrow 2)p(2,t) - w(2 \leftarrow 1)p(1,t)$$

The left side of the equation represents the rate of change of $p(1,t)$ with respect to time t. The right side of the equation represents the net rate of change of $p(1,t)$ due to transitions between the two states. The term '$rate\ in$' refers to the rate at which the system transitions into stae 1 from state 2, and the term '$rate\ out$' refers to the rate at which the system transitions out of state 1 and into state 2. The term $w(1 \leftarrow 2)$ reperesents the trnasiton rate from state 2 to state 1, and the term $w(2 \leftarrow 1)$ represents the transition rate from 1 to state 2. Recall the derivation of Markov chains → It determines what the probabilities $p(k,n)$ are rather than how they change. By analogy:

$$\frac{d}{dt}p(2,t) =' rate\ in' - 'rate\ out' = w(2 \leftarrow 1)p(1,t) - w(1 \leftarrow 2)p(2,t)$$

$$\frac{d}{dt}p(1,t) =' rate\ in' -' rate\ out'$$

$$'rate\ in' = w(1 \leftarrow 2)p(2,t) + w(1 \leftarrow 3)p(3,t) = \sum_{k \neq 1} w(1 \leftarrow k)p(k,t)$$

$$'rate\ out' = w(2 \leftarrow 1)p(1,t) + w(3 \leftarrow 1)p(1,t) = p(1,t)\sum_{k \neq 1} w(k \leftarrow 1)$$

Based on the rates of transition between various states, the equations provided illustrate how the odds of being in two distinct states change with time. The likelihood of being in state 2 is shown by the first equation, while the chance of being in state 1 is shown by the second equation. The names '$rate\ in$' and '$rate\ out$' relate to the probability of transitioning from one state to another, and the rates are determined by those probabilities.

- M-state master equation

$$\frac{d}{dt}p(j,t) =' rate\ in' -' rate\ out' = \sum_{k \neq 1} w(1 \leftarrow k)p(k,t) - p(1,t)\sum_{k \neq 1} w(k \leftarrow 1)$$

According to the equation, the rate of flow into a state (from other states) and the rate of flow out of the state are taken into account when calculating the rate of change in probability for a certain state (to other states).

- Alternative notation

$$\frac{d}{dt}p(j,t) =' rate\ in' -' rate\ out' = \sum_{k \neq 1} w(j \leftarrow k)p(k,t) - p(j,t)\sum_{k \neq 1} w(k \leftarrow j)$$

Here, the term $\sum_{k \neq 1} w(j \leftarrow k)p(k,t)$ represents the sum of the transition rates from all states k other than state j to state j, multiplied by the probability of being in state k.

The term $-p(j,t)\sum_{k \neq 1} w(k \leftarrow j)$ represtns the probability of being in state j multiplied by the sum of the transition rates from state j to all states k other than state k.

We may add and subtract the term $w(j \leftarrow j)p(j,t)$ on the right side.

$$\frac{d}{dt}p(j,t) = \sum_{k \neq 1} w(j \leftarrow k)p(k,t) + w(j \leftarrow j)p(j,t) - w(j \leftarrow j) - p(j,t)\sum_{k \neq 1} w(k \leftarrow j)$$

This yields the following:

$$\frac{d}{dt}p(j,t) = \sum_{k \in \Omega} w(j \leftarrow k)p(k,t) - p(j,t)\sum_{k \in \Omega} w(k \leftarrow j), \qquad \Omega = \{1, \dots, M\}$$

The equation shows the time evolution of a probability distribution $p(j,t)$ in a system with M states. The left side of the equation represents the rate of change of $p(j,t)$ with respect to time t. The right side of the equation represents the net rate of change of $p(j,t)$ due to transitions between the states. The term $\sum_{k\in\Omega} w(j \leftarrow k)p(k,t)$ represents the sum of the transion rates from all states k in the set Ω to stae j, multiplied by the probability of being in state k. The term $-p(j,t)\sum_{k\in\Omega} w(k \leftarrow j)$ represtns the probability of being in states j multiplied by the sum of the transition rates from state j to all states k in the set Ω.

+ **Stationary Case:** The process is time-homogeneous and transition rates are positive in the stationary situation.

- Time homogeneous and $w < 0$ for all (j,k). A process becomes stationary over time (see, e.g., van Kampen, 1981)

The chance of existing in a particular condition approaches a constant value that is independent of time as time increases toward infinity. The stationary probability distribution refers to this.

- $t \to \infty$: $p(k,t) \to p(k)$ independent of time

Stationary distribution: Rate in = rate out for all states

$$\frac{d}{dt}p(j,t) = '\,rate\ in' - 'rate\ out' = \sum_{k\neq j} w(j \leftarrow k)p(k,t) - p(j,t)\sum_{k\neq j} w(k \leftarrow j)$$

$$\frac{d}{dt}p(j,t) = 0 \Rightarrow '\,ratein' = '\,rate\ out'$$

$$\forall j\text{:} \sum_{k\neq j} w(j \leftarrow k)p(k) = p(j)\sum_{k\neq j} w(k \leftarrow j)$$

Here, the symbol $\forall j$ indicates that the equation holds for all states j. The left side of the equation represents the sum of the transition rates from all states k other than state j to state j, multiplied by the probability of being in state k. The right side of the equation represents the probability of being in state j multiplied by the sum of the transition rates from the state j to all states k other than state j.

It denotes that the sum of the transition rates into state j is equal to the probability of being in state j multiplied by the sum of the transition rates out of state j.

- Numerical solution methods

$$\frac{d}{dt}p(j,t) = \sum_{k\neq j} w(j \leftarrow k)p(k,t) - p(j,t)\sum_{k\neq j} w(k \leftarrow j), \qquad \text{Initial probabilities } \mathrm{p(k,0)}$$

The chance of being in state j over time is represented by an equation, which depends on the probabilities of being in other states and the speeds at which those states transition between one another. The equation's solution for a specific set of beginning circumstances can be discovered using numerical solution methods.

M-state model = M-coupled first-order differential equations → use any numerical algorithm for solving ordinary differential equations

- Euler forward method $\frac{d}{dt}p(j,t) = \frac{p(j,t+\tau)-p(j,t)}{\tau}$, τ small

$$p(j,t+\tau) = p(j,t) + \tau\left[\sum_{k\neq j} w(j \leftarrow k)p(k,t) = p(j,t)\sum_{k\neq j} w(k \leftarrow j)\right]$$

It shows the evolution of a probability $p(j,t)$ in a system with a time constant τ. The left side of the equation represents the probability of being in state j at time $t+\tau$. The right side of the equation represents the probability of being in state j at time t plus the change in the probability of being in state j over a small time interval τ. The term in squre brackets '[...]' represents the net rate of change of the probability of being in state '' due to transitions between the states.

Normalization → sum of all $p(k,t)$ equals 1 → solve only M – 1 independent equations Euler forward scheme of master equations and Markov chains

$$p(j,t+\tau) = p(j,t) + \tau\left[\sum_{k\neq j} w(j \leftarrow k)p(k,t) = p(j,t)\sum_{k\neq j} w(k \leftarrow j)\right]$$

This is equivalent to solving the Markov chain model.

$$p_j(t_{n+1}) = p(j \leftarrow j)p_j(t_n) + \sum_{k\neq j} \underbrace{\tau\, w(j \leftarrow k)}_{p(j\leftarrow k)} p_k(t_n)$$

$$p(j \leftarrow j) = 1 - \sum_{k\neq j} \tau\, w(k \leftarrow j)$$

Here, the term $\sum_{k\neq j} \tau\, w(k \leftarrow j)$ represents the sum of the transition rates from the state j to all states k, other than state j multiplied by the time constant τ. The probability of not transitioning out of state j duet to transitions to other states is equal to '1' minus this sum.

Now that you have been exposed to master equations, we can begin to think about various other processes through this deterministic lens. In the following part, we take this approach to two-state transcription and its distribution.

2. Numerical Simulation for Stochastic Systems

We can be healthy (state 1) or sick (state 2). Also, we can speak one language (state 1) or another (state 2). There are scenarios under which something or someone can switch between two states that we name 1 and 2.

$$\frac{d}{dt}p(1,t) = w(1 \leftarrow 2)p(2,t) + w(2 \leftarrow 1)p(1,t)$$
$$p(2,t) = 1 - p(1,t)$$
$$\frac{d}{dt}p(1,t) = w(1 \leftarrow 2)(1 - p(1,t)) + w(2 \leftarrow 1)p(1,t)$$

Here, the term $w(1 \leftarrow 2)(1 - p(1,t))$ represents the transition rate from state 2 to state 1 multiplied by the probability of not being in state 1. The terms $w(2 \leftarrow 1)p(1,t)$ represents the transition rate from state 1 to state 2 multiplied by the probability of being in state 1. This is the time evolution of a probability $p(1,t)$ in at system with two states, labeled 1 and 2.

- Numerical representation

$$\begin{aligned} p_1(t_{n+1}) &= p_1(t_n) + \tau\big(w(1 \leftarrow 2)(1 - p_1(t_n)) + w(2 \leftarrow 1)p_1(t_n)\big) \\ &= \tau\, w(1 \leftarrow 2) + (1 + \tau[w(2 \leftarrow 1) - w(1 \leftarrow 2)])p_1(t_n) \end{aligned}$$

The chance that a process will be in state 1 at time t, represented numerically by p_1, may be found in this equation. The probability at the following time step, t_{n+1}, is computed using the probability at the present time step, t_n, and the transition rates $w(1 \leftarrow 2)$ and $w(2 \leftarrow 1)$ between states 1 and 2. A time step and a linear combination of the probabilities at time t_n and the transition rates are used in the numerical representation, with the time step and the transition rates affecting the coefficients of the linear combination.

- Jumping on a grid

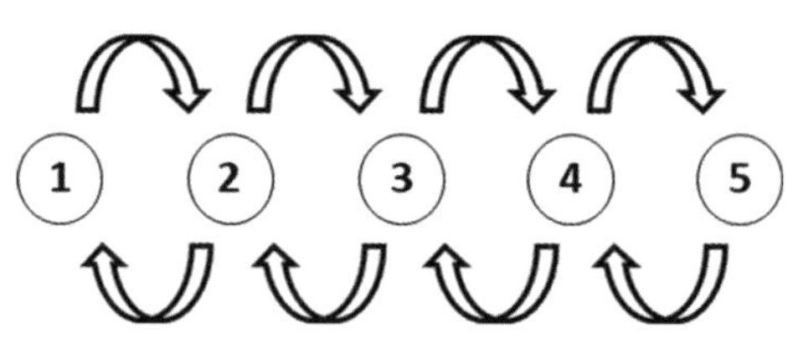

Figure 2.12.2. Schematic representation of jumping on a grid. Arrows denote transitions, and numbers refer to each state k.

A system that jumps on a grid can be in any of the discrete states that are present on the grid, but it can only go to one of its nearby states (Figure 2.12.2). The odds of being in each state at a given moment may be characterized using Markov chains, and the transition rates between states rely solely on the current state of the system and the nearby states.

- Gains and losses

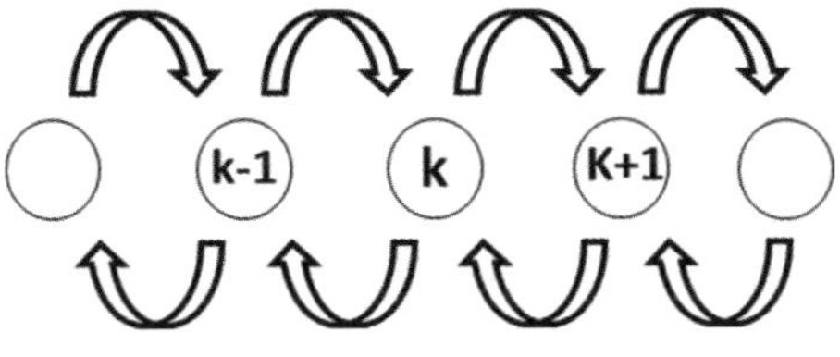

Figure 2.12.3. Schematic representation of gains and losses. Rate in and rate out

A stochastic process with gains and losses is depicted in Figure 2.12.3, in which an entity may experience either a gain or a loss of some kind. Rate in and Rate out, respectively, stand for the rate of gain and loss. A master equation that provides the odds of being in various states over time while accounting for gain and loss rates may be used to explain the process.

+ **Discrete State Transitions (Drift and No-Drift Scenarios):** only two transition rates $[w(left), w(right)]$. Gain increases the occupation of state k [Gains = $w(right)p(k-1) + w(left)p(k+1)$] and Loss decreases the occupation of state k [Losses = $w(right) + w(left) \times p(k)$].

$$\frac{d}{dt}p(k,t) = gains - losses$$

$$\frac{d}{dt}p(k,t) = w(right)p(k-1,t) + w(left)p(k+1,t) - p(k,t)\{w(right) + w(left)\}$$

Here, the equation shows the probability $p(k,t)$ in a system with multiple states that can be labeled as k, $k-1$, and $k+1$. The equation describes how the probability of the system being in state k at time t changes over time due to transitions between the states.

The term $w(right)p(k-1,t)$ represents the transition rate from the state $k-1$ to state k mulitiplied by the probability of being in state $k-1$. The term $w(left)p(k+1,t)$ represents the transition rate from state $k+1$ to state k multiplied by the probability of being in state $k+1$.

The term $p(k,t)\{w(right) + w(left)$ represents the probability of being in state k multiplied by the sum of the transition rates out ot state k.

Boundary states

$$\frac{d}{dt}p(1,t) = w(left)p(2,t) - w(right)p(1,t)$$

$$\frac{d}{dt}p(M,t) = w(right)p(M-1,t) - w(left)p(M,t)$$

In this equation, a state has special properties because it is connected to a smaller number of states. Thus, the state M is a boundary state because it is only connected to state $M-1$ and there is no state labeled $M+1$.

Jumping without drift: $w(left) = w(right)$

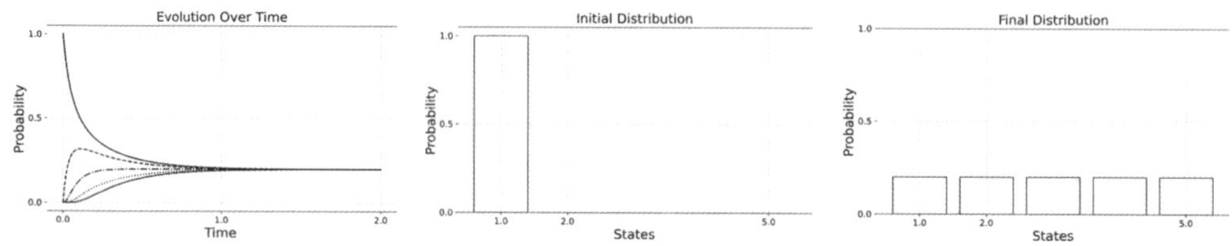

Figure 2.12.4. Simulation of jumping without drift: $M = 5, p_1(0) = 1, w(left) = w(right) = 10$/s. The middle panel shows the initial distribution, and the right panel shows the final distribution.

Jumping with drift: $w(left) \neq w(right)$

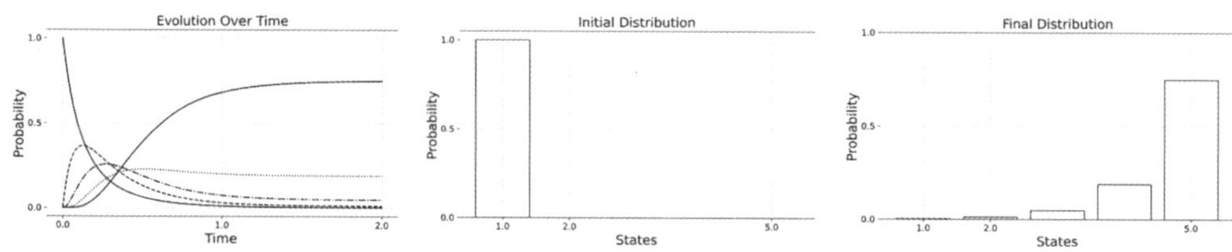

Figure 2.12.5. Simulation of jumping with drift: $M = 5, p_1(0) = 1, w(left) = 2$/s, $w(right) = 8$/s. The middle panel shows the initial distribution, and the right panel shows the final distribution.

The images show simulations of two distinct instances of particles hopping between discrete states. Particles are not drifting in the first scenario (Figure 2.12.4), hence the rate of left and rightward jumps is equal. The pace of leaping to the left and right is not identical in the second scenario (Figue 2.12.5) because particles are drifting as they jump. The figures in both instances display the starting and final probabilities over the states.

+ **Simulation for Jumping on a Grid Models:** We may use programming tools to numerically solve the differential equation in order to mimic the master equation technique. The initial distribution of probabilities across the stages and the transition rates between nearby states, which define how probable it is for a leap to occur in each direction, are among the parameters for the simulation. The behavior of the system may be examined by modifying these parameters, such as the average amount of time spent in each state or the overall diffusion of the particles on the grid. Let us parameterize for the simulation

```
[In]
import numpy as np
M = 5
tau = 0.01
T = 200
p = np.zeros((M, T))
```

The M is an integer representing the number of grid points, the 'tau' is a float representing the single time step, the T is an integer representing the number of time steps and the P is a matrix of size M rows and T columns, initialized to all zeros. For transition coefficients

```
[In]
w_left = 2
w_right = 8
```

For the localized initial condition for three states with their loop functions

```
[In]
p[0,0] = 0.4
p[1,0] = 0.45
p[2,0] = 0.15

for n in range(T-1):
    for k in range(1, M-1):
        gains = w_left*p[k+1,n] + w_right*p[k-1,n]
        losses = (w_left + w_right)*p[k,n]
        p[k, n+1] = p[k, n] + tau*(gains - losses)
```

The code initializes the first column of the p matrix with three values: 0.4, 0.45, and 0.15. Then it enters a loop that will iterate through each time step, n, from 1 to $T-1$. Fro each time step, it enters a second loop that will iterate through each grid point k from 2 to M-1. Inside of these loops, it calculates the gains and losses at each grid point for the current time step using the variables 'w_left', 'w_right', and the value of p at the previous time step. Then, it updates the value of p at the current grid point and time step by adding the product of 'tau' and the difference between gains and losses. To define its first position

```
[In]
    gains = w_left * p[1, n]
    losses = w_right * p[0, n]
    p[0, n+1] = p[0, n] + tau * (gains - losses)
```

The code calculates the gains and losses at the first grid point for a given time step n using the variable 'w_left' and 'w_right' as well as the values of p at the first and second grid points at the previous time step. To define its last position

```
[In]
    gains = w_right * p[M-2, n]
    losses = w_left * p[M-1, n]
    p[M-1, n+1] = p[M-1, n] + tau * (gains - losses)
```

The code, calculates the gains and losses at the last gird point for a given time step n using the variables 'w_left' and w_right' as well as the values of p at the last and second-to-last grid points at the previous time step. It then updates the value of p at the last grid point and the current time step by adding the product of 'tau' and the difference between the gains and losses. Create the probability plot

[In]
```
tvec = np.arange(1, T+1) * tau


import matplotlib.pyplot as plt
plt.figure(figsize=(10, 6))
plt.rcParams.update({'font.size':
20})
for k in range(M):
    plt.plot(tvec, p[k, :])
plt.show()
```

[Out]

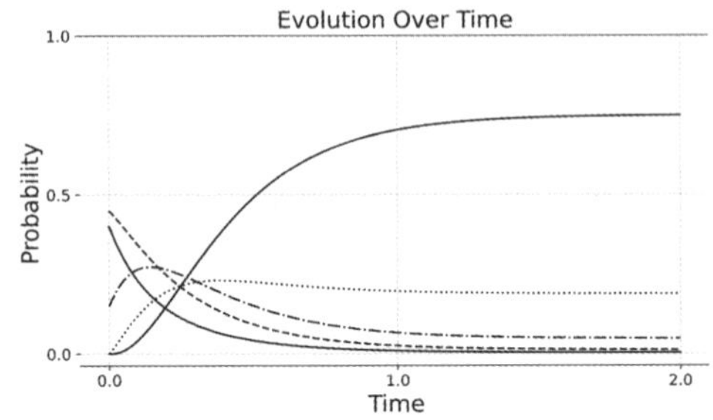

Figure 2.12.6. Simulation of the probabilities

In Figure 2.12.6, we create a 'tvec' as a sequence of integers from 1 to T multiplied by 'tau'. This is represented by the time value used on the x-axis of the plot. Then the code enters a loop that will iterate through each grid point k from 1 to M. For each grid point, it plots the values of p at that grid point (plotted as separate lines on the same figure) over time, represented by the 'tvec' with the lines. Left us create the initial state plot

[In]
```
plt.figure(figsize=(10, 6))
plt.rcParams.update({'font.size':
20})


pinitial = p[:, 0]
plt.bar(np.arange(1, M+1), pinitial)
plt.title('Initial Distribution')
plt.axis([0.5, M+0.5, 0, 1.05])
plt.show()
```

[Out]

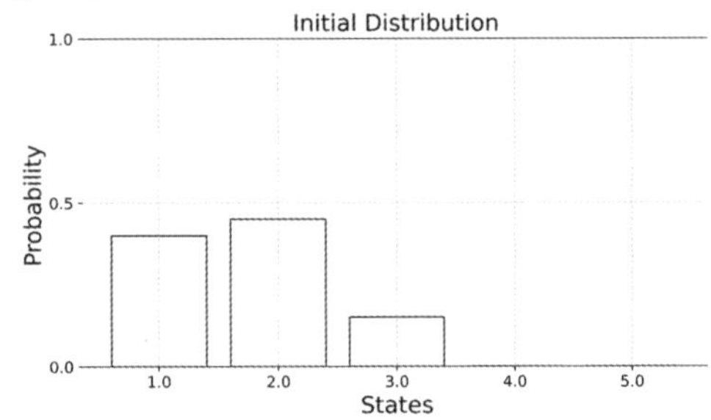

Figure 2.12.7. Visualization of the initial states

We generate a bar plot (Figure 2.12.7) of the initial values of p at each grid point. And adds a plot with the final distribution as below (Figure 2.12.8). Create the final state plot

[In]
```
plt.figure(figsize=(10, 6))
plt.rcParams.update({'font.size':
20})
pfinal = p[:, T-1]
plt.bar(np.arange(1, M+1), pfinal)
plt.title('Final Distribution')
plt.axis([0.5, M+0.5, 0, 1.05])
plt.show()
```

[Out]

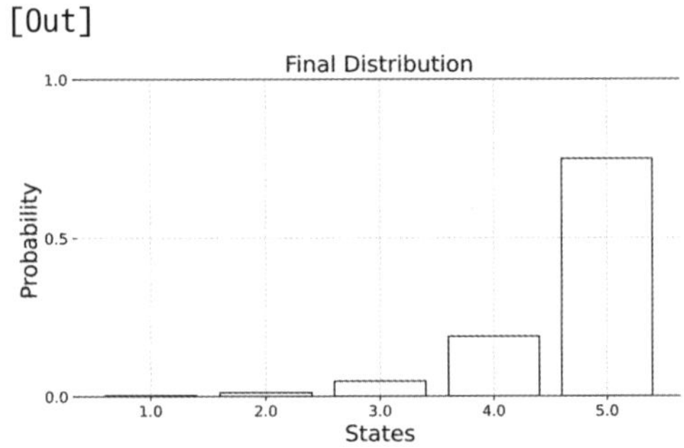

Figure 2.12.8. Visualization of the final states

A summary for Jumping on a grid: When the transition rates to the left and right are equal, there is no drift, and the distribution is uniform.

$w(right) = w(left)$: No drift, Uniform stationary distribution

$w(left) \neq w(right)$:

- Drift to left or right = tendency = deterministic component
- Stationary distribution = skewed distribution
- Preferred states are on the side with the larger transition rate
- The distribution is skewed to the left (left-hand tail) if $w(right) > w(left)$

When the transition rates to the left and right are unequal, there is a drift to one side, resulting in a skewed distribution. The side with the larger transition rate has preferred states, and the distribution is skewed towards that side. This summary suggests that the simulation involving the evolution of values on a grid over time. The values on the grid are evolving according to some rules involving the transition rates 'w_left' and 'w_right'.

3. Practice with Master Equations

In this exercise, we simulate master equations to investigate the behavior of Markov processes. We may learn how different parameters impact the dynamics of complex systems and see phenomena like drift, skewness, and favored states.

Problem 1: Master equation for modeling the progression of nicotine addiction. We consider a system with a discrete state space Ω. Let j denote an integer that is used to label the states of the system. In this case, the master equation that describes the evolution of the probability $p(j,t)$, where t is time, is defined by

$$\frac{d}{dt}p(j,t) = \sum_{k\neq 1} w(j \leftarrow k)p(k,t) - p(j,t)\sum_{k\neq 1} w(k \leftarrow j)$$

Here $w(j \leftarrow k) \geq 0$ denotes the transition rate from k to j. Let us study the progression of nicotine dependency (smoking behavior) as an application. We define four states: nonsmoker, experimental smoker, dependent smoker, and ex-smoker. We code these states j = 1, 2, 3, 4 and have as below (Table 2.12.2);

[Table 2.12.2] States Representations $\Omega = \{1, 2, 3, 4\}$

State	Nonsmoker to experimental	Experimental to nonsmoker	Experimental to dependent	Dependent to ex-smoker
Code	1	2	3	4

The master equation model defined by the equation above comes with 12 different transition rates $w(1 \leftarrow 2)$, $w(1 \leftarrow 3)$, $w(1 \leftarrow 4)$, $w(2 \leftarrow 1)$, …, $w(4 \leftarrow 3)$.

Problem 1a: We consider a model that exhibits only four types of transition. These transitions are formulated in words in the table below, and the corresponding transition rates w are given as well (Table 2.12.3).

[Table 2.12.3] Four Types of Transition

Transition	Transition rate in units of 1,000 people/day
Nonsmoker to experimental	3
Experimental to nonsmoker	9
Experimental to dependent	10
Dependent to ex-smoker	2

Assign these four transitions to the corresponding four transition rates $w(j \leftarrow k)$. That is, state—by writing down the indices j and k—which four rates $w(j \leftarrow k)$ out of all 12 rates are non-zero and state what their values are.

Solution to Problem 1a: Master equation for modeling nicotine addiction progression

[Table 2.12.4] Organized Ω = {1, 2, 3, 4}

State	Nonsmoker to experimental	Experimental to nonsmoker	Experimental to dependent	Dependent to ex-Smoker
Code	1	2	3	4
	↓	↓	↓	↓
Transition rate (1,000 people/day)	3	9	10	2
	$w(k \leftarrow j)$ T1	$w(k \leftarrow j)$ T2	$w(k \leftarrow j)$ T3	$w(k \leftarrow j)$ T4
NS($j \leftarrow k$)		**1 ← 2 9/1,000**	1 ← 3	1 ← 4
ES($j \leftarrow k$)	**2 ← 1 3/1,000**		2 ← 3	2 ← 4
ED($j \leftarrow k$)	3 ← 1	**3 ← 2 10/1,000**		3 ← 4
DE($j \leftarrow k$)	4 ← 1	4 ← 2	**4 ← 3 2/1,000**	

The solution in Table 2.12.4 assigns these transitions to the appropriate transition rates w(j ← k), indicating the non-zero rates and their respective values, which are essential for developing a master equation to model the progression of nicotine addiction.

Problem 1b: Write down the evolution equations for $p(1,t)$, $p(2,t)$, $p(3,t)$, and $p(4,t)$ using the numbers listed in the table for the transition rates (i.e., write down a set of four coupled differential equations: $dp(1,t)/dt = \ldots$, $dp(2,t)/dt = \ldots$, etc.).

Solution to Problem 1b:

$$\frac{d}{dt}p(j,t) = '\text{ rate in}' - '\text{rate out}' = \sum_{k \neq j} w(j \leftarrow k)p(k,t) - p(j,t)\sum_{k \neq j} w(k \leftarrow j)$$

$dp(1,t)/dt = w(1 \leftarrow 2)\, p(2, t) - w(2 \leftarrow 1)\, p(1, t) = 9/1{,}000 \times p(2, t) - 3/1{,}000 \times p(1, t)$
$dp(2,t)/dt = w(2 \leftarrow 1)\, p(1, t) - w(1 \leftarrow 2)\, p(2, t) - p(2, t)\, w(3 \leftarrow 2) = 3/1{,}000 \times p(1, t) - 9/1{,}000 \times p(2, t) - p(2, t) \times 10/1{,}000$
$dp(3,t)/dt = w(3 \leftarrow 2)\, p(2, t) - w(4 \leftarrow 3)\, p(3, t) = 10/1{,}000 \times p(2, t) - 2/1{,}000 \times p(3, t)$
$dp(4,t)/dt = w(4 \leftarrow 3)\, p(3, t) = 2/1{,}000 \times p(3, t)$

The solution provides a set of four coupled differential equations ($dp(1,t)/dt$, $dp(2,t)/dt$, $dp(3,t)/dt$, and $dp(4,t)/dt$) representing the rates of change for each state. These equations are essential for understanding the dynamics of nicotine addiction progression over time.

Problem 1c: The model developed so far does not account for relapse. That is, it does not consider that ex-smokers may become dependent smokers again. (i) To model relapse dynamics in the sense that an ex-smoker becomes a dependent smoker again, which transition rate must become non-zero? Set this transition rate equal to 0.1. (ii) Write down explicitly the evolution equations for $p(1,t)$, $p(2,t)$, $p(3,t)$, and $p(4,t)$ using the numbers listed in the table for the transition rates and the transition rate of 0.1 that you identified in Problem 1c(i).

[Table 2.12.5] Solution to Problem 1c

	$w(k \leftarrow j)$ T1	$w(k \leftarrow j)$ T2	$w(k \leftarrow j)$ T3	$w(k \leftarrow j)$ T4
NS($j \leftarrow k$)		**1 ← 2 9/1,000**	1 ← 3	1 ← 4
ES($j \leftarrow k$)	**2 ← 1 3/1,000**		2 ← 3	2 ← 4
ED($j \leftarrow k$)	3 ← 1	**3 ← 2 10/1,000**		**3 ← 4 100/1,000**
DE($j \leftarrow k$)	4 ← 1	4 ← 2	**4 ← 3 2/1,000**	

$dp(1,t)/dt = w(1 \leftarrow 2)\, p(2, t) - w(2 \leftarrow 1)\, p(1, t) = 9/1{,}000 \times p(2, t) - 3/1{,}000 \times p(1, t)$
$dp(2,t)/dt = w(2 \leftarrow 1)\, p(1, t) - w(1 \leftarrow 2)\, p(2, t) - p(2, t)\, w(3 \leftarrow 2) = 3/1{,}000 \times p(1, t) - 9/1{,}000 \times p(2, t) - p(2, t) \times 10/1{,}000$
$dp(3,t)/dt = w(3 \leftarrow 2)\, p(2, t) - w(3 \leftarrow 4)\, p(4, t) - = w(4 \leftarrow 3)\, p(3, t) = 10/1{,}000 \times p(2, t) + 100/1{,}000 \times p(4, t) - 2/1{,}000 \times p(4, t)$
$dp(4,t)/dt = w(4 \leftarrow 3)\, p(3, t) - = w(3 \leftarrow 4)\, p(4, t) = 10/1{,}000 \times p(3, t) - 100/1{,}000 \times p\,(4, t)$

Table 2.12.5 and the revised set of differential equations incorporates this new transition rate, providing a more comprehensive representation of the nicotine addiction progression that now includes relapse dynamics in the population.

Problem 2: Simulations and analytical considerations for the academic master equation model of a system jumping on a grid. In this problem, consider the grid with $M = 5$ grid points. Only jumps between nearest neighbors are allowed. All simulations can be conducted using the script as above (feel free to write your own script if you want to do so). Conduct all simulations for a time interval from 0 to 2 s. That is, if you use the script, set $T = 200$ and tau = 0.01.

Problem 2a: Consider the case w(left) = w(right) that describes a stochastic process without drift. Select a value for w that is in the interval [2, 6]. Report the value that you have selected. Show by numerical simulation that the process converges to a uniform distribution. To this end, consider two different initial conditions: (i) set $p(1, t = 0) = 1$ and (ii) select for $p(1, 0)$, $p(2, 0)$, and $p(3, 0)$ three values larger than 0. Take the normali-zation condition into account. (iii) What happens when you use a uniform distribution as an initial condition? Report for (i), (ii), and (iii) the following: plot of the initial distribution, plot of the final distribution, plot of trajectories $p(k, t)$. In addition, write for each case (i), (ii), and (iii) a short sentence that summarizes what happens (i.e., what you can see from the simulation).

Solution to Problem 2a:

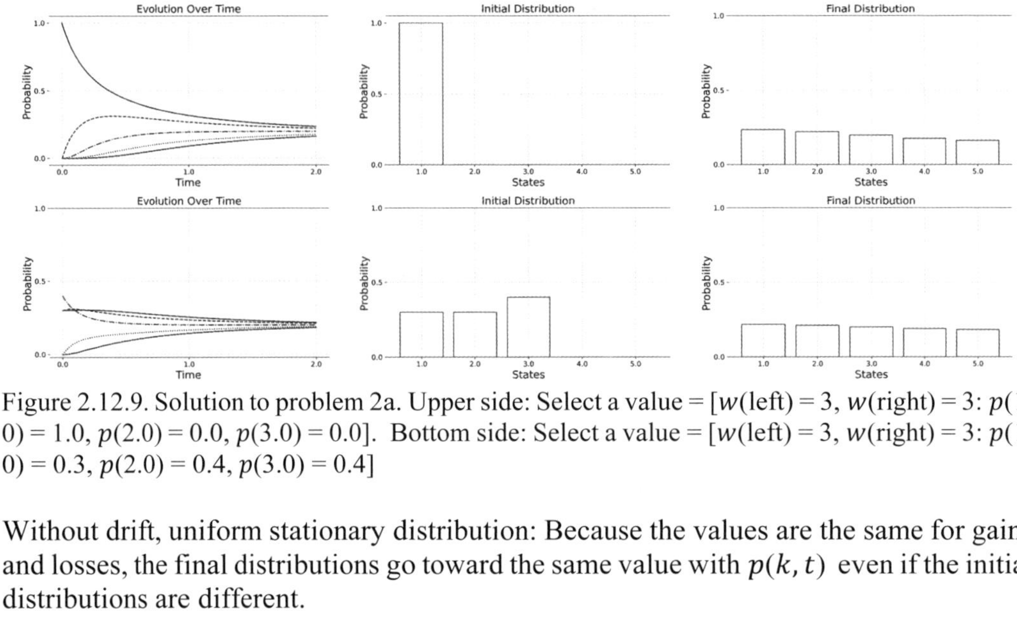

Figure 2.12.9. Solution to problem 2a. Upper side: Select a value = [w(left) = 3, w(right) = 3: p(1, 0) = 1.0, p(2.0) = 0.0, p(3.0) = 0.0]. Bottom side: Select a value = [w(left) = 3, w(right) = 3: p(1, 0) = 0.3, p(2.0) = 0.4, p(3.0) = 0.4]

Without drift, uniform stationary distribution: Because the values are the same for gains and losses, the final distributions go toward the same value with $p(k, t)$ even if the initial distributions are different.

Problem 2b: Assume that $M = 5$ holds, and set $p(k - 1) = 1/5 = 0.2$, $p(k) = 1/5 = 0.2$, and $p(k + 1) = 1/5 = 0.2$.

$$\frac{d}{dt}p(k,t) = gains - losses$$

$$\frac{d}{dt}p(k,t) = w(right)p(k-1,t) + w(left)p(k+1,t) - p(k,t)\{w(right) + w(left)\}$$

$$\frac{d}{dt}p(k,t) = 0.2 * 3 + 0.2 * 3 - 3 * (0.2 + 0.2)$$

$$\frac{d}{dt}p(k,t) = 0$$

Let us simulate a model that drifts to the left. (i) Select two values for $w(\text{left})$ and $w(\text{right})$ such that $w(\text{left}) > w(\text{right})$ and that both values are from the interval [2, 6]. Report the initial distribution, the final distribution, and a figure showing the trajectories $p(k, t)$ for $k = 1, 2, 3, 4$, and 5. Write a short sentence by yourself that describes what happens. (ii) Set $w(\text{right}) = 0$ and run the simulation again. What happens? Can you speculate what would happen if time went to infinity? That is, how would the stationary distribution look in this case? Try to come up with an argument for why the behavior of the stochastic process that you observe for $w(\text{right}) = 0$ is plausible.

Solution to Problem 2b:

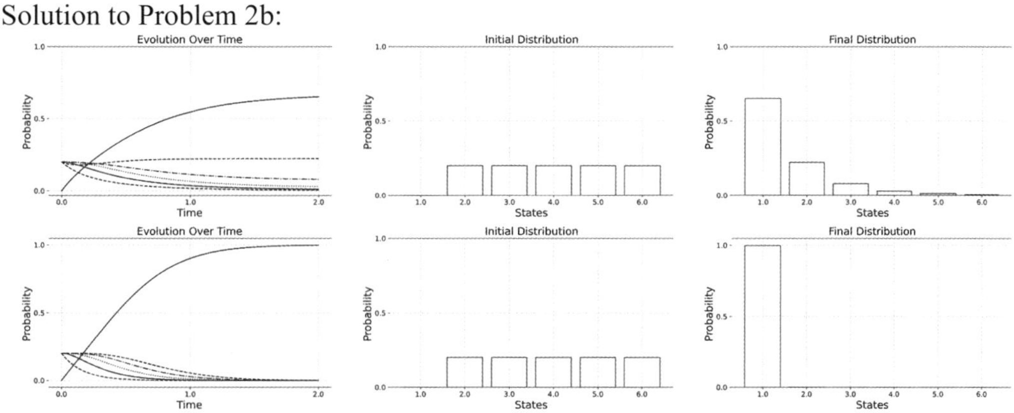

Figure 2.12.10. Solution to problem 2b. Upper side: Select a value = [$w(\text{left}) = 6$, $w(\text{right}) = 2$: $p(1, 0) = 0.2$, $p(2.0) = 0.2$, $p(3.0) = 0.2$, $p(4.0) = 0.2$, $p(5.0) = 0.2$]. Bottom side: Select a value =[$w(\text{left}) = 6$, $w(\text{right}) = 0$: $p(1, 0) = 0.2$, $p(2.0) = 0.2$, $p(3.0) = 0.2$, $p(4.0) = 0.2$, $p(5.0) = 0.2$]

The distribution is skewed because there are more transitions to the left than to the right. In Problem 2b(ii) there are no transitions to the right. Therefore, all realizations will ultimately end up in state $k = 1$, and the stationary distribution is $p(1) = 1$ and $p(k) = 0$ for $k = 2, 3, 4$, and 5.

Remarks: For modeling stochastic processes that entail changes between discrete states, master equations are a helpful tool. These equations explain how the transition rates between states affect the probability of the states as they vary over time. The Master Equation's identification of the rates is an important component in this situation. The two-state Master Equation, which applies to systems having two states and may be resolved analytically or numerically, is one of the most often employed varieties of Master Equations. They can offer insightful information on the kinetics and statistical characteristics of these processes. We can effectively identify the transition rates and run simulations to learn more about the behavior of the system if we have a solid grasp of the underlying concepts.

[2.13]

Markov Diffusion Processes, Part 1

Dynamics of Markov Diffusion Processes, Wiener Process, Practical Practice

Expanding upon our prior investigation of master equations and Markov processes, we have considered the representation of a system's state, X, through a stochastic process. In this section, we will deepen our understanding of stochastic processes with diffusion by focusing on a specific subclass of Markov processes, characterized by almost surely continuous sample paths—an aspect crucial for systems analysis.

Our discussion will encompass systems exposed to Gaussian white noise, which can indeed be described by Markov processes that exhibit almost surely continuous sample paths, thereby broadening our knowledge of these processes and their impact on system behavior.

1. Dynamics of Markov Diffusion Processes

Exploring the realm of Markov processes reveals that these models are governed by the principle where the future state's probability is determined exclusively by the current state, independent of historical states. This attribute, central to Markovian dynamics, is mathematically articulated through partial differential equations [$\partial p(x,t)/\partial t$] in diffusion processes. Such equations enable the precise modeling and prediction of future states based on present conditions, highlighting the efficacy of Markov processes in forecasting variable trajectories.

[Table 2.13.1] Stochastic Processes for Continuous State Spaces in Continuous Time

	Discrete states	Continuous-state variable
Discrete time	Monte Carlo simulations Markov chains	Stochastic iterative maps
Continuous time	Master equations	**Markov diffusion processes**

Table 2.13.1 presents a classification of various types of stochastic processes based on the nature of their state spaces (discrete or continuous) and the time evolution (discrete or continuous). In the realm of discrete time and discrete states, Monte Carlo simulations and Markov chains are the prominent stochastic processes.

Stochastic iterative maps fall under the category of processes having discrete time evolution with a continuous state variable.

For the processes with continuous time evolution, Master equations are used to describe systems with discrete states, while Markov diffusion processes are employed for continuous-state variables. This classification helps to understand the underlying framework for modeling the dynamics of systems in various fields of study.

For the continuous-state, continuous-time, we can consider Markov diffusion processes, represented by the following equation:

- Markov diffusion processes

$$\frac{\partial}{\partial t}P(x,t) = -\frac{\partial}{\partial x}h(x)P(x,t) + \frac{\partial^2}{\partial x^2}D(x)P(x,t)$$

Here, technically, the symbol $\frac{\partial}{\partial t}$ represents the partial derivative of P with respect to t which describes how the value of P changes as t changes. The $\frac{\partial}{\partial x}$ represents the derivative of P with respect to x, which describes how the value of P changes as x changes.

The term $-\frac{\partial}{\partial x}h(x)P(x,t)$ describes how the value of P changes as x changes considering the effect of the function h. And the term $\frac{\partial^2}{\partial x^2}D(x)P(x,t)$ describes how the rate of change of P with respect to x changes as x changes taking into account the effect of the function D.

Understanding this equation is crucial for grasping the diffusion process in Markov processes.

- Another process is given by;

$$\frac{\partial}{\partial t}p(x,t) = -\frac{\partial}{\partial x}h(x)P(x,t) + \frac{\partial^2}{\partial x^2}D(x)P(x,t) + \sum_{k=3}^{\infty}\frac{\partial^k}{\partial x^k}A_k(x)P(x,t), \qquad A_k \neq 0$$

This is the similar to the one in the previous, but it has an additional term at the end. The term $\sum_{k=3}^{\infty}\frac{\partial^k}{\partial x^k}A_k(x)P(x,t)$ represents an infinite sum of terms each representing the k-th partial derivative of P with respect to x multiplied by a function $A_k(x)$ of x and P. The sum starts at $k = 3$ and goes to infinity ∞. Recognizing the differences and implications of this equation is equally essential for understanding the diffusion process, serving as a fundamental concept.

We can now turn our attention to a specific example: The Wiener process.

2. The Wiener Process

The Wiener process was proposed by Norbert Wiener (Masani, 2012). This process characterizes the erratic motion (i.e., diffusion) of a grain of pollen on the water's surface as it is continually being bombarded by water molecules. The resulting motion, which can be viewed as a scaled random walk on any finite time interval, is almost surely continuous.

The Wiener process is considered a series of normally distributed random variables, and for later time points the variances of these normally distributed random variables increase to reflect the fact that it is more uncertain (and thus more difficult) to predict the value of the process after a longer period.

\+ **Mathematical Process:** The evolution equation for $P(x,t)$ is a fundamental mathematical stochastic process that describes how a probability distribution changes over time.

- No deterministic part
- Change in *P* with time is determined entirely by the diffusion term
- Derivation via master equation $w_{up} = w_{down}$
- Change in *P* with time ($\frac{\partial}{\partial t}p$) with diffusion term ($Q\frac{\partial^2}{\partial x^2}P$)

In this process, there is no deterministic part, and the change in *P* over time is determined entirely by the diffusion term ($\frac{\partial}{\partial t}p$). The equation is derived from the master equation, which requires that the rates of upward and downward transitions be equal. The diffusion term is represented by $Q\frac{\partial^2}{\partial x^2}P$, where Q is a constant that represents the intensity of the diffusion process.

$$\frac{\partial}{\partial t}p = Q\frac{\partial^2}{\partial x^2}P \Rightarrow P(x,t) = \frac{1}{\sqrt{2\pi\sigma^2(t)}}exp\left\{-\frac{x^2}{2\sigma^2(t)}\right\}, \qquad \sigma^2(t) = 2Qt$$

Here, the left-hand side of the equation $\frac{\partial}{\partial t}p = Q\frac{\partial^2}{\partial x^2}P$ represents the rate of change of p with respect to time t and the second derivative of P with respect to space x. The right-hand side of the equation $P(x,t) = \frac{1}{\sqrt{2\pi\sigma^2(t)}}exp\left\{-\frac{x^2}{2\sigma^2(t)}\right\}$ defines the functional form of P as a function of x and t. The term $\sigma^2(t) = 2Qt$ is the variance of P as a function of tiem t with Q being a constant.

Initial condition (t = 0): All realizations are at x = 0. Probability density = normal distribution is as below (Figure 2.13.1);

- Mean = 0 for all t
- Variance increases like $\sigma^2(t) = 2Qt,\ Q > 0$

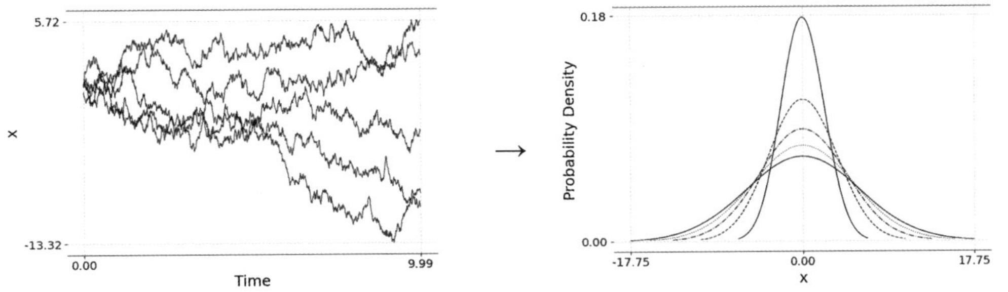

Figure 2.13.1. Schematic representation of density with Wiener processes. Initially ($t = 0$), all realizations are at $x = 0$. The normal distribution becomes the Dirac delta distribution (peak ∞).

Trajectories of the Wiener Process? AR(1) non-oscillatory case (positive a_1)

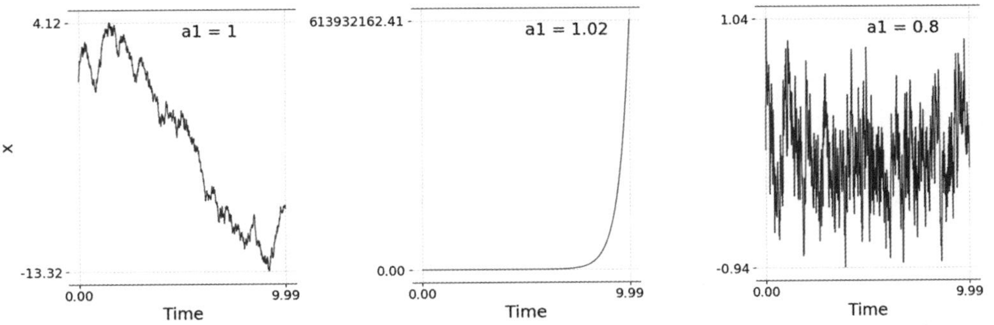

Figure 2.13.2. Representation of the $a_1 = 1$: (left), middle ($a_1 = 1.02$), right ($a_1 = 0.8$)

The trajectories of the Wiener Process in the AR(1) non-oscillatory scenario when a_1 is positive are depicted in Figure 2.13.2. The non-stationary scenario with $a_1 = 1.02$ is shown in the middle panel, while the deterministic example with $a_1 = 1$ is shown in the left panel. The stationary instance with an a_1 value of 0.8 is depicted in the right panel.

- Neither increases nor decays: The phrase 'neither rises nor decays' indicates that the amount under consideration does not change over time. There is no tendency for values to rise or fall.
- No deterministic subprocess: The phrase 'no deterministic subprocess' denotes the absence of any observable trend or pattern in the changes over time. Changes instead happen haphazardly or unexpectedly.

+ **Scaling of Driving Force:** It is a AR(1) process specifically, which implies that each time step's value is dependent on the preceding time step's value with some random variation. When the parameter a_1 equals 1, the process is referred to as a 'unit autoregression,' which governs how strongly it is dependent on the prior value.

- AR(1) for $a_1 = 1$: $X(n+1) = X(n) + \varepsilon(n),\ parameter\ Var(\varepsilon) = s^2$
- Alternative: $X(n+1) = X(n) + s\varepsilon'(n),\ Var(\varepsilon') = 1$
- Use to construct trajectory $X(t)$

These models allow for the simulation of a variable's time-dependent trajectories, which may be helpful for modeling and studying a variety of events. Time step Δt small: $t_n = \Delta t\ n,\ Initial\ time\ t_0 = 0;\ X(0) = 0$

$$X(t_n + \Delta t) = X(t_n) + s\varepsilon(t_n), \qquad s = \sqrt{2Q\Delta t}, \qquad Var(\varepsilon) = 1$$
$$X(\Delta t) = +s\varepsilon(0) \to Var\ \ s(X(\Delta t)) = s^2 = 2Q\Delta t$$

Here, Δt is a small-time interval, $\varepsilon(t_n)$ is a random variable with mean 0 and variance 1, and Q'is a constant.

$$X(2\Delta t) = +s[\varepsilon(\Delta t) + \varepsilon(0)] \to Var\ \ (X(2\Delta t)) = s^2\langle\varepsilon^2(\Delta t) + 2\varepsilon(0)\varepsilon(\Delta t) + \varepsilon^2(0)\rangle$$
$$Var(X(2\Delta t)) = s^2\langle\varepsilon^2(\Delta t) + \varepsilon^2(0)\rangle = 2s^2 = 4Q\Delta t$$

Then, we complete the equation derived the variance of $X(2\Delta t)$ which is the value of X at time $t_n + 2\Delta t$. To derive the variance of $X(2\Delta t)$, we used the equation for $2s^2 = 4Q\Delta t$ and the definition of variance, as well as the properties of the random variable ε. In general

$$Var(X(m\Delta t)) = ms^2 = m2Q\Delta t = 2Qm\Delta t, \qquad \to \sigma^2 = 2Q \cdot t$$

We have derived an expression for the variance of $X(m\Delta t)$ which is the value of X at time $t_n + m\Delta t$ where m is an integer. The expression $ms^2 = m2Q\Delta t = 2Qm\Delta t$ shows that the variance of X increase linearly with the time interval $m\Delta t$ and that the variance is proportional to the squre of the consatn 'Q' which represents the standard deviation of the random variable $\varepsilon(t)$.

- In addition: mean = 0 because of $X(0) = 0$, AR(1) has normal distribution

Conclusion: simulation of Wiener process via $X(t_n + \Delta t) = X(t_n) + \sqrt{2Q\Delta t}\ \varepsilon(t_n)$.

The normal distribution becomes the Dirac delta distribution (peak ∞). Estimating probability density (Figure 2.13.3);

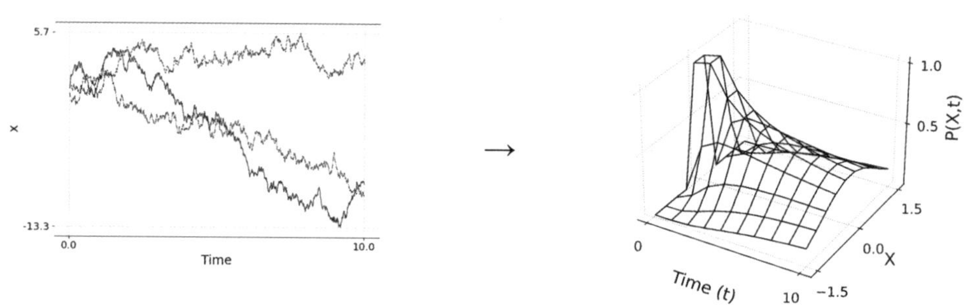

Figure 2.13.3. Estimating density with the Wiener process: (left) the Wiener process with N = 100; (right) a comparison of the analytical solution. Prob. = probability

We employ a continuous-time evolution equation for trajectories to get the derivative of the fluctuating driving force, where continuous-time evolution equation for trajectories given

$$X(t_n + \Delta t) = X(t_n) + \sqrt{2Q\Delta t}\,\varepsilon(t_n)$$

We are looking for $\frac{d}{dt}X(t)$ = fluctuating driving force, defined fluctuating function $\Gamma(t)$ like $X \int_t^{t+\Delta t} \Gamma(t)d\tau = \sqrt{\Delta t}\,\varepsilon(t)$

$$X(t_n + \Delta t) = X(t_n) + \sqrt{2Q}\,X \int_t^{t+\Delta t} \Gamma(\tau)d\tau = X(t_n) + \sqrt{2Q}\Gamma(t_n) + O(\Delta t)$$

In the equation, $X(t_n + \Delta t)$ represents the value of X at time $t_n + \Delta t$ and $X(t_n)$ represents the value of X at the previous time step t_n. The term $\sqrt{2Q}\,X \int_t^{t+\Delta t} \Gamma(\tau)d\tau =$ represents the change in X over the time interval Δt and is given by the integral of a function $\Gamma(\tau)$ with respect to time t. The term $O(\Delta t)$ represents the error in the approximation which is of the order of Δt.

This means that the error decreases as Δt becomes smaller.

- Divide by Δt $\frac{X(t_n+\Delta t)-X(t_n)}{\Delta t} = \sqrt{2Q}\Gamma(t_n) + O(\Delta t)$
- Limiting case $\frac{d}{dt}X(t) = \sqrt{2Q}\Gamma(t)$, Langevin equation of the Wiener process
- $\Gamma(t)$: Langevin force

+ **Langevin Force $\boldsymbol{\Gamma(t)}$:** It is crucial to understand the Langevin force $\Gamma(t)$ in the context of diffusion processes because it enables us to describe the temporal evolution of stochastic variables that are subject to noise or random fluctuations and to take these fluctuations' effects on the dynamics of the system into account.

The Langevin equation offers a means of linking the underlying microscopic, discrete-time stochastic process that causes a system's fluctuations to the macroscopic, continuous-time development of that system.

$$\int_t^{t+\Delta t} \Gamma(t')dt' = u(t_n) = \sqrt{\Delta t}\,\varepsilon(t_n)$$

We have an expression here for the integral of the function $\Gamma(t)$ over the time interval $t, t + \Delta t$. The integral represents the change in X over time interval Δt as a function of the $\Gamma(t)$.

The expression on the right side $u(t_n) = \sqrt{\Delta t}\,\varepsilon(t_n)$ defines the function $u(t_n)$ as the square root of Δt times a random variable $\varepsilon(t_n)$ with mean 0 and variance 1.

- Fluctuating force, Time dependent, Normal distribution

The $\Gamma(t)$ is the continuous-time counterpart of a certain discrete-time fluctuating force $u(t_n)$.

The $u(t_n)$ is defined by normally distributed (appropriately rescaled) random variables: $u(t_n) = \sqrt{(\Delta t)} \cdot \varepsilon(t_n)$.

+ **Summary and General 1D Case:** The Wiener process is

- Fokker-Planck equation $\frac{\partial}{\partial t} p = Q \frac{\partial^2}{\partial x^2} P$
- Langevin equation $\frac{d}{dt} X(t) = \sqrt{2Q}\Gamma(t)$
- Simulation of trajectories via Euler-forward schemas $X(t_n + \Delta t) = X(t_n) + \sqrt{2Q\Delta t}\, \varepsilon(t_n)$

General case (i.e., univariate Markov diffusion process)

$$\frac{\partial}{\partial t} P(x,t) = -\frac{\partial}{\partial x} h(x)P + \frac{\partial^2}{\partial x^2} D(x)P$$

As we defined, the equation above is a partial differential equation that describes how the quantity P changes over time t and space X.

$$X(t_n + \Delta t) = X(t_n) + \sqrt{2Q\Delta t}\, \varepsilon(t_n) + \sqrt{2D}(X(t_n))\Delta t \varepsilon(t_n)$$

This equation describes the evolution of the quantity X over time, using a discrete-time approximation.

$$\frac{d}{dt} X(t) = h(x(t)) + \sqrt{2D(X(t_n))}\Gamma(t)$$

And finally, this equation describes the evolution of X over time in continuous time.

3. Practice with the Wiener Process and Langevin Equations

In this exercise we conduct several simulations of the Wiener process.

Problem 1: We consider both numerical outcomes and analytical solutions. The aim is to answer the following questions:

- (i) How does the number of realizations (sample size) affect how well numerical solutions approximate exact solutions?
- (ii) How does the Wiener process behave qualitatively as time evolves?

You may write your own script to simulate a Wiener process. Alternatively, you may use the script below:

First, let us parameterize the Wiener process and its calculation of a trajectory with the ensemble as given below:

```
[In]
# Define variables
dt = 0.1
Q = 0.5
N = 100
```

We defined three variables. The dt represents the time step which initialized to 0.1, Q is value initialized to 0.5, and the N is the number of elements initialized to 100.

To initialize with time points

```
[In]
import numpy as np

# Create a zero matrix of size N x 1000
x = np.zeros((N, 1000)) timepoints = np.arange(0, dt*1000, dt)
```

The code defined x as 2D array of size N by 100 filled with zeros. The 'timepoints' is an array that start at 0 ends at dt *(1000−1) and increments by dt at each step.

For time loop and its realizations

```
[In]
import numpy.random as npr

# Fill the matrix
for k in range(N):
    x[k, 0] = 0

    for n in range(999):
        # Python uses 0-indexing, so we use 999 instead of 1000-1
        e = npr.normal(0, 1)
        x[k, n+1] = x[k, n] + np.sqrt(2*Q*dt)*e
```

The code iterates over the rows of x form $k = 1$ to $k = N$. For each row k, the inner loop iterates over the columns of x form $n = 1$ to $n = 999$ as follows.

It generates a random normal value with mean 0 and standard deviation 1 using the 'normal' function. It updates the value at position $(k, n + 1)$ in the x array by adding to it the square root of $2 * \text{Q} * dt$ multiplied by the random normal value generated in step 1.

Then, make output

```
[In]
import matplotlib.pyplot as plt

# Create a new figure
plt.figure()

# Plot each row of x
for k in range(N):
    plt.plot(timepoints, x[k, :])
plt.show()
```

→

[Out]

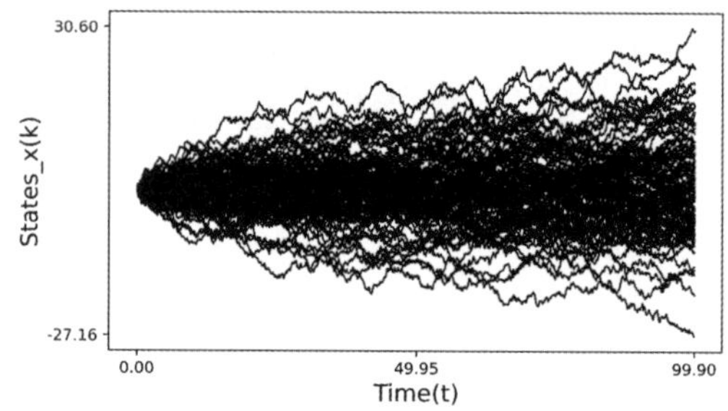

Figure 2.13.4. Output of problem 1 with state x

The created Figure 2.13.4 shows the values in each row of the x array against the 'timepoints' array drawing a line.

Problem 2: Now, let us move on to the simulation of a nonlinear single-variable Langevin equation: example of the incomplete Haken-Kelso-Bunz (1985) model. In this exercise we simulate a single-variable (i.e., univariate) nonlinear Langevin equation and compare the numerical solutions to predictions made by the model. To this end, we use the Haken-Kelso-Bunz (HKB) model. For the sake of simplicity, we take only the first sine term of the model into account and neglect the second term.

Accordingly, the Langevin equation reads

$$\frac{d}{dt}X(t) = h(x) + \sqrt{2Q}\Gamma(t)$$

with the nonlinear (deterministic) force term h defined by

$$h(x) = -\alpha \cdot \sin(x)$$

involving the parameter, $a > 0$.

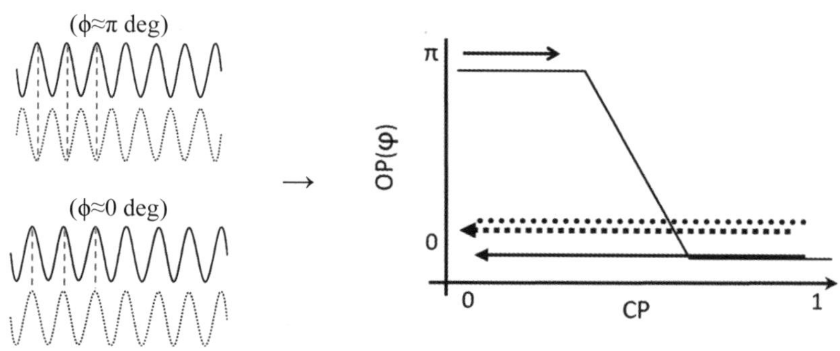

Figure 2.13.5. Reflection of the HKB model

In Figure, 2.13.5, the plot on the left side shows the observed relative phase or phase relation between two oscillators at $\phi \approx 0$ deg (in-phase = upper side) or $\phi \approx \pm 3.14$ deg (anti-phase = bottom side). The plot on the right, the CP denotes the control parameter, the OP(ϕ) denotes the order parameter, and the arrow indicates the attractor.

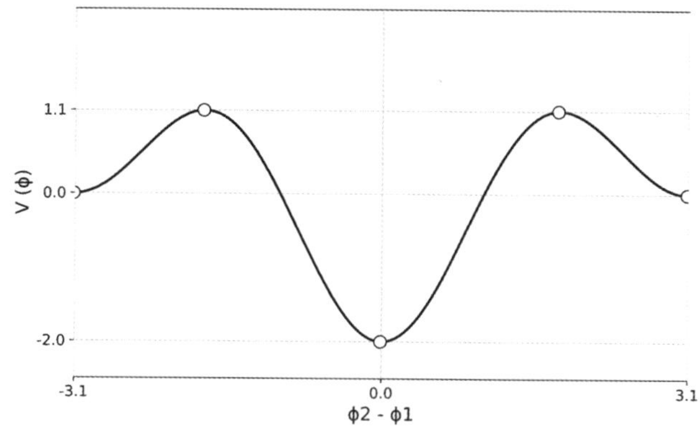

Figure 2.13.6. Dynamics of the HKB model

In Figure 2.13.6, Simulation on the left denotes the energy of the function at each averaged relative phase (bottom line = the vertical-axis). V is the intrinsic dynamics of the potential function, and the balls at the bottom side symbolize stable states (attractors) and the balls on the upper side correspond to unstable states (repellers), $\phi 2 - \phi 1 \approx 0$ denotes a condition of a nearly synchronized in-phase, and $\phi 2$ - $\phi 1 \approx \pi$ indicates a condition of an anti-phase (Haken et al., 1985; Park 2022). The model (Figure 2.13.6) describes the dynamics of the relative phase $X(t)$ between two in-phase oscillating limbs. X is an angular variable with period 2π. The model has a stable fixed point at $x = 0$. The strength of the fixed-point attractor is given by the parameter a. The model makes the following predictions (e.g., Schöner, Haken, and Kelso, 1986):

- (i) In the deterministic case (i.e., for $Q = 0$), for any initial value $X(0)$ different from 0, the phase X relates to $X = 0$ in the long-term limit.
- (ii) In the stochastic stationary case, for a weak fluctuating force (i.e., small Q), the variability in X is given by the ratio of fluctuating strength and attractor strength.
- (iii) In the stochastic stationary case, for a weak fluctuating force, the variable X is distributed like a normal distribution, with mean 0 and variance given by equation

$$Var(x) = Q/a$$

Note that we use the Langevin equation that has the factor 2 in front of the variable Q. For the purpose of this exercise you may write your own script to solve the Langevin equation numerically.

Alternatively, you may use the script below:

Wiener process: calculation of a trajectory, For parameter and time step for simulation

```
[In]
# Parameters
Q = 0.5
dt = 0.1
```

To initialize with its initial value

```
[In]
import numpy as np
# Initialize
x = np.zeros(1000)
timepoints = np.arange(0, dt*1000, dt)
x[0] = 0
```

Define the deterministic function $h(x)$, replace with the actual function of the HKB model

```
[In]
# Define the deterministic function h(x)
def h(x):
    # Replace with the actual function of the HKB model
    return np.sin(x)
```

Generate data with loop

```
[In]
for n in range(len(timepoints) - 1):
    e = np.random.normal(0, 1)
    x[n+1] = x[n] + h(x[n])*dt + np.sqrt(2*Q*dt)*e
```

Create the output (Figure 2.13.7)

```
[In]
import matplotlib.pyplot as plt
fig, ax = plt.subplots(figsize)

ax.plot(timepoints, x)
```

→

[Out]

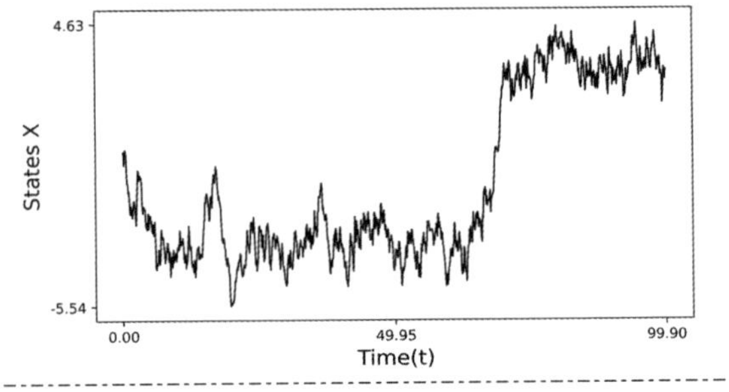

Figure 2.13.7. Results of the calculation trajectory

This simulation incorporates a deterministic function $h(x) = \sin(x)$, which is a part of the nonlinear Langevin equation based on the simplified HKB model. This function contributes to the deterministic part of the system's dynamics.

As in the previous example, a stochastic time series is generated with each new state of the system $x[n+1]$ influenced by both the system's previous state and a random variable e. However, in this simulation, the deterministic $h(x[n])$ component is included in the state update rule $x[n+1]$. The deterministic component $h(x[n]) * dt$ depends on the current state $x[n]$ and represents a 'push' that the system experiences due to the deterministic part of its dynamics. This 'push' is proportional to the sine of the current state, introducing a nonlinearity to the system's evolution over time.

This simulation provides a visual representation of the system's states over time under the influence of both deterministic and stochastic factors. In the resulting plot (Figure 2.13.7), the x-axis represents time, and the y-axis represents the system state. This plot allows us to observe the interplay of deterministic and stochastic influences in the evolution of the system's state. This exercise extends the previous work by introducing a deterministic component, providing further insight into how nonlinear Langevin equations such as the HKB model describe stochastic systems with deterministic influences.

Remarks: Markov Diffusion Processes describe the evolution of a system over time, and these processes are particularly useful in modeling systems that exhibit random behavior, such as stock prices or biological systems. The Wiener Process, also known as Brownian motion, is a continuous-time stochastic process that has stationary and independent increments. It is often used to model the random motion of particles in a fluid or the price movement of a stock. In real-world applications, Wiener Processes and Markov Diffusion Processes can assist academics and practitioners in understanding the behavior of complex systems and forecasting their future evolution. It is crucial to keep in mind that these models are abstractions of actual events and might not always precisely represent the behavior of the system under investigation. Overall, Wiener Processes and Markov Diffusion Processes are effective tools for examining systems in a variety of domains. They serve as a starting point for more sophisticated procedures and have a wide range of applications in both research and industry.

[2.14]

Markov Diffusion Processes, Part 2

The Ornstein-Uhlenbeck Process, (Non)Parametric Analysis, Practical Practice

Having discussed discrete-time Markov chains and their continuous-time counterparts, we now shift our focus to diffusion processes, a significant class of continuous-time, continuous-state Markov processes (i.e., $X \in R$).

Another noteworthy stochastic process is the Ornstein-Uhlenbeck process (Gillespie, 1996). This Gaussian process is stationary and temporally homogeneous. In fact, it stands as the only nontrivial process fulfilling these three conditions, provided linear transformations of the space and time variables are allowed. Over time, the process exhibits a tendency to drift toward its mean function, characterizing it as a mean-reverting process.

1. The Ornstein-Uhlenbeck Process

The Ornstein-Uhlenbeck process is named after Leonard Ornstein and George Eugene Uhlenbeck. It is a continuous-time stochastic process defined as the solution of a special kind of stochastic differential equation called the Langevin equation.

Its original application in analysis was as a model for the velocity of a massive Brownian particle under the influence of friction, also called a damped random walk. Besides the physical sciences, the Ornstein-Uhlenbeck process has applications in fields of science and analysis.

[Table 2.14.1] Stochastic Processes for Ornstein-Uhlenbeck Process

	Discrete states	Continuous-state variable
Discrete time	Monte Carlo simulations Markov chains	Stochastic iterative maps
Continuous time	Master equations	Markov diffusion processes Wiener Process **Ornstein-Uhlenbeck Process**

Table 2.14.1 provides a succinct categorization of various stochastic processes, grouped based on the type of state spaces and the time evolution-both can be either discrete or continuous. As we transition to continuous time processes, discrete states are aptly described by Master equations.

Continuous state variables, on the other hand, are represented by Markov diffusion processes, the Wiener process, and the Ornstein-Uhlenbeck process. These classifications and their corresponding processes serve as fundamental building blocks for modeling complex dynamical systems.

- Limiting case $\gamma \to 0$ corresponds to the Wiener process

$$\frac{d}{dt}X = -\gamma X(t) + \sqrt{2Q}\Gamma(t), \qquad \gamma > 0$$

The equation states that the time derivative $X(t)$ is equal to the negative of a constant $\gamma X(t)$ plus the square root of 2 times Q time s a noise term $\Gamma(t)$.

$$\frac{d}{dt}X = h(X) + \sqrt{2Q}\Gamma(t), \qquad h(X) = -\gamma x$$

The equation is a rearrangement of the above equation where $h(X)$ is defined as the negative of γx

$$\frac{d}{dt}X = -\frac{d}{dt}V(x), \qquad x = X(t) + \sqrt{2Q}\Gamma(t), \qquad V(x) = \gamma x^2/2$$

This equation is a transformation of the first equation where $V(x)$ is defined as the gamma times x squared divided by 2 $\gamma x^2/2$ and x is defined as $X(t)$ plus the square root of t tiems Z tiems a noise term $\sqrt{2Q}\Gamma(t)$.

They state that overdamped motion subjected to a linear restoring force $h(X)$, that is, parabolic potential force with potential $V(X)$; strength of linear force (γ).

- In the case of overdamped stochastic motion near a fixed point that is perturbed by weak noise (small Q), the nonlinear force h can be linearized as $h(X) = -\gamma x$.
- The velocity dynamics of Brownian motion can be formally described by the Ornstein-Uhlenbeck process.

The amplitude of the fluctuating force in the model is denoted by Q and must be greater than or equal to zero ($Q \geq 0$)

This process can be considered a modification of the random walk in continuous time, or the Wiener process, in which the properties of the process have been changed so that there is a tendency for the walk to move back toward a central location, with a greater attraction when the process is further away from the center.

+ Ornstein-Uhlenbeck Process and AR(1): The Ornstein-Uhlenbeck process can also be considered the continuous-time analogue of the discrete-time AR(1) process.

$$x(n+1) = x(n) - \Delta t\, \gamma\, x(n) + \sqrt{2Q\Delta \mathrm{t}}\, \varepsilon(n), \qquad E(\varepsilon^2) = 1$$

Here, the equation states that the value of x at the next time step $n+1$ is equal to the value of x at the current time step n, minus the product of a constant Δt, γ, and x at the current time step n, plus the product of the square root of $\sqrt{2Q\Delta \mathrm{t}}$ and a noise term epsilon at the current time step $\varepsilon(n)$.

Let us simulate a stochastic process with mean reversion using a discrete-time Ornstein-Uhlenbeck process. To ensure accurate simulation, the single time step Δt must be small, such that $\Delta t \cdot \gamma \ll 1$, in order to accurately capture the dynamics of the process over time.

Then, the discrete-time Ornstein-Uhlenbeck process is defined by the equation; Corresponds to AR(1): $0 < a_1 < 1$ (i.e., stable, non-oscillatory case)

$$x(n+1) = a_1 x(n) + \varepsilon^*(n+1), \qquad a_1 = 1 - \Delta t\, \gamma, \;\; E(\varepsilon^{*2}) = 2Q\Delta \mathrm{t}$$

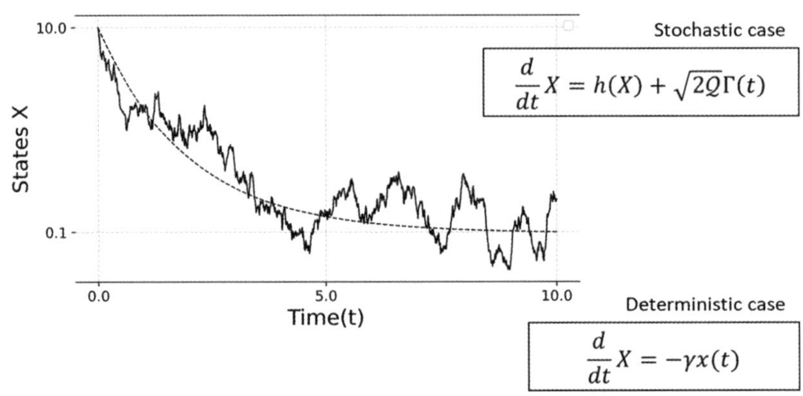

Figure 2.14.1. Representation of the Ornstein-Uhlenbeck process. $X(0) = 10$, γ=0.5, Q=2.0

Figure 2.14.1 depicts the Ornstein-Uhlenbeck process with starting condition $X(0) = 10$, damping coefficient γ of 0.5, and fluctuation strength Q of 2.0.

+ Stationary Process: When the Ornstein-Uhlenbeck process is stationary, its mean, variance, and probability density are well defined and independent of the original circumstances.

Figure 2.14.2 shows the relationships between the mean, variance, and probability density of the Ornstein-Uhlenbeck process. The initial condition for the simulation is $X(t=0) = x0$. The mean of the process can be expressed as $M(t) = x0$ exp(-γ t)), and in the long time limit, the mean converges to zero. The variance of the process is $\sigma^2(t) =$

$Q/2\gamma$ (1-exp(2- γ t)), with an initial variance of $\sigma^2(0) = 0$, and in the long time limit, the variance converges $\sigma^2 = Q/2\gamma$. The probability density of the process is a normal distribution with mean $\mu = 0$ and variance $var = Q/2\gamma$.

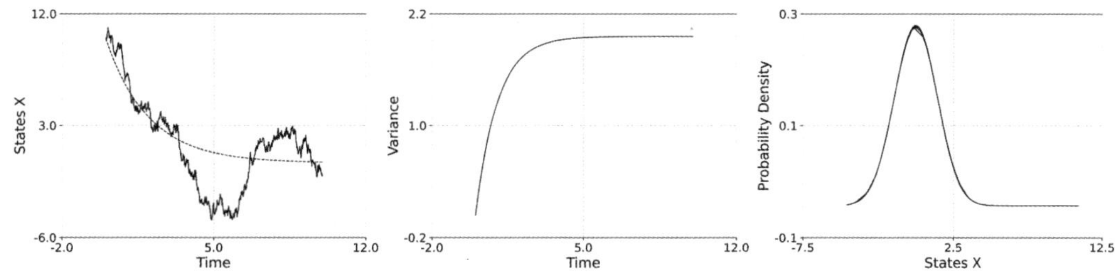

Figure 2.14.2. Some relations with mean, variance, and probability density. Initial condition: $X(t = 0) = x0$; mean: $M(t) = x(0)\ x(0)\exp(-\gamma\ t)$; variance: $\sigma^2(0) = 0, \sigma^2(t) = Q/2\gamma$ (1-exp(2- γ t)); long time limit: M=0, $\sigma^2 = Q/2\gamma$; density = normal distribution $\mu = 0, var = Q/2\gamma$

+ **Simulation for the Ornstein-Uhlenbeck Process:** The Ornstein-Uhlenbeck pro-cess, a stochastic process frequently used to explain mean-reverting behavior, will be simulated using programming in this section.

To achieve this, we must first define a number of parameters that will be utilized in the model before we can start the simulation. We initialize the variables 'ndim', dt and x with their values and the variables 'gamma' and Q are also initialized, respectively.

```
[In]
# Parameters
ndim = 1000
dt = 0.01
gamma = 0.5
Q = 2.0
ndim=1000;
```

For initial value

```
[In]
import numpy as np

# Initialize
x = np.zeros(ndim)
x[0] = 10
```

Give the calculation of the process along with the loop methods, generate data with both deterministic and stochastic components

```
[In]
for n in range(ndim - 1):
    e = np.random.normal(0, 1)
    x[n+1] = x[n] - dt*gamma*x[n] + np.sqrt(2*Q*dt)*e
```

We iterate over the values in the code of n from 1 to 'ndim-1'. At each iteration, it generates a normal distributed random number with mean 0 and standard deviation 1, which it assigns to the variable. Then, it updates the value of x at the next time step $n+1$ using the iterative equation 'x(n)−dt∗gamma∗x(n)+sqrt(2∗Q∗dt)∗e;'.

And display the process including its deterministic time loop

[In]

```
# Plotting
fig, ax = plt.subplots()
ax.plot(np.arange(1, ndim+1)*dt, x)

# Deterministic, time loop
for n in range(ndim - 1):
    e = np.random.normal(0, 1)
    x[n+1] = x[n] - dt*gamma*x[n]

ax.plot(np.arange(1, ndim+1)*dt, x)
```

[Out]

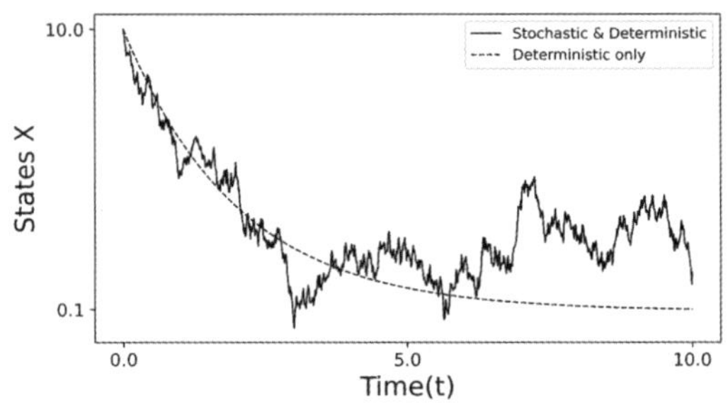

Figure 2.14.3. Visualization of the OU process

In Figure 2.14.3, we generate a plot of the values of x over time where t is an array of time steps equal to dt times the indices of x. After the loop finishes executing, the code uses the 'hold on' command to add another plot using the same data as the first plot, but it is plotted in red without noise term.

2. Parametric Data Analysis

Building upon our earlier exploration of simulations, we now turn our attention to parametric data analysis. This approach refers to the use of methods when the data being studied comes from a sample or population of data that is normally distributed. It assumes that the variance is homogeneous, and the type of data is supposed to be interval or ratio (sometimes referred to as continuous variables/data).

Parametric data analysis: A particular model is given.

$$\frac{\partial}{\partial t}P(x,t) = -\frac{\partial}{\partial x}h(x)P(x,t) + Q\frac{\partial^2}{\partial x^2}D(x)P(x,t)$$

Here, the term $h(x)$ represents a drift that describes how the probability is influenced by the gradient of $h(x)$. The term involving Q and $D(x)$ represents a diffusion that describes how the probability is influenced by the curvature of $D(x)$.

Together, these are describing how the probability density function evolves over time due to both deterministic and stochastic forces.

- Parameters are unknown.
- Example: The Wiener process $h = 0$, $D = Q$ (Q is unknown)

Parameter estimation methods can be a kind of the Maximum likelihood estimation (finance, physics, and mathematics; e.g., Frank et al., 2013).

- Linear regression analysis (e.g., Beek et al., 1995)
- Moment methods (e.g., Frank, 2011)

+ **Non-Parametric Data Analysis:** This is a partial differential equation for the evolution of a probability density function $P(\vec{x}, t)$ as a function fo both time t and a vector of position $\vec{x}$.

$$\frac{\partial}{\partial t}P(\vec{x},t) = -\sum_{k=1}^{M}\frac{\partial}{\partial x_k}h_k(\vec{x})P(\vec{x},t) + \sum_{k=1}^{M}\frac{\partial^2}{\partial {x_k}^2}D_k(\vec{x})P(\vec{x},t)$$

The term on the right-hand side involving the sum over k of the derivative of $h_k(\vec{x})$ with respect to x_k that descrivess how the probability is influenced by the gradient of $h_k(\vec{x})$ in each dimension. The second derivative of $D_k(\vec{x})$ with respect of x_k represents a diffusion term that describes how the probability is influenced by the curvature of $D_k(\vec{x})$ in each dimension.

Nonparametric methods: dimension M of the state space is a priori given.

- Drift h_x and diffusion coefficients D_{kx} are unknown.

Nonparametric time series analysis method: drift diffusion analysis (Friedrich & Peinke, 1997; Stanton, 1997). In time series analysis, it is crucial to understand both parametric and non-parametric analysis methods. Parametric methods, such as linear regression and maximum likelihood estimation, require specific assumptions about the data's distribution and parameters. On the other hand, non-parametric methods like drift diffusion analysis do not rely on such assumptions and can be applied to a broader range of situations. Understanding these methods and their applications helps us make informed decisions when selecting the most appropriate approach for their data, ultimately leading to more accurate predictions and insights

3. Practical Practice (Drift Diffusion Analysis)

Building on the importance of non-parametric methods in time series analysis, drift-diffusion analysis is a notable technique in the univariate, stationary case with additive noise. This model, which is based on sequential sampling with diffusion (Brownian) signals, involves decision-makers accumulating evidence until the process reaches a stopping boundary. At that point, they select the alternative corresponding to that boundary. With applications in psychology, neuroeconomics, and neuroscience, drift-diffusion analysis has proven valuable for explaining observed patterns of choice and response times in various binary-choice decision problems.

$$\frac{d}{dt}X = h(X) + \sqrt{2Q}\Gamma(t) \rightarrow \frac{d}{dt}E(X) = E[h(X)] \rightarrow \frac{d}{dt}M_1(t) = \langle h(X) \rangle$$

Equation of the left side states that the time derivative of X is equal to a deterministic function $h(X)$ plus the square root of 2 times Q ($\sqrt{2Q}$) times a noise term $\Gamma(t)$. The middle equation is reformulation of the left side, where $E(X)$ is the expected value of X and it states that the time derivative of the expected value of X is equal to the expected value of $h(X)$. And the right side of the equation further reformulate of the second equation, where the angle brackets denote the expected value operator.

These equations describe how the expected value and moments of X evolve over time due to both deterministic and stochastic forces.

- Use the conditional expectation, that is, the average under the constraint that $X(t)$ is at t

$$E\left(\frac{d}{dt}X\right)\Big|_{X(t)=x} = E[h(X)]|_{X(t)=x} = h(x) \rightarrow h(x) = \langle \frac{d}{dt}X(t) \rangle, \qquad \textit{evaluated at state } x$$

Here, the equation states that the expected value of the time derivative of X evaluated at a particular state x, is equal to the expected value of $h(x)$, evaluated at the same state x. It also states that the expected value of $h(x)$ is equal to the value of $h(x)$, evaluated at x.

+ **Drift $h(x)$ (Conditional Average of Changes in $X(t)$ at State x):** The conditional average of changes in $X(t)$ at state x is used to calculate the drift $h(x)$ of the Ornstein-Uhlenbeck process. Assume, for instance, that we are given a time step of $\Delta t = 0.1$ and a series of observations with the values $\{x(1) = 3, x(2) = 4, x(3) = 3, (4) = 5\}$, respectively.

By averaging the changes in $X(t)$ that take place when the process is at state $x = 3$, we may get the drift $h(x = 3)$.

- First observation at $x = 3$: change in x is $(4 - 3)/0.1 = 10$

To compute the drift $h(x = 3)$ for the Ornstein-Uhlenbeck process, we use the sequence of observations $\{x(1) = 3, x(2) = 4, x(3) = 3, x(4) = 5\}$ and a time step of $\Delta t = 0.1$. The change in x from the first observation to the second observation at $x = 3$ is $(4 - 3)/0.1 = 10$, and the change in x from the second observation to the third observation at $x = 3$ is $(3 - 4)/0.1 = -10$

- Second observation at $x = 3$: change in x is $(5 - 3)/0.1 = 20$
- Averaging yields: h_{estim}(at $x = 3$) = 15

The change in x from the third observation to the fourth observation at x is (5 – 3)/0.1 = 20. Averaging these changes in x yields h_{estim}(at x = 3) = 15.

+ **Drift Diffusion (Analysis Problem):** How can we implement the conditional average for experimental data? For example, how can we compute $h(x$ = 1.763) when the data are centered at x (e.g., x – 0.2, x + 0.2) to compute the conditional average at x? → Binning method

- Stanton: Use the kernel density estimator method to calculate the conditional average at x

Drift diffusion analysis → Q (univariate, stationary case, additive noise)

$$\frac{d}{dt}X = h(X) + \sqrt{2Q}\Gamma(t)$$

- Mean squared displacement: $MSD(\tau) = E[(X(t+\tau) - X(2))^2]$

Noise amplitude Q can be computed from mean squared displacement: $Q = lim_{\tau\to 0}\, MSD(\tau)/2\tau$, frequently used approximation:

$$Q = \frac{MSD(\Delta t)}{2\Delta t}, \qquad \Delta t = sampling\ interval$$

An often-used approximation to estimate the fluctuation strength Q for the Ornstein-Uhlenbeck process is based on the mean square displacement (MSD) of the process over a small time interval Δt.

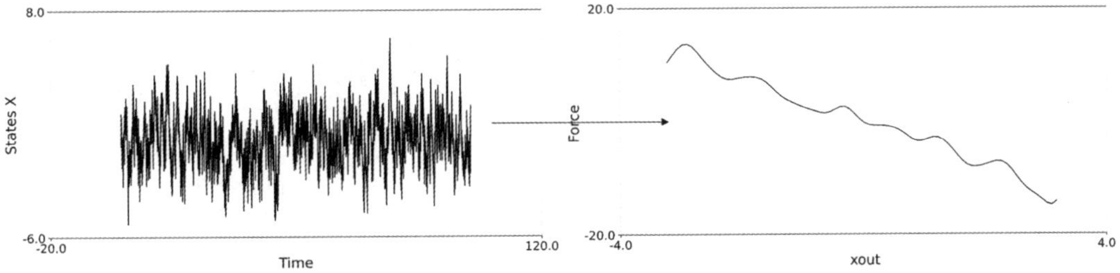

Figure 2.14.4. The Ornstein-Uhlenbeck process from computer-generated data. γ=4, Q=10, N=10000, Δt=0.01. The plot on the left visualizes state transitions over time with stochastic variations, while the plot on the right represents the force exerted with respect to different states, illustrating the transition from initial state fluctuations to a deterministic force pattern.

Specifically, we have Q = MSD(Δt)/(2Δt), where Δt is the sampling interval. As an example, Figure 2.14.4 shows the Ornstein-Uhlenbeck process generated from computer data with parameters γ=4, Q=10, N=10000, Δt=0.01.

The drift-diffusion analysis is crucial for understanding how the expected value and moments of a variable evolve over time due to both deterministic and stochastic forces.

+ Simulation of the Drift-Diffusion-Analysis: Building on the significance of drift-diffusion analysis in understanding the evolution of a variable over time, we can use simulations, such as those implemented in programming, to investigate the effects of varying parameters on the model's predictions and compare them to actual data. To run the model, we must first import the data reflecting stimulus intensity and corresponding reaction times of a subject completing a task in the programming environment.

```
[In]
# Load data from a text file
x = np.loadtxt('outestdata.txt')
x = x - np.mean(x) # Center data
```

The data is stored in a variable called x, and we subtract the mean of the data from each element of x and stores the result back in x for the effect of centering the data around 0. Then we define grid points at which $h(x)$ should be computed, and the sampling interval

```
[In]
dt = 0.01 # Define time step
datasd = np.std(x)  # Define standard deviation of data
xout = np.arange(-2*datasd, 2*datasd, datasd/20)  # Create a new array for plotting
```

Two variables were initialized dt and 'xout'. It sets the value of dt to 0.01 and the value of 'xout' to an array of evenly spaced values ranging from -2 times the standard deviation of the data in x to 2 times the standard deviation of the data in x, with a spacing of 'datasd/20'. Make a plot for the time series

```
[In]
# Get length of the array
tdim = len(x)

# Plotting
fig, ax = plt.subplots()
ax.plot(np.arange(1,  tdim+1)
* dt, x)
```

→ [Out]

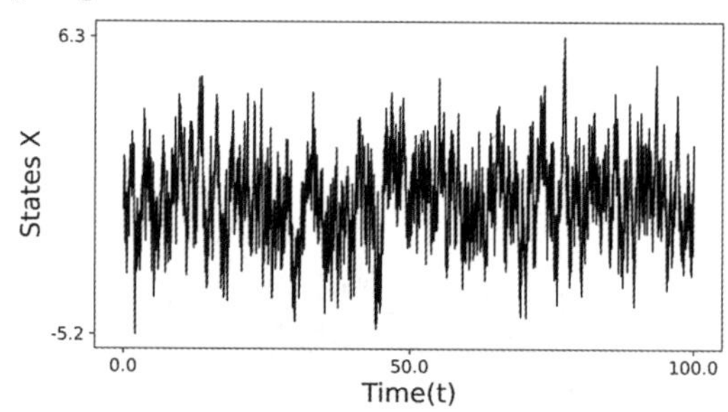

Figure 2.14.5. Visualization of the Ornstein-Uhlenbeck process with State *X*

In Figure 2.14.5, the code generates a plot of the values of x versus time initializing a variable called 'tdim' and sets it to the length of x. Then, let us make a command for the drift analysis corresponding the loaded data.

```
[In]
x = np.loadtxt('outestdata.txt') # Load data from a text file
x = x - np.mean(x)  # Center data
dt = 0.01  # Define time step
datasd = np.std(x)  # Define standard deviation of data
xout = np.arange(-2*datasd, 2*datasd, datasd/20)  # Create a new array for plotting
```

Executes the following tasks, the mean of this data is calculated and subtracted from all elements in x to center the data around 0. A time step dt of 0.01 is defined and the standard deviation of the centered data. Also, an array 'xout' is defined to range from −2 ∗ datasd to 2 ∗ datasd, with a step size of datasd / 20. This creates a grid of points around the mean of the data, covering roughly four standard deviations.

```
[In]
# Define variables
xoutdim = len(xout)
force = np.zeros(xoutdim)
h = datasd / (len(x) ** 0.2)
h2 = h ** 2
```

Then, the code is defining several new variables: 'xoutdim' is set to the length of the array 'xout', which was previously defined. This essentially denotes the number of points at which the force will be calculated. The force is initialized as an array of zeros, with the same length as 'xout'. And the h is a variable calculated by taking the standard deviation of the data (datasd), divided by the total number of data points (len(x)) to the power of 0.2. The $h2$ is simply h squared, which is calculated here for convenience as it is used in the computation of the exponential term in the subsequent force calculation. With them, at each iteration, we initialize the variables 'sum1' and sum2' to 0.0 and sets x to the value of 'xout' at the current index k. Then it enters a loop that iterates over the values of n from 1 to 'tdim−1', where 'tdim' is the length of the array x.

```
[In]
# Loop over xout
for k in range(xoutdim):
    sum1, sum2 = 0.0, 0.0
    for n in range(len(x) - 1):
        expo = -((xout[k] - x[n]) ** 2) / h2
        sum1 += (x[n+1] - x[n]) * np.exp(expo)
        sum2 += np.exp(expo)
    force[k] = sum1 / (sum2 * dt)
```

At each iteration, it calculates an exponent based on the difference between x and the value of x at the current index n, normalized by $h2$. It then adds the product of the difference between the value of x at the next index $n+1$ and the value of x at the current index n to 'sum1', weighted by the exponent; also adds the exponent to 'sum2'. And for the diffusion analysis

```
[In]
# Calculate mean square displacement (lag 1)
diffdata = np.diff(x)
MSDlag1 = np.sum(diffdata**2) / (len(x) - 1)
# Calculate noise amplitude Q
Q = MSDlag1 / (2 * dt)
```

The code calculates a value for 'Q' based on the data in x. First, it calculates the differences between successive elements of x and stores them in an array called 'diffdata'. It then calculates the mean squared displacement 'MSD' of the data over a lag of 1 time step by summing the squares of the differences and dividing the result by the number of differences 'tdim−1'. Then, it divides the 'MSD' by 2 times dt and stores the result in 'Q'. The mean squared displacement is a measure of the variability of the data over time and is related to the variance and standard deviation.

See the output

```
[In]
# Plotting
fig, ax = plt.subplots()
ax.plot(xout, force)
```

[Out]

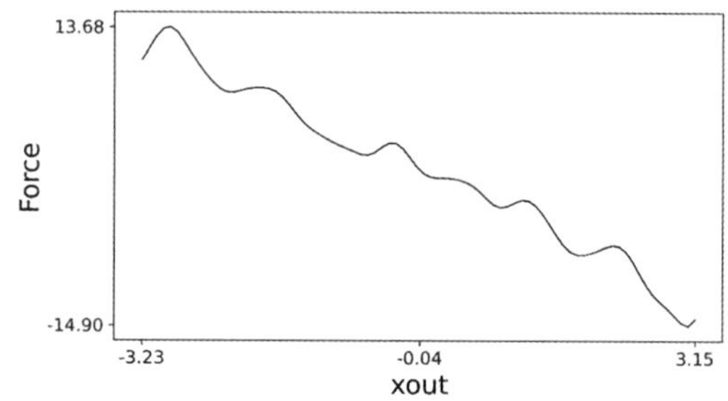

Figure 2.14.6. Visualization of the Ornstein-Uhlenbeck process with Force h(x)

Here we simulate the Ornstein-Uhlenbeck process, which is a solution of the Langevin equation and Friedrich-Peinke-Stanton drift diffusion analysis. As we have proved in Figure 2.14.6, the Ornstein-Uhlenbeck process is stationary, Gaussian, and Markov, which makes it a good candidate for representing stationary random noise. It shows the stochastic evolution of a particle in a fluid under the influence of friction. Notice that the Langevin equation we use in this example is the stochastic differential equation.

Remarks: A popular stochastic model for characterizing the dynamics of a system that is subject to a linear restoring force and a fluctuating force is the Ornstein-Uhlenbeck process. The process may be used to describe the accumulation of noisy sensory input over time that drives decision-making under uncertainty in the context of the drift-diffusion analysis. We may learn more about the mechanisms by comparing the model's predictions to actual data from a two-alternative forced-choice problem. We can also look at how various factors influence the decision-making process. Numerous other disciplines in systems analysis, including as physics, economics, and biology, also make use of the Ornstein-Uhlenbeck process and its variations.

[2.15]

Concluding Remarks

A stochastic process is a system that evolves over time while undergoing chance fluctuations. We describe such a system by defining a family of random variables where X measured at time t is the aspect of the system that is of interest. These obvious methods represent the basis of the construction of a model of a queue that incorporates random variables. It is the often-complex consequences of these idealized models that should be studied. Simple as these models are, by incorporating a little of the randomness of the real world they bring us far closer to understanding the real world than models that neglect such random behavior possibly could.

+ **Trajectories:** The ensemble is depicted schematically in Figure 2.15.1, and depending on whether the states are discrete or continuous, there are two different sorts of paths that can be taken.

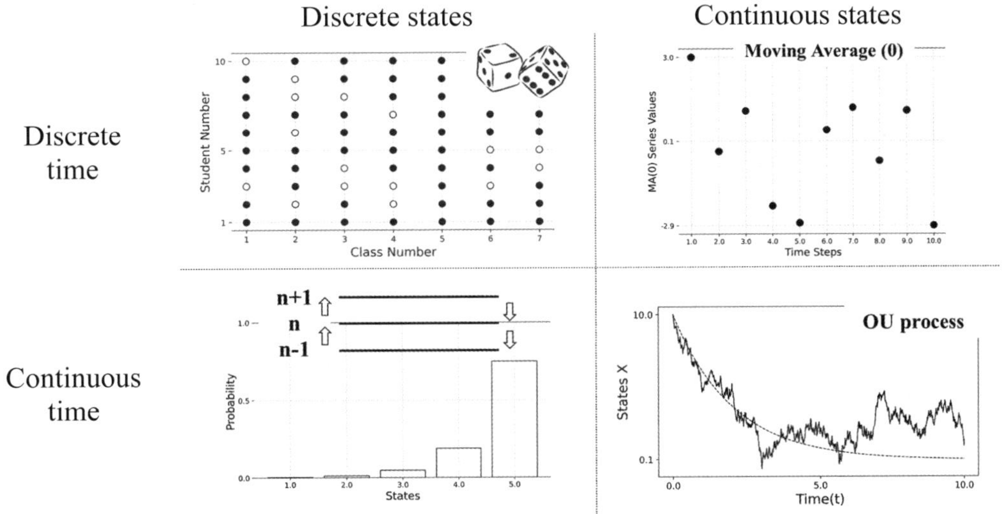

Figure 2.15.1. Schematic representation of the ensemble

The trajectory may be seen as a series of states occurring at distinct time points when the states are discrete. For continuous states, a state space curve that changes continuously over time serves as a representation of the trajectory. For a better understanding of Markov diffusion processes' behavior and the techniques used to evaluate it, it is crucial to distinguish between discrete and continuous trajectories.

A stochastic process is a mathematical model used to describe the behavior of state variables that evolve in time and exhibit fluctuations (evolve in time, exhibit fluctuations). Such processes are defined in terms of an ensemble of trajectories, which represent the possible paths that the state variables can take over time. Each trajectory is associated with a probability, and the process as a whole is described by a hierarchy of joint probabilities and probability densities that describe the likelihood of different sets of trajectories.

- An ensemble of trajectories,
- A hierarchy of joint probabilities and probability densities

In discrete time processes, the evolution of the system is described at discrete time intervals. In discrete state processes, the system can be in one of a finite number of states. In continuous state processes, the system can be in any value within a continuous range of states. Continuous time processes involve the evolution of the system in continuous time. All of these methods can be used to model system that exhibit behavior and uncertainty with different analysis as below.

+ **Probabilities and Probability Densities:** Each process is indexed by each mathematical set and can be defined by either discrete or continuous probability, and index set. The discrete time probability involve evolution at discrete time intervals, while continuous time probability involve evolution in continuous time (Figure 2.15.2).

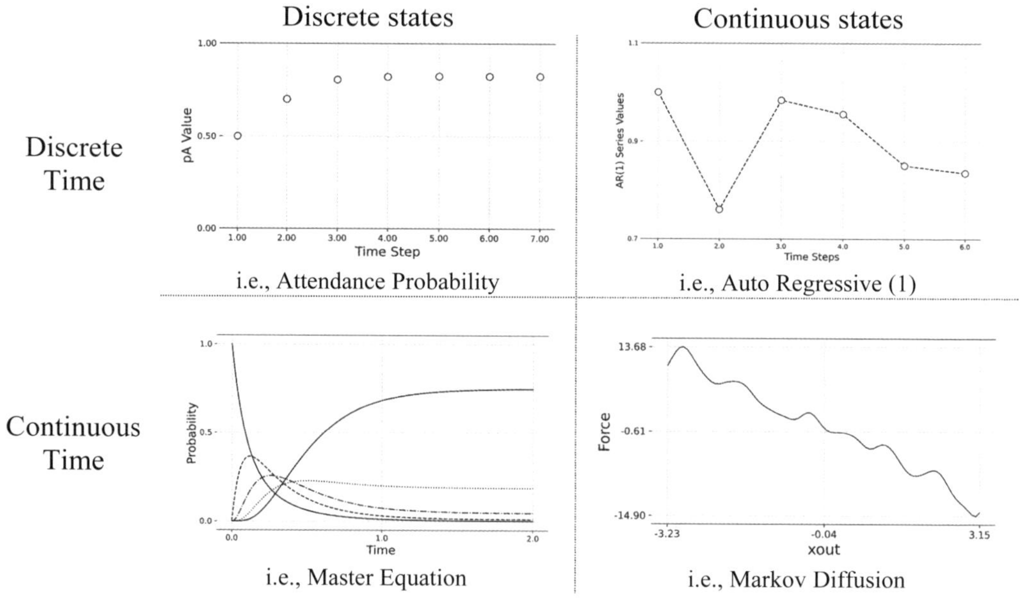

Figure 2.15.2. Schematic representation of the process

At the same time, the discrete state probability involves a finite number of sates, while continuous state probability involves a continuous range of states.

+ Stochastic Models: We provide several types of stochastic models that can be used to analyze (Table 2.15.1).

[Table 2.15.1] Summation of the Stochastic Models

	Discrete states	Continuous states
Discrete time	Monte Carlo simulations Markov chains	Stochastic iterative maps Special case of the Chapman-Kolmogorov equation (integral evolution equation for probability densities)
Continuous time	Gillespie algorithm Master equations	Langevin equations Fokker-Planck equations

In discrete time and states models, Monte Carlo simulations and Markov chains are used. In discrete time and continuous states models, Special case of the Chapman-Kolmogorov equation can be used, which is an integral evolution equation for probability. In continuous time and discreet states models, the Gillespie algorithm, and Master equations are used, while the Langevin equations and the Fokker-Planck equations can be used for the Continuous time and states models.

[Table 2.15.2] Summation of Some connections

	Discrete states	Continuous-state variable
Discrete time	Markov Chain ↑ Transition rate = transition prob./time ↓	Stochastic iterative maps ↑ Numeric Euler forward scheme ↓
Continuous time	Master equation	Langevin equations Fokker-Plank equations
	← Diffusion approximation →	

In Table 2.15.2, time series analysis expands across different domains. A useful time series analysis must be fast and involve relatively easy processes. To handle the increasing types and complexity of time series data, many techniques have been developed for particular applications. A deep understanding of the range of methods depends on many factors. Below are some general inquiries to help you to decide which one is best for your research.

First, define the purpose of the time series analysis, where the research will be used, who the stakeholders are, and how the relevant area works. These factors determine the power required for the analysis governing the selection, as well as in scope and reliability. Next, Understanding the time series type is important for choosing the analysis method (i.e., univariate, multivariate). After understanding the time series type, we can focus on the relevant analysis. Finally, to fit the best method, you need to identify the patterns in the data (time and state) and then choose a method that is able to capture the characteristics properly. Time series features can be described in terms of the systems analysis classes of components and you can find more information on decomposition methods in this book.

Reading List 2.1: Applied Time Series (Probability-Based) Research Example with Elementary Coordination

Purpose: We inquired into whether something akin to attunement to the environmental 24-day/night temperature cycles may be apparent in an experimental setting of bimanual coordination, a setting that has been used to examine self-organization in biological systems. Thus, we asked whether unintentional coordination with an environmental rhythm might increase by assessing whether the dynamics of bimanual coordination is influenced by an overarching temporal structure that is irrelevant to the task—exposing the laws that underlie end-directed capabilities by measuring biological characteristics (motor synchrony) in an environmental cycle (circadian temperature). **Methods:** Participants (n = 32) at the University of Connecticut (Storrs, USA) and Seoul National University (Seoul, Korea) served in the study. In-phase bimanual coordination was examined at different circadian points (allowing for a comparison of day/night temperature effects) in two experimental designs: (1) normal temperature embedded in a 24 h cycle (5:00, 12:00, 17:00, and 24:00), and (2) normal versus abnormal (artificially decreasing or increasing) temperature embedded at two points (5:00, 17:00) during the circadian process. **Results:** Typical bimanual stability varied significantly as a function of the day/night temperature cycle. Although circadian effects under the artificially perturbed temperature manipulation were not straightforward along the day/night temperature cycle, the circadian effect divided by the ordinary circadian seems to be constant along the day/night temperature cycle (Park, 2016).

Background: The focus of intelligence "outside the head" of the organism and environment was invoked in how J. J. Gibson (1950) considered the achievements of perceiving and acting as continuous. Men, earthworms, single-cell organisms, and even plants are aware of their surroundings and organize actions according to their circumstances (Cariani, 1993; Ford, 2008; Strong & Ray, 1975). Exhibits of autonomy and control of functions have been the goal in contemporary research to explain agency more scientifically (Shaw & Kinsella-Shaw, 1988). To understand this directed behavior as a call for the organism-environment system rather than as simply pertaining to the organism (Turvey & Carello, 2012), we seek to expose the laws that underlie intelligent capabilities. One useful strategy looks for cycles at all time scales and aims to show how interacting cyclical processes account for the emergence of new entities, many of which are similarly cyclical (Iberall, 2001).

The central idea is that the earth's cycles—the geophysical, hydrological, meteorology-cal, geochemical, and biochemical—have interacted to create self-replicating living systems abiding by particular cyclicities (Soodak & Iberall, 1978). We inquired into whether something akin to attunement to environmental 24 day/night temperature cycles (Maury et al., 2010) may be apparent in an experimental setting of bimanual coordination, a setting that has been used to examine self-organization in biological systems (Kugler & Turvey, 1987). We asked whether unintentional coordination with an environmental rhythm might increase by assessing whether the dynamics of bimanual coordination is influenced by an overarching tem-poral structure that is irrelevant to the task.

Experiment 1: Experiment 1 embedded a bimanual coordination task in an ordinary 24 h day/night cycle (5:00, 12:00, 17:00, and 24:00). Data were subjected to analysis of variance. We asked, 'Is our system influenced by an ecological feature?' A metronome was used to impose a rhythm reflecting the natural period of a pendulum system. An ordinary temperature condition served as a replication of standard in-phase bimanual coordination tasks. The participant was seated with his or her arms supported on armrests and a pendulum in each hand held firmly to prohibit within-grasp motions. Gaze was elevated to prevent the participant from viewing the pendulum oscillations, which arose from motions strictly about the wrist. Experiment 1 ($N = 8$) used a setting that minimized variability: in-phase coordination of the two pendulums (relative phase $\phi = 0°$) with no detuning (i.e., the two pendulums had the same eigenfrequency). In the experiment, bimanual coordination was performed at a metronome beat (1.21 s) without concern for amplitude or frequency.

Proposed Analysis 1: The mechanism of oscillation on two different (left pendulum and right pendulum) but nearly identical process phases was defined by the following dynamic:

$$\phi = \theta_L - \theta_R$$

Here ϕ is the phase of the strength between the left hand (θ_L) and the right hand (θ_R). The degree of relative phases (0°–180°) depends on the difference in the two oscillators. If each θ is defined as a sine function, as follows,

$$\theta_1 = \omega_1 - \frac{\alpha}{2}\sin(\theta_1 - \theta_2)$$

$$\theta_2 = \omega_2 - \frac{\alpha}{2}\sin(\theta_2 - \theta_1)$$

the logic can simply be rewritten to ϕ as the same dynamic function ($\theta_1 - \theta_2 = \phi$). Then each θ can be specified as the following equation:

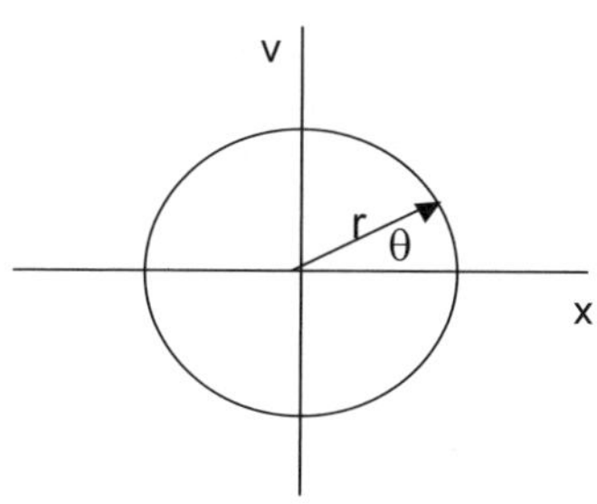

$$\left.\begin{aligned}\dot{r} &= 0\\ \dot{\theta} &= \omega\end{aligned}\right\} \rightarrow \theta(t) = \omega t + \theta(0)$$

$$\theta_1 = \omega_1 = \theta_1(t) = \omega_1(t) + \theta_1(0)$$
$$\theta_2 = \omega_2 = \theta_2(t) = \omega_2(t) + \theta_2(0)$$

$$\phi = \theta_2 - \theta_1 = \omega_2 - \omega_1 = \Delta\omega(t) + \phi(0) = \phi(t)$$

Inspired by the traditional (a)symmetric coupling manipulation between two oscillators, the main experimental variables of the core body temperature phase and the circadian temperature phase in this study were applied as the functional asymmetry expressed through the detuning term [$c\sin(\phi^{℃}) + 2d\sin(2\phi^{℃})$. Such application assumed that homologous but contralateral phase segments of both limbs would not be identical. Logically, the relativity can be denoted through multiplication by –1, and their solutions are linked to their relative values [–1(a) = (b)]. These simulations with the observations reflect the fact that the fixed-point shift can be calculated using the deviation of the average relative phase from the intended phase ($\phi_{ave} - \phi_0$), and the variability can be calculated as the standard deviation of the relative phase ($SD\phi$). This provides us with a basic tool for measuring the system's ability to remain stable in certain circumstances.

Results 1: A variety of measures (e.g., phase shift, variability, entropy) were examined for evidence of entrainment or any influence of the embedding rhythm on stability or attractor location. Absolute differences in parameters (fixed point shift, variability as a function of frequency competition, and entropy production) were found, especially at the circadian 17:00 point.

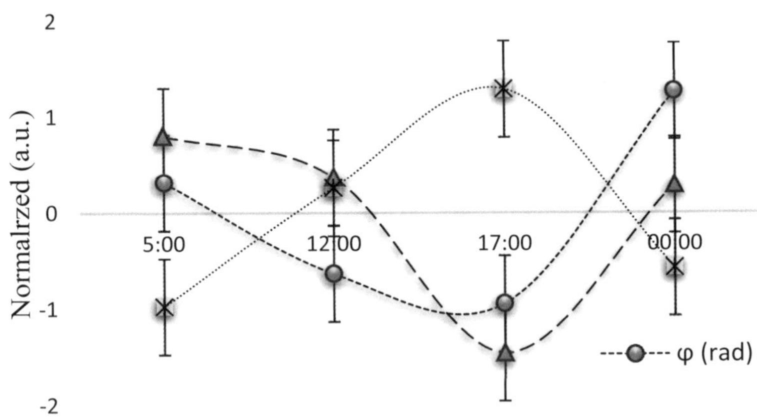

Figure R2.1.1. General tendencies in the normal condition. Normalized = standard score (*Z* calculation), a.u. = arbitrary unit, $\phi_{ave} - \phi_0$ = fixed point shift, $SD\phi$ = variability as a function of the frequency competition, Temp = temperature (Celsius)

In Figure R2.1.1, the parameters changed in the opposite direction from the core body temperature. Whereas the core body temperature rhythm showed a minimum at 5:00 h but had a maximum at about 17:00 h, behavioral performance (Shannon's entropy) showed a maximum at 5:00 h but had a more clearly defined minimum at about 17:00 h in the day/night cycle.

Experiments 2 and 3: Experiments 2 and 3 asked, 'How does our system adapt to regular or irregular thermal structures?' Normal and abnormal day/night circadian temperature effects were compared at dawn (5:00, approximately when core temperature reaches its minimum) and dusk (17:00, approximately when core temperature reaches its maximum; Aschoff, 1983). In-phase coordination without detuning was performed at dawn and at dusk. A metronome was used to impose a rhythm reflecting the natural period of the pendulum system. In addition, a short-term, thermodynamic manipulation was introduced.

Prior to the session midpoint, some participants (N = 8) donned a heat vest for 30 min that increased their core temperature (within 0.5°, SD = 0.2°; Experiment 2), and other participants donned an ice vest for 30 min that reduced their core temperature (within 0.4°, SD = 0.15°; Experiment 3). The expectation was that the influence of the thermal manipulation on coordination would interact with time of day.

Proposed Analysis 2: After establishing some concepts of the different biological factors in terms of the environmental temperature context, we extended the examination to which these features interact with more complex dynamics given a Markov chain process. This helped us to understand how the magnitude of the observation at the point (X_{t+1}) is related to the magnitude of the observation at the preceding point (X_t): such a relationship may be expressed in the general form

$$X_{t+1} = F(X_t)$$

In the equation, $F(X_t)$ is typically referred to as a nonlinear function when the function of F corresponding to an input X's output $F(X_t)$ is not a straight line. However, the investigation has to consider that the spreading of many symptoms is not a continuous variable. It may be of value to switch from continuous time t and a continuous change in mode at time n_{t+1} depending on mode at time n_t via a matrix of switching probabilities (Figure R2.1.2).

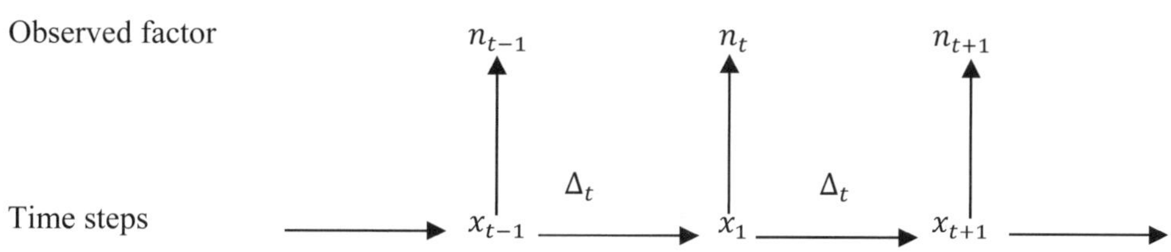

Figure R2.1.2. General process of the stochastic model. n_t is the obtained factor, but x_1 is the unobserved state. The arrows imply the process of the state through time steps.

$$n_{t+1} = n_t + \Delta_t, \qquad \Delta_n = n_{t+1} - n_t$$

Here Δ_n denotes whether the interaction of category ($i = n_{t+1}$) follows the interaction of category ($j = n_t$). This logic considers variables that are defined by a number of events that occur in a given time interval (discrete model = delta t). It can be argued that it makes sense to treat this variable as a discrete one, which may be conceptualized as a Markov process (time continuous, discrete states). As it is, equation can be written as a process of the states depending on the previous states in a non-deterministic manner.

$$\mathrm{H}(\chi) = -\{p_1 \log p_1 + p_2 \log p_2 + \cdots + p_M \log p_M\} = -\sum_{j=1}^{M} p_j \log p_j$$

This simply inverts the expression inside the logarithm while adding a negative sign from equation

$$\sum_{i=1}^{n} p_i \times log_2\left(\frac{1}{p_i}\right)$$

to equation

$$-\sum_{i=1}^{n} p_j \times log_2(p_j)$$

where p_1 has a probability of 1 and p_j is a state given as 1 from the previous character. To estimate this uncertainty from the main experimental dynamics of the relative phase X(t) between two in-phase oscillating limbs, use the following entropy calculations:

$$\text{Entropy} = -\{f_1 \log f_1 + f_2 \log f_2 + \cdots + f_M \log f_M\} = -\sum_{j=1}^{M} f_j \log f_j$$

where f_1 is the relative phase of 1 and f_j is a state given 1 as the previous character. This process specifies how the data are related to the states in the process, considering the estimated entropic force with respect to time series that this application wants to observe with the differential equation below:

$$\frac{d}{dt}H(x) = -\sum_{j=1}^{M} f_j \log f_j + \sqrt{\varrho\xi_t}$$

where $\frac{d}{dt}H(x)$ is the change of uncertainty phase with time determined by the stochastic noise term $\sqrt{\varrho\xi_t}$, a source of fluctuations in H arising from the components' microscopic subsystems. This equation indicates that the common means of defining entropy is based on a Markov model (a stochastic model that assumes that the model is a process whose states depend on previous states in a non-deterministic manner).

Due to this quantitative measure of information, nominal stability is empirically estimable when the form of the possibility distribution of the population is unknown. This provides us with information on the quantitative dynamic energy from the measure of uncertainty disorder, or the system's characteristic in a random variable when we gain information from the changes in variable state over continuous time.

Results 2: Entropy was estimated from the dynamics of the relative phase between the two limbs oscillating at the wrists. Entropy was affected by the temporal locus during the circadian cycle as well as by the introduction of the heat vest and the ice vest; the influence of the thermal manipulation was not identical. Even when the same external temperature perturbations were given, the influence of the vest was negatively exaggerated (increasing entropy) at dawn, but the influence of the vest was positively exaggerated (decreasing entropy) in the evening.

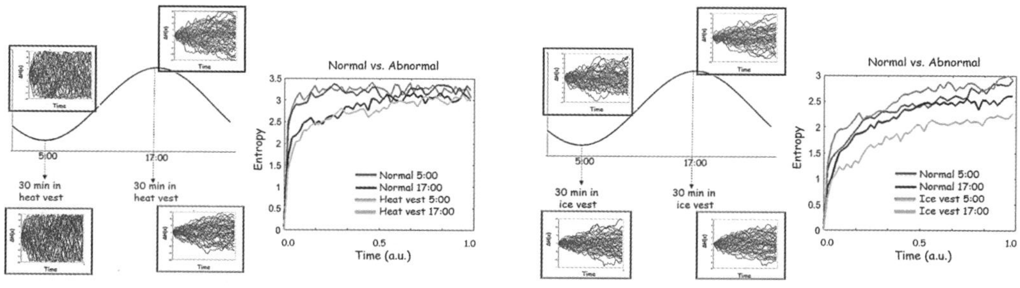

Figure R2.1.3. Entropy maximizations for all uncertainty characteristics. Each set of plots shows the estimated entropy forces according to the time series for each experimental design: The figures of the left set denote the estimated entropy states according to the time series between normal (5:00 and 17:00) vs abnormal (5:00 and 17:00). N = normal, Ab = abnormal in terms of heat-based experimental design. The figures the right set denote the estimated entropy states according to the time series between normal (5:00 and 17:00) vs abnormal (5:00 and 17:00). N = normal, Ab = abnormal in terms of ice-based experimental design.

From the dynamics of the relative phase between the two limbs, we can see that oscillatory bimanual coordination was affected by the temporal locus during the circadian cycle and by the introduction of the both ice or heat vest. The influence of the vest was negatively exaggerated (increasing uncertainty) at dawn and positively exaggerated (decreasing uncertainty) in the evening. The results (Figure R2.1.3) from this biological scale correspond to a theoretical suggestion, that the rate of entropy production varies when a new energy source is accessed via a nonequilibrium phase transition process (Frank, 2011). This reflects the fact that access differs as a function of the circadian cycle and that access can be manipulated by a temporary thermal manipulation.

Reading List 2.2: Applied Time Series (Stochastic) Research Example with Modality Dominance

Purpose: This study measured extrasensory haptic accuracy, that is, judging the position at which a hand-held object is hit by another object. The primary focus was which factors are associated with estimations of the position where contact is made with the grasped implement. **Methods:** Data were collected from 20 participants whose haptic accuracy was analyzed using a discrete numerical state (relative probability densities) as well as the stochastic evolutionary possibility (Monte Carlo simulation based on a Markov chain model). **Results:** The analyses proved that perceived accuracy influenced not only the stimulus magnitudes distinguished by the coefficient of restitution but also the distributions of the encoded impressions by the rotational inertia. In particular, stochastic analysis confirmed that perceptions of the location of the effect of grasped objects are more constrained by kinesthetic properties than by cutaneous-oriented gain in arbitrarily conflicted circumstances (Park, 2019).

Background: The present study aimed to determine which element in a conflicted situation plays the main role. To assess the factors associated with the assumptions, relative probability and the evolution of probability were used, which is an ideal tool for estimating (or likeliness) continuous (or discrete) variables based on the following questions: (i) Is extra perception dynamics (judging the hit position of a rod) affected by the magnitude of a stimulus? (ii) Is extra perception dynamics affected by the distribution of a stimulus? The expectation should be that if these mechanical factors are captured, it could be suggested how the different components contribute to the effects of haptic accuracy in the hand-held object. The results will lead to various estimates of system functions, providing an account of generalization that accommodates a variety of aspects.

Methods: The data used in this experiment were collected from undergraduate students (N = 20, 10 females, 10 male). Their mean age was 20.5 years old (range: 18–23). Participants were enrolled in an introductory course and received credit for their participation. All participants had normal mobility in their right arm. Individuals with self-reported acute or chronic physical and psychiatric disorders (left-handedness: n = 2, sprained ankle: n = 1) were excluded. All participants provided written informed consent. The study was approved by the relevant local ethics committee (SNUIRB No. 1509/002-002) and conformed to the ethical standards of the 1964 Declaration of Helsinki (Collaborative Institutional Training Initiative Program, Report ID No. 20481572).

Analysis of Relative Probability Densities: Mean value and variance are measures of central tendency and variability for continuous (discrete) variables defined on a real line. They may also be applied to variables under certain circumstances. In general, statistics should be applied to variables to determine the center and variability of probability densities and samples. Among several available measures, this analysis focused mainly on the deterministic system, which is described by a discrete numerical state variable (Feindt, 2004). This method explains whether there is a consistent difference between the data. If so, a scientific guess can be made as to which data set belongs to which significant effect in the experimental conditions. First each accuracy was defined as an absolute error (AE $= \sum |\chi_i - \mathrm{T}| / n$) between the actual contact position and the perceived/designated position, then these sizes were converted to a discrete-state unit (1 = distance within 0–1 cm, 2 = distance from 1 to 2 cm). This allows us to see the proportion (percentage) of occurrences of each variable in that outcome in the statistical ensemble as follows:

$$h_j = \sum_{K=1.....N;X^{(k)}=j} 1, \quad \rightarrow \quad f_j = h_j/N$$

Here h_j is the number of occurrences of a repeating event—usually called *frequency*—during the period of time. In statistics, this occurrence of an event (j) is the number (k) of times the event takes place in an experiment. In contrast, relative frequency (f_j) is reflected by how often something happens divided by all outcomes. This event refers to the absolute frequency (h_j) normalized by the total number of events (N). The relative frequency, also known as empirical (experimental) probability, is represented as follows:

$$P_j = \lim_{n\to\infty} \frac{1}{N} \sum_{K=1.....N;X^{(k)}=j} 1$$

P_j is the ratio of the number of outcomes ($\sum_{K=1.....N;X^{(k)}=j} 1$) in which a specified event occurs to the total number of trials ($\lim_{n\to\infty} \frac{1}{N}$) in an actual experiment. This is a measure of the likeliness that an event will occur, and the certainty we adopt can be described in terms of a numerical measure between impossibility (0) and certainty (1). Thus, the higher the probability of an event, the more certain we are that the event will occur, as in the following short form:

$$P_j = \lim_{n\to\infty} f_j$$

Next, to estimate how many realizations per measurement unit occur on a continuous level, a probability function based on the construction of density for these data sets can be used, as follows:

$$P(\chi) = \lim_{\Delta\chi\to 0} \frac{\mathrm{P}(\chi, \Delta\chi)}{\Delta x} = \frac{d}{dx} F(\chi) = \frac{d}{dx} P(\mathrm{X} < x)$$

Analysis of the Evolution of Probability: We are always faced with uncertainty, ambiguity, and variability. Although we now have unprecedented access to information, being able to accurately predict the future is still almost impossible. The Monte Carlo simulation (based on the Markov chain model; Ethier & Kurtz, 2009) allows us to see all possible outcomes and assess the possible impacts of outcomes that may occur in the future using information not only from the current state but also from the sequence of events that preceded it. To address this likelihood, the haptic accuracy state variable was continuously placed as a probability density and variability according to the discrete time series. The relative ratio of the differences between all designated conditions and the ratio of each condition were calculated using joint probability based on the Markov chain model as follows:

$$P_j(n+1) = \sum_{k=1}^{M} P(j \leftarrow k) P_k(n)$$

P is the probability that the system is in state (k) at time step (n) and in state (j) at the next time step. The assumption for probability cannot express how to create probabilities for combined events such as $P[A \cap B]$ or for the likelihood of an event A, given that it is known that event B occurs. For instance, let A be the state of distribution effect accuracy (within ± 1 SD) at one condition and B be the state of non-distribution effect accuracy for the other condition. Does knowledge of haptic accuracy at the one condition change the belief that it will still be accurate for the other condition? That is, $P[B]$ is the probability that the accuracy of the other condition is ignoring information on whether it is accurate at the condition. This differs from $P[B|A]$, the probability that it is accurate at the other condition given that it is accurate at the time step (called the *conditional probability* of B given A). It is quite likely that $P[B]$ and $P[B|A]$ are different.

This probability density and the probabilities of the two variables can be calculated by generating data (X_1) and (X_2) for a participant. Using this information, the time evolution of a system can be modeled to occupy each of a countable number (discrete set) of states about a continuous time step, where switching between states is treated probabilistically. In this case, the master that describes the evolution of the probability $p(j,t)$, where t is time, is defined by

$$\frac{d}{dt} p(j,t) = \text{'rate in'} - \text{'rate out'} = \sum_{k \neq j} w(j \leftarrow k) p(k,t) - p(j,t) \sum_{k \neq j} w(k \leftarrow j)$$

Here $w(j \leftarrow k)$ denotes the transition rate from k to j. This modeling will allow the progression of haptic accuracy dependency (progression of performance probability) as an application. Intuitively, we can make three qualitative statements about this evolutionary probability $\frac{d}{dt} p(j,t)$. First, the probability $p(j,t)$ increases monotonically as a function of time, but whether it increases or decreases depends on the other probabilities, that is, $p(1,t), p(2,t)$ ... $p(n,t)$. Second, the probability $p(j,t)$ increases monotonically as a function of time or does not change with time. That is, $p(j,t)$ does not decay as a function of time, and this is irrespective of the values of the other probabilities. Third, the probability $p(j,t)$ decreases monotonically as a function of time or does not change with time. That is, it does not increase as a function of time, and this is irrespective of the values of the other probabilities. This observation, then, will represent that the dominant state between conditions which influence would be more robust.

Results for Relative Probability: Many inputs evoke a complicated system of orienting reflexes in an organism. These orienting reactions are of course subject to their intensity and roughly correspond to those of the inputs. This gradually changes when these actions have been repeated frequently. It is probable that, the higher the relative frequency of an event, the more certain we are that the event will occur. To measure this, a probability function based on the construction of density for these data sets was calculated.

[Table R2.2.1] Comparative Estimation of the Parameters between Heights and Distances

Condition	Error size				Error variance			
	Height (from the rod)		Distance (from the grasp)		Height (from the rod)		Distance (from the grasp)	
	AE	P(x)	AE	P(x)	AE	P(x)	AE	P(x)
1 (10 cm)	4.74	0.43	4.74	0.44	0.67	0.42	0.67	0.4
2 (20 cm)	3.28	0.31	3.58	0.33	0.42	0.26	0.77	0.45
3 (30 cm)	2.85	0.26	2.51	0.23	0.52	0.32	0.26	0.15

Note: AE = average absolute error (cm), P(x) = probability. Notice the proportion (percentage) of occurrences of each outcome in the statistical ensemble.

These results show the deviation of the performance parameter for the interaction between error (marks) and variance (shade) of the error from the different heights, $F(2, 71) = 26.951$, $\eta^2 = 13.475$ ($p < 0.001$), Pearson correlation $R = -.61$. In addition, these results also reflect the deviation of the performance parameter for the interaction between error (marks) and variance (shade) of the error from the different distances, $F(2, 71) =$ 28,865, $\eta^2 = 14.433$ ($p < 0.001$), Pearson correlation $R = -.63$. The analyses show that observers influenced not only the stimulus magnitudes distinguished by the coefficient of restitution but also the distributions of the encoded impressions by the distance from the hand to the impact (Table R2.2.1). These results were further investigated to address matters such as the conflicted circumstances between the two modalities. Specifically, the participants' judgment of the contact point can also be derived from the mechanical properties of the distribution rather than the magnitude of the stimulus.

[Table R2.2.2] Calculating Conditional Probability Height 1+Distance 3, Height 3+Distance 1

$p(e \mid l)$	$p(x_1 \in e\_s)$	$p(e\_s \times x_1 \in e\_s)/(e\_s \times x_1 \in e\_s + l\_s \times x_1 \in l\_s)$	Conditional p (%)
$e\_s = 0.5$	$p(3 \in e\_s) = 0.26$	$(0.5 \times 0.26)/(0.13 + 0.115)$	0.531 (53)
$l\_s = 0.5$	$p(3 \in l\_s) = 0.23$	$(0.5 \times 0.23)/(0.115 + 0.13)$	0.469 (47)

Note: Error size parameter based (s), e = coefficient of restitution, l = pressure distribution.

$p(e \mid l)$	$p(x_1 \in e\_v)$	$p(e\_v \times x_1 \in e\_v)/p(e\_v \times x_1 \in e\_v + l\_v \times x_1 \in l\_v)$	Conditional p (%)
$e\_v = 0.48$	$p(3 \in e\_v) = 0.32$	$(0.486 \times 0.32)/(0.155 + 0.077)$	0.669 (67)
$l\_v = 0.51$	$p(3 \in l\_v) = 0.15$	$(0.514 \times 0.15)/(0.077 + 0.155)$	0.331 (33)

Note: Error variance parameter based (v), e = coefficient of restitution, l = pressure distribution.

Table R2.2.2 shows which parameter plays a more significant role in the arbitrarily estimated comparisons of the two different combinations. Both densities (error size estimation, error variance estimation) illustrate that the combination height 10 cm and distance 30 cm resulted in better accuracy with less variance than the combination height 30 cm and distance 10 cm. The mechanical foundation of the stimulus distribution (conditioned by rotational inertia as more kinesthetic-oriented gain = relatively many scores at the extreme ends) makes its influence more significant than its impulse magnitude (conditioned by the coefficient of restitution as more cutaneous-oriented gain = relatively few scores at the extreme ends) for accuracy in arbitrarily conflicted circumstances between the two modalities. Likewise, the most critical characteristic of the hand-held object's perceived size-weight ratio was not the object's weight but rather the rotational inertia in the conflicted circumstance (Charpentier, 1891). Localized discrete contact points in the grasped implement can also be derived from the same mechanical property of the distribution rather than from the magnitude of the stimulus.

Results for the evolution of probability: In the discrete time series of a Markov process based on the model, the probability of a particular state in the future behavior depends only on the current state and not on the past. Because we can iterate a Markov process from an initial state numerically, we might be observing an estimation of parameters. By repeating the steps in previous procedures and making panels for *P*(distribution, not distribution), we can state whether we get the impression that the graphs do not change much when we look at different sample sizes. If this does not change much, then we can say that we have good approximations for the *N* to infinity limiting case. These estimates are maximum likelihood estimators for those parameters where the likelihood of observing a given data set is maximal.

[Table R2.2.3] Comparative Estimation of the Parameters (Heights and Distances)

	Error size		Error variance	
Condition	Height $P(x_j) = H$	Distance $P(x_j) = D$	Height $P(x_j) = H$	Distance $P(x_j) = D$
1 (10 cm)	0.22	0.22	0.21	0.21
2 (20 cm)	0.16	0.17	0.13	0.23
3 (30 cm)	0.13	0.11	0.16	0.07

Note: The proportion (percentage) of occurrences of the outcome in the statistical ensemble is given using the vector matrix and component notation. Proving for the error variance (30 cm) cases: P(x = H) = 49% P(30cm|H) = 49% = 0.32 = P(30cm|H) × P(H) = 0.32 × 0.49 = 0.16, P(x = D) = 51% P(30cm|D) = 51% = 0.15 = P(30cm|D) × P(D) = 0.15 × 0.51 = 0.07. Component notation: $P_j(n+1) = \sum_{k\in a,b}^{M} P(j \leftarrow k)P_k(n),\ j \in [a,b]$. $P(x_j)$ = probability Markov Chain model.

Table R2.2.3 shows that especially the error variance becomes more significant based on conditional probability, $F(1, 15) = 17{,}030$, $\eta^2 = .549$ ($p < 0.001$). As it is, the time evolution of a system was analyzed to occupy each of a countable number (discrete set) of states about a continuous time step, where switching between states is treated probabilistically.

[Table R2.2.4] Calculation of Transition Probability

E_S	Height: $\frac{d}{dt}p(j,t) = .008$	Rate in =	$(.06 \times .16) + (.06 \times .09)$	= .015
		Rate out =	$(-.06 \times 0.22) + (.09 \times .22)$	= .006
	Distance: $\frac{d}{dt}p(j,t) = .007$	Rate in =	$(.05 \times .17) + (.11 \times .11)$	= .021
		Rate out =	$(-.05 \times .22) + (.11 \times .22)$	= .132
E_V	Height: $\frac{d}{dt}p(j,t) = .024$	Rate in =	$(.08 \times .13) + (.05 \times .16)$	= .018
		Rate out =	$(-.08 \times .21) + (.05 \times .21)$	= –.006
	Distance: $\frac{d}{dt}p(j,t) = .03$	Rate in =	$(-.02 \times .23) + (.14 \times .07)$	= .005
		Rate out =	$(.02 \times .21) + (-.14 \times .21)$	= –.025

Note: Model for the analysis. Calculated rate in = w(1←2)P(2,t) + w(1←3)P(3,t), and calculated rate out = w(2←1)P(1,t) + w(3←1)P(1,t). E_S = error size parameter based. E_V = error variance parameter based. The model exhibits only three transitions, where t is time and j is the discrete state defined by (1 = 10 cm, 2 = 30 cm, 3 = 50 cm). H = conditional height for the coefficient of restitution, D = conditional distance for the pressure distribution. The left table denotes modeling for progression on the basis of conditional probability {Ω [P(x1,x2)] = 1, 2, 3}.

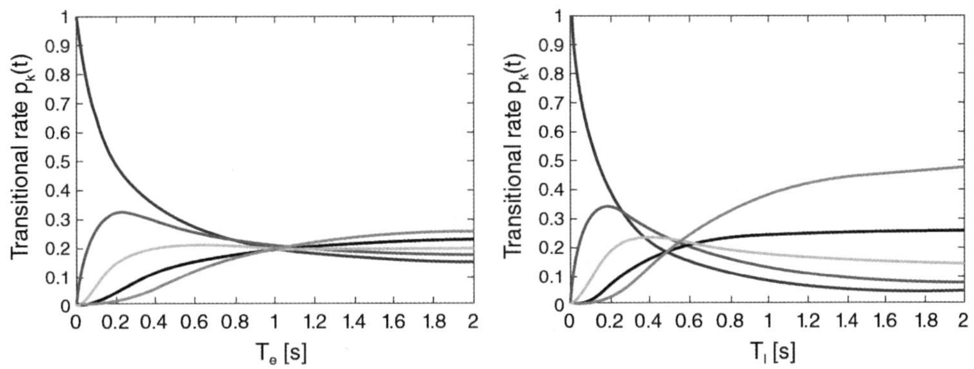

Figure R2.2.1. The transitional rate (probability) of the continuous-time process: (left) the transitional rates of states with height [(e): $\frac{d}{dt}p(j,t)$=.008] according to the continuous-time process (S = state, T = time step) calculated by the master equation function, (right) the transitional rates of states with distance [(I): $\frac{d}{dt}p(j,t)$=.03] according to the continuous-time process (S = state, T = time step) calculated by the master equation function. Both images reflect how much the final states differed from their initial states in each condition. The probability-weighted function $P(x_1,x_2)$ was used as the initial value for the simulation as follows: three height levels: level 1 = 0.21, level 2 = 0.13, level 3 = 0.15; three distance levels: level 1 = 0.21, level 2 = 0.23, level 3 = 0.07.

This additional analysis shows which influence on the conditions is most robust. The results, in conjunction with a prior analysis on relative probability, suggest that the standard property (rotational inertia) of the hand-held object is the crucial feature in perceived position accuracy as well (see Table R2.2.4). Note that both mechanical sources of information could be perceptually independent and combine additively, although it could also be the case that one source or the other dominates (Figure R2.2.1).

Part 3.

Applied Systems Analysis

Contents

Part 3. Applied Systems Analysis

[3.1]

Fundamentals of Systems Analysis

Overview of Systems Analysis, Vectors and Scalars, Vector Operation

Systems analysis is fundamental to almost all aspects of life (Schwill, 1994). It is a method of identifying and solving problems that looks at each component in the overall system for the purpose of achieving specific goals. This approach can inform the decision-making process or enhance efficient use of time, property, and resources in business and in life. It includes defining process steps toward an end result.

In biology, there is the act of obtaining and extrapolating information on biological samples. This is derived from statistics, which is a form of mathematical analysis. Systems analysis attempts to make sense of the (statistics-based) information we acquire from and about our surroundings to model behavior and information into concrete visualizations and understandable images (Steel, 2007). In psychology, systems analysis manifests in the form of mathematical psychology, in which human behavior is seen through the lens of formulas and laws (Downs & Stea, 2017). Even a quick Google search pulls up scientific journals on theology and mathematics. In physics we see the actual physical embodiment of systems analysis in the way we fall to the ground or in how the movement of heavenly bodies can be predicted from simple formulas (Skyttner, 2005).

1. Overview of Systems Analysis

We are continuously reminded of the importance of systems analysis by life itself. We enjoyed systems analysis in our school days but enjoyed it slightly less as postgraduates, because somehow systems presented themselves in every corner of anything we studied.

There was always some representable idea in any topic we encountered. All of life seems to be interconnected through systems, which in some ways explains why we can see systems in our surroundings (Johnson, 2002). Although we were fascinated with all the numbers and figures in research, we often struggled with analysis: At the last minute we would pull a solution out of nowhere to solve a problem, or we would forget how to derive a certain analysis. We often made arithmetic mistakes and would rely on our lecture notes to let it pass by and consider the methods we would use. Remembering all the concepts and stochastics was not easy for us. We knew that our systems journey would not just end with calculation, that it would continue back in our college department. We would soon be introduced to the idea of taking advantage of the systems analysis around us and using computers to solve life's problems. That programs would be able to assist us with analysis was a plus.

+ **Unveiling the Depths of Applied Systems Analysis:** At the heart of this dynamic field lies an adept use of a plethora of mathematical tools. These tools not only enable us to unravel the underlying principles of modern advanced systemic methods but also seamlessly integrate theoretical concepts with practical applications. Our journey transcends a mere introduction to systems analysis; it is an endeavor to cultivate a profound understanding of its broad spectrum of influences and implications, including the crucial process of system dynamics modeling. This involves defining problems, establishing dynamic hypotheses based on causal loops, constructing and testing models, and evaluating these models in terms of policy-making and real-world impact. In this era, characterized by an unprecedented surge in data availability and computational power, systems analysis stands as a beacon, illuminating paths previously unseen. It offers a platform that converts the simplicity of raw data into complex structures through advanced modeling techniques. These models, mirroring real-world systems, provide an interactive playground to experiment, adjust, and analyze system behavior.

As we delve deeper, you will discover a multitude of applications born at the crossroads of systems analysis, data, and models (Figure 3.1.1).

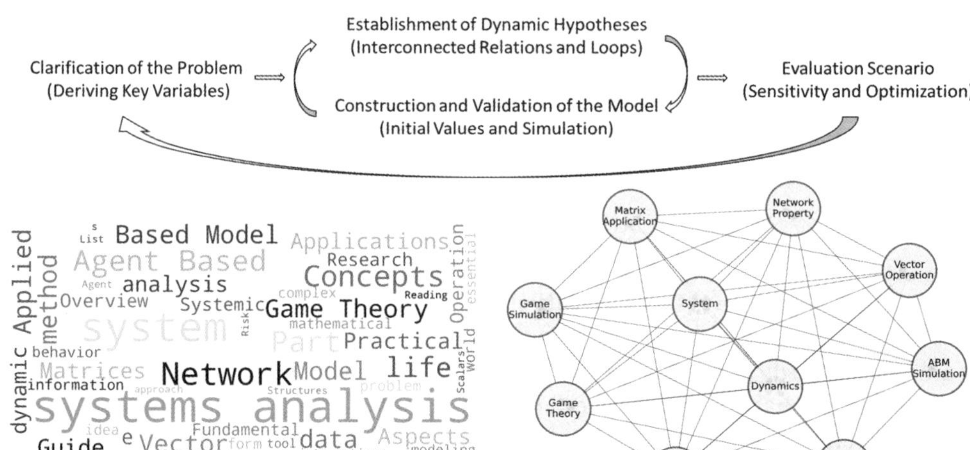

Figure 3.1.1. Schematic representation of the core themes and concepts. The upper section depicts the modeling procedures, outlining the systematic approach used in systems analysis. The bottom left features a word cloud, visually representing the most prevalent and significant terms associated with systems analysis. The bottom right illustrates the interconnected relationships within systems analysis, with a special focus on system dynamics modeling, highlighting how different elements within the system interact and influence each other.

+ **Mastering Essentials Concepts:** Our journey begins in the multivariate vector universe, vital for capturing magnitude and direction in systems, and moves on to matrices, the essential pieces for handling multidimensional data. We then explore the complex realm of networks to understand the interdependent interactions in complex systems. The intriguing world of agent-based modeling follows, allowing a bottom-up examination of system dynamics through the interactions of individual agents. This method enables us to simulate complex systems, test our hypotheses, and deepen our understanding of their dynamics. Lastly, we delve into game theory, providing insights into systems where outcomes depend on both individual and collective choices. As we traverse these interconnected realms, this section will offer a comprehensive overview of applied systems analysis, including the vital stages of system dynamics modeling. It serves as a fundamental roadmap for your exploration into the captivating world of systems analysis.

2. Vectors and Scalars

Linear algebra, akin to high school algebra, operates not just on conventional single numbers, but expands its domain to vectors. This progression into vector calculus permits traditional algebraic operations, including addition, subtraction, and multiplication, to be generalized for vectors.

Vectors and scalars form the bedrock of system analysis because of their inherent properties. Vectors carry both magnitude and direction, making them essential for understanding multidimensional systems. On the other hand, scalars offer a simplified perspective, aiding in breaking down complex system components. Together, their interaction provides an intricate insight into system behavior and functionality, solidifying their importance in system analysis. This section gives an overview of the operations (a couple of basic examples are given). Let us start by defining what we mean by 'scalars' and 'vectors.'

+ **Identifications:** A scalar is a number (Bernstein, 2018). Examples of scalars are temperature, distance, speed, and mass—all quantities that have a magnitude but no direction, other than perhaps positive or negative. You are familiar with scalars. In fact, we could go so far as to describe the algebra you learned in school as scalar algebra, or the calculus many of us learned in high school as scalar calculus, because both deals almost exclusively with scalars. This is to be contrasted with vector calculus or vector algebra, which most of us only got in college if at all. However, what is a vector?

A vector is a list of numbers (Diestel & Faires, 1974). There are (at least) two ways to interpret what this list of numbers means. One way is to think of the vector as being a point in space. Then this list of numbers is a way of identifying that point in space, where each number represents the vector's component of that dimension. Another way to think of a vector is as a magnitude and a direction (e.g., a quantity-like velocity: the fighter jet's velocity is 250 mph north-by-northwest). In this way of thinking, a vector is a directed arrow pointing from the origin to the end point given by the list of numbers. Usually, we denote vectors by including a small arrow above the symbol, like this: $\vec{a}$.

+ **Structures:** Scalars are single numbers. They can be represented by an italic lowercase letter. It does not matter whether they are integers or doubles—they are all scalars.

$$n \in N, \ \ r \in R$$

n is an element of the natural numbers (*N*), and *r* is an element of the real numbers (*R*). Vectors are arrays of numbers. Mathematically speaking, they can be thought of as directions from an origin or as points in space. The individual elements of a vector are scalars. The space of the vector is denoted as $\mathbb{R}^m$. This vector is made up of real numbers, and there is *m* of them. Also, sets are important in vectors. Here is the proper notation for understanding vectors with sets:

$$\vec{v}_s = \{1, 2, 4\}$$

One thing that is important here is that in a set order does not matter, whereas in a vector it does. Note that the quantities of scalars are magnitude (temperature, speed) while those of vectors are magnitude + direction (force, velocity) according to quantity as follows:

$$\vec{v}_i = \|v_i\| * \vec{d}_i, \qquad \|v_i\| = \sqrt{v_{i_x}^2 + v_{i_y}^2}, \qquad \vec{d}_i = \frac{\left(v_{i_x}, v_{i_y}\right)}{\sqrt{v_{i_x}^2 + v_{i_y}^2}}$$

where the $\vec{v}_i$ is the vector [i.e., individual's (i) velocity ($\vec{v}$)] represented by the length of the magnitude ($\|v_i\|$) with the direction ($\vec{d}_i$).

3. Vector Operation

Now let us dive into the world of vectors in more detail. We will look at operations that can be done with vectors and scalars. One thing to pay basic attention to is what types of objects (vector or scalar) each operation applies to and what type of object each operation produces. The first key operation is scalar multiplication (multiplying a scalar and a vector). If k is a scalar and u is a vector, their product can be defined as ku. The second key operation is vector addition (adding one vector to another). If we have two vectors u and v in R^2, it can be defined as $u + v$.

- Scalar multiplication (ku), Vector addition ($u + v$)

The third key operation is vectors in R^n. The set of all *n*-tuples of real numbers, denoted R^n, is called the n-*space*. A particular *n*-tuple in R^n is called a *point* or *vector*.

- $R^n = u = (a_1, a_2, \dots, a_n)$

Sometimes a vector in *n*-space R^n is written vertically rather than horizontally. A column vector is a type of matrix with only one column (it is also called a column matrix).

- Row vector = [1, 2], column vector $\begin{bmatrix}1\\2\end{bmatrix}, \begin{bmatrix}3\\4\end{bmatrix}, \begin{bmatrix}1.5\\2/3\\-15\end{bmatrix}$

This is the simplest column vector with 1 column and 1 row. The first element of the matrix is a_1, the second element is a_2, and so on until the las element, a_n.

See some column vector examples below:

$$u = (2, 4, -5)\ and\ \ v = (1, -6, 9)$$

$$u + v = (2 + 1, 4 + (-6), -5 + 9) = (3, -2, 4)$$

$$7u = (7(2), 7(4), 7(-5)) = (14, 28, -35)$$

With column vector

$$u = \begin{bmatrix}2\\3\\-4\end{bmatrix}, v = \begin{bmatrix}3\\-1\\-2\end{bmatrix}$$

$$2u - 3v = \begin{bmatrix}2*2\\2*3\\2*(-4)\end{bmatrix} + \begin{bmatrix}(-3)*3\\(-3)*-1\\(-3)*-2\end{bmatrix} = \begin{bmatrix}4\\6\\-8\end{bmatrix} + \begin{bmatrix}-9\\3\\6\end{bmatrix} = \begin{bmatrix}-5\\9\\-2\end{bmatrix}$$

Here, the given vector operation involves performing arithmetic on two vectors, u and v. The operation begins by multiplying vector u by 2 and vector v by $(-3$, and then adding the resulting vectors together. The resulting vector has components $(-5, 9, -2)$.

Theorem

Basic properties: This the standard properties of vector addition and scalar multiplication is a vector space. These are known as the axioms of a vector space and they define how vectors and scalars behave in the space.

- Vectors: $u, v, w \in R^n$, Scalars: $k, k' \in R$

In a vector space, especially in R^n, the accompanying equations describe certain characteristics of vectors and scalars (the n-dimensional real space).

(i)	$(u + v) + w = u + (v + w)$	(v)	$k(u + v) = ku + kv$
(ii)	$u + 0 = u$	(vi)	$(k + k')u = ku + k'u$
(iii)	$u + (-u) = 0$	(vii)	$(kk')u = k(k'u)$
(iv)	$u + v = v + u$	(viii)	$1u = u$

(i) The associative feature of vector addition states that adding vectors in any particular sequence has no effect on the outcome and that $(u+v)+w=u+(v+w)$.

(ii) The identity condition of vector addition states that any vector will result in the original vector when $u+0=u$.

(iii) The inverse property of vector addition is $u+(-u)=0$, which states that for any vector u, there is a corresponding vector $-u$ such that combining the two vectors yields the zero vector.

(iv) The property of vector addition known as $u+v=v+u$ ensures that the outcome is unaffected by the order in which two vectors are added.

(v) The property of $k(u+v)=ku+kv$, states that there are two possible ways to multiply a scalar with the sum of two vectors: either by multiplying the scalar with each vector separately and adding the results, or by adding the two vectors first and multiplying the result with the scalar.

(vi) The property of scalar addition over vector multiplication $(k+k')u=ku+k'u$, states that there are two ways to multiply the sum of two scalars with a vector: either by multiplying each scalar separately with the vector and then adding the two results, or by adding the two scalars first and then multiplying the result with the vector.

(vii) The associative feature $(kk')u=k(k'u)$, means that the outcome is unaffected by the order in which two scalars are multiplied.

(viii) When a vector is multiplied by the scalar 1, the output is the original vector, which is known as the identity property of scalar multiplication $1u=u$.

Note: Spam is the collection of all possible linear coordination of given vectors and scalars (i.e., all possible vectors you get by scaling each of them in some way and adding them together).

+ **Dot (Inner) Product:** Consider arbitrary vectors u and v in R^n, say, $u=(a_1,a_2,\dots,a_n)$ and $v=(b_1,b_2,\dots,b_n)$. The dot product or inner product or scalar product u and v is denoted and defined by

$$u*v=a_1b_1+a_2b_2+\cdots+a_nb_n$$

The dot product is an operation on vectors that enables us to easily find the angle between two vectors. That is, $u*v$ is obtained by multiplying corresponding components and adding the resulting products. The vectors u and v are said to be orthogonal (or perpendicular) if their dot product is 0 (i.e., if $u*v=0$). For examples, let:

$$u=(1,-2,3),\qquad v=(4,5,-1),\qquad w=(2,7,4),$$
$$u*v=1(4)-2(5)+3(1)=4-10-3=-9,\qquad u*w=2-14+12=0,$$
$$v*w=8+35-4=39$$

Thus, u and w are orthogonal: If, and then

$$u = \begin{bmatrix} 2 \\ 3 \\ -4 \end{bmatrix}, \quad v = \begin{bmatrix} 3 \\ -1 \\ -2 \end{bmatrix}, \qquad u * v = 6 - 3 + 8 = 11$$

By computing the dot products of the pairs (u, v), (u, w), and (v, w), we obtain the values -9, 0, and 39, respectively. The dot product of u and w is zero, which indicates that vectors u and w are orthogonal, meaning they are perpendicular to each other. The dot product of u and v is calculated as 11, showing that they are not orthogonal.

Theorem

For any vectors $u, v, w \in R^n$, and any scalar $k \in R$. In a vector space, especially in Rn, the accompanying equations describe certain characteristics of vectors and scalars (the n-dimensional real space).

(i)	$(u+v)+w = uv + vw$	(iv)	$uv = vu$
(ii)	$(ku)v = k(uv)$	(v)	$uu \geq 0, and\ uu = 0\ if\ u = 0$
(iii)	$u(kv) = (kv)u = k(vu) = k(uv)$		

(i) The associative feature of vector addition states that adding vectors in any particular sequence has no effect on the outcome and that $(u+v)+w = uv + vw$.

(ii) A scalar can be multiplied with the product of two vectors in one of two ways: This is known as the distributive property of scalar multiplication over vector multiplication, or $(ku)v = k(uv)$.

(iii) The commutative property of scalar multiplication with vector multiplication is $u(kv) = (kv)u = k(vu) = k(uv)$, which means that the outcome is unaffected by the order in which a scalar and a vector are multiplied.

(iv) The equation $uv = vu$ asserts that two vectors u and v is commutative, which means that the outcome is unaffected by the order of the vectors in the dot product.

(v) The definition of a vector's norm or magnitude, a non-negative scalar that counts the length of a vector, is $uu \geq 0, and\ uu = 0\ if\ u = 0$ A vector's squared magnitude is never negative and only equals zero if and only if the vector is the zero vector (i.e., has a length of zero).

+ **Norm (length) of a Vector:** The length of the vector is referred to as the vector norm or the vector's magnitude. The norm or length of a vector $u \in R^n$, denoted $\|u\|$, is the nonnegative square root of $u * u$. In particular, if $u = (a_1, a_2, \ldots, a_n)$, then

$$\|u\| = \sqrt{u * u} = \sqrt{a_1^2 + a_2^2 + \cdots + a_n^2}$$

That is, $\|u\|$ is the square root of the sum of the squares of the components of u.

- Thus, $\|u\| \geq 0$, and $\|u\| = 0$ if and only if $u = 0$

A vector u is called a unit vector if $\|u\| = 1$ or, equivalently, if $u * u = 1$. For any non-zero vector $v \in R^n$, the vector $\hat{v} = \frac{1}{\|v\|}v = \frac{v}{\|v\|}$ is the unique unit vector in the same direction as v. The process of finding $\hat{v}$ from v is called *normalizing* v .

Examples: Suppose $u = (1, -2, -4, 5, 3)$. To find $\|u\|$, we can first find $\|u\|^2 = u * u$ by squaring each component of u and adding $\|u\|^2 = 1^2 + (-2)^2 + (-4)^2 + 5^2 + 3^2 = 1 + 4 + 16 + 25 + 9 = 55$. Then $\|u\| = \sqrt{55}$.

- Let

$$v = (1, -3, 4, 2), \qquad w = \left(\frac{1}{2}, -\frac{1}{6}, \frac{5}{6}, \frac{1}{6}\right)$$

- Then

$$\|w\| = \sqrt{\frac{9}{36} + \frac{1}{36} + \frac{25}{36} + \frac{1}{36}} = \sqrt{\frac{36}{36}} = \sqrt{1} = 1$$

The w is a unit vector, but v is not a unit vector. However, we can normalize v as follows:

$$\hat{v} = \frac{1}{\|v\|} = (\frac{1}{\sqrt{30}}, \frac{-3}{\sqrt{30}}, \frac{4}{\sqrt{30}}, \frac{2}{\sqrt{30}})$$

This is the unique unit vector in the same direction as v. It demonstrates the process of normalizing vector v to obtain a unit vector $\hat{v}$. By dividing each component of v by its magnitude (√30), we obtain a new vector $\hat{v}$ with a length of 1, which maintains the original direction of v while being scaled to a unit length.

Definition: We can simply define each operation as below:

- Addition

$$\begin{bmatrix} a_1 \\ a_2 \\ \vdots \\ a_n \end{bmatrix} + \begin{bmatrix} b_1 \\ b_2 \\ \vdots \\ b_n \end{bmatrix} = \begin{bmatrix} a_1 + b_1 \\ a_2 + b_2 \\ \vdots \\ a_n + b_n \end{bmatrix}$$

Here, the vectors are represented as arrays of number enclosed in brackets, and the vectors are added component-wise, meaning that the first element of the first vector is added to the first element of the vector, the second element of the first vector is added to the second element of the vector, and so on.

The scalar multiplication is an operation that multiplies a scalar (a number) by a matrix, and the resulting matrix is called the product matrix. Scalar multiplication:

$$c\begin{bmatrix} a_1 \\ a_2 \\ \vdots \\ a_n \end{bmatrix} = \begin{bmatrix} ca_1 \\ ca_2 \\ \vdots \\ ca_n \end{bmatrix}$$

Here, given a matrix A and a scalar c, the product matrix cA is found by multiplying each element in matrix A by c. the resulting matrix cA then have the same dimensions as A. Dot product:

$$\begin{bmatrix} a_1 \\ a_2 \\ \vdots \\ a_n \end{bmatrix} * \begin{bmatrix} b_1 \\ b_2 \\ \vdots \\ b_n \end{bmatrix} = a_1b_1 + a_2b_2 + \cdots + a_nb_n, (i.e., \begin{bmatrix} 2 \\ 5 \end{bmatrix}, \begin{bmatrix} 7 \\ 1 \end{bmatrix} = 2 * 7 + 5 * 1 = 19)$$

The doc product is an operation that takes two equal-length sequences of numbers (usually coordinate vectors), and returns a single number. It is often used in Euclidean geometry to find the angle between two vectors, as well as their magnitude, and data analysis to determine the similarity between two vectors. Length:

$$\|\vec{a}\| = \sqrt{a_1^2 + a_2^2 + \cdots + a_n^2}, \qquad [i.e., \sqrt{2^2 + 5^2} = \sqrt{29}\ (= lenth\ of\ the\ vector)]$$

The length of a vector is the square root of the sum of the squares of its components, which the length of a vector 'a' is represented by the equation above.

+ **Distance, Angles, and Projections:** Sometimes we need to calculate the distance between vectors or points. Given some points as vectors in R^n, we can denote the distance between the two points as the distance between vectors $u = (a_1, a_2, \dots, a_n)$ and $v = (b_1, b_2, \dots, b_n)$ in R^n and defined by

$$d(u, v) = \|u - v\| = \sqrt{(a_1 - b_1)^2 + (a_2 - b_2)^2 + \cdots + (a_n - b_n)^2}.$$

We can show that this definition agrees with the usual notion of distance in the Euclidean plane R^2 or space R^3. The angle θ between non-zero vectors u, v in R^2 is defined by

$$cos\theta = \frac{uv}{\|u\|\|v\|}.$$

The definition is well defined, because, by the Schwarz inequality,

$$-1 \leq \frac{uv}{\|u\|\|v\|} \leq 1$$

The position of a vector u onto a non-zero vector v is the vector denoted and defined by

$$proj(u, v) = \frac{uv}{\|v\|^2} v = \frac{uv}{vv} v$$

We show below that this agrees with the usual notion of vector projection. The projection is calculated by taking the dot product of u and v, and dividing it by the squared magnitude of v, then multiplying the result by vector v. This process yields a vector that represents the component of u in the direction of v.

Example: Suppose $u = (1, -2, 3)$ and $v = (2, 4, 5)$ then

$$d(u, v) = \|u - v\| = \sqrt{(1-2)^2 + (-2-4)^2 + (3-5)^2} = \sqrt{1 + 36 + 4} = \sqrt{41}$$

To find $cos\theta$, where θ is the angle between u and v

$$uv = 2 - 8 + 15 = 9, \qquad \|u\|^2 = 1 + 4 + 9 = 14, \qquad \|v\|^2 = 4 + 16 + 25 = 45$$

Then

$$cos\theta = \frac{uv}{\|u\|\|v\|} = \frac{9}{\sqrt{14}\sqrt{45}}$$

Also

$$proj(u, v) = \frac{uv}{\|v\|^2} v = \frac{9}{45}(2, 4, 5) = \frac{1}{5}(2, 4, 5) = (\frac{2}{5}, \frac{4}{5}, \frac{5}{5})$$

Here, the projection of the vector u onto the vector v can be calculated by using the formula $proj(u, v) = \frac{uv}{\|v\|^2}$. In this case, the v is equal to $\frac{1}{5}(2, 4, 5)$, which is a unit vector with a magnitude of 1. Thus, the projection of u onto v can be calculated as above.

Cauchy-Schwarz Inequality

The Cauchy-Schwarz inequality has an equality situation where the vectors are linearly dependent (i.e., one vector is a scalar multiple of the other). For any two vectors in a real inner product space, the Cauchy-Schwarz inequality says that the following inequality holds.

$$|\vec{x} \cdot \vec{y}| \leq \|\vec{x}\|\|\vec{y}\|$$

This result is crucial to linear algebra. A scalar called the dot product shows how one vector is projected onto another.

According to the Cauchy-Schwarz inequality, the dot product of the two vectors is always bigger than or equal to the absolute value of the product of the vectors' magnitudes.

In Figure 3.1.2, both sides are regular numbers because the dot product of two vectors must be a regular number and the length of the vectors must be scalar, but the dot product of the multiplication of two vectors has to be less than or equal to the length of the vectors.

$$\vec{x}\cdot\vec{y} \in R^n \rightarrow |\vec{x}\cdot\vec{y}| \leq \|\vec{x}\|\|\vec{y}\| = (\textit{dot product of the vectors} \leq \textit{length of the vectors}) \longrightarrow$$

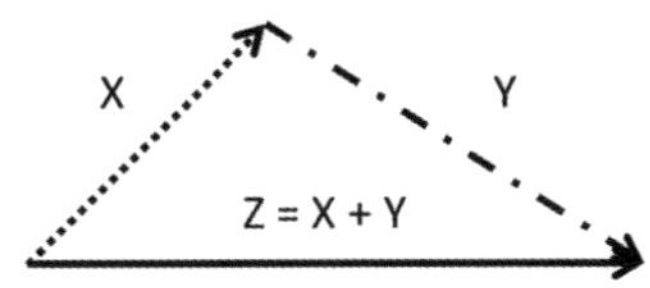

$\|Z\| = \|X + Y\| < \|X\| + \|Y\|$

Figure 3.1.2. Schematic representation of the vector

Numerous fields, including optimization, signal processing, and statistics, make use of the Cauchy-Schwarz inequality.

+ **Cross-product:** The cross product (or vector product [occasionally directed area product, to emphasize its geometric significance]) is a binary operation on two vectors in a three-dimensional-oriented Euclidean vector space, and is denoted as (only) in R^3.

$$\vec{a} = \begin{bmatrix} a_1 \\ a_2 \\ a_3 \end{bmatrix}, \vec{b} = \begin{bmatrix} b_1 \\ b_2 \\ b_3 \end{bmatrix}, \qquad \vec{a}\cdot\vec{b} \begin{bmatrix} a_2b_3 - a_3b_2 \\ a_3b_1 - a_1b_3 \\ a_1b_2 - a_2b_1 \end{bmatrix}$$

The cross product of two vectors '$\vec{a}$' and '$\vec{b}$' is denoted as '$\vec{a}\cdot\vec{b}$' and is a vector that is perpendicular to both '$\vec{a}$' and '$\vec{b}$'. The direction of the cross product is given by the right-hand rule. That is,

$$\vec{a} = \begin{bmatrix} 1 \\ -7 \\ 1 \end{bmatrix}, \quad \vec{b} = \begin{bmatrix} 5 \\ 2 \\ 4 \end{bmatrix}, \quad \vec{a}\cdot\vec{b} = \begin{bmatrix} -7\cdot 4 - 1\cdot 2 \\ 1\cdot 5 - 1\cdot 4 \\ 1\cdot 2 - 9(-7)\cdot 5 \end{bmatrix} = \begin{bmatrix} -30 \\ 1 \\ 37 \end{bmatrix}$$

In this example, given vectors $\vec{a}$ = (1, −7, 1) and $\vec{b}$ = (5, 2, 4), the cross product $\vec{a}\cdot\vec{b}$ is calculated as (−30, 1, 37), which is a vector perpendicular to both $\vec{a}$ and $\vec{b}$.

Remarks: The main purpose of this section is to explain the reasoning behind the use of vectors to represent data. The motive behind systems analysis is to determine patterns of observations behind the curtain and relationships between objects. Understanding data as vectors and using vector algebra to manipulate them solves a lot of problems in any field of research and systems analysis. Thus, please feel free to explore vectors in more detail including its applications in data science to build on the basics of the concepts mentioned here.

[3.2]

Matrices

A Guide to Concepts, Matrix Applications, Implementing Matrices

Like vectors, it is not difficult to grasp the key concepts of matrices. Early in basic algebra we are taught linear equations: These self-explanatory equations include two variables that give a line when plotted on a graph (i.e., Cartesian coordinates; $u + v = z$). Systems of such equations tangle up these lines, giving multiple equations with missing pieces that can only be found by untangling (augmenting and combining forms of) these equations.

$$vector\ \vec{z} = \vec{u} + \vec{v}$$

$$matrix \begin{bmatrix} 3 \\ 1 \end{bmatrix} + \begin{bmatrix} 1 \\ 2 \end{bmatrix} = \begin{bmatrix} 3+1 \\ 1+2 \end{bmatrix} \quad \longrightarrow$$

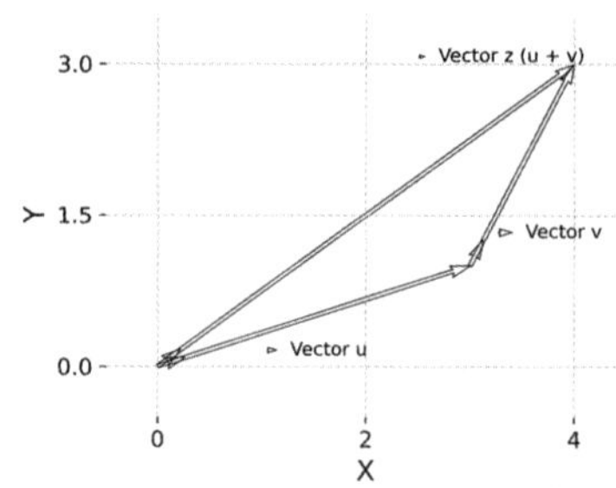

Figure 3.2.1. Vector and matrix operation in a Cartesian coordinate system

In Figure 3.2.1, illustrates vector and matrix operations in a 2×2 Cartesian coordinate system. The equation 'vector' represents the addition of two vectors $\vec{u}$, and $\vec{v}$, resulting in a new vector $\vec{z}$. The equation 'matrix' represents the addition of two matrices resulting in a new matrix.

In systems analysis these equations can be in the form of vectors, and these vectors can be combined into matrices, which can be represented in most programming languages using arrays. When we have the matrices, we can solve these systems of linear equations in many ways.

1. A Guide to Concepts

We might be wondering how a matrix differs from a vector from the perspective of data analysis. Simply put, a vector is a single column (attribute) of a data set, whereas a matrix is a collection of all columns. A matrix, like a vector, is a collection of numbers (Berry et al., 1999). The difference is that a matrix is a table of numbers rather than a list. Many of the same rules we just outlined for vectors apply equally well to matrices (in fact, you can think of vectors as matrices that happen to have only one column or one row). Matrices are usually denoted with a capital letter, like this: A.

+ **Identification**: A matrix A over a field k or, simply a matrix A (when k is implicit) is a rectangular array of scalars usually presented in the following from:

$$A = \begin{bmatrix} a_{11} & a_{12} & \cdots & a_{1n} \\ a_{21} & a_{22} & \cdots & a_{2n} \\ \vdots & \vdots & \vdots & \vdots \\ a_{m1} & a_{m2} & \cdots & a_{mn} \end{bmatrix}$$

The rows of such matrix A are the m horizontal lists of scalars. A matrix A when the field is implicit, is a rectangular array of scalars usually presented in the form:

$$(a_{11}, a_{12}, \dots, a_{1n}), (a_{21}, a_{22}, \dots, a_{2n}), \dots, (a_{m1}, a_{m2}, \dots, a_{mn})$$

The rows of this matrix A are the m horizontal lists of scalars, and each element in the matrix is denoted by a two-subscript notation, where the first subscript represents the row, and the second represents the column.

The columns of A are the n vertical lists of scalars.

$$\begin{bmatrix} a_{11} \\ a_{21} \\ \vdots \\ a_{m1} \end{bmatrix}, \begin{bmatrix} a_{12} \\ a_{22} \\ \vdots \\ a_{m2} \end{bmatrix}, \dots, \begin{bmatrix} a_{1n} \\ a_{2n} \\ \vdots \\ a_{mn} \end{bmatrix}$$

Note that the element a_{ij}, called the ij -entry or ij -element, appears in row i and column j . We frequently denote such a matrix by simply writing

$$A = [a_{ij}]$$

+ **Matrix with (m) Rows and (n) Columns**: A matrix with rows and columns is called an m by n matrix, written as

$$m \times n$$

The pair of numbers m and n is called the size of the matrix. Two matrices A and B are equal, written $A = B$, if they have the same size and if corresponding elements are equal. Thus, the equality of two $m \times n$ matrices is equivalent to a system of mn equalities, one for each corresponding pair of elements.

A matrix with only one row is called a row matrix or row vector, and a matrix with only one column is called a column matrix or column vector. A matrix whose entries are all zeros is called a zero matrix and is usually denoted by 0. A matrix whose entries are all real numbers is called a *real matrix* and is said to be a matrix over R.

2. Matrix Applications

In linear algebra, matrix multiplication is a binary operation that produces a matrix from two matrices. For matrix multiplication, the number of columns in the first matrix must be equal to the number of rows in the second matrix.

+ **Addition and Scalar Multiplication:** Let $A = [a_{ij}]$ and $B = [b_{ij}]$ be two matrices of the same size, say $m \times n$ matrices. The sum of A and B, written $A + B$, is the matrix obtained by adding corresponding elements from A and B.

$$A + B = \begin{bmatrix} a_{11} + b_{11} & a_{12} + b_{12} & \cdots & a_{1n} + b_{1n} \\ a_{21} + b_{21} & a_{22} + b_{22} & \cdots & a_{2n} + b_{21} \\ \vdots & \vdots & \vdots & \vdots \\ a_{m1} + b_{m1} & a_{m2} + b_{m2} & \cdots & a_{mn} + b_{mn} \end{bmatrix}$$

The resulting matrix, known as the matrix product, has the number of rows of the first matrix and the number of columns of the second matrix. The resulting matrix has the number of rows of the first matrix and the number of columns of the second matrix. As you can see, the number of columns in the first matrix must be equal to the number of rows in the second matrix; the product of matrices A and B is denoted above.

The product of matrix A by a scalar k, written kA or simply kA, is the matrix obtained by multiplying each element of A by k.

$$kA = \begin{bmatrix} ka_{11} & ka_{12} & \cdots & ka_{1n} \\ ka_{21} & ka_{22} & \cdots & ka_{2n} \\ \vdots & \vdots & \vdots & \vdots \\ ka_{m1} & ka_{m2} & \cdots & ka_{mn} \end{bmatrix}$$

Observe that $A + B$ and kA are also $m \times n$ matrices. We also define

$$-A = (-1)A \ and \ A - B = A + (-B)$$

The matrix $-A$ is called the negative of the matrix A, and the matrix $A - B$ is called the difference of A and B

+ **Vector Spaces:** A vector space is a set of objects called vectors that can be added together and multiplied (scaled) by numbers called scalars. It has two operations, vector addition and scalar multiplication. Here, V is called a vector space (over the field k) if the following axioms hold for any vectors:

$$u, v, w \in V$$

Set Notation

Common mathematical symbols used to describe elements and their connections inside sets are enumerated in the following set notation.

- $a \in A$: element a belongs to set A
- $a, b \in A$: element a and b belong to A

When you see the sign ∈ it means that the element to the left of the symbol is a member of the set to the right.

- $\forall x \in A$: for every x in A

The symbol ∀ stands for the universal quantifier, which means that the statement that comes after it is true for each and every member of the given set.

- $\exists x \in A$: there exists an x in A

The existential quantifier is represented by the symbol ∃ which denotes the existence of at least one element in the given set that fulfills the condition.

- $A \subseteq B$: A is a subset of B
- $A \cap B$: intersection of A and B
- $A \cup B$: union of A and B

The set on the left is a subset of the set on the right, as shown by the set inclusion symbol ⊆ The intersection of two sets, or the set of items that are present in both sets, is represented by the symbol ∩, ∪ stands for the union of two sets, or the collection of components that may be found in any of the two sets.

- $\emptyset$: empty set

The empty set, or set without any items, is represented by the character ∅ In the context of vectors and scalars.

- u, v, w: vectors in V
- k: the given number fields
- $a, b, c, or\ k$: scalars in k

Here, 'u, v, w' denotes vectors in a vector space V whereas k denotes a number field. Scalars are indicated in the field k by '$a, b, c, or\ k$' (Vector addition: $u + v \in V$, Scalar multiplication: $ku \in V$).

+ **Linear Combination:** Let V be a vector space over a field k. A vector v in V is a linear combination of vectors $u_1, u_2, \dots, u_m$ in V if there exist scalars $a_1, a_2, \dots, a_m$ in k such the $v = a_1u_1, a_2u_2, \dots, a_mu_m$ at. Alternatively, v is a linear combination of $u_1, u_2, \dots, u_m$ if there is a solution to the vector equation $v = x_1u_1, x_2u_2, \dots, x_mu_m$, where $x_1, x_2, \dots, x_m$ are unknown scalars.

Spanning sets: A set of vectors is called a spanning set for a vector space if every vector in the space can be written as a linear combination of the vectors in the set. Let V be a vector space over k. Vectors $u_1, u_2, \ldots, u_m$ in V are said to span V or to form a spanning set of V if every v in V is a linear combination of the vectors $u_1, u_2, \ldots, u_m$ that is, if there exist scalars $a_1, a_2, \ldots, a_m$ in k such that $v = a_1u_1, a_2u_2, \ldots, a_mu_m$.

Subspace: A subspace is a subset that satisfies conditions as follow. Let V be a vector space over a field k, and let W be a subset of V. W is a subspace of V if W is itself a vector space over k with respect to the operations of vector addition and scalar multiplication on V. The way in which one shows that any set W is a vector space is to show that W satisfies the eight axioms (set notations) of a vector space. However, if W is a subset of a vector space V, then some of the axioms automatically hold in W, because they already hold in V. Simple criteria for identifying subspaces follow. Suppose W is a subset of a vector space. W is a subspace of V if the following two conditions hold: The zero vector 0 belongs to W.

- For every $u, v \in W, k$, (i) the sum $u + v \in W$ and (ii) the multiple $ku \in W$.

Property (i) in states that W is closed under vector addition, and property (ii) in states that W is closed under scalar multiplication. Both properties may be combined into the following equivalent single statement:

- For every $u, v \in W, a, b \in k$, the linear combination $au, bv \in W$.

The intersection of subspaces: In the context of matrices, the intersection of subspaces is the set of all matrices that belong to the subspaces. Let U and W be subspaces of a vector space V. We can show that the intersection $U \cap W$ is also a subspace of V. Clearly, $0 \in U$ and $0 \in W$, because U and W are subspaces; hence $0 \in U \cap W$.

Now suppose u and v belong to intersection $U \cap W$. Then $u, v \in U$ and $u, v \in W$. Furthermore, because U and W are subspaces, for only scalars $a, b \in k$ ($au, bv \in U$, and $au, bv \in W$). Thus, $au, bv \in U \cap W$, therefore $U \cap W$ is a subspace of V. The intersection of any number of subspaces of a vector space is a subspace of V.

+ **Square Matrices:** A square matrix is a matrix with the same number of rows and columns. An $(n \times n)$ square matrix, which is said to be of order n, is sometimes called an *n-square matrix*. Not every two matrices can be added or multiplied; however, if we only consider square matrices of given order n, then this inconvenience disappears.

The operations of addition, multiplication, scalar multiplication, and transposition can be performed on any $n \times n$ matrices, and the result is again an $n \times n$ matrix. For example, the following are square matrices of order 3 (T = transpose [row and column]):

$$A = \begin{bmatrix} 1 & 2 & 3 \\ -4 & -4 & -4 \\ 5 & 6 & 7 \end{bmatrix}, \quad B = \begin{bmatrix} 2 & -5 & 1 \\ 0 & 3 & -2 \\ 1 & 2 & -4 \end{bmatrix}, \quad A + B = \begin{bmatrix} 3 & -3 & 4 \\ -4 & -1 & -6 \\ 6 & 8 & 3 \end{bmatrix}$$

$$A^T = \begin{bmatrix} 1 & 2 & 3 \\ -4 & -4 & -4 \\ 5 & 6 & 7 \end{bmatrix}, \quad B^T = \begin{bmatrix} 2 & 0 & 1 \\ -5 & 3 & 2 \\ 1 & -2 & -4 \end{bmatrix}, \quad A^T + B^T = \begin{bmatrix} 3 & 2 & 4 \\ -9 & -1 & -2 \\ 6 & 4 & 3 \end{bmatrix}$$

In this example, we can see that the addition, multiplication, scalar multiplication, and transposition operations can be performed on square matrices, and the resulting matrices are also square matrices of the same order.

For practice with using the matrices

$$A = \begin{bmatrix} 3 & 1 & -2 \\ 4 & -2 & 3 \\ -2 & 1 & -1 \end{bmatrix}, \quad B = \begin{bmatrix} 2 & 0 & -1 \\ -4 & 1 & 2 \\ 1 & -1 & 0 \end{bmatrix}$$

Here, we are given two matrices A and B. We can perform matrix operations like addition, subtraction, and multiplication on these matrices. To add two matrices, simply add the corresponding elements in each matrix:

$$A + B = \begin{bmatrix} 3+2 & 1+0 & -2-1 \\ 4-4 & -2+1 & 3+2 \\ -2+1 & 1-1 & -1+0 \end{bmatrix} = \begin{bmatrix} 5 & 1 & -3 \\ 0 & -1 & 5 \\ -1 & 0 & -1 \end{bmatrix}$$

To subtract two matrices, subtract the corresponding elements in each matrix:

$$A - B = \begin{bmatrix} 3-2 & 1-0 & -2+1 \\ 4+4 & -2-1 & 3-2 \\ -2-1 & 1+1 & -1-0 \end{bmatrix} = \begin{bmatrix} 1 & 1 & -1 \\ 8 & -3 & 1 \\ -3 & 2 & -1 \end{bmatrix}$$

To multiply two matrices, multiply the elements in the rows of the first matrix by the corresponding elements in the columns of the second matrix and then sum the results:

$$AB = \begin{bmatrix} (3)(2)+(1)(-4)+(-2)(1) & (3)(0)+(1)(1)+(-2)(-1) & (3)(-1)+(1)(2)+(-2)(0) \\ (4)(2)+(-2)(-4)+(3)(1) & (4)(0)+(-2)(1)+(3)(-1) & (4)(-1)+(-2)(2)+(3)(0) \\ (-2)(2)+(1)(-4)+(-1)(1) & (-2)(0)+(1)(1)+(-1)(-1) & (-2)(-1)+(1)(2)+(-1)(0) \end{bmatrix} = \begin{bmatrix} 0 & 3 & -1 \\ 19 & -5 & -8 \\ -9 & 2 & 4 \end{bmatrix}$$

Similarly, to multiply BA, multiply the elements in the rows of matrix B by the corresponding elements in the columns of matrix A and then sum the results:

$$BA = \begin{bmatrix} (2)(3)+(0)(4)+(-1)(-2) & (2)(1)+(0)(-2)+(-1)(1) & (2)(-2)+(0)(3)+(-1)(-1) \\ (-4)(3)+(1)(4)+(2)(-2) & (-4)(1)+(1)(-2)+(2)(1) & (-4)(-2)+(1)(3)+(2)(-1) \\ (1)(3)+(-1)(4)+(0)(-2) & (1)(1)+(-1)(-2)+(0)(1) & (1)(-2)+(-1)(3)+(0)(-1) \end{bmatrix} = \begin{bmatrix} 8 & 1 & -3 \\ -12 & -4 & 9 \\ -1 & 3 & -5 \end{bmatrix}$$

In this example, we demonstrate how to perform addition, subtraction, and multiplication operations on matrices A and B. The resulting matrices display the outcome of these operations.

+ **Identity Matrices and Scalar Matrices:** The n-square identity or unit matrix, denoted by I_n, or simply I, is the n-square matrix with I's on the diagonal.

The identity matrix I is similar to the scalar I in that, for any n-square matrix A,

$$AI = IA = A$$

More generally, if B is an $m \times n$ matrix, then

$$BI_n = I_m B = B$$

For any scalar k, the matrix kI that contains k's on the diagonal and elsewhere is called the *scalar matrix* corresponding to the scalar k . Observe that

$$(kI)A = k(IA) = kA$$

That is, multiplying a matrix A by the scalar matrix kI is equivalent to multiplying A by the scalar k.

+ **Simplify the Matrix Operation:** Let us simplify a basic application concept of matrices as below. Linear system equation (simply)

$$\begin{matrix} 2x + 5y + 3z = -3 \\ 4x + 0y + 8z = 0 \\ 1x + 3y + 0z = 2 \end{matrix} \rightarrow \underbrace{\begin{bmatrix} 2 & 5 & 3 \\ 4 & 0 & 8 \\ 1 & 3 & 0 \end{bmatrix}}_{\substack{matrix\ A \\ constant}} \underbrace{\begin{bmatrix} x \\ y \\ z \end{bmatrix}}_{\substack{variable \vec{x} \\ vectors}} = \underbrace{\begin{bmatrix} -3 \\ 0 \\ 2 \end{bmatrix}}_{\vec{v}} \rightarrow A\vec{x} = \vec{v}$$

The equation represents a system of linear equations. This system can be represented in the form $A\vec{x} = \vec{v}$, where A is a matrix representing the coefficients of the variables, y, and z, $\vec{x}$ is a column vector representing the variables x, y, and z, and $\vec{v}$ is a column vector representing the constant terms in the equations.

- Dot product

$$\begin{bmatrix} 2 \\ 7 \\ 1 \end{bmatrix} * \begin{bmatrix} 8 \\ 2 \\ 8 \end{bmatrix} = \begin{bmatrix} 2 \cdot 8 \\ 7 \cdot 2 \\ 1 \cdot 8 \end{bmatrix} = 2 \cdot 8 + 7 \cdot 2 + 1 \cdot 8$$

$$\begin{bmatrix} x_1 & y_1 \end{bmatrix} \begin{bmatrix} x_2 \\ y_2 \end{bmatrix} = \begin{bmatrix} x_1 x_2 \\ y_1 y_2 \end{bmatrix} = x_1 \cdot x_2 + y_1 \cdot y_2$$

The dot product is a mathematical operation that takes two vectors and returns a scalar value. It is also known as the scalar product or inner product.

The dot product of two vectors is calculated by multiplying the corresponding components of the vectors and then summing the results.

- Cross-product

$$\begin{bmatrix} v_1 \\ v_2 \\ v_3 \end{bmatrix} * \begin{bmatrix} w_1 \\ w_2 \\ w_3 \end{bmatrix} = \begin{bmatrix} v_2 \cdot w_3 - w_2 \cdot v_3 \\ v_3 \cdot w_1 - w_3 \cdot v_2 \\ v_1 \cdot w_2 - w_1 \cdot v_2 \end{bmatrix} = \vec{p} = \vec{v} \times \vec{w}$$

$$\hat{\imath}(v_2 \cdot w_3 - w_2 \cdot v_3) + \hat{\jmath}(v_3 \cdot w_1 - w_3 \cdot v_2) + \hat{k}(v_1 \cdot w_2 - w_1 \cdot v_2)$$

The Cross-product is an operation between two vectors that results in a third vector that is perpendicular to the plane defined by the first two vectors. Given two vectors $\vec{v}$ and $\vec{w}$, their cross-product $\vec{p}$ is calculated as above.

This can also be expressed using the unit vectors $\hat{\imath}$, $\hat{\jmath}$, and $\hat{k}$. The resulting vector $\vec{p}$ is perpendicular to both $\vec{v}$ and $\vec{w}$ and has a magnitude equal to the area of the parallelogram defined by $\vec{v}$ and $\vec{w}$.

+ **Eigenvalues and Eigenvectors:** Eigenvalues are a special set of scalars associated with a matrix equation that are sometimes also known as characteristic values. Eigenvectors are vectors that do not change their orientation when multiplied by the transition matrix but just scale by a factor of the corresponding eigenvalues (Fox & Kapoor, 1968).

Some applications of eigenvalues and eigenvectors are very useful when handling data in matrix form because you can decompose them into matrices that are easy to manipulate. Matrix-vector multiplication $A\vec{v}$ gives the same result as just scaling the eigenvector $\vec{v}$ by value λ as given:

$$A\vec{v} = \lambda\vec{v}$$

The left side represents matrix-vector multiplication, and the right side is scalar-vector multiplication (i.e., $k\vec{v}$).

$$\underbrace{\begin{bmatrix} \lambda & 0 & 0 \\ 0 & \lambda & 0 \\ 0 & 0 & \lambda \end{bmatrix}}_{scaling\ by\ \lambda} \rightarrow \lambda \underbrace{\begin{bmatrix} 1 & 0 & 0 \\ 0 & 1 & 0 \\ 0 & 0 & 1 \end{bmatrix}}_{\substack{identity\ matrix\ (I)\ with \\ 1s\ down\ the\ diagonal}} \rightarrow A\vec{v} = (\lambda I)\vec{v}$$

Here, the symbol λ (lambda) is often used to represent the eigenvalues of a matrix. In this case, left side of the matrix is a diagonal matrix with λ as its main diagonal elements. When this matrix is multiplied with another vector $\vec{v}$, the resulting vector $A\vec{v}$ is obtained by scaling each element of $\vec{v}$ by a factor of $\vec{v}$. The identity matrix I is a square matrix with 1s on the main diagonal and 0s everywhere else.

In the expression given, the matrix λI is being multiplied with the vector $\vec{v}$ to obtain the vector $A\vec{v}$.

$$A\vec{v} = (\lambda I)\vec{v} \rightarrow \begin{matrix} A\vec{v} - (\lambda I)\vec{v} = 0 \\ (A - \lambda I)\vec{v} = 0 \end{matrix} \rightarrow \det(A - \lambda I) = 0$$

This equation is from $A\vec{v} - (\lambda I)\vec{v} = 0$, where I is the identity matrix. The roots of this equation are the eigenvalues of the matrix A. In other words, the eigenvalues of A are the value of λ that make the determinant of equal to zero $\det(A - \lambda I) = 0$.

Vector Spaces and Eigenvalue-Eigenvector

The following is an underlying assumption behind diagonalization and eigen decomposition: Let us assume that the matrix you are trying to decompose is a square matrix and has linearly independent eigenvectors (different eigenvalues).

Linear equations are often visually presented in the Cartesian coordinate system, a system that can plot coordinates in a theoretical two-dimensional (2D) space with the center at the origin (that has coordinates (0, 0)). Vectors can be placed on this graph, arrows pointing to a certain coordinate in the 2D plane. Linear transformations fascinate us greatly, and one way to describe them is as functions that convert one vector to another in a 2D space. The set of vectors that a linear transformation produces can be visualized as points, where the points represent the new endpoints when the vectors are put into the linear transformation. A simple way to visualize this is to make a grid out of these points.

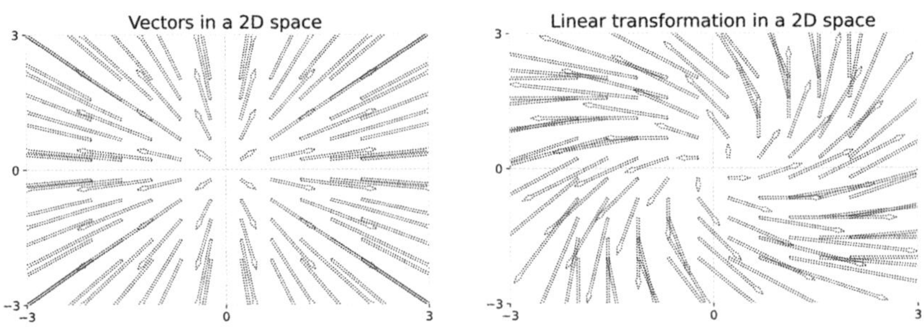

Figure 3.2.2. Vectors in a 2D space. The plot of the left side is represented as points and then as a grid. The plot of the right side is represented as linear transformation in a 2D space.

In Figure 3.2.2, linear transformations can be quite self-explanatory, as they transform the lines that vectors make (right side of the figure). A set of vectors can be simplified by two basic vectors, which we can call *basis vectors*. These are the two vectors that point in the x and y directions, or i-hat and j-hat. A linear transformation at its most basic can be described by the change in the two basis vectors in a 2D space. The determinant is a scalar that can be computed from the elements of a square matrix. As mentioned previously, vectors can be turned into matrices; thus, if we have a square matrix, we can turn it back into a set of vectors. A square matrix can represent two vectors, so in this case two vectors can give us a determinant. As shown earlier, the two basis vectors of a linear transformation give a determinant. And it just so happens that the determinant of a square matrix is the area formed by its basis vectors.

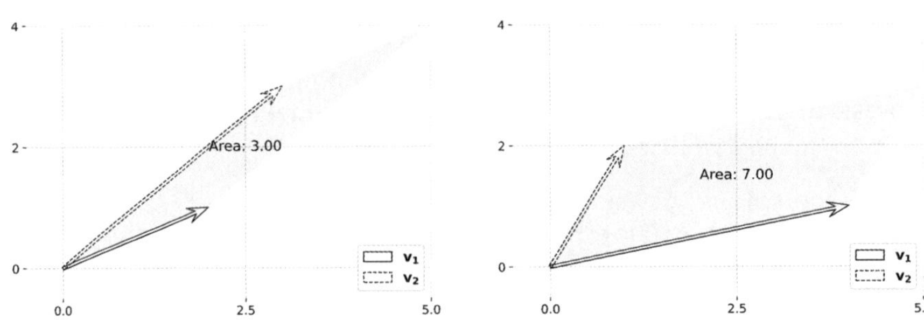

Figure 3.2.3. Determinant of two vectors (a square matrix); the area formed by the vectors (horizontal axis = x, vertical axis = y)

Linear transformations transform sets of vectors into a new set of vectors. In all linear transformations there exist vectors in the set that scale in size only but practically remain the same vector. These vectors are called *eigenvectors*. They remain on their same span and merely scale by a certain scalar. These scalars are called the *eigenvalues* of the eigenvectors.

Visual representations (Figure 3.2.3) of these concepts show how we can see math in many things, and these concepts can be extended and applied to so many things in life.

+ **Derivative and Integral Mappings:** This is mathematical operations used to describe how a function changes over time or distance. In matrix, the derivative describes the rate of change of the function, while the integral describes the accumulation of the function over a given interval.

Consider the vector space $A = p(t)$ of polynomials over the real field R

- Let $u(t)$ and $v(t)$ be any polynomials in V, and let k be any scalar.
- Let $D\colon V \to V$ be the derivative mapping.

One proves in calculus that

$$\frac{d(u+v)}{dt} = \frac{du}{dt} + \frac{dv}{dt}, \qquad and \ \frac{d(ku)}{dt} = k\frac{du}{dt}$$

The first equation states that the derivative of the sum of two functions is equal to the sum of the derivatives of those two functions, and the second equation states that the derivative of a constant times a function is equal to the constant times the derivative of the function. That is

$$D(u+v) = D(u) + D(v) \ and \ D(ku) + kD(u)$$

Thus, the derivative mapping is linear.

- Let $J : V \to R$ be an integral mapping, say,

$$J(f(t)) = \int_0^1 f(t)dt$$

Here, the $J(f(t))$ is an integral mapping that takes in a function $f(t)$ and returns the definite integral of $f(t)$ from t=0 to t=1. This means that the for any function $f(t)$, $J(f(t))$ will return the area under the curve of $f(t)$ from t=0 to t=1.

- One also proves in calculus that

$$\int_0^1 [u(t) + v(t)]\,dt = \int_0^1 u(t)dt + \int_0^1 v(t)dt$$

This is a property of integration known as the linearity of integration which states the integral of the sum of two function $u(t)$ and $v(t)$ from 0 to 1 is equal to the sum of the integrals of $v(t)$ and $v(t)$ from 0 to 1.

- And

$$\int_0^1 ku(t)dt = k\int_0^1 u(t)dt$$

- That is

$$J(u+v) = J(u) + J(v) \ \ and \ \ J(ku) + kJ(u)$$

The equation states that the integral mapping J applied to the sum of two functions is equal to the sum of the integral mappings applied to each function individually. And it also states that the integral mapping applied to a function multiplied by a scalar k is equal to the scalar k multiplied by the integral mapping applied to the function.

These statements hold true for any integral mapping and J and functions u and v, as long as J is a linear operator. Thus, the integral mapping is linear.

3. Implementing Matrices

With respect to mathematical conceptualization, let us dive into some examples of actual coding. Here, we will show how each object (or application) will be implemented in practice.

+ **Matrix Addition:** Matrix addition (or subtraction) is really similar to the addition we did with vectors earlier. The only difference is that there are multiple columns instead of just one. The general idea remains the same: You only need to add up the corresponding components.

In the general formula, we use a and b for placeholders, and you can see how each component is added up:

$$\begin{pmatrix} a_1 & a_2 \\ a_3 & a_4 \end{pmatrix} + \begin{pmatrix} b_1 & b_2 \\ b_3 & b_4 \end{pmatrix} = \begin{pmatrix} a_1+b_1 & a_2+b_2 \\ a_3+b_3 & a_4+b_4 \end{pmatrix}$$

Although this is fairly simple to grasp, here is a simple example of two-matrix addition:

$$A = \begin{pmatrix} 3 & 5 \\ 1 & 0 \end{pmatrix}, \qquad B = \begin{pmatrix} 2 & -3 \\ 1 & 2 \end{pmatrix}$$

$$A + B = \begin{pmatrix} 3+2 & 5+(-3) \\ 1+1 & 0+2 \end{pmatrix} = A + B = \begin{pmatrix} 5 & 2 \\ 2 & 2 \end{pmatrix}$$

In the given example, two matrices A and B are added together. Matrix A consists of elements 3, 5, 1, and 0, while matrix B consists of elements 2, −3, 1, and 2. When the corresponding elements of A and B are added together, we get a new matrix $A + B$ with elements 5, 2, 2, and 2.

This demonstrates the process of matrix addition, where the resulting matrix is formed by adding the corresponding elements of the two input matrices.

Matrix addition is really simple to do in programing (Jupyter; NumPy). As with vectors, you can use the addition sign:

```
[In]
A = np.matrix([
      [3, 5],
      [1, 0]])
B = np.matrix([
      [2, -3],
      [1, 2]])
print(A + B)
```

→

```
[Out]
[[5 2]
 [2 2]]
```

→

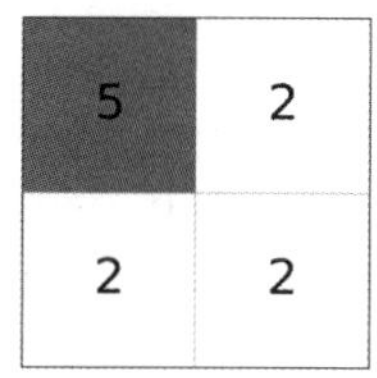

The code given in the prompt is using the 'numpy' library and 'numpy.matrix()' to define two matrices A and B and then adding them together using the '+' operator. This creates a 2×2 matrix called A (B) with the values '[,]' in the first row and '[,]' in the second row.

\+ **Scalar Multiplication:** This concept is more or less the same as with vectors. Every number in the matrix will be multiplied by scalar n. The formula is also very similar:

$$n = \begin{pmatrix} a_1 & a_2 \\ a_3 & a_4 \end{pmatrix} = \begin{pmatrix} na_1 & na_2 \\ na_3 & na_4 \end{pmatrix}$$

For this example, we use an arbitrary matrix, and I set the scalar n to 2:

$$A = \begin{pmatrix} 3 & 5 \\ 1 & 0 \end{pmatrix}$$

$$2A = \begin{pmatrix} 2 * 3 & 2 * 5 \\ 2 + 1 & 2 * 0 \end{pmatrix} = 2A = \begin{pmatrix} 6 & 10 \\ 2 & 0 \end{pmatrix}$$

Implementation in programing (Python; NumPy). As with vectors as well, you can use the multiplication sign:

```
[In]
A = np.matrix([
    [3, 5],
    [1, 0]])
print(2 * A)
```

→

```
[Out]
[[6 10]
 [2  0]]
```

→

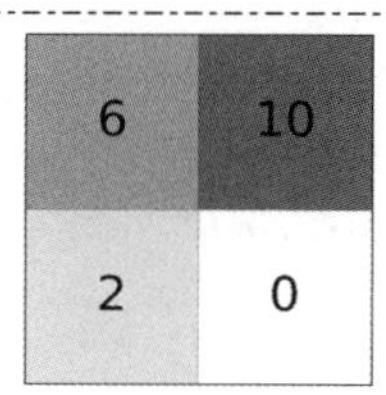

Everything is pretty much identical to vectors. The code creates a 2×2 matrix A with the elements 3, 5, 1, and 0. The matrix is then multiplied by the scalar value 2 and the resulting matrix is printed.

+ **Matrix Multiplication**: This topic is slightly more complex to grasp than the others encountered so far. It is not as hard as it might seem at first, but you will need to solve a couple of examples to get the full gist.

For the examples in this section on matrix multiplication, consider two matrices:

- Matrix A —has dimensions of m by n (m rows, n columns)
- Matrix B —has dimensions of n by p (n rows, p columns)

Multiplying A and B yields a new matrix that has dimensions of m by p (m rows, p columns). In general, the resulting matrix has the number of rows that matrix A has and the number of columns that matrix B has. We will probably need to read the concept a couple of times before we understand it fully.

To help us out, here is everything we have said so far presented visually:

$$\underbrace{A}_{mn} * \underbrace{B}_{np} = \underbrace{AB}_{mp}$$

As you can see, the two n in the middle need to match. If they are not equal, matrix multiplication cannot be performed. Most programming languages will give you an error if the dimensions do not match. Once you understand the basic rule of matrix multiplication, you are ready for the general formula:

$$\begin{pmatrix} a_1 & a_2 \\ a_3 & a_4 \end{pmatrix} + \begin{pmatrix} b_1 & b_2 \\ b_3 & b_4 \end{pmatrix} = \begin{pmatrix} a_1 + b_1 & a_2 + b_3 & a_1 + b_2 & a_2 + b_4 \\ a_3 + b_1 & a_4 + b_3 & a_3 + b_2 & a_4 + b_4 \end{pmatrix}$$

Let us state this in the most general way (please embed this into your brain): Matrix multiplication is performed by calculating the dot product of the corresponding row of matrix A and the corresponding column of matrix B.

If you understand that sentence, you understand matrix multiplication. If not, let us drive the point home with a simple example:

$$A = \begin{pmatrix} 3 & 4 \\ 1 & 0 \end{pmatrix}, \quad B = \begin{pmatrix} 2 & 2 \\ 1 & 2 \end{pmatrix}$$

Matrix multiplication is performed by calculating the dot product of the corresponding row of matrix A and the corresponding column of matrix B. In this example, the resulting matrix AB is calculated as follows:

$$AB = \begin{pmatrix} (3*2) & + & (4*1) & (3*2) & + & (4*2) \\ (1*2) & + & (0*1) & (1*2) & + & (0*2) \end{pmatrix}$$

When the calculations are completed, we get:

$$AB = \begin{pmatrix} 6+4 & 6+8 \\ 2+0 & 2+0 \end{pmatrix} = AB = \begin{pmatrix} 10 & 14 \\ 2 & 2 \end{pmatrix}$$

This illustrates the concept of matrix multiplication, where the elements of the resulting matrix are formed by calculating the dot product of the corresponding rows and columns from the input matrices.

As with vectors, you can use the dot function to perform multiplication with programing (NumPy):

```
[In]
A = np.matrix([
    [3, 4],
    [1, 0]])
B = np.matrix([
    [2, 2],
    [1, 2]])
print(A.dot(B))
```

→

```
[Out]
[[10 14]
 [2  2]]
```

→

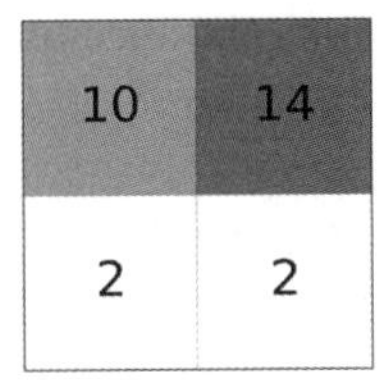

The matrix multiplication is a hard concept to immediately understand, but what will really help is doing it by hand. The code is using the 'dot()' method to perform matrix multiplication on A and B. It is a way to combine two matrices by multiplying the elements in each row of the first matrix by the corresponding elements in each column of the second matrix, and then summing the results.

+ **Matrix Transposition:** Matrix transposition sounds fancy, particularly if you are not familiar with what transposition means, but the idea is really simple: You merely exchange the rows and columns of a matrix. The transpose operator is in most cases denoted with a capital letter T, and the notation can be put either before the matrix or as an exponent. Either way, here is the general formula:

$$T = \left(\begin{bmatrix} a_1 & a_2 \\ a_3 & a_4 \end{bmatrix}\right) = \begin{bmatrix} a_1 & a_3 \\ a_2 & a_4 \end{bmatrix}$$

As you can see, the diagonal elements stay the same, and the off-diagonal elements switch their position. Here is a simple example with a 2×2 matrix:

$$A = \begin{pmatrix} 3 & 4 \\ 1 & 0 \end{pmatrix}$$

Matrix A is a 2×2 matrix with elements 3, 4, 1, and 0. When finding the transpose of a matrix, the diagonal elements remain the same, while the off-diagonal elements switch their positions. In this example, the transpose of matrix is calculated as follows:

$$T(A) = \begin{pmatrix} 3 & 1 \\ 4 & 0 \end{pmatrix}$$

The transpose operation has swapped the off-diagonal elements (4 and 1) of the original matrix A, while the diagonal elements (3 and 0) have remained the same.

Implementation in programming really could not be any simpler:

```
[In]
A = np.matrix([
      [3, 4],
      [1, 0]])
```

→

```
[Out]
[[3 1]
 [4 0]]
```

→

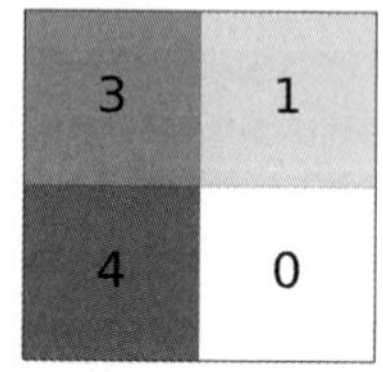

In the code, A, T is the transpose of the matrix A. The transpose of a matrix is a new matrix that is obtained by swapping the rows and columns of the original matrix printed as the statement 'print(A, T)'.

+ **Identity Matrices:** Just like transposition, identity matrices are really simple to grasp. An identity matrix is a matrix in which every diagonal element is 1 and all other elements are 0. An identity matrix is usually denoted with a capital letter I, and the number representing its size is a subscript.

Here is what an identity matrix of size 3 would look like:

$$I_3 = \begin{pmatrix} 1 & 0 & 0 \\ 0 & 1 & 0 \\ 0 & 0 & 1 \end{pmatrix}$$

There will not be any example of an identity matrix (for now).

Let us just see how to create them in programming:

```
[In]
A = np.eye(3)
    print(A)
```

→

```
[Out]
[[1. 0. 0.]
 [0. 1. 0.]
 [0. 0. 1.]]
```

→

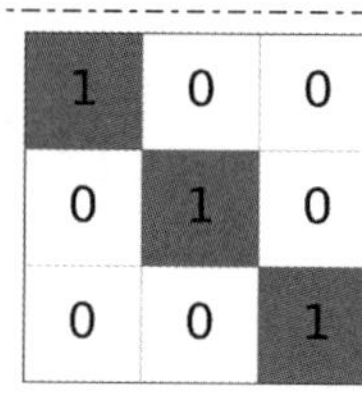

The 'np.eye(3)' is a function that returns a 3×3 identity matrix with ones on the diagonal and zeros elsewhere. The identity matrix is defined as above I assigned the value of the 3×3 identity and the dimension of the array (number of columns) is equal to the number passed as an argument to the function (in this case, 3).

+ **Determinants:** The determinant is a scalar value that can be computed from the elements of a square matrix and encodes certain properties of the linear transformation described by the matrix. The determinant of a matrix A is denoted

$$det(A), det\,A, or\ |A|$$

Geometrically, it can be viewed as the volume scaling factor of the linear transformation described by the matrix. The calculation process is simple for a 2×2 matrix, gets a little more difficult for 3×3 matrices, and should not be attempted by hand for larger ones.

The goal here is to develop an understanding; computers were made to do the calculation. Here is the general formula for calculating the determinant of a 2×2 matrix:

$$det = \begin{pmatrix} a & b \\ c & d \end{pmatrix} = ad - bc$$

And to drive the point home, here is the most basic example of calculation by hand:

$$A = \begin{pmatrix} 3 & 1 \\ 2 & 6 \end{pmatrix}$$

$$|A| = (3 * 6) - (1 - 2)$$

$$|A| = 18 - 2 = 16$$

To find the determinant of matrix A, denoted as $|A|$, we follow these steps:

- Multiply the diagonal elements: $3 * 6 = 18$
- Multiply the off-diagonal elements: $1 * 2 = 2$
- Subtract the product of the off-diagonal elements from the product of the diagonal elements: $18 - 2$

Implementation in code:

```
[In]
A = np.matrix([
    [3, 2],
    [1, 6]])
print(np.linalg.det(A))
```

→

```
[Out]
16.00000000
0000007
```

→

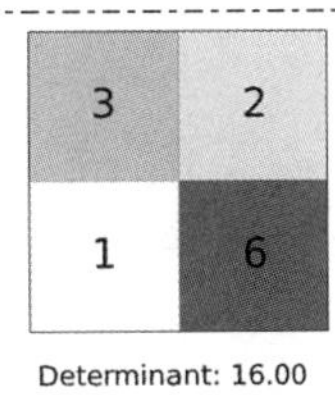

The determinant of a matrix is a scalar value that can be calculated from the elements of a matrix. It is denoted by the symbol $|A|$. Because the A is the matrix in the provided above, the determinant of A was calculated as $|A| = (ad) - (bc) = (3 * 6) - (2 * 1) = 18 - 2 = 16$ as printed.

\+ **Inverse Matrices:** A square matrix is called invertible (or nonsingular) if multiplying the original matrix by its inverse results in an identity matrix. From this statement, you can conclude that not all matrices have inverses.

For a matrix to be invertible, it has to satisfy the following conditions:

- It must be square.
- The determinant cannot be 0

A matrix that is not invertible is called a singular matrix. Logically speaking, for a square matrix to be singular, its determinant must be equal to 0. Let us see why by exploring the general formula:

$$A^{-1} = \begin{pmatrix} a & b \\ c & d \end{pmatrix}^{-1} = \frac{1}{ad-bc}\begin{pmatrix} d & -b \\ -c & a \end{pmatrix}$$

As you can see, the inverse matrix is denoted by the superscript −1. The formula might already look really familiar to you—you have previously seen the $ad - bc$ term (the determinant). You can see here why the determinant cannot be 0: Division by 0 is undefined. This term is then multiplied by a slightly rearranged version of the original matrix. The diagonal elements are switched, and the off-diagonal elements are multiplied by −1. Here is a simple example of calculating the inverse of a 2 × 2 matrix:

$$A = \begin{pmatrix} 4 & 3 \\ 5 & 4 \end{pmatrix}$$

In this example, the inverse of a 2×2 matrix A is calculated by first determining the determinant of A, which is found to be 1.

$$A^{-1} = \begin{pmatrix} 4 & 3 \\ 5 & 4 \end{pmatrix}^{-1} = \frac{1}{(4 * 4) - (3 * 5)}\begin{pmatrix} 4 & -3 \\ -5 & 4 \end{pmatrix}$$

then swapping the diagonal elements of A and negating the off-diagonal elements to create a new matrix,

$$A^{-1} = \frac{1}{16 - 15}\begin{pmatrix} 4 & -3 \\ -5 & 4 \end{pmatrix} = A^{-1} = \frac{1}{1}\begin{pmatrix} 4 & -3 \\ -5 & 4 \end{pmatrix}$$

and finally multiplying this new matrix by the inverse of the determinant (1/1) to obtain the inverse of matrix A, given as.

$$A^{-1} = \begin{pmatrix} 4 & -3 \\ -5 & 4 \end{pmatrix}$$

Implementation in code:

```
[In]
A = np.matrix([
    [4, 3],
    [5, 4]])
print(np.linalg.inv(A))
```

→

```
[Out]
[[ 4. -3.]
 [-5.  4.]]
```

→

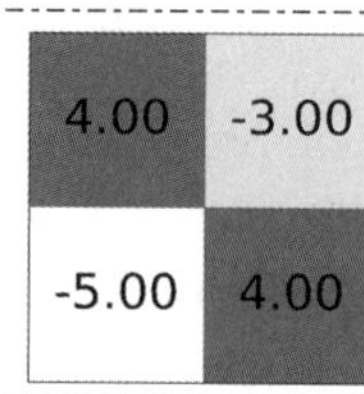

Because the determinant of A would be calculated as $|A| = (4 * 4) - (3 * 5) = 16 - 15 = 1$, the inverse of A would be

$$A^{-1} = 1/1 * [[4, -3], [-5, 4]] = [[4, -3], [-5, 4]]$$

This is confirmed by the 'inv()' function which computes the multiplicative inverse of a matrix. Now let us verify the earlier claim, that multiplying the original matrix by its inverse yields an identity matrix:

$$A^{-1}A = I_n$$

Here is the example calculated by hand, and the statement holds true

$$A^{-1} = \begin{pmatrix} 4 & 3 \\ 5 & 4 \end{pmatrix}^{-1}, \quad A = \begin{pmatrix} 4 & 3 \\ 5 & 4 \end{pmatrix}$$

$$A^{-1}A = \begin{pmatrix} (4*4) & + & (-3*5) & & (4*3) & + & (-3*4) \\ (-5*4) & + & (4*5) & & (-5*3) & + & (4*4) \end{pmatrix}$$

Specifically, when $A^{-1}A$ is computed,

$$A^{-1}A = \begin{pmatrix} 16-15 & 12-12 \\ -20+20 & -15+16 \end{pmatrix} = A^{-1}A = \begin{pmatrix} 1 & 0 \\ 0 & 1 \end{pmatrix}$$

The obtained matrix is $\begin{pmatrix} 1 & 0 \\ 0 & 1 \end{pmatrix}$, confirming the property that the product of a matrix and its inverse yields the identity matrix.

```
[In]
print(A.dot(np.linalg.inv(A)))
```

→

```
[Out]
[[1.00000000e+00 1.77635684e-15]
[0.00000000e+00 1.00000000e+00]]
```

This code printed the doc product of matrix A and the inverse of matrix A. The 'np.linalg.inv()' function computes the inverse of a matrix, and the 'dot()' function computes the dot product of two matrices which printed as above.

Remarks: Maybe not all of the ideas discussed here are directly applicable to systems analysis (or maybe they are not yet applicable from your perspective), but in general linear algebra is worth knowing. It is something that will probably be asked about in data science, so knowing the basics is a must. The main takeaway here is that although mathematics is notoriously disliked by many researchers, as it presents a very technical, detailed, and objective lens that many find hard to look through, linear algebra appears in all software, and in algorithms, and using mathematics to our advantage creates infinite possibilities and improvements in the realm of technology. Current systems already use these mathematical methods, and hopefully soon we may contribute to the innovation and creation of a better future.

Supplemental Information 3.1: Math Symbols with Code

Summations, factorials, matrices, and more are simple when we look at them with our favorite languages (Manning & Schutze, 1999). If you are interested in pursuing a career or research in data science, the day will come when it will be time for you to move beyond the programing libraries (e.g., Python, R, MATLAB) to follow your curiosity into the math behind it all.

This will typically take you to the vast, open collection of papers detailing how it all works. The deeper you go to understand the core mathematics, the closer you may come to the flash of insight needed to create a new method. Everything seems to be alright on the first paper until you come across something like this:

$$Q^*(s,a) = \sum_{s',r} P(s',r|s,a)\left[r + \gamma \max_{a'} Q^*(s',a')\right]$$

For anyone who has studied math for years or worked at the math level of machine learning, such an equation can be carefully parsed into meaning and code. For many others, though, this may look like hieroglyphics. The truth is that it almost seems like the ancient math leaders chose the most interesting-looking symbols to describe rather intuitive methods (Kramer, 1982). The result? Equations and variables that look much more complex than they really are.

Code can be used for more than writing programs; it is also a globally agreed-upon language for explaining the complex. When learning the math behind everything data science, the best way to gain a universal understanding of math is to write code snippets to describe equations. Eventually these symbols become understood to the point that they can almost be read as text in a typical paper.

Here are some examples of how simple math can be when described with code.

+ **Summation and Product:** The summation symbol is one of the most useful and commonly used symbols in iterative mathematics.

$$\sum_{i=0}^{5} x^i$$

Despite its complex design, its implementation is rather simple, yet incredibly useful.

```
[In]
x = [1, 2, 3, 4, 5, 6]
result = 0
for i in range(6):
    result += x[i]
    print(result)
```

→

```
[Out]
1
3
6
10
15
21
```

The code defines a list x with 6 elements and initializes a variable 'result' to 0. It then defines a 'for' loop that iterates over the element in x. On each iteration, it adds the current element to 'result' and then prints the value as above.

All this symbol represents is a for-loop in the range of the number on top, starting from the number on the bottom. The variable set on the bottom becomes the index variable, and any result per loop is added to an overall value. Less commonly, the following may be used:

$$\prod_{i=0}^{5} x^i$$

Typically called the product operator, this symbol functions in the same manner, but instead of adding each result it multiplies them.

```
[In]
x = [1, 2, 3, 4, 5, 1]
result = 1
for i in range(6):
    result *= x[i]
    print(result)
```

→

```
[Out]
1
2
6
24
120
120
```

This is similar to the previous one, but with a few differences. The list x has the same elements, but the initial value of 'result' is 1 instead of 0. In addition, the 'for' loop now updates 'result' by multiplying it by the current element of x instead of adding it. This process continues until all elements of x have been processed, at which point 'result' is equal to $120(1 * 2 * 3 * 4 * 5 * 1)$ as output of this printed as above.

+ **Factorial:** Factorial is the '!' symbol that exists on almost every calculator.

$$n!$$

To many, this one may be a bit more obvious, but it is still worth it to write some code to understand the mechanics. 5! would be represented as follows:

```
[In]
x = [1, 2, 3, 4, 5, 1]
result = 1
for i in range(1, 6):
    result *= i
    print(result)
```

→

```
[Out]
120
```

Here, the list x is not used at all and the 'for' loop iterates over the range (1, 6) instead of the elements of x. The value of 'result' starts at 1 and is updated on each iteration of the loop by multiplying the current element of the range. On the first iteration, 'result' becomes $2(1 * 2)$. On the second iteration, 'result' becomes $6(2 * 3)$. This process continues until all elements of the range have been processed, at which point 'result' is equal to 120 ($1 * 2 * 3 * 4 * 5$).

+ **Conditional Brackets:** Conditional brackets are used to divert the flow of an equation based on a set of conditions.

$$x = \begin{cases} 0 & i < 0, i \notin y \\ 1 & i < 0, i \notin y \\ \sum y & i \in y \end{cases}$$

For coders, this is simply the common 'if' statement. The above conditional can be represented as follows:

```
[In]
i = 3
y = [-2, 3, 4, 1]
result = 0
if i in y:
    result = sum(y)
elif i > 0:
    result = 1
else:
    result = 0
print(result)
```

→

```
[Out]
3
```

The right notation of each row of the bracket dictates what each path should execute. We threw the extra 'contains' symbol into each condition as well to add more insight. As seen above, we checked whether the i value was in the y list. Recognizing that it was, we returned the sum of the array. If the i value had not been in the array, we would have returned 0 or 1 based on the value. To be specific, the code defines an 'if' statement with two 'elif' clauses and an 'else' clause. The 'if' clause checks if i is an element in the list y. If it is, then the 'result' variable is updated to the sum of all elements in y. the 'elif' clause checks if i is greater than 0. If it is, then 'result' is updated to 1. Finally, the value of 'result' is printed as above.

+ **Pointwise and Cartesian Matrix Multiplication:** We want to quickly cover an operation that is typically performed for a data scientist by his or her favorite language library: matrix multiplication. The easiest form to understand is the pointwise operation. This is simply written as follows:

$$x = \begin{bmatrix} 2 & 1 \\ 4 & 3 \end{bmatrix} \odot \begin{bmatrix} 1 & 2 \\ 3 & 4 \end{bmatrix}$$

Note that the first requirement is that each matrix have the same shape (i.e., #rows= and #Columns=). The code looks like this:

```
[In]
y = [[2,1],[4,3]]
z = [[1,2],[3,4]]
x = [[0,0],[0,0]]
for i in range(len(y)):
    for j in range(len(y[0])):
        x[i][j] = y[i][j] * z[i][j]
        print(x)
```

→

```
[Out]
[[2, 0], [0, 0]]
[[2, 2], [0, 0]]
[[2, 2], [12, 0]]
[[2, 2], [12, 12]]
```

Finally, let us take a look at a typical matrix multiplication process most commonly used in data analysis. In complex terms, this operation finds the dot product of each primary row and each secondary column. The main takeaway from this is the following requirement: assume [#rows, #columns] →matrices i x j requires #columns(i) == #rows(j) → with a final product of shape [#rows(i), #columns(j)]. This may seem confusing. Our best suggestion would be to take a look at images for some great visualizations of these requirements.

$$x = \begin{bmatrix} 1 & 2 \\ 3 & 4 \end{bmatrix} \times \begin{bmatrix} 2 \\ 1 \end{bmatrix}$$

The code for this equation looks like this (using the Jupyter; NumPy dot method):

```
[In]
y = [[1,2],[3,4]]
z = [[2], [1]]
# x has shape [2, 1]
x = [[0], [0]]for i in
range(len(y))
    for j in range(len(z):
        x[i][j] = np.dot(y[i],
z[:, j])
print(x)
```

→

```
[Out]
[[4], [10]]
```

The code defines two 2D lists: y and z, and a 2D list x with shape [2, 1]. It then defines a nested 'for' loop to iterate over the rows of y and the columns of z. The outer loop iterates over the rows of y, and the inner loop iterates over the columns of z. On each iteration, the code uses the 'np.dot()' function to compute the dot product of the current row of y and the current column of z. The result is then assigned to the corresponding element of x as printed result above.

These are just a few examples but understanding this simple code can allow any programming effort to take on the initially ominous world of mathematics. These methods can all be consolidated for efficiency and usually have a library (i.e., NumPy, Panda) method readily available. The point of writing them in code is to see how much sense they make when written out in the form of their true operations (Table S3.1.1).

[Table S3.1.1] Basic Arithmetic and Comparison Operators

	Name	Sign	Example	Description
Arithmetic	Addition	+	$a + b$	Adds values on either side of the operator
	Subtraction	–	$a - b$	Subtracts right hand operand from left hand operand
	Multiplication	*	$a * b$	Multiplies values on either side of the operator
	Division	/	b / a	Divides left hand operand by right hand operand
	Exponential	**	$a ** b$	Performs exponential calculation on operators
Comparison	Equal	==	$x == y$	If the values of two operands are equal, then the condition becomes true
	Not equal	!=	$x! = y$	If the values of two operands are not equal, then the condition becomes true
	Greater than	>	$x > y$	If the value of the left operand is greater than that of the right operand, then the condition becomes true
	Less than	<	$x < y$	If the value of the left operand is less than that of the right operand, then the condition becomes true
	Greater than or equal to	>=	$x >= y$	If the value of the left operand is greater than or equal to that of the right operand, then the condition becomes true
	Less than or equal to	<=	$x <= y$	If the value of the left operand is less than or equal to that of the right operand, then the condition becomes true

Note: Operators are the constructs that can manipulate the value of operands (Python). Consider the expression 4 + 5 = 9. Here, 4 and 5 are called operands and + is called the operator.

Supplemental Information 3.2: Differential, Derivative, and Integral

A function is one of the basic concepts in mathematics (Råde & Westergren, 1995). It defines a relationship between a set of inputs and a set of possible outputs where each input is related to one output. One variable is the independent variable, and the other variable is the dependent variable. Functions are essential to defining physical relationships. The statement 'y is a function of x' means that something related to y is directly related to x by a formula. Let us say that the input is 6 and the function is to add 5 to input 6. The result will be 6 + 5 = 11, which is the output. A few exceptions in mathematics cannot be solved by ordinary methods of geometry and algebra alone. A branch of mathematics known as calculus is used to solve these problems. To better understand them, we need to understand the following concepts.

+ **What Is the Differential?** Differential calculus is a fundamental division of calculus, along with integral calculus (Courant, 2011). It is a subfield of calculus that deals with infinitesimal change in some varying quantity. The world we live in is full of interrelated quantities that change periodically. For example, the area of a circular body changes as the radius changes, or a projectile change with its velocity. These changing entities are called variables, and the rate of change in one variable with respect to another is called the derivative. Let us consider a first-order differential equation, ($dy/\mathrm{dx} = x + y$). For example,

$$dy/dx = 2x + 3y$$

In this equation, the rate of change of the dependent variable y with respect to the independent variable x is given by 2 times x plus 3 times y. To get a clearer understand-ding, we consider a specific point $(x, y) = (1, 1)$. At this point, we can calculate the rate of change:

$$dy/dx = 2(1) + 3(1) = 2 + 3 = 5$$

So, at the point (1, 1), the rate of change of y with respect to x is 5. This means that if x increases by a small amount from 1, y will increase approximately 5 times that amount. This equation describes the rate of change of the dependent variable y with respect to the independent variable x.

```
[In]
def dy_dx(x, y):
    return 2 * x + 3 * y
# Calculate the rate of change at point (1, 1)
x_pt, y_pt = 1, 1
rate_of_change = dy_dx(x_pt, y_pt)
```

→

[Out]

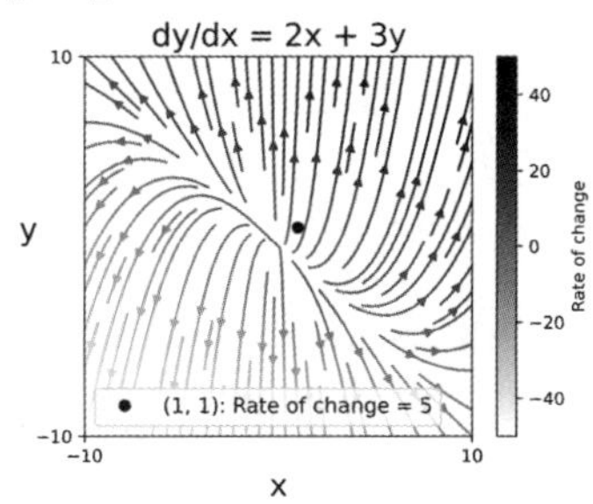

Figure S3.2.1. Define the differential equation: $dy/dx = 2x + 3y$

The rate of change is given by the sum of x and y (Figure S3.2.1). Solving this equation provides a function that represents the relationship between x and y, giving insight into how these variables change in relation to one another. Differential equations contain unknown functions and some of their derivatives.

+ **What Is the Derivative?** The derivative of a function is one of the most powerful concepts in mathematics (Liu, 2009). It is usually a new function called the derivative function or the rate function. The derivative of a function represents an instantaneous rate of change in the value of a dependent variable with respect to the change in value of an independent variable. Let us consider a simple example of a quadratic function $[f(x) = x^2 + 3x + 2]$, and then calculate its derivative and evaluate it at a specific point. First, we find the derivative of the function with respect to x:

$$f'(x) = 2x + 3$$

The derivative function $f'(x)$ represents the instantaneous rate of change in the value of $f(x)$ with respect to x. Now, let us evaluate the derivative at a specific point, say $x = 1$:

$$f'(1) = 2(1) + 3 = 5$$

At $x = 1$, the instantaneous rate of change of the function $f'(x)$ with respect to x is 5. This means that, at this point, the slope of the tangent line to the graph of $f(x)$ is 5. So, if x increases by a small amount from 1, the value of $f(x)$ will increase approximately 5 times that amount.

```
[In]
def f(x):
    return x**2 + 3 * x + 2
def f_derivative(x):
    return 2 * x + 3
# Evaluate the derivative at x = 1
x_pt = 1
f_derivative_pt = f_derivative(x_pt)
```

→ [Out]

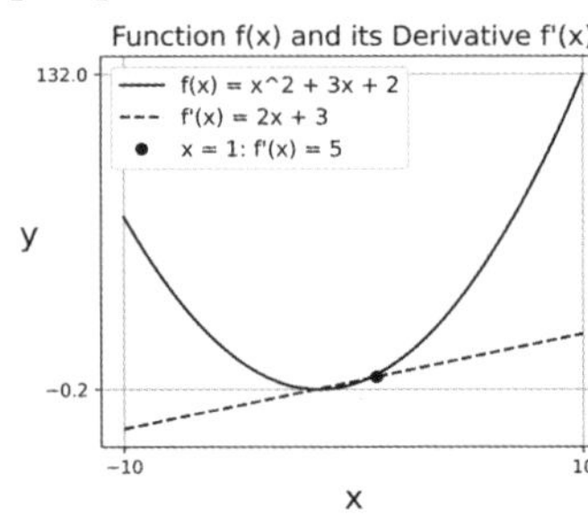

Figure S3.2.2. Define the function $f(x) = x^2 + 3x + 2$ and its derivative $f'(x) = 2x + 3$

This fundamental tool of calculus can also be interpreted as the slope of the tangent line (Figure S3.2.2). It measures how steep the graph of a function is at some given point on the graph. In simple terms, the derivative is the rate at which a function changes at some particular point.

+ **What Is the Integral?** The integral of a function can be interpreted geometrically as the area under the curve of the mathematical function $f(x)$ plotted as a function of x. You can see yourself drawing a large number of blocks to approximate the area under a complex curve, getting a better answer if you use more blocks. The integral gives you a mathematical way of drawing an infinite number of blocks and getting a precise analytical expression for the area. Let us consider a simple example of a linear function and calculate its integral over a specific interval to find the area under the curve [Function: $f(x) = 2x + 1$]. We want to find the area under the curve of this function between $x = 0$ and $x = 2$. To do this, we first find the integral of the function with respect to x:

$$f(x) = \int (2x + 1)\ \mathrm{d}x = x^2 + x + \mathrm{C}$$

Now, we can find the area under the curve between $x = 0$ and $x = 2$ by evaluating the integral at these points:

$$Arear = f(2) - f(0) = (2^2 + 2) - (0^2 + 0) = 6\ square\ units$$

In this example, the area under the curve of the function $f(x) = 2x + 1$between $x = 0$ and $x = 2$ is 6 square units. This is very important for geometry—and profoundly important for the data sciences, where the definitions of many realistic entities can be cast in a mathematical form like the area under a curve.

```
[In]
def f(x):
    return 2 * x + 1
# Define the integral function F(x)
def integral_f(x):
    return x**2 + x
# Calculate the area between x = 0 and x = 2
area, _ = quad(f, 0, 2)
```

→

[Out]

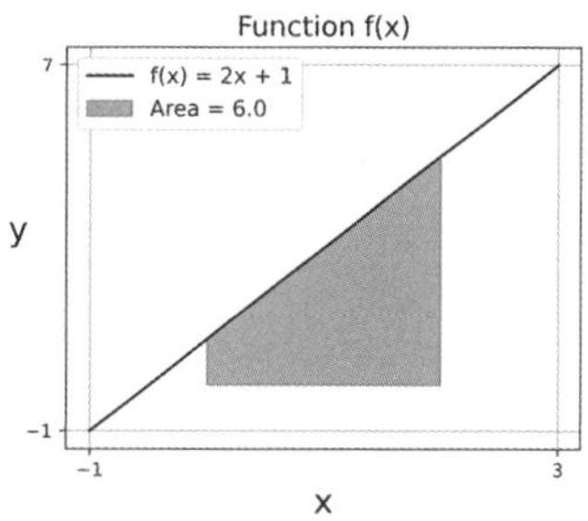

Figure S3.2.3. Define the integral function $f(x) = x^2 + x + C$

The area of a little block under the curve can be thought of as the width of the strip weighted by (i.e., multiplied by) the height of the strip (Figure S3.2.4). Notice the function we provided, $f(x) = 2\,x + 1$, is a linear function, which is why the plot shows a straight line, not a curve. If you want to plot a curve, we need to change the function.

Here is an example where we plot a quadratic function $g(x) = x$^2

```
[In]
def g(x):
    return x**2


# Calculate the area under the curve with g(x)
area, _ = quad(g, 0, 2)
```

→

[Out]

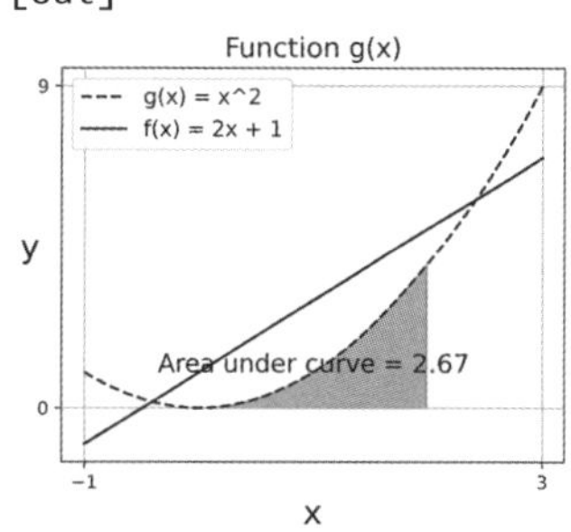

Figure S3.2.4. Define the integral function $g(x) = x$^2

Many properties of continuous bodies depend on weighted sums, which to be exact must be infinite weighted sums—a problem tailor-made for the integral (Clauser et al., 1969), a vast number of problems involve such infinite sums in their solutions, which makes the integral an essential tool for the systems analysis.

+ **Practical Practice with Differential Equations:** Let us consider a theoretical simulation using a fundamental model called The Predator-Prey Model. The differential equations are very sneaky, as they can be found in many facets of life, providing an indirect solution to many problems. In data analysis, differential equations are often used to simulate real-world scenarios. They are also used to analyze many different things, whether it be data structures and algorithms or traffic in networks. The applications of differential equations are truly endless.

One example of a differential equation is a Lotka-Volterra equation, which models the relationships between two species involved in a predator-prey biological system (Hofbauer & Sigmund, 1998). Proposed in 1910, this model describes change in the populations of two species over time, which is useful in the field of mathematical biology, as the dependency of two species can be modeled mathematically and therefore solved. These equations also have surprising use in economic systems, as presented by the Goodwin (1967) model, a nonlinear model that demonstrates the cyclical relationship between employment and wage share in a working economy.

$$\frac{dx}{dt} = \alpha x - \beta xy, \qquad \frac{dy}{dt} = \delta xy - \gamma y,$$

The Lotka-Volterra equations above model predator-prey relationships. In these equations, x and y are the number of prey and predators, respectively; t represents time; dx/dt and dy/dt represent the rate of growth of the two populations; and α, β, γ and δ are positive real parameters describing the interactions between the two species. For example:

- α is the natural growth rate of prey when there is no predator
- β is the natural death rate of prey due to predation
- γ is the natural death rate of predators when there are no prey
- δ is a factor describing how many caught prey let produce a new predator

When programing this defining of parameters:

```
[In]
a = 1.
b = 0.1
c = 1.5
d = 0.75
def dX_dt(X, t=0):
    return array([ a*X[0] -   b*X[0]*X[1] ,
                  -c*X[1] + d*b*X[0]*X[1] ])
```

We defined the dynamics of the predator-prey system, where the first element of x represents the prey population and the second element represents the predator population. The parameters a, b, c, and d are used to specify the growth and death rate of the population. Defined function of 'dx_dt' takes in two argument x and t. the x is a list or array with two elements, and t is a time parameter with a default value of 0. The function then returns an array of two elements, which are the growth rate of the prey 'a∗X[0] - b∗X[0]∗X[1]' and predator '−c∗X[1] + d∗b∗X[0]∗X[1]' populations. When graphed, the two equations form a graph similar to that of simple harmonic motion, with one trailing behind the other (as a decrease in one population does not immediately affect the other population). When programing this for calculation:

```
[In]
from scipy import integrate
t = linspace(0, 50,  1000)                 # time
X0 = array([10, 5])                        # initials: 10 preys and 5 predators
X, infodict = integrate.odeint(dX_dt, X0, t, full_output=True)
infodict['message']                        # >>> 'Integration successful.'
```

We used the 'odeint' function from the 'scypy.integrate' module to solve a system of ordinary differential equations (ODE). The dx_dt represents the system of 'ODE' and the initial conditions are given by the array 'X0'. The 'odeint' integrates the ODE over a given time interval specified by the array t and returns the solution at each time point in the array x. for visualization

```
[In]
rabbits, foxes = X.T
f1 = p.figure()
p.plot(t, rabbits, 'b-.', label='prey')
p.plot(t, foxes   , 'r-.', label='predator')
```

↓

[Out]

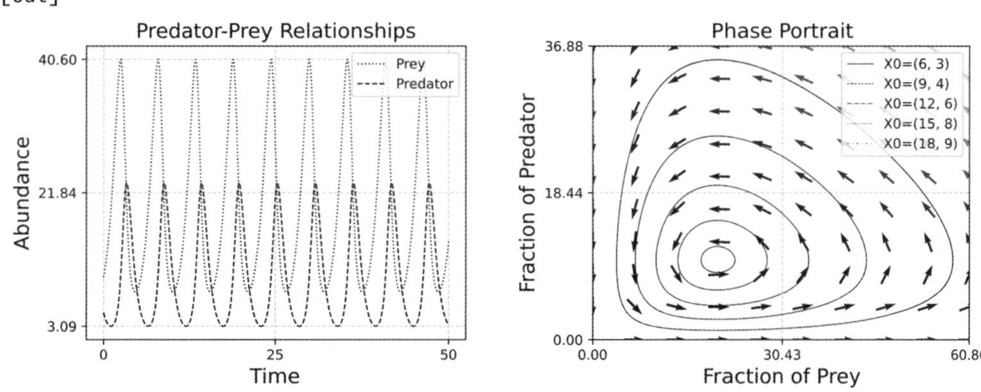

Figure S3.2.5. Linearization of the two equations (visualizing the simple harmonic motion of two populations). The left plot describes the relationship between the two species. The right plot shows how changing either the predator or the prey population can have an unintuitive effect.

In Figure S3.2.5, it shows a plot of the solutions x(t) and y(t) for $t \in [0, 50]$ for the set (α, β, γ and δ) = (2/3, 4/3, 1, 1), $x(0) = 1$, y(0) = 1. Lotka-Volterra equations provide insight and a foundational basis for the cyclical relationship between two populations, and they are in general for use with real-world situations (Park, 2023). The information provided only goes so far, and additional application or context may be needed to provide a more accurate representation of a real-world predator-prey relationship.

+ **Practical Practice with Derivatives:** Let us practice with a fundamental derivative model called SEIR concerning an example of the spread of a virus (i.e., COVID-19). The population is set to N, and the results are represented in fractions of the population. At each time, the population is divided into four categories that sum to a total of 1 (S = susceptible, E = exposed, I = infected, R = recovered or dead). The fractions of the population evolve over time as follows:

$$\frac{dS}{dt} = \beta_t \frac{S_t}{N} I_t,\ \frac{dE}{dt} = \beta_t \frac{S_t}{N} I_t - \sigma E_t,\ \frac{dI}{dt} = \sigma E_t - \gamma I_t,\ \frac{dR}{dt} = \gamma I_t$$

The parameter γ is the rate per day at which infected people either recover or die from the virus. It is considered a fixed parameter (set to $\gamma = 1/18$) reflecting an estimated duration of illness of 18 days (Atkeson, 2020). The parameter σ is the rate at which those exposed to the virus become infected. It is also a fixed parameter (set to $\sigma = 1/4$) reflecting an estimated incubation period of 4 days (i.e., COVID-19). When programing:

```
[In]
γ = 1 / 18 # to reflect an average incubation period of 4 days.
σ = 1 / 4 # to match an average illness duration of 18 days.
def F(x, t, R0=2.2):
    s, e, i, r = x
    β = R0(t) * γ if callable(R0) else R0 * γ # New exposure of susceptible
    ne = β * s * i
    ds = - ne                                 # Time derivatives
    de = ne - σ * e
    di = σ * e - γ * i
    dr = γ*i
    return ds, de, di, dr
t_length = 365*2 # 2 years investigated
grid_size = 1000
t_vec = np.linspace(0, t_length, grid_size)
```

The code defines a function F that represents a simple mathematical model of the infectious disease. It takes in three arguments x, t, and 'R0'. The x is a list of arrays with four elements representing the proportion of the population that is susceptible, exposed, infectious, and recovered at a given time t. the 'R0' is the basic reproductive number of the disease which is a measure of how contagious the disease is (as constant). It also specifies the average incubation period of the disease 'γ' and the average duration of illness 'σ'. Then, it returns the time derivatives of the four variables in x which represent the rates of change of the four proportions. The rates of change are calculated using a set of differential equations and 't-length' represents the length of time, in this case set to 2 years.

```
[In]
R0_vals = np.linspace(1.5, 3.0, 3)
i_paths, c_paths = [], []
for r in R0_vals:
    i_path, c_path = solve_path(r, t_vec)
    i_paths.append(i_path)
    c_paths.append(c_path)
def plot_paths(paths, labels,
times=t_vec):
    fig, ax = plt.subplots()
    for path, label, ls in zip(paths,
labels):
        ax.plot(times, path)
```

→ [Out]

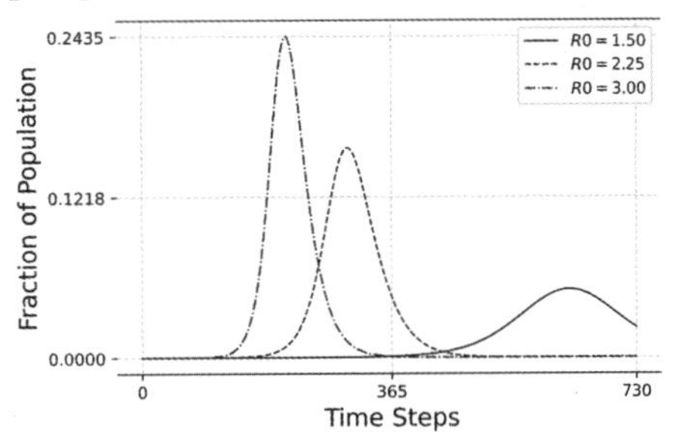

Figure S3.2.6. Simulation with transmission rates (R0): the fraction of the population (left line: transmission rate = 1.5, middle line: transmission rate = 2.2, right line: transmission rate = 3.0)

In Figure S3.2.6, β_t is the transmission rate. Of the people that one who is infected bumps into, fraction S/N are susceptible and hence transition to being exposed. R_t is the ratio of this meeting rate β_t and recovery + death rate γ on day t. This parameter governing the ratio of the rates at which those who are susceptible become infected and those who are ill either recover or die varies over time and is controlled by social distancing and quarantining. Let us parameterize this rate R_t for an initial period of intense application of social dis-tancing followed by a relaxation of these measures to allow economic activity to resume.

$$R_{1t} = R_{1,0} \exp(-\eta_1 t) + (1 - \exp(-\eta_1 t)\, \bar{R}_1, \qquad R_{2t} = R_{2,0} \exp(-\eta_2 t) + (1 - \exp(-\eta_2 t)\bar{R}_2$$

$$R_t = (R_{1t} + R_{2t})/2$$

Here $R_0 = (R_{1,0} + R_{2,0})/2$ is the initial value of R_t representing the spread of the disease in its initial phase. The parameters $\bar{R}_i$ for $i = 1,2$ indicate the long-term values to which R_{1t} converges. Thus, in the long run R_t converges to $(\bar{R}_2 + \bar{R}_2)/2$.

To get a U-shaped pattern for R_t, we make R_{1t} a rapidly declining function and R_{2t} a slowly rising function. The parameter η_1 governs the rate at which R_{1t} falls toward $\bar{R}_1$. The parameter η_2 governs the rate at which R_{2t} rises toward $\bar{R}_2$. When programing:

```
[In]
def R0_mitigating(t, r0=2.2, η=1, r_bar=1.5):
    R0 = r0 * exp(- η * t) + (1 - exp(- η * t)) * r_bar
    return R0
η_vals = 1/5, 1/10, 1/20, 1/50, 1/100
```

Generate the results given;

```
[In]
fig, ax = plt.subplots()
for η, label in zip(η_vals, labels):
    ax.scatter(t_vec,
R0_mitigating(t_vec, η=η))
i_paths, c_paths = [], []
for η in η_vals:
    R0 = lambda t: R0_mitigating(t, η=η)
    i_path, c_path = solve_path(R0,
t_vec)
    i_paths.append(i_path)
    c_paths.append(c_path)
```

→

[Out]

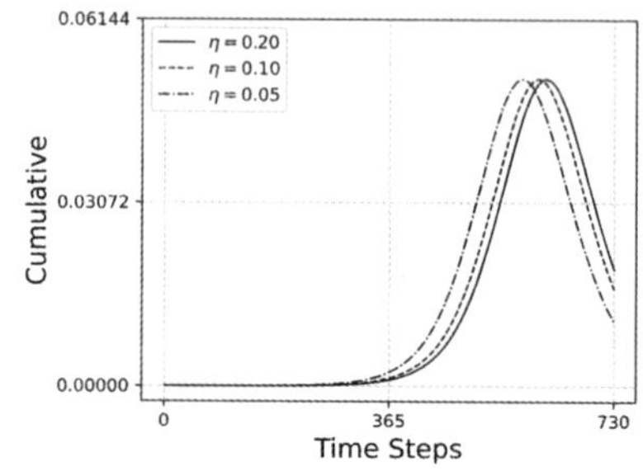

Figure S3.2.7. Simulation with social distancing (η) when the transition rate equals 3: the fraction of the population (left line: social distancing = 0.2, middle line = 0.1, right line = 0.05)

The code defines three arguments time t, initial value of the basic reproductive number 'R0', and how quickly it decreases over time 'η'. The 'r_bar' represents the value of 'R0' is reached asymptotically as t approaches infinity. And the function calculates the value of 'r0' at a given time t using an exponential decay function. The value of 'R0' is initialized to 'r0' at $t = 0$ and it decrease exponentially towards 'r_bar' as t increase. And the rate of decay is controlled by the parameter 'η' which the values correspond to different rates of decay for R0. To get the image in Figure S3.2.7, we ran the simulation. We may infer from the simulations how different sorts of interventions (such as social distance) affect the quantity of instances and how those numbers evolve over time.

+ **The Relationship between the Differential and Derivative:** The differential and derivative are intimately connected to each other because they are interrelated (Wahlquist & Estabrook, 1975). In mathematics, entities that change are called variables, and the rate of change in one variable with respect to another is called the derivative. Equations that define the relationship between variables and their derivatives are called differential equations. Differentiation is a method of computing a derivative, which is the rate of change in the output y of a function with respect to change in the variable x. In simple terms, the derivative is the rate of change in y with respect to x. This relationship is expressed as $y = f(x)$, which means that y is a function of x. The derivative of the function $f(x)$ is the function whose value generates the slope of $f(x)$ where it is defined, and $f(x)$ is differentiable. It refers to the slope of the graph at a given point. When comparing changes in related quantities where y is the function of x, the differential dy can be written as $[dy = f'(x)\,dx]$. The derivative of a function is the slope of the function at any point and is written as d/dx. For example, the derivative of $\sin(x)$ can be written as $[d/dx \sin(x) = \cos(x)]$. The equation sates that the derivative of the sine function with respect to x is equal to the cosine function. Because the rate of change of the sine function is equal to the cosine function, the function is a measure of how the changes as its input x changes.

[3.3]

Networks, Part 1

A Guide to Concepts, Network Applications, Network Structures, Measuring Centralities

Someone needed a way of keeping track of the order of things, so we played around with and created different data structures until finding one that worked best for that specific problem (McGinn, 1989). Someone else needed a good way of storing data and so played around with different number systems until identifying one that worked best for the situation.

In this section, we look at network from the viewpoint of systems analysis. We will not be covering proofs because we do not take a mathematics or computer science perspective; instead, we will be looking at the implementation of basic algorithms and a couple of fundamental applications with code. Because network can be used to represent almost any problem, they pop up everywhere (Liben-Nowell & Kleinberg, 2007).

1. A Guide to Concepts

The basic idea of network was first introduced in the 18th century by the Swiss mathematician Leonhard Euler, one of the most eminent mathematicians of the 18th century. His work on the famous Seven Bridges of Königsberg problem is commonly credited as the origin of graph theory (Euler, 1956). The city of Königsberg in Prussia (now Kaliningrad, Russia) was set on both sides of the Pregel River and included two large islands—Kneiphof and Lomse—that were connected to each other, and to the two mainland portions of the city, by seven bridges (as illustrated in the figure below at left).

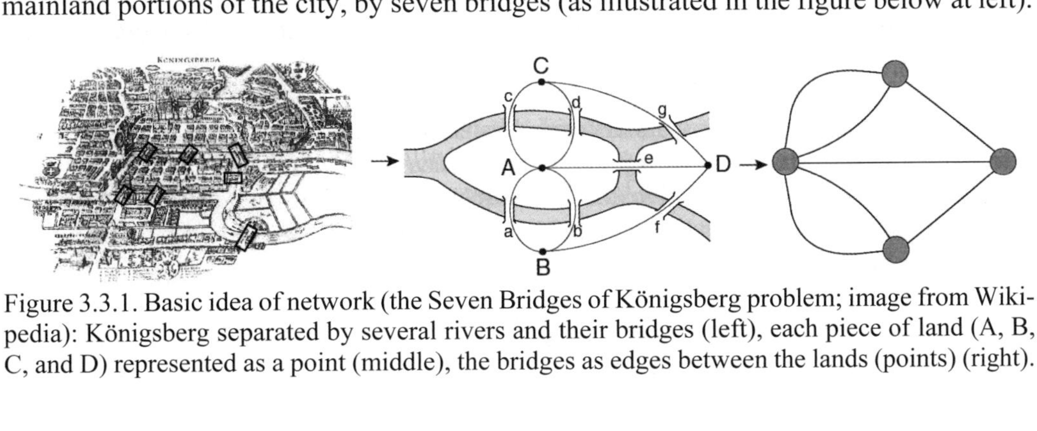

Figure 3.3.1. Basic idea of network (the Seven Bridges of Königsberg problem; image from Wikipedia): Königsberg separated by several rivers and their bridges (left), each piece of land (A, B, C, and D) represented as a point (middle), the bridges as edges between the lands (points) (right).

The problem was to devise a walk through the city that would cross each of those bridges once and only once (Figure 3.3.1).

+ **Graph Theory:** Euler, recognizing that the relevant constraints were the four bodies of land and the seven bridges, drew out the first known visual representation of a modern graph. A modern graph, seen in the image below at right, is represented by a set of points, known as vertices or nodes, connected by a set of lines known as edges.

This abstraction of a concrete problem about a city and bridges to a graph makes the problem tractable mathematically, as it includes only the information important for solving the problem. Euler actually proved that this specific problem has no solution. However, the difficulty he faced was in the development of a suitable technique for analysis and of subsequent tests that established this assertion with mathematical rigor. Thus, the branch of math known as graph theory lay dormant for decades (Heckmann et al., 2015). A simple problem that can be expressed as a graph theory problem might be the following: Given the constraints in a certain picture, how many different sets of objects can we make by choosing an article from each category (Krishnaa, 2018)? This problem could also be expressed and solved using only mathematics, but the advantage of using graph theory is that it allows us to visualize the problem using nodes to represent items and edges to represent relationships between them.

In essence, graph theory is the study of the properties and applications of graphs, which common folk refer to as networks. This is fundamental, because our goal in this section is to apply graph theory to systems analysis.

+ **Identification**: People needed a good way of labeling and processing tasks, so they found a way to build on the tools they had and created a way to juggle all the things that one single system needed to do at any given time. To truly know why we use graphs and what they are, we need to go deep down to the very roots of something that stems from discrete mathematics: trees (West, 2001).

One of the most fundamental characteristics of nonlinear structures is that their data do not follow a particular order—at least, not an obvious numerical one, like in arrays or linked lists. Trees start with a root node and might connect to other nodes, which means that they can contain subtrees within them. Trees are defined by a certain set of rules: One root node may or may not connect to others, but ultimately it all stems from one specific place (Moretti, 2003).

Some trees have even more specific rules, like binary search trees, which can only ever have two links to two nodes at any given time. However, what if we did something kind of crazy and just threw these rules out the window? As it turns out, we can do that, and it is just that we would not be dealing with trees anymore—we would be dealing with something called a graph (Bondy & Murty, 1976).

Trees are nothing more than restricted graphs, just with many more rules to follow. A tree is always a graph, but not all graphs are trees. What is it that makes a tree different from the larger umbrella of graphs? For one thing, a tree can only flow in one direction, from the root node to either leaf nodes or child nodes. A tree can also only have one-way connections—a child node can only have one parent, and a tree cannot have any loops or cyclical links.

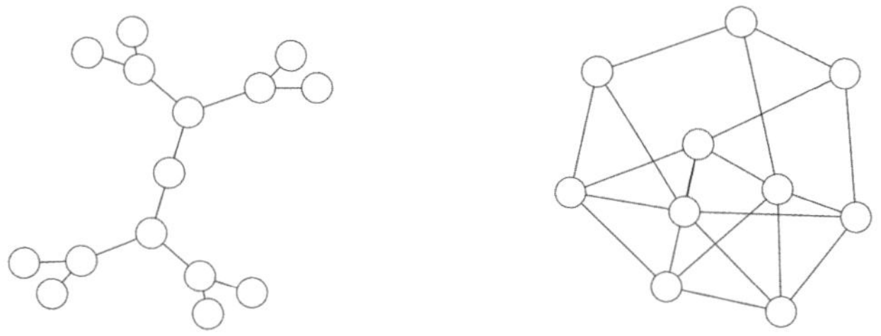

Figure 3.3.2. Tree data structures (left) compared to graph data structures (right)

With graphs (Figure 3.3.2.), all of these restrictions go straight out the window. Graphs do not have any concept of a root node. And why would they? Nodes can be connected in any way possible. One node might be connected to five others. Graphs also do not have any notion of one-directional flow—instead, they might have direction, or they might have no direction whatsoever. Or, to complicate matters further, they could have some links that have direction and others that do not, but we will not get into that here. Let us stick with the simple stuff to start.

+ **Elements:** Graph theory is ultimately the study of relationships. Given a set of nodes and connections, which can abstract anything from city layouts to computer data, graph theory provides a helpful tool for quantifying and simplifying the many moving parts of dynamic systems (Bollobás, 2013). Studying graphs through a framework provides answers to many problems about arrangement, networking, optimization, matching, and operations.

The primary structure of a graph shows how different things are connected, sometimes in a structure called a network (Newman & Girvan, 2004). These are called vertices (fundamental elements drawn as nodes), and the lines over there are called edges (a set of two elements drawn as a line connecting two vertices).

Figure 3.3.3. Nodes and links of a graph

In Figure 3.3.3, nodes are the parts of graphs that store data, and edges are the connections between them, but edges can store data too. The edges between nodes can describe any connection between them (called *adjacent*). Nodes contain any amount of data, like each agent's capital and strategy, and edges include data about the strength of a connection.

+ **Connectivity:** Another essential property of a graph is connectivity. A disconnected graph has some vertices (nodes) that cannot be reached by other vertices.

Figure 3.3.4. Connectivity of a graph

A disconnected graph might have one vertex off to the side with no edges. It could also have two so-called connected components, which are graphs that are connected on their own but have no connection between them as you can see in the above graph (Figure 3.3.4). Thus, a connected graph has no disconnected vertices. Depending on the information in the graph, which is usually controlled by (n, p), we use connectivity for an initial condition.

+ **Representation List:** An object-oriented language can be used to create vertices and edge objects and give them each a property (i.e., capital and strategies). A vertex can have a list of edges it is connected to and vice versa. However, operations that graph can be inconvenient if we need to search through vertices and edge objects.

Thus, we represent connections on simple graphs that simply use a list of edges.

Figure 3.3.5. List of edges in a graph

The edges themselves are represented by a list of two elements: [1, 2]. These elements are normally identification numbers for vertices. Thus, this list shows two nodes that have an edge between them (Figure 3.3.5). An edge list encompasses all smaller lists. Because the edge list contains all other lists, it is sometimes called a *two-dimensional list*.

+ **Adjacency List:** We represent the edge list of a graph as an adjacency list. The vertices normally have a number that corresponds to the index in an array.

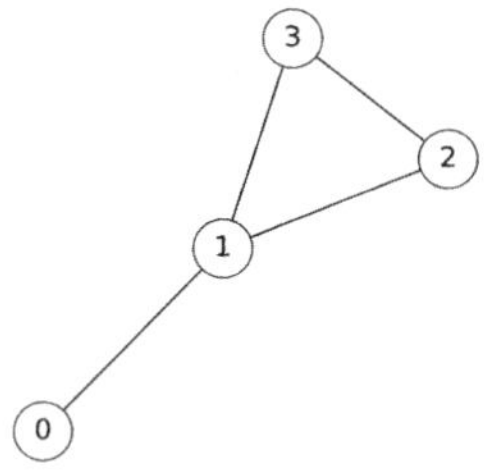

Adjacency list (two dimensional)
[0 = [0, 1], 1 = [1, 2],
2 = [1, 3], 3 = [2, 3]]

Figure 3.3.6. Adjacency of a graph

In Figure 3.3.6, each space in the array is used to store a list of nodes to which the primary node is adjacent. For instance, the opening at index 0 represents a vertex with an identification number of 0. That vertex shares an edge with one node, so a reference to that node is stored in the first spot in the array. Because it contains other lists, the adjacency list is two dimensional, and thus it allows us to use an adjacency matrix that is essentially a two-dimensional array, but the lists inside are all the same length. The adjacency list represents the graph as a map of nodes to a list of edges. The idea is that each node tracks all of its outgoing edges.

+ **Basic Calculations:** Let us create a graph based on the fundamental properties. We start by randomly adding edges between pairs of nodes one at a time. Now we can calculate the degree based on the adjacency list (Figure 3.3.7). For example, consider an undirected network that has four nodes (vertices) with possible edges (lines).

Network (undirected)

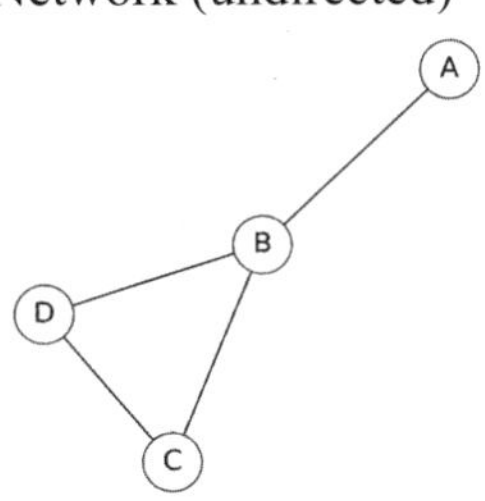

Label

0 = A,
1 = B,
2 = C,
3 = D

Pairs of nodes

AB = BA 1
AC = CA 0
AD = DA 0
BC = CB 1
BD = DB 1
CD = DC 1

Figure 3.3.7. Describing a graph with an adjacency matrix

Consider an adjacency matrix whose rows and columns are labeled by vertices (nodes). Here we have 1, 2, 3, and 4 vertices. The actual labels we use can be anything, but let us call our adjacency matrix A. This is the actual definition of our matrix. An entry in row m column n will be equal to either 1 or 0. It will be equal to 1 if there is an edge between m and n, and it will be equal to 0 if m is not connected to n.

$$Am \times n = \begin{matrix} & 1 \quad 2 \quad 3 \quad 4 \\ \begin{matrix}1\\2\\3\\4\end{matrix} & \begin{bmatrix} & & & \\ & & & \\ & & & \\ & & & \end{bmatrix} \end{matrix}, \quad A = \begin{cases} 1 & if\ mn \in edge \\ 0 & otherwise \end{cases}$$

The diagonal elements of the adjacency matrix are 0 for graphs without loops.

$$A = \begin{array}{c} \\ 1 \\ 2 \\ 3 \\ 4 \end{array}\!\!\begin{array}{c} \begin{array}{cccc} 1 & 2 & 3 & 4 \end{array} \\ \begin{bmatrix} 0 & 1 & 0 & 0 \\ & & & \\ & & & \\ & & & \end{bmatrix} \end{array}, \quad A = \begin{array}{c} \\ 1 \\ 2 \\ 3 \\ 4 \end{array}\!\!\begin{array}{c} \begin{array}{cccc} 1 & 2 & 3 & 4 \end{array} \\ \begin{bmatrix} 0 & 1 & 0 & 0 \\ & 0 & & \\ & & 0 & \\ & & & 0 \end{bmatrix} \end{array}, \quad A = \begin{array}{c} \\ 1 \\ 2 \\ 3 \\ 4 \end{array}\!\!\begin{array}{c} \begin{array}{cccc} 1 & 2 & 3 & 4 \end{array} \\ \begin{bmatrix} 0 & 1 & 0 & 0 \\ 1 & 0 & 1 & 1 \\ 0 & 1 & 0 & 1 \\ 0 & 1 & 1 & 0 \end{bmatrix} \end{array}$$

In an adjacency matrix, rows and columns represent vertices (nodes), so the sum of each row or column is the degree of the related node. In the example above, vertex 2 has three neighbors. For the columns of the matrix:

$$A = \begin{array}{c} \\ 1 \\ 2 \\ 3 \\ 4 \end{array}\!\!\begin{array}{c} \begin{array}{cccc} 1 & 2 & 3 & 4 \end{array} \\ \begin{bmatrix} 0 & 1 & 0 & 0 \\ 1 & 0 & 1 & 1 \\ 0 & 1 & 0 & 1 \\ 0 & 1 & 1 & 0 \end{bmatrix} \end{array} \quad \rightarrow$$

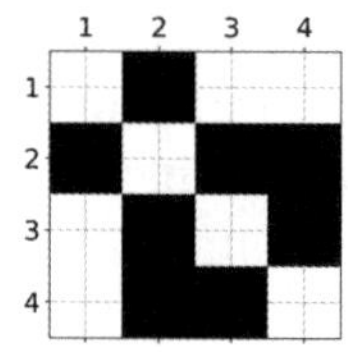

$$\rightarrow \quad \deg(v_{2m}) = 3, \quad \deg(v_{2n}) = 3$$

If we look at the column that represents vertex 2, we get 3, which represents its degree. Keep in mind that the adjacency matrix of a graph has all of the same information that is contained in the graph. Notice that the graph has been shown to us in any random way the computer generates the visuals. The resulting $m \times n$ matrix is obtained by $[A = G(n,p)]$. Any node can be randomly exposed to the other nodes, which creates random connections.

$$A = \begin{bmatrix} a_{11} & a_{12} & \cdots & a_{1n} \\ a_{21} & a_{22} & \cdots & a_{2n} \\ \cdots & \cdots & \vdots & \cdots \\ a_{m1} & a_{m2} & \cdots & a_{mn} \end{bmatrix}$$

Given the collection of nodes influenced by the connection probability $[p \in (0,1)]$, we can find what distribution of the connection in the network is the probability of the degree. There are two parameters: the number of nodes (n) and the probability that an edge is present (p). For a graph shown as $G(n,p)$, $n = 5$ and $p = 0.6$

- The number of possible edges is $n(n-1)/2 = 5 * 4/2 = 10$.

The expected average node degree is $p(n-1) = 0.6 * 4 = 2.4$. To quantify the probability that a node has degree d for all $[0 \le d \le (n-1)]$, notice the following:

- A node has degree 0 if nothing is connected to it.
- A node has degree $(n-1)$ if all nodes are connected to it.

For a node to have degree d, in a network with n nodes there must be d connections and $(n-1-d)$ nodes that are not connected to it. Because the probability of connecting to a node is p, the probability of not connecting to a node is $(1-p)$. The outcome d heads and $(n-1-d)$ tails occurs with probability $p^k(1-p)^{n-1-d}$, but there are choose $(n-1)$ ways in which this outcome can occur (the order of the flip results does not matter). The probability that a given node has degree d is given by the binomial distribution $\binom{n-1}{d} p^k (1-p)^{n-1-d}$. The binomial distribution is $\mu = p(n-1)$, which is also the average node degree.

The Graph theory is a strong tool in many domains because of its various and varied concepts and rules. Let us now discuss a few of the numerous applications of graph theory.

Properties of Three Matrices Associated with a Graph

Examining the properties of any matrices associated with a graph. Thus, we explore most common (significant) matrices here. First, the adjacency matrix is more intuitive and easier to understand. The adjacency matrix $A(m, n)$, whose entries are also determined by $a_{i,j}$

$$a_{i,j} = \begin{cases} 1 & if\ \{i,j\} \in E \\ 0 & otherwise \end{cases} \rightarrow$$

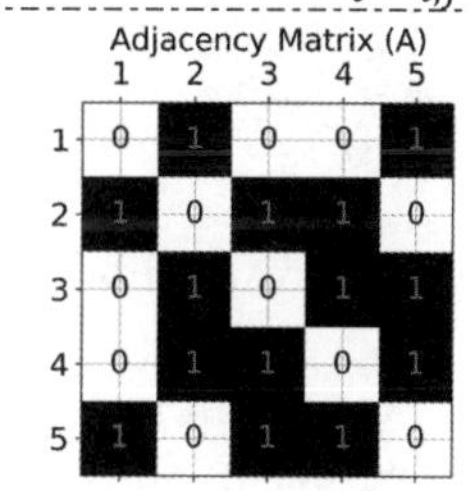

Next, we can also provide a similar definition for the degree matrix d in the context of graph theory, which is a diagonal matrix that encompasses all the information regarding the local connectivity of a node.

$$d_{i,j} = \begin{cases} d(i) & if\ i = j \\ 0 & otherwise \end{cases} \rightarrow$$

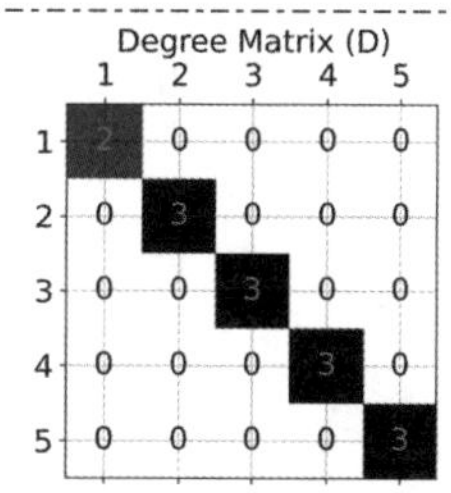

where $d(i)$ denotes the degree of node i, i.e., the number of its neighbors. On the other hand, lastly, we can obtain our initial definition of the Laplacian matrix l by employing a similar approach, defining each entry as

$$l_{i,j} = \begin{cases} -1 & if\ \{i,j\} \in E \\ d(i) & if\ i = j \\ 0 & otherwise \end{cases} \rightarrow$$

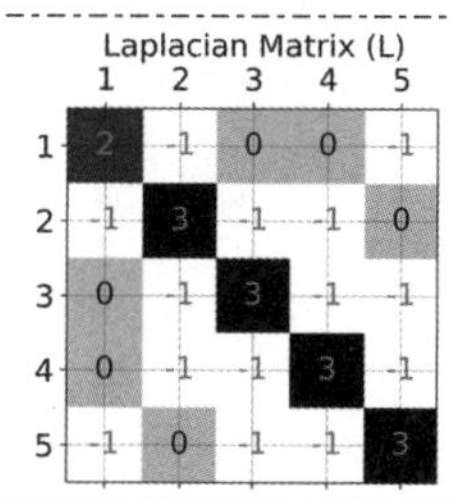

Notably, the Laplacian matrix can simplify in terms of the previous element-wise definition of adjacency and degree matrices as L = D – A.

2. Network Applications

A network is a collection of nodes that are connected by links (i.e., a graph). Informally speaking, a graph is a set of nodes joined by a set of lines or arrows. A node can represent an individual or any agent, and that agent's relationships can be defined by links between nodes (e.g., parenthood: direct link, friendship: undirect link). The structure of a system of interacting agents is represented as a network. Notice that almost every complex system is graph-like (Figure 3.3.8). Network theory provides a set of techniques for analyzing these graphs (Scott, 1988).

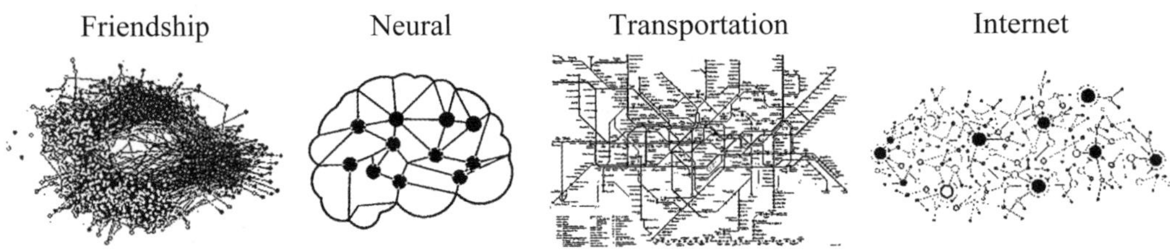

Figure 3.3.8. Representation of many networks

+ **Links:** Each agent can have a number of links. The number of links an agent has is called its degree. In some networks the degree is the same for each agent, whereas in others the degree is different for each agent (Figure 3.3.9). The distribution of degrees in a population of agents is an important characteristic of a network.

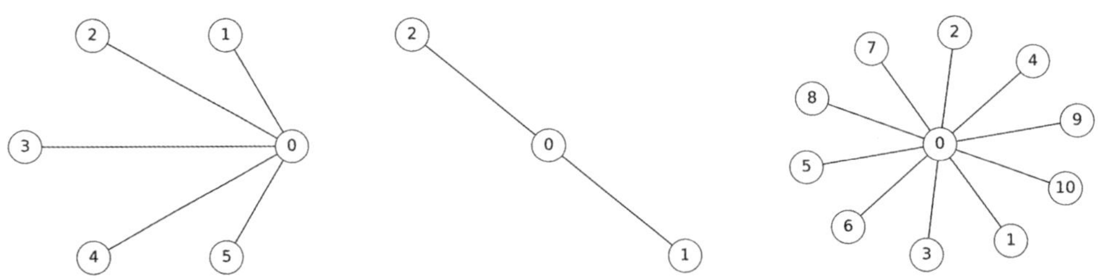

Figure 3.3.9. The number of links among agents: (left) a central node of degree 5, (middle) a central node of degree 2, (right) a central node of degree 10

+ **Degree and Density:** Degree is the number of links incident on a node. The number of links entering a node is the indegree, and the links leaving the node is the outdegree.

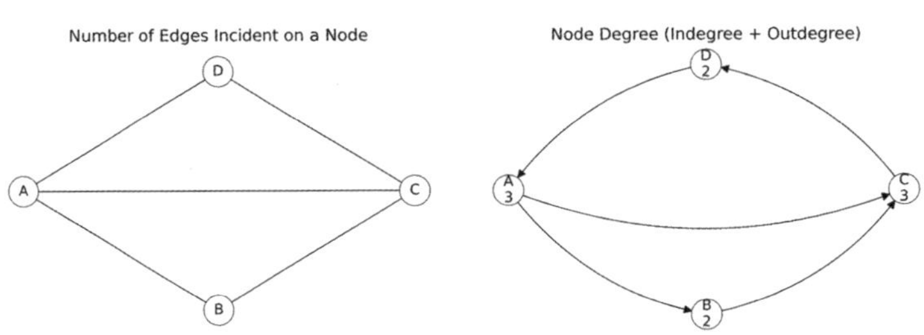

Figure 3.3.10. Representation of the degrees

In Figure 3.3.10, if G is a graph with m edges, $\sum \deg(v) = 2m = 2|E|n$. In a sense, the size of a network is defined by the number of nodes (n) and edges (m) in it. Thus, the density of a network $p(G)$ can be the portion of all possible edges actually realized in the network (i.e., fraction between 0 and 1) given by;

$$undirected\ p(G) = \frac{m}{\frac{n(n-1)}{2}} = \frac{2m}{n(n-1)}, \qquad directed\ p(G) = \frac{m}{n(n-1)}$$

+ **Paths and Clustering Coefficient:** The paths between nodes are another characteristic of a network. Various paths are possible, and the shortest paths of links between nodes may also differ. The network structure affects the length of the shortest path and is an important indicator for comparing networks.

$$path\ length\ L = \frac{\sum_{i,j} d(i \to j)}{n(n-1)}$$

This formula works for both undirected and directed cases and calculates the average length of shortest paths for all possible node pairs in the network.

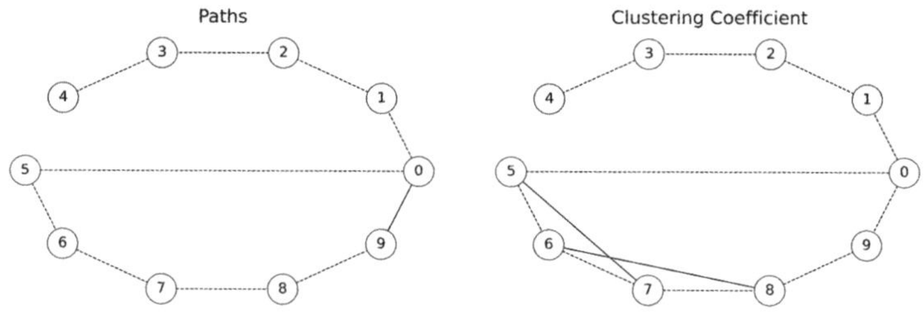

Figure 3.3.11. Different paths and clustering.

In Figure 3.3.11, the left graph demonstrates different paths between nodes, with the black solid line representing the shortest path between node 0 and node 9. The right graph represents the clustering in a network, where the black solid lines show extra connections forming a cluster among nodes. The dashed lines in both graphs indicate other connections in the network. The clustering coefficient indicates how frequently links of a node are linked with each other. The more links a node has, the more clustering there is. The clustering coefficient is the average calculation for all nodes of a network. If there is no connection, the clustering coefficient is 0.

$$C = \frac{\sum_i C(i)}{n}, \qquad C(i) = \frac{|\{\{j,k\}|d(i,j) = d(i,k) = d(j,k) = 1\}|}{\deg(i)(\deg(i)-1)/2}$$

The formula calculates the probability for its neighbors to be each other's neighbors (average clustering coefficient). Then, $C(i)$ is the total number of possible node pairs within node (i)'s neighbors, while the numerator is the number of actually connected node pairs among them.

Degree Distribution

When we represent the degree of a node as a distribution, it gives us another essential property as given $P(k) = |\{i|deg(i) = k\}|/n$. The formula calculates a probability distribution of network which the probability for a node to have degree k, which can be obtained as follows;

```
[In]
g = nx.karate_club_graph()
hist(g.degree().values(), bins = 20)
nx.degree_histogram(g)
```

→

```
[Out]
[0, 1, 11, 6, 6, 3, 2, 0, 0, 1,
1, 0, 1, 0, 0, 0, 1, 1]
```

This list contains the value of (unnormalized) $P(k)$, for $k = 0,1, \ldots kmax$, in this order. For larger networks, it is often more useful to plot (i.e., histogram) list in a log-log scale.

3. Network Structures

There are many different types of graph representations, and it is important to be able to recognize what type of graph, especially when we are programming and want to solve a particular problem. Graph structures are subsets of nodes and edges in a graph that possess certain characteristics or relate to each other in particular ways (i.e., regular, random, small-world, and scale-free; Broder et al., 2011). Such structures are used to isolate interesting or important sections of a graph because they form a significant domain-specific structure or contribute significantly to the properties of the graph.

+ **Regular Network:** In a regular graph, each vertex has the same number of neighbors (i.e., every vertex has the same degree or valence). A regular directed graph must also satisfy the stronger condition that the indegree and outdegree of each vertex are equal to each other. A regular graph with vertices of degree d is called a d-regular graph or regular graph of degree d. Also, a regular graph of odd degree will contain an even number of vertices (Table 3.3.1).

$$A = G[n, d(p)], \qquad d(p) \in (0,1) \;\rightarrow\; Am \times n$$

[Table 3.3.1] Regular Network Property

Define	Parameterize	Range	Created regular network
Number of individuals	$n = \text{dim*dim}$	$\in (0, \infty)$	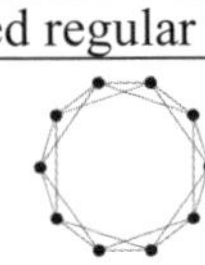
Connection probability	$d(p)$	$\in (0,1)$	
Linear dimension(matrix)	Dim	row, col	

Code example of the regular network; the module requires an import, and create object and illustrate the graph

```
[In]
Import networkx
import matplotlib.pyplot as plt
G = networkx.random_regular _graph(n = 4, d = 2)
networkx.draw(G, node_color = 'k', node_size = 500)
```

```
Parameters    ----------    n : int (Number of nodes),    d : int (Number of
degrees of each vertex is equal to existing nodes)
```

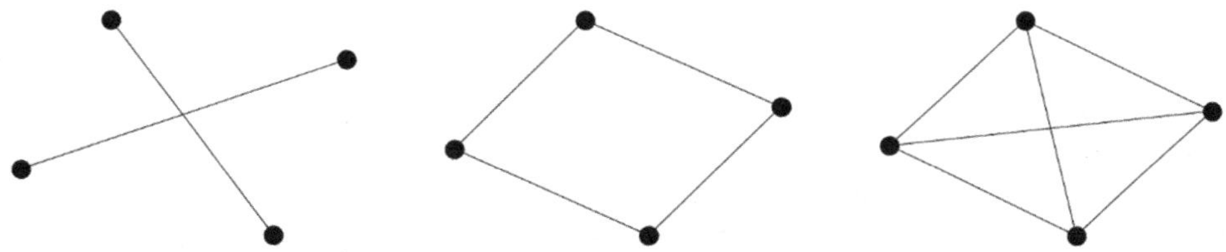

Figure 3.3.12. Regular graph. All nodes have the same degree: (left) n = 4, d = 1; (middle) n = 4, d = 2; (right) n = 4, d = 3

In Figure 3.3.12, the output is a visual representation of three different regular graphs, where all nodes have the same degree, with the number of nodes and degrees varying in each case.

+ **Random Network:** In a random network (Erdős & Rényi, 1959), there is a population of n nodes by randomly picking one of the nodes and linking them randomly with another node (Table 3.3.2).

$$A = G[n, d], \qquad p \in (0,1) \;\rightarrow\; Am \times n$$

[Table 3.3.2] Random Network Property

Define	Parameterize	Range	Created random network
Number of individuals	n	$\in (0, \infty)$	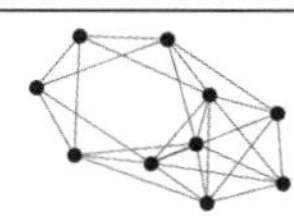
Connection probability	d	$\in (0,1)$	
Linear dimension(matrix)	m*n	row, col	

Code example of the random network; the same module requires the same import, and create object and illustrate graph

```
[In]
G = netwrokx.Erdos_Renyi_graph(n = 10, d = 0.2, directed=False)
networkx.draw(G, node_color = 'k', node_size = 500)
```

```
Parameters    ----------    n : int (Number of nodes), m : int (Number of edges to
attach from a new node to existing nodes)
```

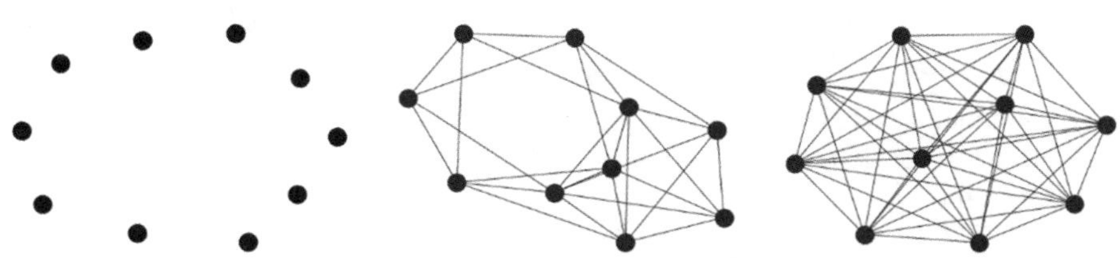

Figure 3.3.13. Random graph. The links and turtles are the connected components of the network. (left) n = 10, d = 0; (middle) n = 10, d = 0.5; (right) n = 10, d = 1

In Figure 3.3.13, The plots show three examples of random graphs, where the density of edges increases from left to right. The links and turtles in the figures refer to the connected components of the network.

+ **Small-World Network:** A small-world network is a type of graph in which most nodes are not neighbors of one another, but the neighbors of any given node are likely to be neighbors themselves as well, and most nodes can be reached from every other node by a small number of hops or steps (Table 3.3.3). Consider a network of friends. Since friends of a person are also friends of each other in a social network but not all of possible friendships (Watts & Strogatz, 1998), a small world network would capture both a high level of clustering and a short average path length.

$$A = G[n, p, \beta], \qquad p \text{ and } \beta \in (0,1) \;\rightarrow\; Am \times n$$

[Table 3.3.3] Small-World Network Property

Define	Parameterize	Range	Small-world network
Number of individuals	n	$\in (0, \infty)$	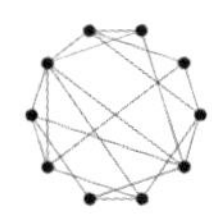
Connection probability	p	$\in (0,1)$	
Rewiring probability	β	$\in (0,1)$	
Linear dimension (matrix)	m*n	row, col	

Code example of the small-world network; the same module requires the same import, and create object and illustrate graph

```
[In]
G = networkx.watts_strogatz_graph(n = 10, p = 4, β = 0.5)
pos = nx.circular_layout()
networkx.draw_networkx(G, pos, node_color = 'k', node_size = 500)
```

Parameters ---------- n: int (Number of nodes), p: int (Number of edges to attach from a new node to existing nodes), β: int (Rewiring probability to the edge)

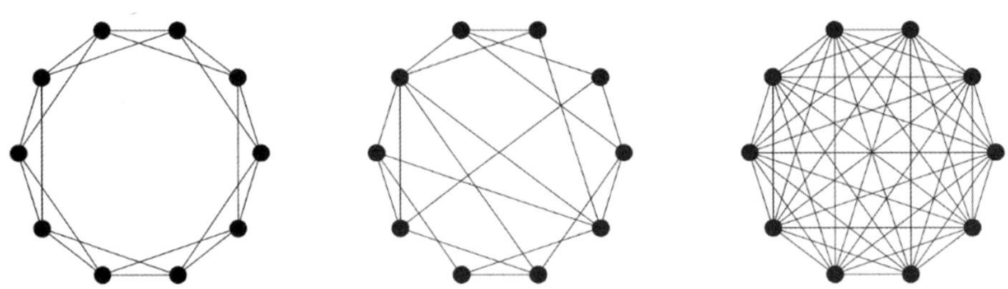

Figure 3.3.14. Small-world network: Different values from the network parameter create various topologies of the network. n = 10, p = 0.4, β = 0 (left); n = 10, p = 4, β = 0.5 (middle); n = 10, p = 10, β = 0.9 (right)

In Figure 3.3.14, the output shows three small-world networks. The left network is a regular network with β=0. The middle network is a small-world network with β=0.5.

The rightmost network is a small-world network with β=0.9. If we relink many nodes, the average path length will initially drop quickly. Just a little relinking lead to networks with high clustering and a low average path length. This is a characteristic of small-world networks.

+ **Scale-Free Network:** Scale-free networks are a type of network characterized by the presence of large hubs. It has a power-law degree distribution, at least asymptotically. Some nodes do not link to others, but others have many links to other nodes. Thus, in scale-free networks (Barabási & Bonabeau, 2003), degree is not normally distributed; these networks are considered hubs (Table 3.3.4).

$$A = G[n, p(k)], \qquad p(k) \in (0,1) \rightarrow Am \times n$$

[Table 3.3.4] Scale-Free Network Property

Define	Parameterize	Range	Scale-free network
Number of individuals	n	$\in (0,1)$	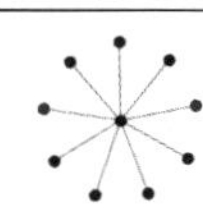
Connection probability	$p(k)$	$\in (0,1)$	
Linear dimension (matrix)	m*n	row, col	

Code example of a scale-free network; the same module requires the same import, and create object and illustrate graph

```
[In]
G = networkx.barabasi_albert_graph(n = 10, p(k) = 5, seed = None)
networkx.draw_networkx(G, node_color = 'k', node_size = 500)
```

Parameters ---------- n: int (Number of nodes), $p(k)$: int (Number of edges to attach from a new node to existing nodes)

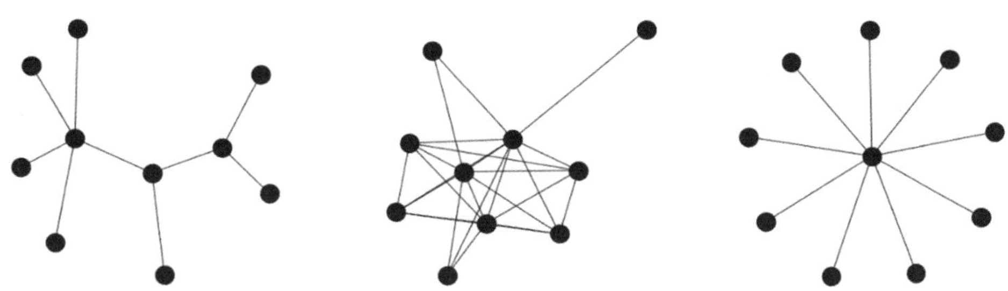

Figure 3.3.15. Scale-free network: (left) a typical scale-free network, (right) the distribution of degree in the World Wide Web. (left) n = 10, $p(k)$ = 1; (middle) n = 10, $p(k)$ = 5; (right) n = 10, $p(k)$ = 9

In Figure 3.3.15, the output shows three examples of scale-free networks, including a typical one on the left and the distribution of degree in the World Wide Web on the right, with different values of the number of edges to attach from a new node to existing nodes. In a scale-free network (Strogatz, 2001), there are more outliers than would be in a normal distribution. Nodes with a high degree have a higher probability of getting more nodes connected. This is also called the rich become richer phenomenon.

4. Measuring Network Centrality

Network analysis sounds like a clandestine, cyberpunk job tasked only to top National Security Agency personnel. However, this idea of mathematically dissecting networks is far from new; in fact, its logical origins extend back to the times of the eminent Leonhard Euler (18th century). The branch of math behind network analysis, graph theory, is far from new, yet its applications in the digital world are undeniably exciting.

From social media to sociology studies to more traditional networks, graph theory has resurfaced as network analysis (Tichy et al., 1979). Graph theory, a subbranch of discrete mathematics, is at the highest level simply the study of connections between things. These things, which are more formally referred to as vertices, vertexes, or nodes, have varying connections between themselves, referred to as edges or links.

Graphs are visually captivating and are typically displayed as follows (Figure 3.3.16):

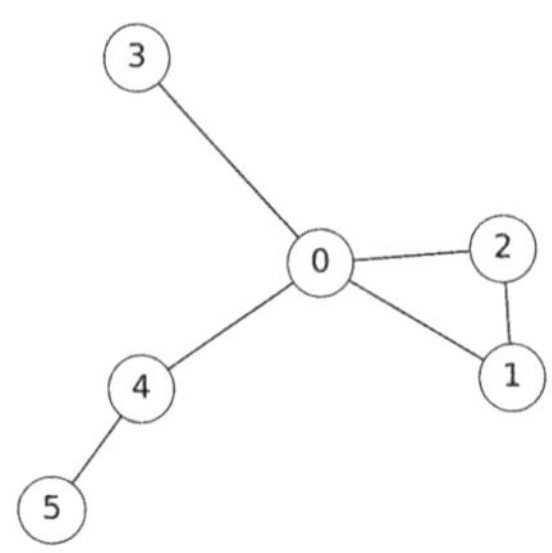

V: [0, 1, 2, 3, 4, 5]

E: [(0, 1), (0, 2), (0, 3), (0, 4), (1, 2), (4, 5)]

D: [(0, 4), (1, 2), (2, 2), (3, 1), (4, 2), (5, 1)]

Figure 3.3.16. A simple graph with its properties. V = vertices, E = edges, D = degrees

+ **From Graph Theory to Network Analysis:** The simple graph above is one of the most basic graphs possible. It is simple because it is lacking additional properties. Graphs, like the dynamic systems of the objects they represent, take on an unfathomable number of shapes and sizes; it is therefore helpful to create a set of properties to specify unique graph attributes. Let us quickly review the most common properties encountered in graph theory:

- Undirected versus directed edges
- Unweighted versus weighted edges
- Including/excluding multiple edges and loops

The clearest and largest means of classifying graphs begins with the type of edges within the graph. Two main types of edges exist: those with direction and those without. An undirected graph, like the simple graph above, is composed of undirected edges. In a directed graph, or a digraph, every vertex has a minimum of one incoming edge and one outgoing edge, signifying the strict direction of each edge relative to its two connected vertices (Figure 3.3.17).

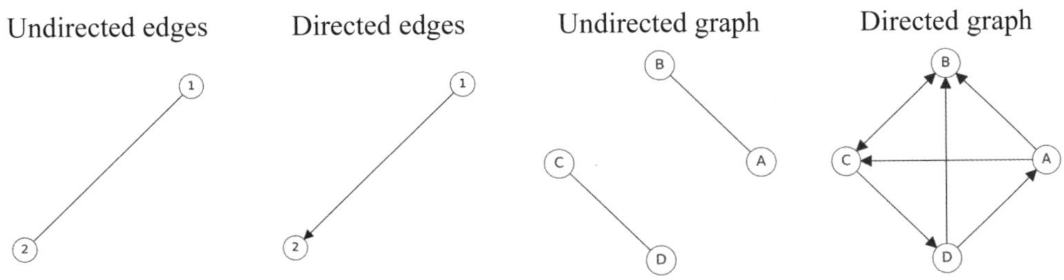

Figure 3.3.17. The directed and undirected relations (representing an arrow)

Consider a social media network (e.g., Facebook). On the social network, there is no direction; regardless of who sent the original request, once it is accepted, two people are friends. On the network, users can follow another user without that user reciprocating. The examples above of undirected and directed graphs highlight this difference. This difference is important because, as you will see shortly, it greatly affects how we carry out a main type of network analysis: centrality. For the rest of this section, we deal with undirected graphs.

+ **Centrality:** In the analysis of graphs, or network analysis, indicators of centrality identify the most important nodes within a graph (Gómez, 2019), in other words, how central a node is within its parent graph. However, different nodes can be considered important depending on how 'importance' is defined. Centrality comes in different flavors, and each flavor defines the importance of a node in a different way, which leads to a unique way of measuring centrality. Centrality is often the first measurement introduced to those learning about network analysis because of its wide application. Calculating it can identify the most influential person(s) in a social media, identify key infrastructure nodes in an intranet network, or even pinpoint super-spreaders of disease. Centrality algorithms use graph theory to calculate the importance of any given node in a network. They cut through noisy data, revealing parts of the network that need attention—but they all work differently. Here we cover the three most common measures of network centrality:

- Degree centrality; Closeness centrality; Betweenness centrality

Centrality indices answer the question 'what characterizes an important node?' The answer is given in terms of a real-valued function on the vertices of a graph, where the values produced are expected to provide a ranking that identifies the most important nodes.

Degree Centrality: Degree centrality is by far the simplest and most intuitive measurement. To understand it, let us first explore the concept of the degree of a node in a graph. In an undirected graph, the degree of a node is the number of direct connections the node has with other nodes, or how many direct, one-hop connections each node has to other nodes in the network. Degree centrality assigns a score based simply on the number of links held by each node—the higher the degree of a node, the more important it is in the graph. Example of degree centrality; import module and create object

```
[In]
import networkx as nx
G = nx.watts_strogatz_graph(n = 10, m = 4, p = 0.5)
pos = nx.circular_layout()
```

The function takes in three argument n, m, and p. The n is the number of nodes in the graph, the m is the number of neighbors each node is connected to, and p is the probability of rewiring each edge. Calculate the degree centrality

```
[In]
deg_centrality = nx.degree_centrality(G)
nx.draw_networkx_labels(G, pos)
plt.bar(deg_centrality.keys(), deg_centrality.values())
```

[Out]

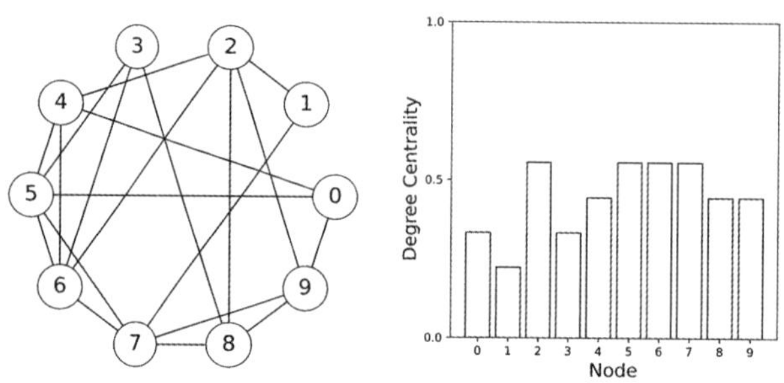


```
Degree centrality
  {node: value}
    0: 0.333
    1: 0.222
    2: 0.556
    3: 0.333
    4: 0.444
    5: 0.556
    6: 0.556
    7: 0.556
    8: 0.444
    9: 0.444
```

Figure 3.3.18. Output of degree centrality

The theory behind this is that the more connected a node is, the more likely that node is to hold the most information or individuals who can quickly connect with the wider network (Figure 3.3.18). Mathematically speaking, degree centrality is defined as D(i) for a node i as follows:

$$C_D(p_i) = \sum_{k=1}^{n} a(p_i, p_k)$$

The calculation is easier than the complex notation above implies—for each node, simply count how many other nodes it is connected to. Not all marketers are mathematically aware of this measurement, yet there is no doubt that this is the idea behind common social media marketing strategies such as influencer marketing. This is the answer to the arithmetic of why celebrities are paid to say or promote specific products or services.

Closeness Centrality: Another measure of centrality is closeness centrality. The closeness centrality (or closeness) of a node is a measure of its centrality in a network that is calculated as the normalized average of all of its geodesic distances. This measure calculates the shortest paths between all nodes, then assigns each node a score based on its sum of shortest paths. To understand this measurement more deeply, we first need to talk about the geodesic distance between two nodes in a graph. Example of closeness centrality; import module and create object

```
[In]
import networkx as nx
G = nx.watts_strogatz_graph(n = 10, m = 4, p = 0.5)
pos = nx.circular_layout()
```

calculate the degree centrality

```
[In]
clos_centrality = nx.closeness_centrality(G)
nx.draw_networkx_labels(G, pos)
plt.bar(clos_centrality.keys(), clos_centrality.values())
```

[Out]

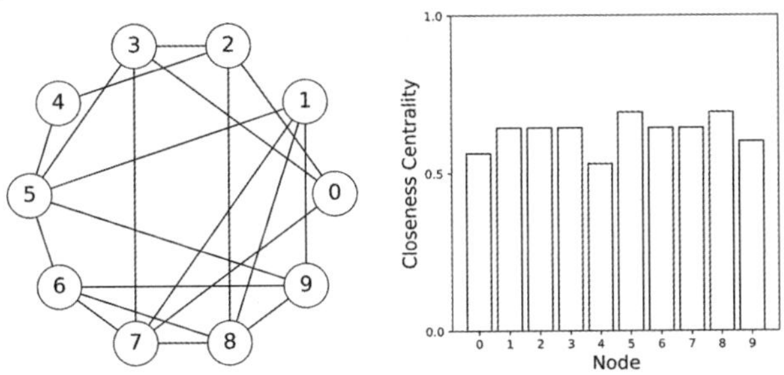


```
Closeness centrality
{node: value}
0: 0.562
1: 0.643
2: 0.643
3: 0.643
4: 0.529
5: 0.692
6: 0.643
7: 0.643
8: 0.692
9: 0.6
```

Figure 3.3.19. Output of closeness centrality

The geodesic distance d between two nodes a and b is the number of edges or links between these two nodes on the shortest path between them (i.e., the path with the minimum number of edges) (Figure 3.3.19). Mathematically speaking, geodesic distance can be defined as follows:

- $d(a, b)$ = number of edges between a and b on the shortest path from a to b if a path exists from a to b
- $d(a, b)$ = 0 if $a = b$
- $d(a, b) = \infty$ (infinity) if no path exists from a to b

Closeness centrality scores each node based on its closeness to all other nodes in the network. Thus, for each node, we simply find the normalized, average geodesic distance to all other nodes. The final formula for the closeness centrality (C_C) of node i in a graph is defined as follows:

$$C_C(p_i) = \frac{N-1}{\sum_{k=1}^{n} d(p_i, p_k)}$$

Why use this alternative definition of centrality at all? Because importance within a graph—what we are attempting to capture with measurements of centrality—can mean very different things. Having the most connections in a network, which is what degree centrality measures, intuitively translates to a power-by-concentration concept; the node with the highest degree is undeniably powerful. However, what if we cared about power not in terms of concentration but in terms of information speed?

In other words, what if we were looking not for the most connected node but rather for the node that can spread information the fastest? It is surprising that the node with the highest degree concentration is rarely the node with the shortest average geodesic distance. Let us suppose we want to send a piece of specific information to each node of the graph—we would need to select a node in the graph that can transmit it quickly to all of the nodes in the network. In this particular case, we would calculate closeness centrality as described above to find the best broadcasters in a given network.

Betweenness Centrality: This measurement of centrality, betweenness centrality, also ties neatly into the concept of geodesic distance. This measure shows which nodes are bridges between nodes in a network; it is calculated by identifying all shortest paths (geodesic distances) between all pairs of nodes and then by counting how many times each node falls on one of these paths. Instead of aggregating the geodesic distance from one node to all others, we instead count the number of times a node shows up in between the shortest path. Example of betweenness centrality; import module and create object

```
[In]
import networkx as nx
G = nx.watts_strogatz_graph(n = 10, m = 4, p = 0.5)
pos = nx.circular_layout()
```

calculate the degree centrality

```
[In]
btw_centrality = nx.betweenness_centrality(G)
nx.draw_networkx_labels(G, pos)
plt.bar(btw_centrality.keys(), btw_centrality.values())
```

[Out]

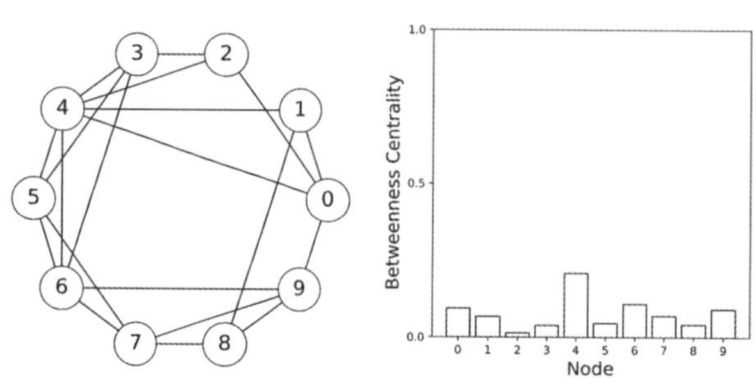


```
Betweenness centrality
   {node: value}
     0: 0.094
     1: 0.067
     2: 0.014
     3: 0.039
     4: 0.208
     5: 0.046
     6: 0.108
     7: 0.069
     8: 0.042
     9: 0.091
```

Figure 3.3.20. Output of betweenness centrality

Betweenness centrality measures the number of times a node lies on the shortest path between other nodes; it represents the degree to which nodes stand between each other (Figure 3.3.20). Mathematically speaking, the betweenness centrality $B(i)$ of node i in a graph is defined as follows:

$$C_B(n_i) = \textstyle\sum_{j<k} g_{jk}(n_i)/g_{jk}$$

Nodes with high betweenness may have considerable influence within a network not because of their concentration or the speed at which they transfer to other nodes but rather by virtue of their control over information passing between other nodes. Nodes with high betweenness centrality scores are also the ones whose removal from a network will most likely disrupt communications between other nodes. Betweenness centrality has some pretty realistic applications. For example, imagine analyzing a network of global terrorists with the goal of maximal disruption. To achieve this, we would start by removing the node with the highest betweenness centrality. Probably a more relatable example is using this same process to help identify and correct possible bottlenecks in logistic processes. Both scenarios provide us with insight into the application of this measure of centrality. It defines and measures the importance of a node in a network based on how many times it occurs in the shortest path between all pairs of nodes in a graph. The field of graph analytics is vast and has immense practical applications (Keim et al., 2008). Here we considered one fundamental metric of network analysis: centrality. This is simple but not easy, as there are multiple starting points for centrality. This means we must carefully consider which type of centrality is right for any given scenario (Table 3.3.5).

[Table 3.3.5] Comparison of Common Measures of Centrality

Type	Formula	Description	Translates to
Degree	$C_D(p_i) = \sum_{k=1}^{n} a(p_i, p_k)$	Number of nodes connected	Most connected node
Closeness	$C_C(p_i) = \frac{N-1}{\sum_{k=1}^{n} d(p_i, p_k)}$	Shortest average geodesic distance	Fastest communicating node
Betweenness	$C_B(n_i) = \sum_{j<k} g_{jk}(n_i)/g_{jk}$	Number of times found in geodesic distances	Most influential node

Let us close with a necessary disclaimer that slight caveats exist for each of these indices. Not only is it critical to choose the right type of centrality, but it is likely that the node-ranking system that will result from each calculation will be a bit deceptive. Any ranking system is clearly relative to other nodes, which heavily overlooks absolute measurements. Maybe the difference in importance between one node and the preceding or following node is insignificant, or perhaps it is the largest difference in the list of nodes. Either way, it is exhilarating to apply math to something as humanly natural as group dynamics. Yet it is clear that this is just a basic introduction to graph theory. Centrality is but one of many important metrics to consider when analyzing networks.

Eigenvector Centrality with Data-Driven Example

Eigenvector centrality quantifies node v_i as influential if it is connected to other influential nodes within a network. Let $\lambda_1, \lambda_2, \ldots, \lambda_n$ denote the eigenvalues of the adjacency matrix $A = \{a_{ij}\}$ of network G. Then the largest eigenvalue of matrix A is λ_{max} with corresponding eigenvector $e = [e_1, e_2, \ldots, e_n]^T$ such that

$$\lambda_{max} e_i = \sum_{j=1}^{N} a_{ij} e_j$$

Then, the eigenvector centrality for node v_i, denoted as $C_E(v_i)$, is defined as

$$C_E(v_i) = \lambda_{max}^{-1} \sum_{j=1}^{N} a_{ij} e_j$$

where a_{ij} defines the tie between nodes v_i and e_j , $a_{ij}e_j$ is the eigenvector value of node e_j and λ is a constant. Thus, centrality is measured as a function of the centrality of all adjacent nodes; this is a kind of extension of degree—it looks at a combination of a node's edges and the edges of that node's neighbors. Eigenvector centrality cares if you are a hub, but it also cares how many hubs you are connected to. It is calculated as a value from 0 to 1: the closer to one, the greater the centrality. Nodes with more links have higher eigenvector centralities (Table 3.3.6).

[Table 3.3.6] Numerical Results for Eigenvector Centrality

Nodes(n)	0	1	2	3	4	5	6	7	8	9
Eigen_cent	0.32	0.08	0.27	0.27	0.20	0.44	0.43	0.36	0.29	0.32
Degrees	5	1	4	4	3	9	8	6	5	5

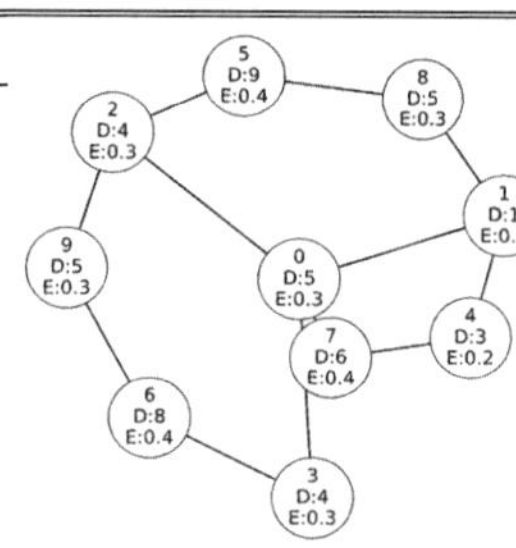

Note: Scale-free graph of $n = 10$ nodes with expected degrees of m

The importance of the node is not only dependent on the number of connections but also on the connectivity of the adjacent nodes, as proposed by Bonacich (1972). Thus, the metric assumes that not all connections are equally important following the belief that influential adjacent nodes increase one's own influence (Newman & Girvan, 2004). In particular, we observed that the increasing complexity of functional roles means elements or players temporarily occupy different positions and areas on the pitch and fulfill different tasks, which can be acknowledged as part of the role repertoire of the different tactical positions.

[Table 3.3.7] Data-Driven (Football) Network Property with Eigenvector Centrality

	DM	GK	LCD	LD	LF	LM	OM	RCD	RD	RF	RM
In Degree	0.12	0.03	0.09	0.08	0.09	0.10	0.11	0.09	0.10	0.09	0.11
Out Degree	0.13	0.06	0.10	0.10	0.06	0.08	0.10	0.11	0.12	0.06	0.09
Centrality	0.11	0.06	0.09	0.09	0.08	0.09	0.10	0.10	0.10	0.08	0.10
Eigen_Cent	0.49	0.13	0.26	0.26	0.21	0.32	0.37	0.27	0.26	0.21	0.32

Note: The centrality (Eigen_Cent = Eigenvector Centrality) is estimated form the sum of weighted edges of a node; GK = Goalkeeper, LCD = Left Central Defender, LD = Left Defender, LF = Left Forward, LM = Left Midfielder, OM = Offensive Midfielder, RCD = Right Central Defender, RD = Right Defender, DM = Defensive Midfielder, RF = Right Forward, RM = Right Midfielder. It shows the DM is significantly more central ($p < .05$) than other positions. Players represent nodes such that the number of passes between them defines the edge weight (Sources: www.fifa.com/worldcup/archive/brazil2014).

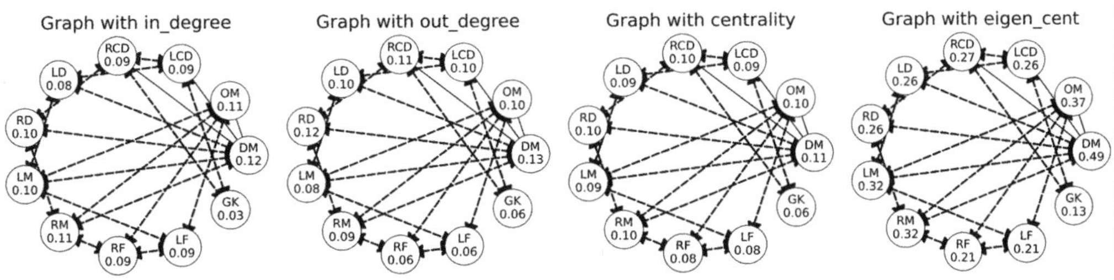

Figure 3.3.21. Representation of the different network attributes: in-degree, out-degree, centrality, and eigenvector centrality. Each node in the graphs represents a different entity, and the labels within each node provide the attribute values.

In Table 3.3.7 and Figure 3.3.21, the represented characteristics and connection structures are composed of data-driven statistics. Here, nodes are basic elements of the graph [encoded by links between control elements the number of members of a specific team], and lines form edges by connecting two individuals, denoting the degree of their connections. In the context of their scores, these capture the successful influence and positively contribute to the element's performance more frequently than others. Especially, for more detail, metrics of weighted in-degree referred to the sum of the incoming weighted edge values of a node. Weighted out-degree also refers to the sum of outgoing weighted edges of a node. Centrality calculations show quantification of the influence of tactical positions on their interplay as well as the balance of influence between players overall.

Eigenvector centrality is useful for understanding which nodes can get information to many other nodes quickly. If you know a lot of well-connected people, you could spread a message very efficiently (i.e., Google, PageRank algorithm).

Remarks: The foundational framework for investigating the characteristics of networks is provided by graph theory, which serves as the basis for analyzing network structures, such as random, regular, small-world, and scale-free arrangements. The key instruments we treated here that enable us to comprehend and evaluate the behavior of complex systems are the identification of network topologies, measurement centralities like degree or betweenness, and modeling of network dynamics, which involves studying changes in structure and behavior over time.

Networks, Part 2

Practical Aspects of Networks, Simulation of Networks Model

Networks and graphs are immensely useful tools for modeling complex relational problems. Networks are composed of two main entities: nodes and edges. Nodes are commonly represented as circles. In the academic literature, nodes are also known as vertices. Edges are commonly represented as lines between circles. Edges denote the relationships between nodes.

According to John Quackenbush of the Harvard School of Public Health, the heart of a graph lies in its edges, not in its nodes (Silverman et al., 2020). Let us consider a few key concepts mentioned above and then implement them using the handy Package (i.e., Networkx).

1. Practical Aspects of Network

A graph $G(V, E)$ is a data structure that is defined by a set of vertices (V) and a set of edges (E).

- A vertex (v) or node is an indivisible point represented by lettered components.
- An edge (vu) connects vertex v and vertex u together (i.e., a complete graph of n vertices is a graph in which every vertex shares an edge with every other vertex and therefore contains the maximum number of edges).

The size of the maximum possible edge set of any simple graph of n vertices is equal to $(n \times [\mathrm{n} - 1])/2$. This is because it is equivalent to the number of vertex pair combinations, that is, $(n\ choose\ 2) = n!/((n-2)!\,(2!))$.

+ **Creating a Simple Graph:** The first thing you will need to do is install the package (i.e., Networkx) on your machine.

Using pip or (conda), it is as easy as follows:

```
[In]
pip install network
```

The command of the code installs a Python package. The 'networkx' package is a library for working with graphs and networks in the Python.

Once installed, import the package and initialize a graph object.

```
[In]
import networkx as nx
G = nx.Graph()
```

The 'Graph' class is the most basic class and it represents an undirected graph with no self-loops or parallel edges. Thus, it creates an empty graph G with no nodes or edges.

Let us add the first two nodes and an edge between them. At this point our graph is just two connected nodes (Figure 3.4.1).

```
[In]
G.add_node(1)
G.add_node(2)
G.add_edge(1, 2)
```

→ [Out]

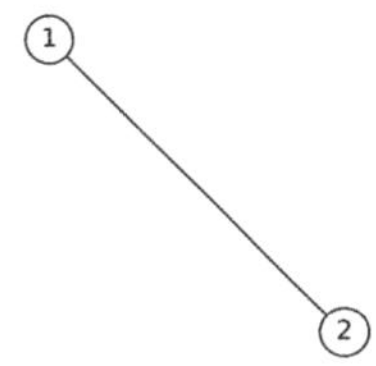

Figure 3.4.1. A graph with two vertices

Adding edges one at a time is a pretty slow process, but we can also add lists of nodes and lists of edges in which each edge is represented by a node tuple.

Our graph should now look something like this (Figure 3.4.2):

```
[In]
G.add_nodes_from([2,3,4,5,6])
G.add_edges_from([(1,2),(4,5),(3,5),(2,3),(5,6)])
```

→ [Out]

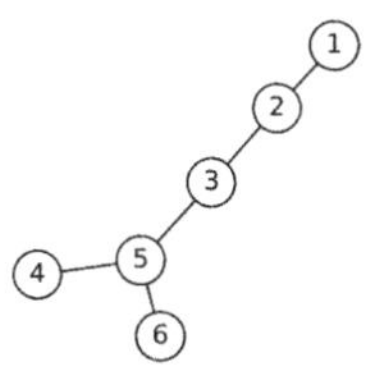

Figure 3.4.2. A created graph with seven vertices

\+ **Accessing and Storing Information:** We can see a list of nodes by printing these attributes of our graph.

```
[In]
print(G.nodes())
```

→

```
[Out]
[1, 2, 3, 4, 5, 6]
```

We can see the edges by printing these attributes of our graph.

```
[In]
print(G.edges())
```

→

```
[Out]
[(1, 2), (2, 3), (3, 4), (3, 5),
(5, 6)]
```

It is also possible to define nodes as strings (Figure 3.4.3).

```
[In]
# Add nodes
G.add_nodes_from([3, 'ib', 'A'])
# Add edges
G.add_edges_from([(3, 'ib'), (3, 'A')])
```

→

[Out]

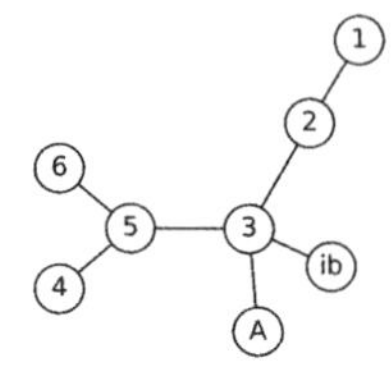

Figure 3.4.3. Graph with edges

Most important, each node can be assigned any number of attributes that are then stored in dictionaries.

```
[In]
G.nodes[1]['color'] = 'black'
G.nodes[1]['count'] = 10
print(G.nodes.data())
```

→

```
[Out]
[(1, {'color': 'black', 'count': 10}), (2, {}),
(3, {}), (4, {}), (5, {}), (6, {}), ('ib', {}),
('A', {})]
```

To make these data more manageable, feed the output of nodes.data() into a dict() function to get it into a nested dictionary where each node is a key.

```
[In]
print(dict(G.nodes.data()))
```

→

```
[Out]
{1: {'color': 'black', 'count': 10}, 2: {}, 3:
{}, 4: {}, 5: {}, 6: {}, 'ib': {}, 'A': {}}
```

In a network, if two nodes are joined together by an edge, then they are neighbors of each other. There are generally two types of networks (as you can see in the figure below): directed and undirected. In undirected networks, edges do not have a directionality associated with them. In directed networks, they do (Figure 3.4.4). A directed graph can be created as follows:

```
[In]
import networkx as nx
G=nx.Graph()
G.add_node(“A”)
G.add_node(“B”)
G.add_node(“C”)
G.add_edge(“A,””B”)
G.add_edge(“B,””C”)
G.add_edge(“C,””A”)
print(“Nodes: “ + str(G.nodes()))
print(“Edges: “ + str(G.edges()))
```

→

```
[Out]
Nodes: ['A', 'B', 'C']
Edges: [('A', 'B'), ('B', 'C'),
('C', 'A')]
```

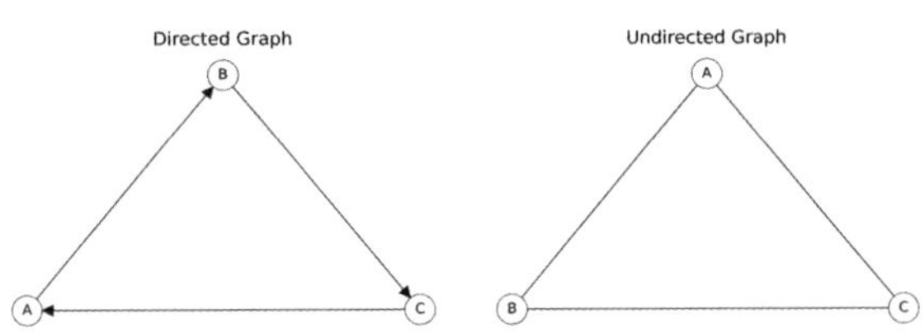

Figure 3.4.4. Different types of graphs with direction: (left) directed, (right) undirected

An example of an undirected network is Facebook: Individuals are nodes, and edges are drawn between individuals who are Facebook friends with each other. An example of a directed network is an air traffic network: Airports are nodes, and flights between airports are the edges.

+ Symmetric and Asymmetric Networks: The next network that we create below is a symmetric network because the relationship 'working together in a field' is a symmetric relationship. If A is related to B, B is also related to A. Let us create the network we saw above in 'Networkx'. We will use the Graph() method to create a new network and add_edge() to add an edge between two nodes.

```
[In]
import networkx as nx
G_symmetric = nx.Graph(seed=None)
G_symmetric.add_edge('A','B')
G_symmetric.add_edge('A','C')
G_symmetric.add_edge('A','D')
G_symmetric.add_edge('A','E' )
G_symmetric.add_edge('B','C')
G_symmetric.add_edge('B','D')
G_symmetric.add_edge('B','E')
G_symmetric.add_edge('E','C')
```

Now, visualize the network we just constructed (Figure 3.4.5-1):

```
[In]
plt.figure()

# Draw nodes
nx.draw_networkx_nodes(G_symmetric, pos)
```

→

[Out]

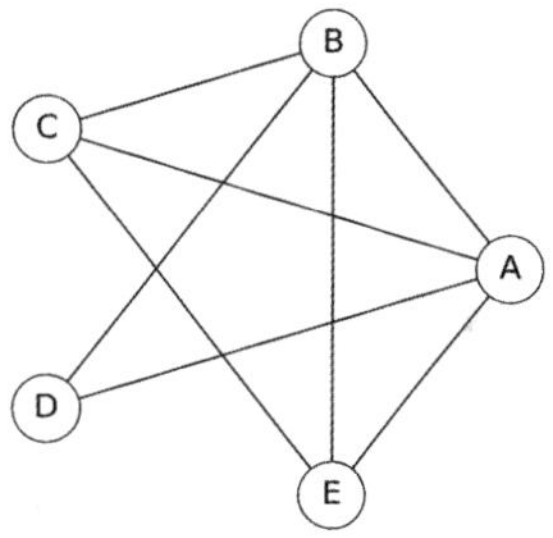

Figure 3.4.5-1. Created networks with symmetry

What if the relationship between nodes is not symmetric? A network in which the relationship is asymmetric (A is related to B, but this does not necessarily mean that B is associated with A) is called an asymmetric network. We can build an asymmetric network in 'networkx' using the DiGraph method, which is short for directional graph (make an asymmetric graph).

```
[In]
G_asymmetric = nx.DiGraph()
G_asymmetric.add_edge('A','B')
G_asymmetric.add_edge('A','D')
G_asymmetric.add_edge('C','A')
G_asymmetric.add_edge('D','E')
```

Now we can visualize it. We can use the draw_networkx() function as before. However, it is possible that nodes do not separate out and are distinctly visible in the network drawn. To address this, we can use the function to force a layout, which positions the nodes in a manner that we can distinctly see them. We can accomplish this using the spring _layout() function followed by the draw_networkx() function (Figure 3.4.5-2).

```
[In]
plt.figure()

# Draw nodes
nx.draw_networkx_nodes(G_asymmetric, pos)
```

→

[Out]

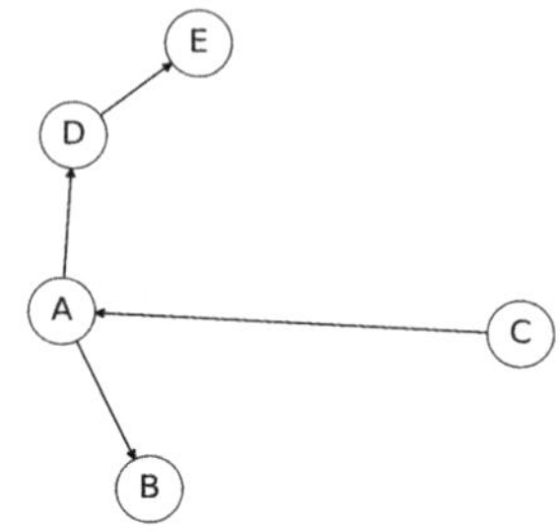

Figure 3.4.5-2. Created networks with asymmetry

+ **Network Connectivity:** Now that the network is made, can we know more about a particular node in the network?

Degree: The degree of a node defines the number of connections the node has. The 'networkx' has the function degree, which we can use to determine the degree of a node in the network.

```
[In]
print(nx.degree(G_symmetric, 'D'))
```

→

```
[Out]
2
```

People who share connections in a social network tend to form associations. In other words, there is a tendency in social networks to form clusters. We can determine the clusters of a node, or the local clustering coefficient, which is the fraction of pairs of the node's friends (i.e., connections) that are connected with each other.

To determine the local clustering coefficient, we use the nx.clustering(Graph, Node) function. In the symmetric network, the average clustering coefficient (the sum of all local clustering coefficients divided by the number of nodes) is 0.867. We can obtain this as follows:

```
[In]
print(nx.average_clustering(G_symmetric))
```

→

```
[Out]
0.8666666666666666
```

Distance: We can also determine the shortest path between two nodes and its length in 'networkx' using the nx.shortest_path(Graph, Node1, Node2) and nx.shortest_path_length(Graph, Node1, Node2) functions, respectively.

```
[In]
print(nx.shortest_path(G_symmetric, 'D',
'E'))
```

→

```
[Out]
['D', 'A', 'E']
```

The code shows the output of the 'nx.shortest_path' function when applied to a graph 'G_symmetric' with nodes labeled D and E. Cleary, the output indicates that the shortest path between nodes D and E in the graph is D, A, E.

2. Simulation of Networks Model

In conducting network simulations, it is crucial to incorporate elements that mirror the real world. To facilitate this, we study into techniques for importing (or exporting) data to and from external sources in our analysis environment, such as Python. Indeed, we are capable of crafting our own datasets using common tools like Excel, saving them in formats like 'csv' or 'txt' [e.g., g = nx.read_adjlist('Your_Network_Data.csv')].

Moreover, we can streamline the process by employing straightforward commands in the programming interface. To illustrate, our analysis begins with data extracted directly from Facebook's 'combined ego networks'. This dataset is a composite of the friends lists from 10 individuals, offering a comprehensive perspective of their interconnected social networks. To commence, let us obtain the file, either from the Supplemental dataset or via Stanford University's website (https://snap.stanford.edu/data/). For researchers interested in a more tailored analysis, it is possible to extract personal data from Facebook or Twitter using their respective APIs. These APIs allow for the automated collection of user-specific data, facilitating personalized or comparative studies in the domain of social network analysis. We read in the file (use the Supplement data set) and construct the graph:

```
[In]
G_fb = nx.read_edgelist("…/facebook_combined.txt," create_using = nx.Graph(),
nodetype=int)
```

The network consists of 4,039 nodes connected via 88,234 edges, it is a big network. You get the information using the 'info()' function.

[In]		[Out]
`print(nx.info(G_fb))`	→	Number of nodes: 4039 Number of edges: 88234 Average degree: 43.6910

We can also visualize the network such that the node color varies by degree and node size by betweenness centrality and degree centrality (Figure 3.4.6). The code for doing this is as follows:

```
[In]
pos = nx.spring_layout(G_fb)
betCent = nx.betweenness_centrality(G_fb, normalized=True, endpoints=True)
node_color = [20000.0 * G_fb.degree(v) for v in G_fb]
node_size =  [v * 10000 for v in betCent.values()]
plt.figure(figsize=(20,20))
nx.draw_networkx(G_fb, pos=pos, with_labels=False,
node_color=node_color,node_size=node_size )
nx.draw_networkx_edges(G_fb, pos, edge_color='k')
plt.axis('off')
plt.show()
```

[Out]

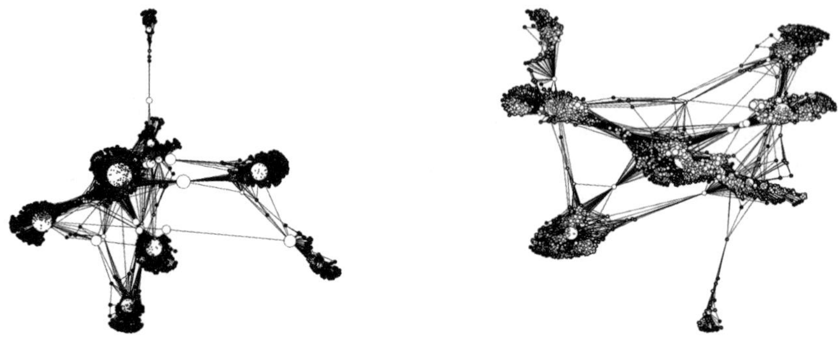

Figure 3.4.6. Visualized social (Facebook) networks with between centrality (left) and degree centrality (right). Node sizes are determined by their degree centrality, which reflects the number of direct connections a node has in the network. The color of the nodes is uniform (white), edged by black.

In Figure 3.4.6, we implemented Facebook social network graph consisting of 4,039 nodes and 88,234 edges, with node color representing degree centrality and node size representing betweenness centrality, using the spring layout. Then we plot the betweenness centrality against the degree centrality for the network data as follows. We calculates the degree centrality using the 'nx.degree_centrality' and betweenness centrality using the 'nx.betweenness_centrality' function of the nodes in a graph 'G_fb', sorting the centrality measures by their keys, and then creating a scatter plot with the sorted centrality measures. The sorted centrality values are extracted and stored in the lists 'deg_cent_sorted' and 'bet_cent_sorted'. The created scatter plot is using 'matplotlib' with 'deg_cent_sorted' on the x-axis and 'bet_cent_sorted' on the y-axis.

[In]

```
degCent = nx.degree_centrality(G_fb)
betCent =
nx.betweenness_centrality(G_fb,
normalized=True, endpoints=True)
deg_cent_sorted = [i[1] for i in
sorted(zip(degCent.keys(),
degCent.values()))]
bet_cent_sorted = [i[1] for i in
sorted(zip(betCent.keys(),
betCent.values()))]
plt.scatter(deg_cent_sorted, bet_cent_so
rted)
```

→

[Out]

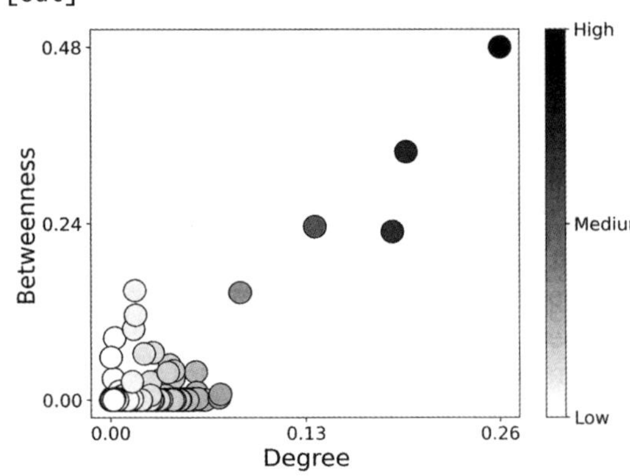

Figure 3.4.7. A created network with betweenness centrality referenced by its degree

In Figure 3.4.7, scatter plot of betweenness centrality (y-axis) versus degree centrality (x-axis) for the Facebook social network graph, illustrate the relationship between the two centrality measures for each node in the network. The plot provides insight into how well-connected nodes (high degree centrality) are in terms of their role as bridges within the network (high betweenness centrality). Nodes with both high degree and betweenness centrality may have more influence on the overall network structure and information flow.

The analysis of a network (graph) is one of the best analytical methods for producing a potential new framework for the analysis of any data. Such an analysis examines the structure of systems characterized by connections (or edges) between individuals (nodes). Every individual can be symbolized by a range of objects, from individuals and species to patches in a landscape, whereas an edge signifies interactions between individuals or connectivity between patches.

This tool has successfully contributed to many fields of research, such as ecology, social interaction, geographical space use, behavior and movement patterns, and even fisheries. Please feel free to use network analysis (various platforms and modules) to help develop, guide, and measure using the basic modeling techniques we practiced above.

General Guide for Deciding Which Network Analysis to Run

The key to network analysis lies in the steps of network structure estimation (to construct the network), network description (to characterize the network), and network stability analysis (to assess the robustness of results).

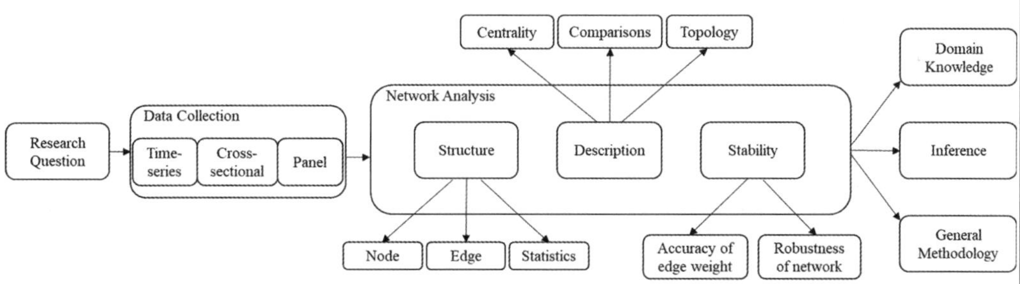

Figure 3.4.8. Schematic representation of network analysis (Source: Borsboom et al., 2021)

In Figure 3.4.8, these steps are informed by substantive research questions and data-collection procedures. The output of network analysis combines with general methodological considerations and domain-specific knowledge to support scientific inference.

Remarks: Network architectures are used in many different applications, including power grids, transportation systems, and biological and social networks. The most significant nodes in the network, which frequently play a major role in the network's overall function, must be identified through the assessment of centralities. We can forecast how a network will change over time and in response to environmental changes by simulating its dynamic behavior. When dealing with networks, there are several practical considerations to keep in mind, as we covered modelling and estimation above.

Supplemental Information 3.3: Modules, Packages, and Libraries in Programming

Programming, with its inherent complexity, often demands efficient approaches to manage intricate tasks. Even constructing a basic dice roll simulator from the ground up could entail extensive coding. To streamline this process, a common strategy involves segmenting large programming tasks into more manageable subtasks. This is precisely why programming environments like Python, R, MATLAB, Mathematic, and so on are equipped with modules, packages, and libraries.

+ **Modules:** A module is essentially a bundle of related code, usually encapsulated in a file with a '.py' extension. As an example, let us consider a function crafted to welcome new students to a specific course:

```
[In]
def systems_message(course):
    print("Thank you for joining our"+ course + "course. You will receive all the
details promptly.")
```

To encapsulate this function in a module named 'systems', one would save the code in a file named 'systems.py'. To employ this function in a different program, the relevant module must be imported using the 'import' statement. The function can then be invoked with the syntax 'module.function()'

[In]		[Out]
import systems systems.systems_message ("Systems Dynamic Part 3")	→	Thank you for joining our Systems Dynamic Part 3 course. You will get all the details easily.

Given that a module may contain numerous distinct items, there are instances when we might need to import just a specific function instead of the entire module. To achieve this, we can use the syntax:

```
[In]
from systems import systems_message
```

This approach not only simplifies your workflow but also reduces the likelihood of errors.

+ **Packages:** A package is essentially a directory containing a collection of related modules, allowing for a hierarchical organization of the module namespace. Similar to organizing files on a hard drive into folders and subfolders, modules can be structured into packages and sub-packages. To import specific modules from a package, use dot notation. For instance, to import the 'dynamic' module from the 'systems' sub-package within the 'my_model' package, one of the following code can be:

```
[In]
import my_model.systems.dynamic
```

Or

```
[In]
from my_model.systems import dynamic
```

There are many built-in and open-source packages that we have already been using in this book. For example:

- NumPy is the fundamental Python package for scientific computing.
- Pandas is a Python package for fast and efficient processing of tabular data, time series, matrix data, etc.

As your application grows larger and uses many different modules, the packages become a crucial component for optimizing your code structure.

+ **Libraries:** In the context of programming, a library represents an extensive collection of reusable code. This typically includes a variety of related modules and packages. While a package is essentially a collection of modules, a library is generally perceived as a more expansive aggregation of packages, each serving a specific functionality or a set of related functionalities. There are numerous indispensable libraries available in the programming landscape today. For example, consider the following imports:

```
[In]
import matplotlib.pyplot as plt
import numpy as np
import pands as pd
```

The Matplotlib library, as demonstrated above, is a highly regarded tool for creating a wide array of data visualizations in Python. It offers a comprehensive suite of plotting capabilities, ranging from simple bar charts and line graphs to complex scatter plots and three-dimensional data visualizations. Similarly, the NumPy and Pandas packages, which have been previously discussed, are often referred to as libraries. This is due to their extensive range of applications and their comprehensive nature. NumPy is pivotal in scientific computing, renowned for its powerful array and matrix processing capabilities. Pandas, on the other hand, is essential for data manipulation and analysis, especially renowned for its effectiveness in dealing with structured data like time series and tabular data.

The concepts of modules, packages, and libraries are fundamental building blocks in the realm of programming, especially in systems analysis. Modules provide the basic structure, encapsulating individual functionalities. Packages organize these modules into a coherent and structured hierarchy, facilitating better management and accessibility. Libraries, encompassing a collection of these packages, offer a broad spectrum of functionalities, making them indispensable tools for programmers to efficiently and effectively develop complex, scalable, and robust systems.

[3.5]

Agent-Based Model, Part 1

A Guide to Concepts, Comparing Agent-Based Model to Other Methods, Implementation

Agent-based modeling is a long-standing but underused method that allows system analysts to simulate artificial worlds for hypothesis testing and theory building (Jackson et al., 2017). Agent-based models offer unprecedented control and statistical power by allowing researchers to precisely specify the behavior of any number of agents and observe their interactions over time.

Agent-based models are especially useful for investigating group behavior or evolutionary processes and can uniquely reveal nonlinear dynamics and emergence—the process whereby local interactions aggregate into often surprising collective phenomena, such as spatial segregation and relational homophily (DeAngelis & Diaz, 2019). This section provides an overview of agent-based modeling. The key features of an agent-based models are presented along with main considerations when modeling. Approaches to agent-based modeling are briefly outlined, and an example is given.

1. A Guide to Concepts

There is no universal agreement among researchers on the precise definition of the term agent, with researchers continually debating whether definition should be by an agent's application or environment; however, definitions do tend to agree on more points than they disagree on.

+ **Identification**: In the realm of agent-based modeling, agents are distinguished by their complexity and behavioral patterns, and their identification often hinges on a set of common features, regardless of their specific application. These features include autonomy, heterogeneity, and activity, each playing a critical role in defining the nature and functionality of the agents.

- Minimalistic agents: Simple and autonomous, these agents follow basic rules, show predictable behavior, and operate independently with low heterogeneity and activity levels, ideal for modeling overall behavior rather than detailed complexities.

- Brownian agents: Characterized by random, autonomous movement similar to Brownian motion, they make limited decisions independently, with constrained heterogeneity and activity, useful in uncertain, random environments like some physical or biological systems.
- Complex agents: These agents display advanced, rule-governed behavior, high autonomy, and intelligence in interactions. They exhibit significant heterogeneity, representing diverse traits, and form groups through natural aggregation of similar individuals.

These highly active agents exert substantial independent influence, making them suitable for detailed models in social, economic, or ecological systems where individual behaviors and interactions are key.

In Agent-Based Modeling, the environment, or system's space, crucially influences the dynamics and outcomes. This space varies in forms like lattice, geographic, and network, each impacting agent behavior and system analysis distinctively.

- In a lattice space, agents are placed on a grid, interacting with their immediate neighbors, thus emphasizing local interactions and neighborhood effects.
- Geographic spaces introduce real-world spatial dimensions, where agents navigate and respond to topographical or ecological variables, making it ideal for models that require a high degree of environmental realism.
- Network spaces, on the other hand, allow agents to be connected in complex patterns, mirroring social, biological, or technological networks, where the focus shifts to the influence of network structure on agent behavior.

These diverse environments, when integrated with detailed agent characteristics, enable a comprehensive exploration of systems, ranging from simple simulations to complex, real-world scenarios, enriching the understanding of applied systems analysis.

+ **Modeling:** Agent-based models are computational simulations in which artificial entities interact over time within customized environments. These entities (agents) are programmed to represent humans who behave in precisely specified ways.

Agents are adaptive in that they respond to their environment through learning and evolution and are autonomous in that they control their own goals, states, and behaviors. They are also intentionally simplified, usually following only one or two basic rules (representing habits, norms, or preferences) throughout a simulation.

A well-programmed model offers insight into how local interactions between agents can lead to complex group- and system-level phenomena. Consider how a single bird's tendency to align and remain close (but not too close) to its peers can create a swirling flock that appears to be moving with a collective mind (Reynolds, 1987) or how predator-prey interdependence can explain animal species' resurgence following near extinction.

Agent-based models are uniquely equipped to shed light on such phenomena and countless other applications involving interacting individuals. Perhaps because of their ability to simulate large-scale dynamics with bottom-up processes, Agent-based models are popular in economics, sociology, political science, and some applied sciences (e.g., artificial intelligence).

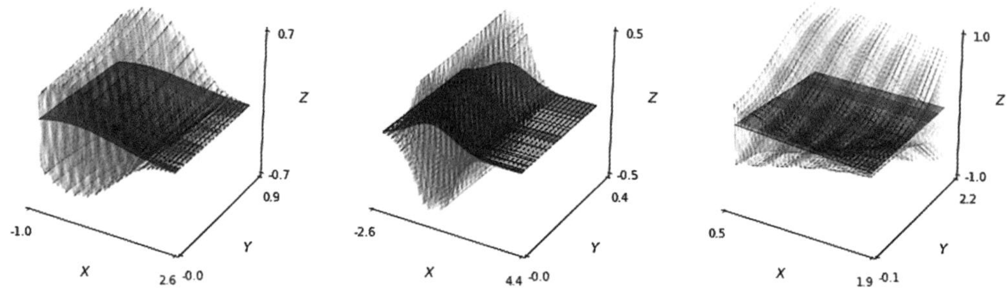

Figure 3.5.1. Prototypical representation of the Agent-Based Model simulation. Different behavioral patterns with distributions were achieved by the basic features of the system in R^3 (x, y and z); dots = individuals; mesh background = density. Note: the system becomes highly sensitive to the individuals' conditions (considered social ties in this simulation) at which the model is set as an initial value.

2. Comparing Agent-Based Modeling to Other Methods

Agent-based modeling has several main advantages over traditional modeling techniques, such as top-down techniques like nonlinear dynamical systems modeling (Table 3.5.1), in which related state variables are aggregated (e.g., through differential equations).

[Table 3.5.1] Differences between Agent-Based Modeling and Traditional Modeling

	Traditional modeling	Agent-based modeling
Modeling techniques	Deterministic (one future)	Stochastic (multiple futures)
	Allocative (top down)	Aggregative (bottom up)
	Equation-based formulas	Adaptive agents
	Do not give explanations	Explanatory power
	Few parameters	Many parameters
	Spatially coarse	Spatially explicit
	Environment given	Environment created
	You react to them	You learn from them

Source: Crooks & Heppenstall, 2011

+ **Model Features:** Compared to laboratory experiments, field studies, and archival investigations (including the analysis of Big Data), agent-based models offer a unique combination of experimental control and massive scale along with the ability to capture nonlinearities and underlying mechanisms. However, like any tool, agent-based models come with limitations, of which external validity is the most notable. This drawback is mitigated by supplementing agent-based models with other tools—such as laboratory or field experiments—in multimethod investigations (Table 3.5.2).

[Table 3.5.2] Comparing Agent-Based Models to Other Methods

	Field studies	Lab experiments	Archival studies	ABMs
Control and realism	Low control; high realism	Medium control; medium realism	Low control; medium realism	High control; low realism
Scale	Medium to high scale	Low to medium scale	High scale	High scale
Nonlinear dynamics	Medium visibility	Low visibility	Medium visibility	High visibility
Mechanism	Medium clarity	High clarity	Low clarity	High clarity

Source: O'Sullivan et al., 2012

With their high controllability, agent-based models are often able to isolate and directly manipulate the discrete systematical processes underlying complex social phenomena. Systematical mechanisms can take many forms and can exist on many levels of analysis.

Agent-based models are best suited for studying how manifestations of individual (or dyadic) behavior influence larger scale group-level phenomena, such as when a slight individual desire for similarity catalyzes neighborhood segregation (Schelling, 1971). One question is whether the mechanism provided by agent-based modeling is the same in the real world: Just because a mechanism sufficiently generates some outcome does not mean it necessarily or always generates that outcome. However, revealing even likely mechanisms is valuable for both basic research and policy decisions.

+ **Building a Model:** After being inspired by the rich history and unique methodology of agent-based modeling, you might want to try your hand at model building. Although training in agent-based model is absent from most college (graduate) programs, many works have linked agent-based model to specific research questions (Schelling, 1971; Axelrod, 1997; Smith & Conrey, 2007; Tesfatsion and Judd, 2006; Gilbert and Troitzsch, 2005) provide a broader overview of agent-based modeling. They include a discussion of the benefits of agent-based modeling over other methodologies (Nowak and Sigmund, 2004) give an in-depth overview of emergence in agent-based models and the utility of simple models for simulating complex processes.

A number of websites provide courses, videos, and code libraries of previous models from which researchers can adapt code (e.g., in Python, Mathematica, MATLAB, R, and C). To augment these resources, we discuss basic models here (i.e., the artificially driven money model, the boids model, and Schelling's segregation model, etc.). These examples offer insight into how data analysts can translate their research questions into simulations. Some of these steps do not apply to all models or all research questions, so you should feel free to adapt them to your own needs. Nevertheless, they provide a useful guide for exploring systemic processes and for creating simulated worlds with the potential for collective emergence.

3. Implementation

In this part, we propose a very basic agent-based models using computer programing. Jupyter notebook is a famous platform in the Python that offers many tools for building a framework for agent-based modeling (Pal et al., 2020). We initialize a population of agents with a set of behaviors (rules), living in an environment governed by a set of laws (again rules), and then let the agents behave without intervening. The interesting fact about this modeling technique is that just by observing the agents behave according to their inner features (i.e., idiosyncrasies), it is possible to anticipate emergent phenomena that would not be very intuitive if considered at a macro-level.

+ **Defining Framework (Agent-Based Model):** Before diving into the code, we will briefly consider the framework of the agent-based models. The primary goal of this practice is to observe what happens when information about a product (useful in a marketing campaign), an event, or something similar is spread among a population of individuals. In this case, we will stick to the example of a product, let us say one just introduced on the market. We will initialize a grid space in which the agents can move around, one cell at a time, in a neighboring cell (according to common definition of a neighborhood). In this scenario, agents are simple (i.e., elements, objects, and individuals) who are endowed with only two rules: (i) they can move in a neighboring cell, and (ii) they can influence their neighbors with their knowledge with probability p. The agents' ability to influence others depends on their 'knowledge' feature: This feature equals 1 if the agent is aware of the product (in this case, the agent can influence others) and 0 otherwise. Thus, the idea is initializing in a grid space a population of agents. Each agent has an internal state of knowledge that can be either 1 (if aware of the product) or 0 (if not aware). Agents with product knowledge can influence those with whom they have more contact (probably their neighbors) about the existence of the product, and this influence will have an impact with a given probability p.

+ **Implement:** As mentioned above, for our purpose we are going to use a modular framework for building, analyzing, and visualizing agent-based models (you can easily install it via pip).

```
[In]
pip install mesa          # or (via conda: install -c conda -forge mesa)
```

Notice the Mesa is a Python library designed for agent-based modeling. Installing Mesa via pip means using the pip package manager to download and install the Mesa library and its necessary dependencies, making the library available for use in Python projects.

Then, let us first import the necessary packages:

```
[In]
from mesa import Agent, Model
from mesa.time import RandomActivation
from mesa.datacollection import DataCollector
from mesa.space import MultiGrid
```

Now we will initialize the class of my Agent (agent with fixed initial wealth) as follows:

```
[In]
class Agent(Agent):
    def __init__(self, unique_id, model):
        super().__init__(unique_id, model)
        self.knowledge = 0
```

The code has four methods as follows. First, the '__init__(self, unique_id, model)' is the class constructor, which is called when an 'Agent' object is created. It takes two arguments 'unique_id' which is a unique identifier for the agent, and 'model' which is a reference to the model that the agent is a part of. The '__init__' method calls the superclass's '__init__' method using the 'super().__init__(unique_id, model' statement and initializes the 'knowledge' attribute of the agent to 0.

```
[In]
    def spread_news(self):
        if self.knowledge == 0:
            return
        neighbors = self.model.grid.get_neighbors(self.pos,moore = True, 
include_center=True)
        neig_agents = [a for n in neighbors  for a in 
self.model.grid.get_cell_list_contents(n.pos)]
        for a in neig_agents:
            if random.random()<0.3:
                a.knowledge = 1
```

Next, the 'spared_new(self)' method is used to spread the knowledge of an agent to its neighbors. It first checks if the agent has any knowledge (i.e., 'self.knowledge' is not equal to 0). If the agent has knowledge, it gets the neighbors of the agent's current position using the 'self.model.grid.get_neighbors' method. It then iterates through the neighbors and with probability 0.3, sets their 'knowledge' attribute to 1.

```
[In]
    def move(self):
        possible_steps = self.model.grid.get_neighborhood(self.pos, moore=True, 
include_center=False)
        new_position = self.random.choice(possible_steps)
        self.model.grid.move_agent(self, new_position)
    def step(self):
        self.move()
        self.spread_news()
```

Then, the 'move(self)' method is used to move the agent to a new position. It gets a list of possible steps (i.e., positions in the grid) using the 'self.model.grid.get_neighborhood' method, and then chooses a random step from this list using the 'random.choice' function.

It then moves the agent to the new position using the 'self.model.gird.move_agent' method. Fourth, the 'step(self)' method is called on each agent at each time step of the simulation. It calls the 'move' and 'spread_news' methods to move the agent and have it spread its knowledge to its neighbors, respectively. As anticipated, our agents are endowed with two main rules:

- spread_news(): This function allows those agents who have knowledge = 1 to spread information about the product. Neighbors get influenced with probability.
- move(): With this function, agents can move in a neighboring cell, according to Moore's definition of a neighborhood.

The function step() allows the agent to perform both spread_news() and move() at each step of the simulation. Now let us define the model: Define a function that is able to count, at each step, how many agents are aware of the product def compute_informed(model):

```
[In]
def compute_informed(model):
    return  sum([1 for a in model.schedule.agents if a.knowledge == 1])
```

The code uses a list comprehension to iterate through the agents in the model accessed using 'model.schedule.agents' and counts the number of agents that have knowledge, and returns this count using the 'sum' function. Define the model

```
[In]
class News_Model(Model):
    def __init__(self, N, width, height):
        self.num_agents = N
        self.grid = MultiGrid(width, height, True)
        self.schedule = RandomActivation(self)
        self.running = True
```

The 'News_Model(Model)' has a single method '__init__(self. N. width, height)' which is the class constructor and takes three arguments: First, the 'N' is the number of agents, the 'width' is the width of the grid on which the agents will be placed, and the 'height' is the height of the grid on which the agents will be placed. This method creates a multi-grid using the 'MultiGrid', then creates 'RandomActivation' schedule using the 'RandomActivation' class from the library and sets the 'running' attribute of the model True. Create the agents

```
[In]
        for i in range(self.num_agents):
            a = Agent(i, self)
            self.schedule.add(a)
            x = self.random.randrange(self.grid.width)
            y = self.random.randrange(self.grid.height)
            self.grid.place_agent(a, (x, y))
            l = [1,2,3,4,5]
```

The code iterates over the range ‘range(self.num_agents)’ and in each iteration creates a new ‘Agent’ object ‘a’ with a unique identifier ‘i’ and a reference to the ‘News_Model’ object ‘self’. It then adds the agent to the model’s schedule using the ‘add’ method of the ‘schedule’ attribute (which is an instance of the ‘RandomActivation’ class), and places the agent on the model’s grid at a random position using the ‘place_agent’ method of the ‘grid’ attribute (which is an instance of the ‘MultiGrid’ class). Finally, the code creates a list ‘1’ with the elements ‘[1,2,3,4,5]’. This code does not appear to be related to the rest of the code snippet, and it is not clear what purpose it serves. Make the agents aware of the product

```
[In]
            if i in l: #only agents with id in the list are aware of the product
                a.knowledge = 1
```

The code above creates elements ‘[1,2,3,4,5]’ then iterates over the range of self-number-agents, for each agent ‘a’ with an identifier ‘i’ that is in ‘1’, sets the agent’s ‘knolwedge’ attribute to 1.

```
[In]
        self.datacollector = DataCollector(
            model_reporters = {“Tot informed”: compute_informed},
            agent_reporters={“ “: “knowledge”})
```

The ‘DataCollection’ is initialize with two arguments as follows. First, the ‘model reporters’ is a dictionary that specifies what model-level data to collect at each time step. The key is the name of the data, and the value is a function that returns the data. In this case, the data collector will collect the total number of informed agents (i.e., agents with knowledge) at each time step, using the ‘compute_informed’ function defined earlier.

```
[In]
    def step(self):
        self.datacollector.collect(self)
        self.schedule.step()
```

Next, the ‘agent_reporters’ is a dictionary that specifies what per-agent data to collect at each time step. The key is the name of the data, and the value is the name of the attribute of the agent that stores the data. In this case, the data collector will collect the value of the ‘konwledge’ attribute of each agent at each time step. Finally, the code defines a ‘step’ method that is called at each time step of the simulation. It first collects the data using the ‘collec’ method of the ‘DataCollector’ object, and then advances the schedule by one time step using the ‘step’ method of the ‘schedule’ attribute (which is an instance of the ‘RandomActivation’ classs).

Now we can run the model. We will initialize it with 100 agents in a grid with height = width = 50. Hence, 2,500 will be available to place agents. Plus, we will also inspect the output of the model: We are interested in seeing whether the number of aware agents increases after many steps.

```
[In]
model = News_Model(100, 50, 50)
for i in range(100):
    model.step()
```

Finally, inspect the results

```
[In]
out = model.datacollector.get_agent_vars_dataframe().groupby('Step').sum()
out
```

It seems that the number of aware agents actually increases.

[Out]

Step (range 100)	0	1	2	3	4	…	95	96	97	98	99
Influenced (knowledge)	5	5	5	6	6	…	89	89	89	89	89

Let us also visualize it:

```
[In]
import matplotlib.pyplot as
plt
import numpy as np

fig, ax = plt.subplots()

plt.scatter(out.index, out.val
ues)
out.plot(ax=ax)
```

→ [Out]

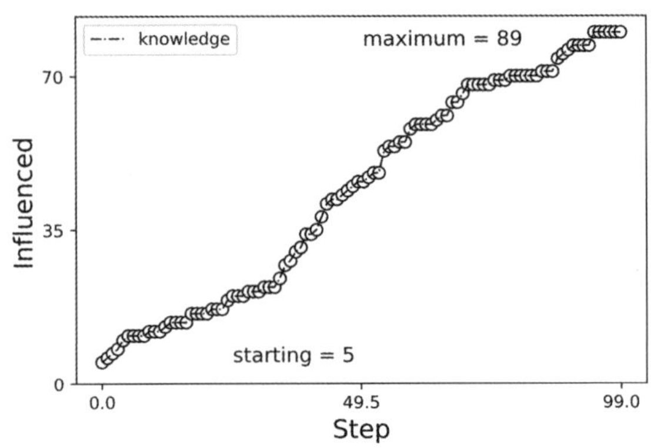

Figure 3.5.2. Results of the ABM simulation with 100 steps

In the plot (Figure 3.5.2), notice that more than 80% of the initial population was influenced, starting with only five agents aware of the product.

Remarks: The agent-based model is a versatile approach prevalent in systems theory, complexity theory, and general computer science. The model framework simulates the spread of product information among individuals in a grid space, where agents representing humans can move and influence neighbors with varying knowledge levels and probabilities, analyzing, and visualizing. The model, thus provides accurate representations of many actual complex phenomena.

Agent-Based Model, Part 2

Practical Aspects of Agent-Based Model, Simulations of Agent-Based Model, Cellular Automata

Agent-based models are powerful modeling techniques, and they leave room for the modeler to be creative. We can make them as complex as we please, play with some sensitivity analysis with parameters, model intervention strategies, and so forth. With practice, this becomes easier and more intuitive.

In a more complex framework for the same scenario, we might be interested in inspecting whether the position of the initial agents aware of the product is relevant in terms of the speed of diffusion. This scenario would be very well modeled with the agent-based model framework as an underlying structure (i.e., network): We could inspect which agents it is most important to target first for knowledge of the product to spread faster.

1. Practical Aspects of Agent-Based Model

In Schelling's (1971) model, agents were paired in a two-dimensional space and interacted in a multidimensional space in which one agent's position did not impede another agents' movement. This is a classic agent-based model that demonstrates how even a mild preference for similar neighbors can lead to a much higher degree of segregation than we would intuitively expect.

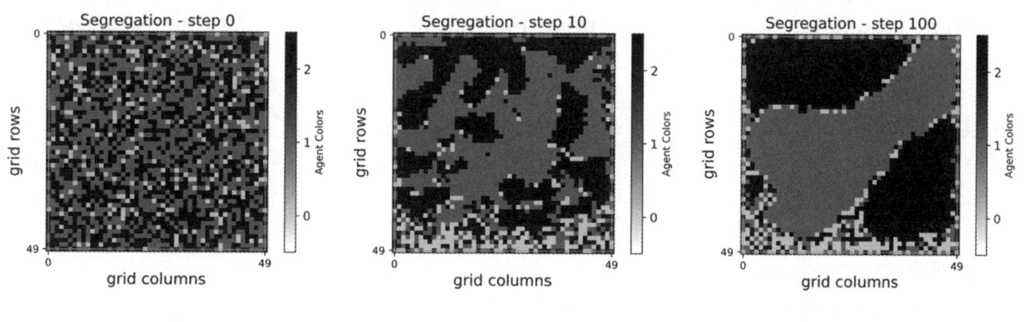

Figure 3.6.1. A prototype of Schelling's segregation (unhappy agent = dark ←→ white = happy agent). Recreated simulation result for 50 % happiness threshold and time steps t = 0 (left), t = 10 (middle), t = 100 (right) (Source: Dynamic Models of Segregation, Schelling, 1971)

The model consists of agents on a square grid, where each grid cell can contain at most one agent. Agents come in two states. They are happy if a certain number of their eight possible neighbors are of the same color and unhappy otherwise. Unhappy agents will pick a random empty cell to move to each step until they are happy (Figure 3.6.1). By default, the number of similar neighbor's agents need to be happy is set to three. This means that agents will be perfectly happy when the majority of their neighbors are a different color (e.g., a white agent will be happy with five dark neighbors and three white ones). Despite this, the model consistently leads to a high degree of segregation, with most agents ending up with no neighbors of a different color.

Because Schelling's segregation model focused on neighborhood dynamics, he programmed agents to interact only with their next-door neighbors. In expend, the agents could also interact with any other agent in the model, although they were more likely to interact with friends than with enemies—and they only interacted with one partner per round.

- In Schelling's model, agents decided whether to stay in their neighborhood or to move to another vacant space on a grid. Also, the agents played a game (i.e., the prisoner's dilemma).
- Agents received no payoff initially, as there was no interactive decision making. However, the agents received a payoff that depended on their decisions.
- Agents changed via movement, moving to a randomly selected grid space or moving closer to those who treated them nicely (i.e., their friends).
- Models ran until a point of equilibrium.

In Schelling's model, this equilibrium was the point at which agents were no longer moving across neighborhoods. Also, the model equilibrium represented the point at which agents had all formed groups or group formation was impossible. Schelling's central parameter was agents' desired similarity, whereas his dependent measures were agents' positions at the conclusion of the simulation. His finding was that a relatively low rate of similarity seeking could produce relatively homophilous agent distributions at the conclusion of the simulation.

+ **Practice with Schelling's Segregation Model:** This demonstration runs the model in a Jupyter notebook. Here we implement the model class, instantiate it, run it, and plot the time series of the number of happy agents.

First, we need to import a couple of necessary packages.

```
[In]
from mesa import Model, Agent
from mesa.time import RandomActivation
from mesa.space import SingleGrid
from mesa.datacollection import DataCollector
```

We create a Schelling agent with unique id, their location (x, y), and types of the agents (minority = 1, majority = 0)

```
[In]
class SchellingAgent(Agent):
    def __init__(self, pos, model, agent_type):
        super().__init__(pos, model)
        self.pos = pos
        self.type = agent_type
```

The class of the 'SchellingAgent' has a method that takes three arguments: 'pos', 'model', and 'agent_type'. The code uses the 'super()' function to call the 'agent' class. It then sets the 'pos' and 'type' attributes of the 'SchellingAgent' corresponding to the arguments.

```
[In]
    def step(self):
        similar = 0
        for neighbor in self.model.grid.neighbor_iter(self.pos):
            if neighbor.type == self.type:
                similar += 1
```

The 'step' method counts the number of similar agents in the neighborhood. The 'neighbor_iter' method is used to iterate over the neighbors of the 'SchellingAgent'. The 'type' attribute of each neighbor is compared to the 'type' attribute of the 'Schelling-Agent', and if they are equal, the 'similar' variable is incremented.

```
[In]
        if similar < self.model.homophily:
            self.model.grid.move_to_empty(self)
        else:
            self.model.happy += 1
```

Here, if the number of similar neighbors 'similar' is less than some parameter 'homophily' in the model, the agent moves to an empty location in the model's grid. Otherwise, the 'happy' attribute of the model is incremented. And, create the Schelling model's class

```
[In]
class Schelling(Model):
    def  __init__(self,  height=20,  width=20,  density=0.8,  minority_pc=0.2,
homophily=3):

        self.height = height
        self.width = width
        self.density = density
        self.minority_pc = minority_pc
        self.homophily = homophily
        self.schedule = RandomActivation(self)
        self.grid = SingleGrid(width, height, torus=True)
```

The code has several attributes including 'height', 'width', 'density', 'minority_pc', and 'homophily' which are set during the initialization. The model also define a 'schedule' which appears to be an instance of the 'RandomActivation' class, and a 'gird' which is an instance of the 'SingleGrid' class. The 'SingleGrid' is initialized with the 'width', 'height', and the optional parameter 'torus' se to 'True'. For testing purposes, agent's individual x and y

[In]
```
        self.happy = 0
        self.datacollector = DataCollector(
            {"happy": "happy"},  # Model-level count of happy agents
            {"x": lambda a: a.pos[0], "y": lambda a: a.pos[1]},)
```

The code initializes the 'happy' of the model to 0. It also creates a 'datacollector' which appears to be an instance of the 'DataCollector' which is initialized with two arguments. And it maps the strings x and y to lambda functions that return the elements of an agent's 'pos' respectively. Then we need to set up agents using a grid iterator that returns the coordinates of a cell as well as its contents (coord_iter)

[In]
```
        for cell in self.grid.coord_iter():
                    x = cell[1]
                    y = cell[2]
                    if self.random.random() < self.density:
                         if self.random.random() < self.minority_pc:
                            agent_type = 1
                        else:
                            agent_type = 0
                        agent = SchellingAgent((x, y), self, agent_type)
                        self.grid.position_agent(agent, (x, y))
                        self.schedule.add(agent)
                self.running = True
                self.datacollector.collect(self)
```

The code iterates through all the cells in the 'gird' and determines if an agent should be placed in the cell based on the 'density' of the model. If an agent should be placed in the cell, it randomly assigns the agent a type of either 0 or 1 based on the 'minority_pc'. It then creates an instance of the 'SchellingAgent' class and places it in the cell, adding it to the model's schedule. The 'running' is then set to 'True' and the 'datacollector' collects data from the model.

Finally, let us run one step of the model. If all agents are happy, it will halt the model.

[In]
```
        def step(self):
                self.happy = 0  # Reset counter of happy agents
                self.schedule.step()
```

```
        # collect data
        self.datacollector.collect(self)
        if self.happy == self.schedule.get_agent_count():
            self.running = False
```

The code resets the 'happy' to 0, then advances the model's schedule by one step. It then collects data using the 'datacollector' and checks if the number of happy agents is equal to the number of agents in the schedule. If this is the case, the 'running' attribute of the model is set to 'False'. Now we instantiate a model instance: a 10×10 grid with a 70% chance of an agent being placed in each cell, approximately 30% of agents set as minorities, and agents wanting at least five similar neighbors.

```
[In]
import matplotlib.pyplot as plt
%matplotlib inline
model = Schelling(10, 10, 0.7, 0.3, 5)
```

We want to run the model until all agents are happy with where they are. However, there is no guarantee that a given model instantiation will ever settle down. Thus, let us run it either for 100 steps or until it stops on its own, whichever comes first:

```
[In]
while model.running and model.schedule.steps <
100:
    model.step()
print(model.schedule.steps)
```

→

```
[Out]
100
```

This simulation has a 'dataCollector' object that stores how many agents are happy at the end of each step. It can also generate a pandas 'dataframe' of the data it has collected:

```
[In]
model_out =
model.datacollector.get_model_vars_dataframe()
model_out.head()
```

→

```
[Out]
 Steps  0  1   2   3   4
 happy  0  32  41  44  42
```

Then we can plot the 'happy' series:

```
[In]
model_out.happy.plot(color='b',label='')
plt.xlabel('steps(t)')
plt.ylabel('happy(agents)')
plt.title('Schelling model')
plt.grid(True)
plt.show()
```

→

[Out]

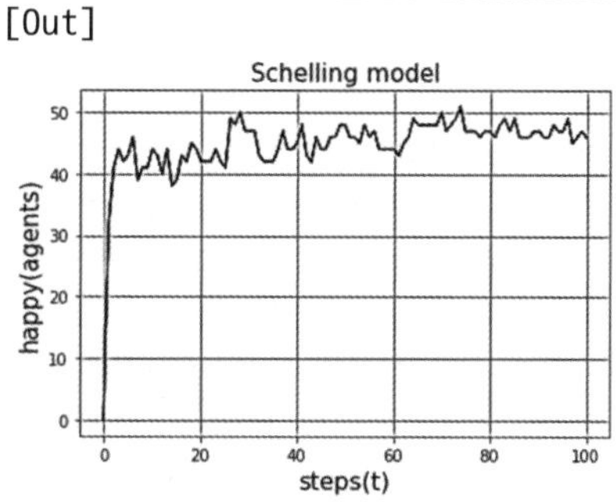

Figure 3.6.2. The number of happy agents in Schelling's model over time

In Figure 3.6.2, the plot depicts the number of happy agents in Schelling's model over time, illustrating the change in agent satisfaction as the model progresses through a maximum of 100 steps. The graph provides insights into the dynamics of the Schelling model, showing how the happiness of agents evolves as they seek neighborhoods with a desired level of similarity. For testing purposes, here is a table giving each agent's x and y values at each step.

```
[In]
x_positions =
model.datacollector.get_agent_vars_dataf
rame()
x_positions.head()
```

→

[Out]

Step	0	1	2	3	4
Agent	0,2	0,3	0,4	0,5	0,6
x	6	7	9	3	3
y	9	3	4	9	6

The result table is presenting the x and y positions of each agent at each step during the model's execution. The table provides a detailed view of agent movement across the grid, allowing for a deeper understanding of their behavior and decision-making within the Schelling model. Now we can do a parameter sweep to see how segregation changes with homophily. First, we create a function that takes a model instance and returns the fraction of agents that are segregated—that is, have no neighbors of the opposite type.

```
[In]
from mesa.batchrunner import BatchRunner
```

Let us find the % of agents that only have neighbors of their same type.

```
[In]
def get_segregation(model):
    segregated_agents = 0
    for agent in model.schedule.agents:
        segregated = True
        for neighbor in model.grid.neighbor_iter(agent.pos):
            if neighbor.type != agent.type:
                segregated = False
                break
        if segregated:
            segregated_agents += 1
    return segregated_agents / model.schedule.get_agent_count()
```

The function named 'get_segregation' takes a 'model' as an argument. The function iterates through all the agents in the model's schedule and checks if each agent has neighbors with different types. If an agent has no neighbors with different types, it is considered 'segregated' and the count of segregated agents is incremented. The function returns the proportion of segregated agents by dividing the number of segregated agents by the total number of agents in the schedule. Now, we set up the batch run, with a dictionary of fixed and changing parameters. The mechanism holds everything fixed except for Homophily.

```
[In]
fixed_params = {"height": 10, "width": 10, "density": 0.7,
"minority_pc": 0.3}
variable_parms = {"homophily": range(1,9)}
model_reporters = {"Segregated_Agents": get_segregation}
param_sweep =
BatchRunner(Schelling,variable_parameters=variable_parms,
fixed_parameters=fixed_params, iterations=10,
max_steps=200, model_reporters=model_reporters)
param_sweep.run_all()
```

→

```
[Out]
80it
[00:10,  7.31it/s]
```

The 'BatchRunner' creates multiple simulations with different parameter combinations. The 'fixed_params' contains several parameters for the 'Schelling' model and their fixed values: the 'height' and 'width' of the grid are both set to 10, the 'density' is set to 0.7, and the 'minority_pc' is set to 0.3. The 'variable_params' contains a single parameter, 'homophily' which will vary across the simulations and take on values in the range 1 to 8 (inclusive). The 'model_reperters' specifies that the 'get_segregation' function should be used to collect data from the model. The 'run_all' method is called on the 'BatchRunner' to run all the simulations, and the 'BatchRunner' is given the 'Schelling' model, the number of iterations, and maximum number of steps for each simulation.

Then, see what happened with the segregation simulation

```
[In]
df =
aram_sweep.get_model_vars_dataframe()
plt.scatter(df.homophily,
df.Segregated_Agents,)
plt.xlabel('homophily', fontsize=15)
plt.ylabel('segregated(agents))
plt.title('Schelling Segregation')
plt.grid(True)
plt.show()
```

→

[Out]

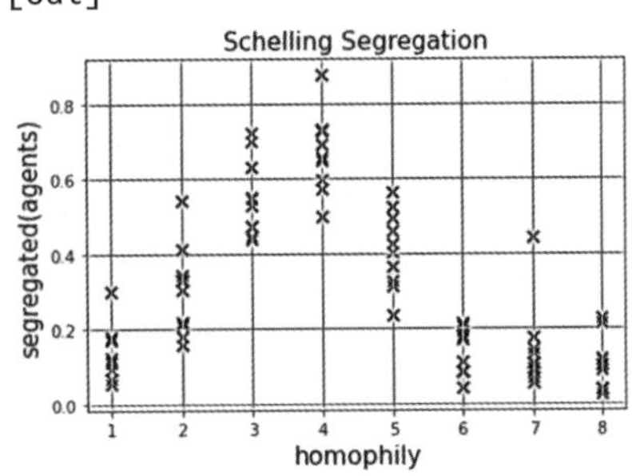

Figure 3.6.3. The degree of segregation in Schelling's model over time

In Figure 3.6.3, the models we have seen so far might be characterized as rule based in the sense that they involve systems governed by simple rules. We next explore agent-based models. Since the 1970s, this modeling has become an important tool in economics and other social sciences and in some natural sciences. Agent-based models are useful for modeling the dynamics of systems that are not in equilibrium (although they are also used to study equilibrium). And they are particularly useful for understanding relationships between individual decisions and system behavior. Agent-based models include agents that are intended to model people and other entities that gather information about the world, make decisions, and take actions. The agents are usually situated in space and interact with each other locally. They usually have imperfect, local information. Often there are differences among agents, unlike in models in which all components are identical. And agent-based models often include randomness, either among the agents or in the world.

2. Simulation

Here we develop and implement a limited information diagnostic strategy for assessing the plausibility of monetary models (Jenkins, 2020). The primary strategy focuses on a model's ability to reproduce empirical estimates of an actual individual's response to monetary policy.

+ **Monetary Policy Model:** A key input to this diagnostic is a univariate time series representation of the response of the individual in monetary policy. Its primary effect is to signal future movements in the money supply.

- Identify the agent types and other objects along with their attributes: There is some number of agents, and all agents begin with 1 unit of money. At every step of the model, an agent gives 1 unit of money to some other agent.
- Each agent has only one variable: how much wealth it currently has.
- Each agent has a unique identifier: unique_id and an agent with fixed initial wealth = 1.

```
[In]
from mesa import Agent, Model
class MoneyAgent(Agent):
    def __init__(self, unique_id, model):
        super().__init__(unique_id, model)
        self.wealth = 1
```

Define the environment the agents will live in and interact with: There is only one model-level parameter, how many agents the model contains (a model with some number of agents).

```
class MoneyModel(Model):
    def __init__(self, N):
        self.num_agetns = N
        # Create agents
        for i in range(self.num_agents):
            a = MoneyAgent(i, self)
```

When the model is started, it populates itself with the given number of agents.

- Specify the methods by which agent attributes are updated in response: Every agent (a = MoneyAgent i) is expected to have a step method = step.

- Control which agents interact, when they interact, and how they interact: At each step of the model, one or more of the agents (all of them) are activated and take their own step, changing internally and/or interacting with one another or the environment.

Random activation activates all of the agents once per step, in random order = self.schedule. Import modules

```
[In]
from mesa import Agent, Model
from mesa.time import RandomActivation
```

An agent with fixed initial wealth

```
[In]
class MoneyAgent(Agent):
    def __init__(self, unique_id, model):
        super().__init__(unique_id, model)
        self.wealth = 1
```

The code defines a class 'MoneyAgent' which has an '__init__' method, and sets the 'wealth' attribute to 1. The agent's step will go here

```
[In]
def step(self):
        pass
```

A model with some number of agents

```
[In]
class MoneyModel(Model):
    def __init__(self, N):
        self.num_agents = N
        self.schedule = RandomActivation(self)
```

The code takes a parameter 'N' and the '__init__' sets the 'num_agents' attribute to the value of 'N'. It also creates a 'RandomActivation' object and assigns it to the 'schedule'.

Create agents

```
[In]
        for i in range(self.num_agents):
            a = MoneyAgent(i, self)
            self.schedule.add(a)
```

The 'for' loop iterates 'self.num_agetns' where 'self' refers to the current instance of the 'MoneyModl' class. On each iteration, it creates a new object, passing 'i' and 'self' as arguments to the 'MoneyAgent' which object is then added to the 'schedule' using the 'add' method.

Advance the model by one step

```
[In]
    def step(self):
        self.schedule.step()
```

Then, let us create the model object, and run it for one step

```
[In]
empty_model = MoneyModel(10)
empty_model.step()
```

Implement the agent model computationally: Now, run a few steps of the model to see how the agent activation order is shuffled each step. Check their wealth. If they have money, give 1 unit of it away to another random agent. To pick an agent at random, it needs a list of all agents. It rewrites the agent's step method as follows:

```
[In]
import random
class MoneyAgent(Agent):
    """An agent with fixed initial wealth."""
    def __init__(self, unique_id, model):
        super().__init__(unique_id, model)
        self.wealth = 1
    def step(self):
        if self.wealth == 0:
            return
        other_agent = random.choice(self.model.schedule.agents)
        other_agent.wealth += 1
        self.wealth -= 1
```

The 'step' method first checks if the 'wealth' of the 'MoneyAgent' instance is 0. If it is, the method returns without doing anything else. If the 'wealth' is not 0, the method selects a random agent from the 'schedule' of the 'model' attribute of the 'MoneyAgent' and increase the agent's 'wealth' by 1.

Create a model with 10 agents, and run it for 10 steps.

```
[In]
model = MoneyModel(10)
for i in range(10):
    model.step()
```

The model construct with the argument 10 which sets the 'num_agents' of the 'model' instance to 10. Then a loop iterates 10 times on each iteration, the 'step' method of the 'model'. This method in in the 'schedule' attribute of the 'model' stance. Create a plot to visualize the data in a histogram.

```
[In]
import matplotlib.pyplot as plt
agent_wealth = [a.wealth for a in model.schedule.agents]
plt.hist(agent_wealth)
```

We import the 'pyplot' module from the 'matplotlib' library and use it to create a histogram of the wealth of 'MoneyAgent' in the 'schedule' attribute of the 'model'. The 'agent_wealth' list generates a list of the 'wealth' of each 'MoneyAgent' in the 'schedule' attribute of the 'model' and the 'hist' function generates a histogram of the values in the 'agent_wealth' list.

[Out]

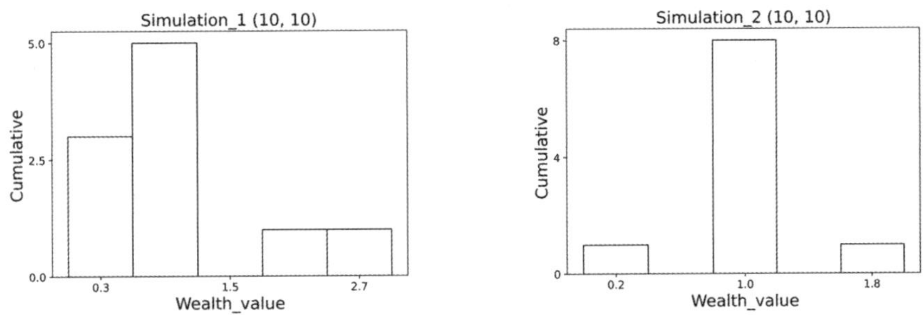

Figure 3.6.4. Simulations of the money model with 10 agents and 10 steps

In Figure 3.6.4, the two simulations are slightly different, as each run of the model is random. To see something like the distribution, create multiple model runs and see what emerges from all of them. Run the model

```
[In]
all_wealth = []
for j in range(100):
    model = MoneyModel(10)
    for i in range(10):
        model.step()
```

The code sets the 'num_agents' attribute of the 'model' instance to 10. Then a loop iterates 10 times on each iteration, the 'step' method of the 'model' called the 'step' method of each 'MoneyAgent' objects in the 'schedule' attribute of the 'model'.

Store the results

```
[In]
    for agent in model.schedule.agents:
        all_wealth.append(agent.wealth)
```

Display results simply with histogram

```
[In]
plt.hist(all_wealth, bins=range(max(all_wealth)+1))
```

[Out]

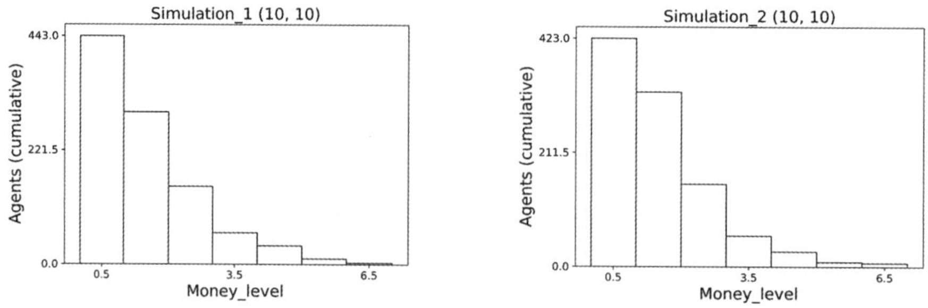

Figure 3.6.5. Simulations of the money model produced by running the model 100 times

In Figure 3.6.5, despite the fact that all agents on average give and receive 1 unit of money every step, the model converges to a state where most agents have a small amount of money and a small number of agents have a lot of money. Let agents move around and interact with nearby neighbors (grid) and allow them to have any arbitrary position in continuous space (continuous). Instead of giving their unit of money to any random agent, they will give it to an agent in the same cell. A model with some number of agents

```
[In]
from mesa.space import MultiGrid
class MoneyModel(Model):
    def __init__(self, N):
        self.num_agents = N
        self.grid = MultiGrid(width, height, True)
        self.schedule = RandomActivation(self)
```

Create agents

```
[In]
        for i in range(self.num_agents):
            a = MoneyAgent(i, self)
            self.schedule.add(a)
```

Add the agent to a random grid cell

```
[In]
            x = random.randrange(self.grid.width)
            y = random.randrange(self.grid.height)
            self.grid.place_agent(a, (x, y))
```

Each agent's position is stored in the cell it is currently in (grid), and the agent has a variable with an (x, y) coordinate tuple (pos). Let the agent move around and only give money to other agents in the same cell. Now put it all together:

```
[In]
class MoneyModel(Model): #A model with some number of agents
    def __init__(self, N, width, height):
```

```
        self.num_agents = N
        self.grid = MultiGrid(width, height, True)
        self.schedule = RandomActivation(self)
        for i in range(self.num_agents):
            a = MoneyAgent(i, self)
            self.schedule.add(a)
            x = random.randrange(self.grid.width)
            y = random.randrange(self.grid.height)
            self.grid.place_agent(a, (x, y))
    def step(self):
        self.schedule.step()

class MoneyAgent(Agent): #An agent with fixed initial wealth
    def __init__(self, unique_id, model):
        super().__init__(unique_id, model)
        self.wealth = 1
    def move(self):
        possible_steps = self.model.grid.get_neighborhood(
            self.pos,
            moore=True,
            include_center=False)
        new_position = random.choice(possible_steps)
        self.model.grid.move_agent(self, new_position)
    def give_money(self):
        cellmates = self.model.grid.get_cell_list_contents([self.pos])
        if len(cellmates) > 1:
            other = random.choice(cellmates)
            other.wealth += 1
            self.wealth -= 1
    def step(self):
        self.move()
        if self.wealth > 0:
            self.give_money()
```

Create a model with 50 agents on a 10 × 10 grid and run it for 20 steps.

```
[In]
model = MoneyModel(50, 10, 10)
for i in range(20):
    model.step()
```

The loop iterates 20 times on each iteration, called 'step' method of each of the 'MoneyAgent' in the 'schedule' attribute of the 'model' instance. Visualize the number of agents residing in each cell. The 'agent_counts' array is intialzied to a 2D array of zeros with dimensions 'model.grid.width, model.grid.height'. The loop, iterates over the cells in the grid and counts the number of agents in each cell which the number of agents in each cell is stored in the corresponding element of the 'agent_counts' array.

```
[In]
import numpy as np
agent_counts = np.zeros((model.grid.width, model.grid.height))
for cell in model.grid.coord_iter():
    cell_content, x, y = cell
    agent_count = len(cell_content)
    agent_counts[x][y] = agent_count
plt.imshow(agent_counts, interpolation='nearest')
plt.colorbar()
```

[Out]

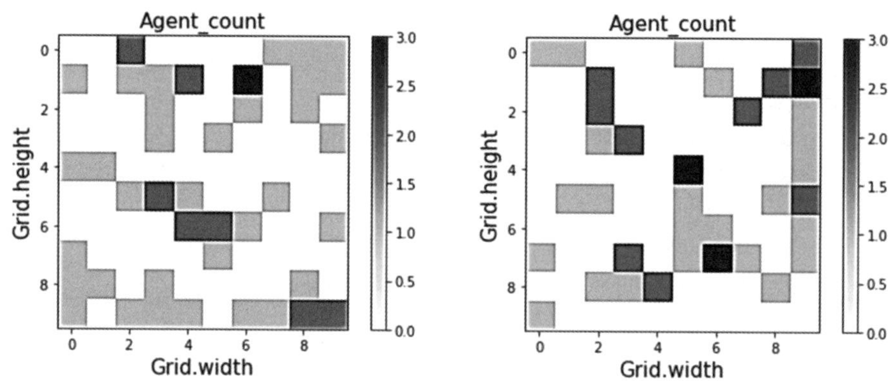

Figure 3.6.6. Simulation of the money model with 50 agents on a 10 × 10 grid, run for 20 steps

In Figure 3.6.6, both plots show how many agents are residing in each cell. Although the positions are different, their distributions look similar.

+ **Distributed Behavioral Model:** Maybe you have been captivated by the mesmerizing flow of a flock of birds flying against the setting sun. Hundreds, or even thousands, of birds flying together, forming endless shapes as if they were one entity are an incredible sight (Getchell, 2008). Here we implement something similar in programing, taking one small step toward forming an understanding of complex systems. In 1987 Craig Reynolds published 'Flocks, Herds and Schools: A Distributed Behavioral Model,' which describes an agent-based model of herd behavior. Agents in this model are called boids, which is both a contraction of birdoid and an accented pronunciation of bird (although boids are also used for modeling the behavior of fish and herding land animals). His model of artificial life followed three simple rules:

- Separation: Steer to avoid crowding local flock mates.
- Alignment: Steer toward the average heading of local flock mates.
- Cohesion: Steer to move toward the average position of local flock mates.

With only these rules you can see emergent complexity arise out of almost nothing. First, we need to import a couple of necessary packages.

```
[In]
import numpy as np
from mesa import Agent
```

Then we create some boids as our birds. We refer to these as boids rather than birds because birds are just one possible model—it could be fish or any other flocking pattern. For this purpose, we create a class. Boids have a vision that defines the radius in which they look for neighbors to flock with. Their speed (a scalar) and heading (a unit vector) define their movement. Separation is their desired minimum distance from any other boid. Let us create the class

```
[In]
class Boid(Agent):
    def __init__(self, unique_id, model, pos, speed=5, heading=None,
                 vision=5, separation=1):
        super().__init__(unique_id, model)
        self.pos = pos
        self.speed = speed
        if heading is not None:
            self.heading = heading
        else:
            self.heading = np.random.random(2)
            self.heading /= np.linalg.norm(self.heading)
        self.vision = vision
        self.separation = separation
```

It is clear that we need a position for each boid, so, we create a new boid flocker agent; unique_id: Unique agent identifyer, pos: Starting position, speed: Distance to move per step, heading: numpy vector for the boid's direction of movement, vision: Radius to look around for nearby boids, separation: Minimum distance to maintain from other Boids. The '__init__' method starts by calling the 'Agent' class. The 'super()' function returns a temporary object of the superclass, which allows you to call its methods. The code sets the 'pos', 'speed', and 'separation' attributes of the boid using the provided parameters. It then checks if a 'heading' which sets the attribute to a random 2D vector and normalizes it (i.e. scales it to have a length of 1). Then, the code returns 'None' which is standard behavior for the 'init' method. Let us return the vector toward the center of mass of the local neighbors.

```
[In]
def cohere(self, neighbors):
        center = np.array([0.0, 0.0])
        for neighbor in neighbors:
            center += np.array(neighbor.pos)
        return center / len(neighbors)
```

The code creates a new 2D array called 'center' with all zeros. It then iterates over each 'neighbor' and adds the 'pos' attribute of the 'neighbor' to 'center'. After all of the neighbors have been processed, the function returns the average position of the neighbors by dividing 'center' by the length of the 'neighbors'. this function is implemented by a 'cohesion' behavior, where a boid tries to move towards the center of mass of its neighbors. This is a common behavior in simulations of flocking or swarming, as it can produce the appearance of cohesive groups of agents moving together. Return a vector away from any neighbors closer than separation dist.

```
[In]
def separate(self, neighbors):
        my_pos = np.array(self.pos)
        sep_vector = np.array([0, 0])
        for neighbor in neighbors:
            their_pos = np.array(neighbor.pos)
            dist = np.linalg.norm(my_pos - their_pos)
            if dist < self.separation:
                sep_vector -= np.int64(their_pos - my_pos)
        return sep_vector
```

The code starts by creating two arrays, 'my_pos' and 'sep_vector' which represent the position of the current boid and the separation vector that will be returned by the function, respectively. Both arrays are initialized to all zeros. Next, the function iterates over each 'neighbor' which creates a new array called 'their_pos' that represents their position. It then calculates the distance between the two 'boid' using the 'np.linalg.norm' function and checks if the distance is less than the 'separation' of the current boid. If the distance is less than 'self.separation', the function updates the 'sep_vector' array by subtracting the difference between 'their_pos' and 'my_pos' which is a vector pointing. After all of the neighbors have been processed, the function returns the 'sep_vector' which represents the desired separation from all of the neighbors. This is another common behavior in simulations of flocking or swarming, as it can produce the appearance of agents avoiding collisions with each other. Return a vector of the neighbors' average heading.

```
[In]
def match_heading(self, neighbors):
        mean_heading = np.array([0, 0])
        for neighbor in neighbors:
            mean_heading += np.int64(neighbor.heading)
        return mean_heading / len(neighbors)
```

The code creates a new array called 'mean_heading' with all zeros. It then iterates over each 'neighbor' and adds the 'heading' to 'mean_heading'. After all of the neighbors have been processed, the function returns the average heading of the neighbors by dividing 'mean_heading' by the length of the 'neighbors' list. This function is implemented by a 'matching heading' behavior, where a boid tries to align its heading with that of its neighbors as it can produce the appearance of cohesive groups of agents moving in the same direction.

And, we get the boid's neighbors, compute the new vector, and move accordingly.

```
[In]
def step(self):
        neighbors = self.model.space.get_neighbors(self.pos, self.vision, False)
        if len(neighbors) > 0:
            cohere_vector = self.cohere(neighbors)
            separate_vector = self.separate(neighbors)
            match_heading_vector = self.match_heading(neighbors)
            self.heading += (cohere_vector +
                             separate_vector +
                             match_heading_vector)
            self.heading /= np.linalg.norm(self.heading)

        new_pos = np.array(self.pos) + self.heading * self.speed
        new_x, new_y = new_pos
        self.model.space.move_agent(self, (new_x, new_y))
```

The 'step' function starts by calling the 'get_neighbors' method of the 'space' of the boid, passing it the 'pos', 'vision', and 'False' as arguments. This returns a list of 'boid' that are within the 'vision' distance of the current boid. If the boid has at least one neighbor (determined by checking the length of the 'neighbors' list), the function calls the 'cohere', 'separate', and 'match_heading' of the boid. These methods return arrays that represent the desired cohesion, separation, and heading-matching behavior, respectively.

The function then adds these vectors together and normalizes the result (i.e. scales it to have a length of 1) using the 'np.linalg.norm' function. The resulting vector is then added to the 'heading' of the boid. Finally, the function calculates the new position of the boid by adding the 'heading' to its 'pos', multiplied by the 'speed' of the boid. It then calls the 'move_agent' method of the 'space' of the 'model' attribute of new position as arguments. This updates the position of the boid in the model's space. Overall, the function is responsible for updating the position and heading of the boid based on its neighbors, combined the cohesion, separation, and heading-matching behaviors.

Now let us address agent creation, placement, and scheduling. We need to create a new model with N (number of boids), width, height (the size of the space), speed (how fast the boids move), vision (how far each boid looks for its neighbors), separation (the minimum distance each boid will attempt to stay from any other).

Here, the 'schedule' is an instance of the 'RandomActivation' class which is used to schedule the steps of the boid. The 'space' is an instance of the 'ContiuousSpace' which is used to represent the 2D space in which the boid move. And the 'running' is a flag indicating whether the model is currently running or not.

```
[In]
class BoidModel(Model):
    def __init__(self, N, width, height, speed, vision, separation):
        self.N = N
        self.vision = vision
        self.speed = speed
        self.separation = separation
        self.schedule = RandomActivation(self)
        self.space = ContinuousSpace(width, height, True)
        self.make_agents()
        self.running = True
```

This method initializes the attributes of the 'BoidModel' using the provided arguments and then creates the individuals using the 'make_agents' added to the 'schedule', which is an instance of the 'RandomActivation' class. Overall, the class is responsible for creating and managing a group of individuals that move within 'contiuousSpace' class to represent the 2D space in which the boid move. Then create N agents with random positions and starting headings.

```
[In]
def make_agents(self):
        for i in range(self.N):
            x = random.random() * self.space.x_max
            y = random.random() * self.space.y_max
            pos = (x, y)
            heading = np.random.random(2) * 2 - np.array((1, 1))
            heading /= np.linalg.norm(heading)
            boid = Boid(i, self, pos, self.speed, heading, self.vision,
                        self.separation)
            self.space.place_agent(boid, pos)
            self.schedule.add(boid)
    def step(self):
        self.schedule.step()
```

This method uses a loop to create 'self.N Boid'. For each boid, it generates random starting positions within the bounds of the model's space (x between 0 and 'self.space.x_max', and y between 0 and 'self.space.y_max'), and a random heading (a 2D array with values between −1 and 1). The heading is then normalized using the 'np.linalg.norm' function. The boid is then created and added to the 'space' and 'schedule' attributes of the model using the 'place_agent' and 'add' methods, respectively. The 'make_agents' and 'step' of the model class are responsible for creating and updating the boid in the model. And the 'step' method is called repeatedly to advance the model by one step at a time. Do the following to set the number of agents residing in each cell.

```
[In]
import matplotlib.pyplot as plt
%matplotlib inline
def draw_boids(model):
    x_vals = []
    y_vals = []
    for boid in model.schedule.agents:
        x, y = boid.pos
        x_vals.append(x)
        y_vals.append(y)
    fig = plt.figure(figsize=(5, 5))
    ax = fig.add_subplot(111)
    rng = np.random.RandomState(0)
    colors = rng.rand(100)
    sizes = 1000 * rng.rand(100)
    ax.scatter(x_vals, y_vals)
```

We define two empty lists 'x_vals' and 'y_vals', iterate through each boid in 'model.schedule.agents', and retrieves its x, and y position. These positions are appended to the 'x_vals' and 'y_vals' lists, respectively. The function then creates a figure on the subplot with 'x_vals' and 'y_vals' as arguments. Finally, let us visualize (Figure 3.6.7) what happened with the parameters given.

```
[In]
model = BoidModel(100, 100, 100, speed=5,
vision=5, separation=1)

for i in range(50):
    model.step()
    draw_boids(model)
```

→

[Out]

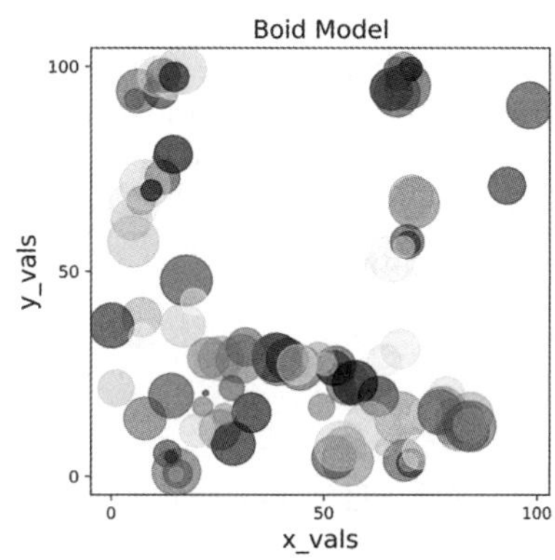

Figure 3.6.7. Simulation results of the Boid Model. The sizes and colors of the agents (circles) are random

Agent-based modeling is not a new technique, but its promise and power are often overlooked. We believe that two assumptions have hindered their use. The first is that agent-based models are difficult to learn or understand. However, good models should be easy to understand conceptually, and the resources discussed above should make their implementation easier.

The second is that agent-based models fail to generate new knowledge. Yet good agent-based models harness the power of emergence, in which higher level phenomena derive from the simple behavior of agents. Like any method, agent-based modeling is imperfect, but it does offer a powerful way of implementing precise hypotheses and exploring emergence. Not only can researchers build whole worlds to examine social processes, they can also sample from these worlds over thousands of generations to yield unprecedented insight into collective behavior. Now, let us dive into the agent-based model a bit further (adding more rules) with another famous simulation.

3. Cellular Automata

Cellular automata are mathematical models that contain a grid of cells, usually (but not always) two dimensions. The concept was discovered by John von Neumann while studying self-replicating machines at the Los Alamos Laboratory in New Mexico in the 1940s. von Neumann conceived of self-replicating machines as a thought experiment in the context of colonizing other planets and developed cellular automata to express these ideas in a mathematically rigorous way. The field of cellular automata remained largely theoretical and academic throughout the next 20 years. Conway's Game of Life: After years of experimentation, in 1970 John Conway finished his cellular automaton, called the Game of Life. He chose the governing ruleset according to four guidelines:

- There should be no explosive growth, and exist small initial patterns with chaotic, unpredictable outcomes.
- There should be potential for von Neumann universal constructors, and the rules should be as simple as possible.

To dive further into the Game of Life, we first need to understand the concepts of the cell and the neighborhood. In cellular automata, the neighborhood describes the cells immediately surrounding a particular cell. There are different types of neighborhoods, but for this example we refer to the four cardinal and four diagonal cells, or eight neighbors total.

Figure 3.6.8. The cellular automata cell

If a cell is black (■) it is considered live, and if it is white (□) it is considered dead (Figure 3.6.8). Here are the four simple rules for the Game of Life:

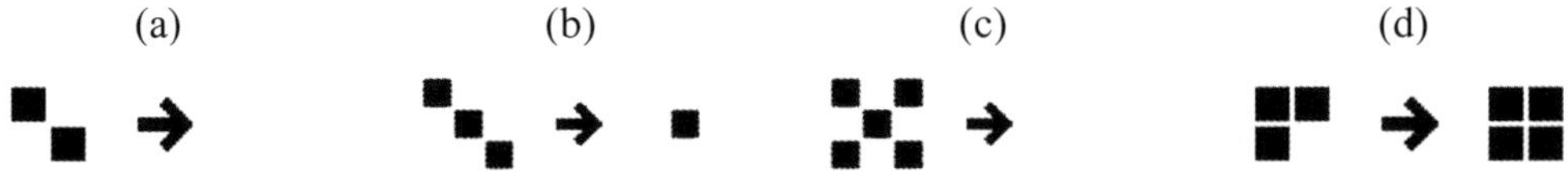

Figure 3.6.9. The dynamics of cellular automata cells

In Figure 3.6.9, (a) any live cell with fewer than two live neighbors die. (b) Any live cell with two or three live neighbors lives on to the next generation. (c) Any live cell with more than three live neighbors dies. (d) Any dead cell with exactly three live neighbors becomes a live cell.

Let us look at an example. In this case the structure on the left will turn into the structure on the right in the next generation, or the next step, of the simulation (Figure 3.6.10).

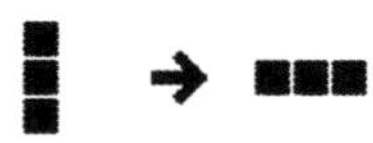

Figure 3.6.10. The dynamics of cellular automata cells (turn structure)

For this to happen, we must apply the four rules outlined above to every single cell in the simulation. This process is shown below for the relevant cells. Dead cell, only two neighbors, no change; live cell, only one neighbor, dies; dead cell, only two neighbors, no change; dead cell, three neighbors, becomes a live cell; live cell, two neighbors, no change; dead cell, three neighbors, becomes a live cell; dead cell, only two neighbors, no change; live cell, only one neighbor, dies; dead cell, only two neighbors, no change (Figure 3.6.11).

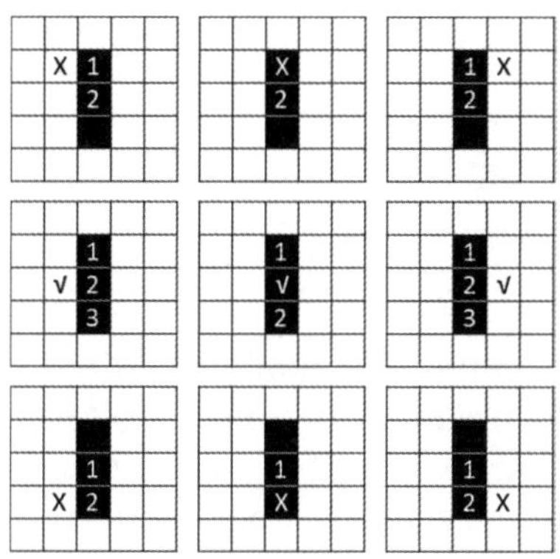

Figure 3.6.11. Cellular automata cells with neighbors

If you have been paying attention, you may have noticed that this structure will return to the first structure on the next generation and will continue to repeat.

This is a stable pattern called an oscillator.

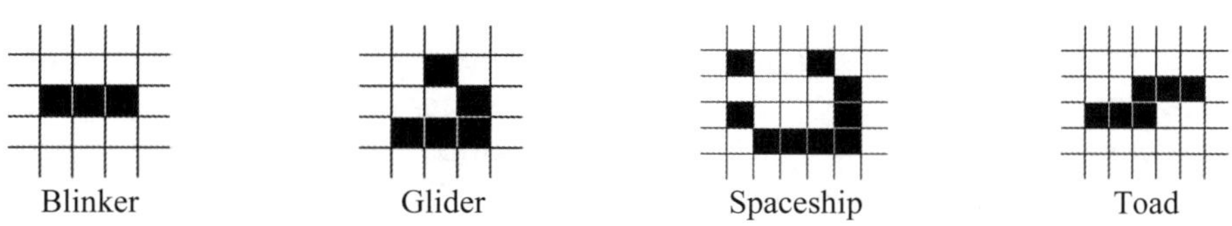

Figure 3.6.12. Repetition of cellular automata cells

In Figure 3.6.12, we can see that from these simple rules we get incredibly complex, unpredictable behaviors. The next thing researchers started to find were interesting shapes and structures and their ensuing generations. A good example of this is Bill Gosper's early glider gun (Figure 3.6.13).

Figure 3.6.13. Schematic representation of the cellular automata cell

+ **Practice with the Model:** In general, cellular automata can be defined in any number of dimensions. A famous two-dimensional example is Conway's Game of Life, in which cells live and die, sometimes producing beautiful patterns. Thus, let us implement the features that evolve in time on a grid. Each grid cell has a finite state, such as 0 (white cell = □) or 1 (black cell = ■), which is updated based on a certain set of rules. A specific cell uses information about the surrounding cells, called its *neighborhood*, to determine what changes should be made.

First, we import a package and arrange the cells in a random array of 10 bits.

```
[In]
import numpy as np
rng = np.random.RandomState(42)
data = rng.randint(0, 2, 10)
print(data)
```

→

```
[Out]
[0 1 0 0 0 1 0 0 0 1]
```

The code calls the 'randint' on the 'RandomState' with the arguments 0, 2, and 10. The 'randint' method returns random integers from a discrete distribution within a specified range. In this case, the method returns 10 random integers from the range that are either 0 or 1. To update the state of the cells, we need to define a function or rule: A given cell C only knows about the state of its left and right neighbors, labeled L and R, respectively (L, C, R).

Because our input cells are binary, there are 2^3 = 8 possible inputs into the function.

```
[In]
for i in range(8):
    print(np.binary_repr(i, 3))
```

→

```
[Out]
000, 001, 010, 011, 100, 101,
110, 111
```

Here, the code 'binary_repr' fnction within a 'for' loop iterates over the rand (0, 8). For each iteration, the function is called with the loop variable 'i' as the input and return the binary representation of 'i' as a string, with a minimum width of 3 bits. For each input, we can assign 0 or 1 to the output. The output of (L, C, R) is the value that will replace the current cell in the next time step. It updates rules in which each rule is represented by an 8-bit binary number.

For example, 'rule 30' could be constructed by first converting to binary and then building an array for each bit.

```
[In]
rule_number = 30
rule_string = np.binary_repr(rule_number, 8)
rule = np.array([int(bit) for bit in rule_string])
print(rule)
```

→

```
[Out]
[0 0 0 1 1 1 1 0]
```

The 'binary_repr' function returns the binary representation of 'rule_number' as a string with a minimum width of 8 bits. Then is passed to a list comprehension, which iterates over the string and converts each character to an integer using the 'int' function. After this code is run, the 'rule' will be an array with the following values; [0, 1, 1, 1, 1, 0, 1, 1]. The code associates the leading bit with 111 and the final bit with 000.

For rule 30 the relationship between the input, rule index, and output is as follows:

```
[In]
for i in range(8):
  triplet = np.binary_repr(i, 3)

  print(f'input:{triplet},   index:{7-i},   output
{rule[7-i]}")
```

→

```
[Out]
input:000, index:7, output 0
input:001, index:6, output 1
input:010, index:5, output 1
input:011, index:4, output 1
input:100, index:3, output 1
input:101, index:2, output 0
input:110, index:1, output 0
input:111, index:0, output 0
```

The code defines a 'for' loop that iterates over the range (0, 8). For each iteration, it calls the 'binary_repr' function with the loop variable 'i' and a width of '3' arguments. The function returns the binary representation of 'i' as a string with a minimum width of 3 bits assigned to the variable 'triplet'. The code then prints a string with three placeholder 'input', 'index', and 'output'. The 'input' is replaced with the value of 'triplet', the 'index' is replaced with the value of '7 – i', and the 'output' is replaced with the value of 'rule [7 – i] as the results appeared at [Out].

We can define a function that maps the input cell information with the associated rule index, converting the binary input to a decimal and adjusting the index range, and take in any input and look up the output based on our rule. For example:

```
[In]
def rule_index(triplet):
  L, C, R = triplet
  index = 7 - (4*L + 2*C + R)
  return int(index)
rule[rule_index((1, 0, 1))]
```

→

```
[Out]
0
```

Then we can create a data structure containing all the triplets for our state array and apply the function across the appropriate axis to determine our new state.

```
[In]
all_triplets = np.stack([
    np.roll(data, 1),
    data,
    np.roll(data, -1)])
new_data = rule[np.apply_along_axis(rule_index, 0,
all_triplets)]
print(new_data)
```

```
[Out]
[0 1 1 0 1 1 1 0 1 1]
```

This is the process for a single update of cellular automata (CA). To do many updates and record the state over time, we will create a function.

```
[In]
def CA_run(initial_state, n_steps, rule_number):
    rule_string = np.binary_repr(rule_number, 8)
    rule = np.array([int(bit) for bit in rule_string])
    m_cells = len(initial_state)
    CA_run = np.zeros((n_steps, m_cells))
    CA_run[0, :] = initial_state

    for step in range(1, n_steps):
        all_triplets = np.stack(
            [
                np.roll(CA_run[step - 1, :], 1),
                CA_run[step - 1, :],
                np.roll(CA_run[step - 1, :], -1),
            ]
    )
        CA_run[step, :] = rule[np.apply_along_axis(rule_index, 0, all_triplets)]
    return CA_run
```

Here, the simulation takes three arguments as follows. The 'initial_state' is a binary array representing the initial state of in the cellular automation. The 'n_steps' is an integer representing the number of time steps to run, and the 'rule_umber' represent the rule to use for the simulation which is converted to a binary string to an array of integers, where each integer corresponds to a specific condition for the state of a cell. It returns a 2D array where each row represents the state of the cells at a particular time step, and each column represents a cell.

At each time step, it calculates the state of each cell based on the states of its neighbors in the previous time step using the 'rule' array. The states of the neighbors are obtained by 'rolling' the previous time step' array which shifts the elements by one index to the left or right. The simulation then uses 'np.apply_along_axis' to apply the 'rule_index' function to each triplet of neighbors, which returns an index into the 'rule' array based on the states of the three cells. The resulting array of indices is used to index into the 'rule array to obtain the new states of the cells.

Let us see the data in numerical form:

```
[In]
initial = np.array([0, 1, 0, 0,
 0, 1, 0, 0, 0, 1, 0, 0, 0, 0,
1, 0, 1, 1, 1, 0])
data = CA_run(initial, 10, 30)

print(data)
```

→

```
[Out]
[[0. 1. 0. 0. 0. 1. 0. 0. 0. 1. 0. 0. 0. 0. 1. 0. 1. 1. 1. 0.]
 [1. 1. 1. 0. 1. 1. 1. 0. 1. 1. 1. 0. 0. 1. 1. 0. 1. 0. 0. 1.]
 [0. 0. 0. 0. 1. 0. 0. 0. 1. 0. 0. 1. 1. 1. 0. 0. 1. 1. 1. 1.]
 [1. 0. 0. 1. 1. 1. 0. 1. 1. 1. 1. 1. 0. 0. 1. 1. 1. 0. 0. 0.]
 [1. 1. 1. 1. 0. 0. 0. 1. 0. 0. 0. 0. 1. 1. 1. 0. 0. 1. 0. 1.]
 [0. 0. 0. 0. 1. 0. 1. 1. 1. 0. 0. 1. 1. 0. 0. 1. 1. 1. 0. 1.]
 [1. 0. 0. 1. 1. 0. 1. 0. 0. 1. 1. 1. 0. 1. 1. 1. 0. 0. 0. 1.]
 [0. 1. 1. 1. 0. 0. 1. 1. 1. 1. 0. 0. 0. 1. 0. 0. 1. 0. 1. 1.]
 [0. 1. 0. 0. 1. 1. 1. 0. 0. 0. 1. 0. 1. 1. 1. 1. 1. 0. 1. 0.]
 [1. 1. 1. 1. 1. 0. 0. 1. 0. 1. 1. 0. 1. 0. 0. 0. 0. 0. 1. 1.]]
```

Now, visualize of this interesting pattern (Figure 3.6.14). To do this, we simulate using an 'matshow' function

```
[In]
import matplotlib.pyplot as plt
plt.rcParams['image.cmap']    =
'binary'
rng = np.random.RandomState(0)
data = CA_run(rng.randint(0, 2,
300), 150, 30)
fig, ax = plt.subplots()
ax.matshow(data)
```

→

[Out]

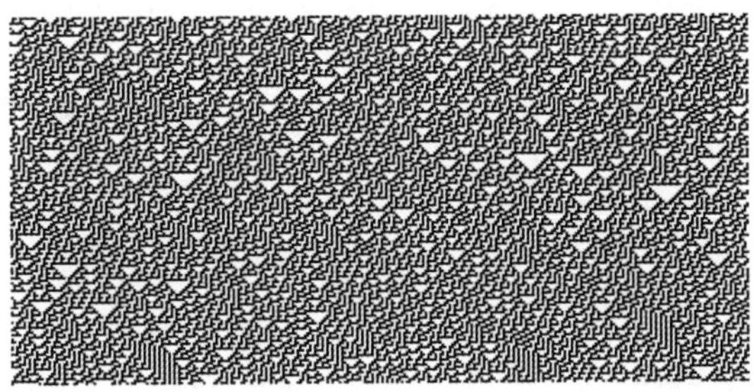

Figure 3.6.14. Simulation results of the cellular automata model

For larger simulations, interesting patterns start to emerge. To visualize our simulation results we used the 'ax.matshow' function.

With the code set up to produce the simulation, we can now start to explore the properties of these different rules. Programing separated the rules into four classes which are outlined below.

Cellular automata which rapidly converge to a repetitive or stable state.

```
[In]
def plot_CA_class(rule_list, class_label):
    rng = np.random.RandomState(seed=0)
    fig, axs = plt.subplots(1, len(rule_list))
    initial = rng.randint(0, 2, 100)

    for i, ax in enumerate(axs.ravel()):
        data = CA_run(initial, 100, rule_list[i])
        ax.set_title(f'Rule {rule_list[i]}')
        ax.matshow(data)
    return fig, ax
_ = plot_CA_class([60, 106, 150], 'Class with different rules')
```

[Out]

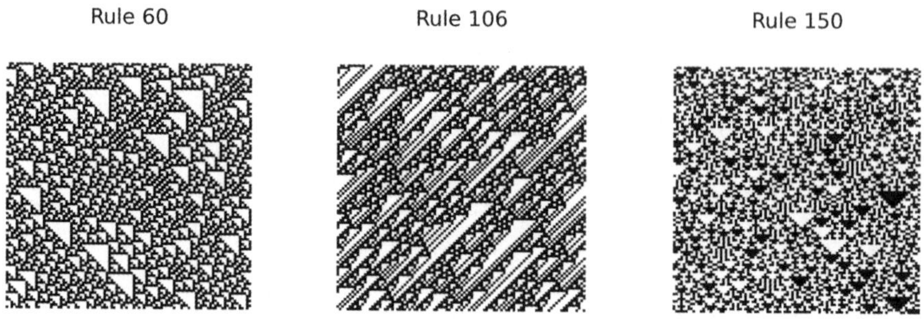

Figure 3.6.15. Simulation results of the cellular automata model with different rules

Cellular automata models have been used with different computer architectures than the ones we use today (called *von Neumann architecture*), such as magnetism and quantum vibration. CA have also been proposed as a means of generating pseudorandom numbers in cryptography as well as modeling one-way functions. It is possible to construct NOT, AND, and OR gates in the Game of Life and to simulate a Turing machine.

Systems Analysis and Complexity: A Symbiotic Exploration

In the context of systems analysis, complexity can be defined as the study of multi-faceted interactions within a system that lead to emergent properties and behaviors, which are not predictable from the individual components alone. 'Applied Systems Analysis' provides an in-depth examination of this concept, emphasizing how a myriad of interconnected elements can self-organize to create coherent, large-scale structures and behaviors. Employing tools such as network analysis, agent-based modeling, and game theory, this approach unravels the intricate dynamics of complex systems. Network analysis sheds light on the dense web of interactions, agent-based modeling simulates individual entity behaviors, and game theory explores strategic decision-making processes. These methodologies illuminate key phenomena like self-organization and adaptation. By integrating computational techniques, this interdisciplinary perspective enhances our understanding of complexity, providing a comprehensive view of the dynamic interplay and unpredictability inherent across various domains.

Remarks: There are likely interesting applications with regard to simulating agents and individuals developing autonomous intelligence. Scholars have even put forward the idea that our physical universe is an immensely complex automaton. Successfully replicating the processes and dynamics that occur within systems models is highly challenging. A potentially infinite number of individual components are linked together by often unknown interconnected processes that play out at different spatial and temporal scales. The bottom-up modeling advocated by agent-based modeling allows local phenomena to be understood and measured at a systemic level. Whereas established methods, such as spatial interaction modeling, treat populations as aggregate homogeneous components, agent-based models potentially allow every individual to be assigned its own characteristics. This powerful paradigm holds great promise for facilitating greater understanding of systems.

Supplemental Information 3.4: High Performance Computing (HPC) via Terminals

High-performance computing (HPC) uses computer clusters to solve advanced computation problems. HPC technologies are the tools and systems used to implement and create high-performance computing systems. Thus, it integrates a multidisciplinary field that combines system software, programming languages, algorithms, and computational techniques (Brazell & Michael, 2005).

The term is most commonly associated with computing used for scientific research or computational science. HPC as a term arose after the term 'supercomputing' (Collette et al., 2004) but is sometimes used as a synonym for it; however, in other contexts, 'supercomputer' is used to refer to a more powerful subset of 'high-performance computers,' and the term supercomputing becomes a subset of high-performance computing since networking clusters and grids use multiple processors and computers. A few examples of commercial HPC technologies include applications in various fields, such as simulating more accurate short- and long-term weather forecasts, improving performance, safety, and reliability in engineering projects, and conducting complex scientific simulations.

- Fast internal network
- High-speed storage managed by a scheduler
- Remote access used as a single machine

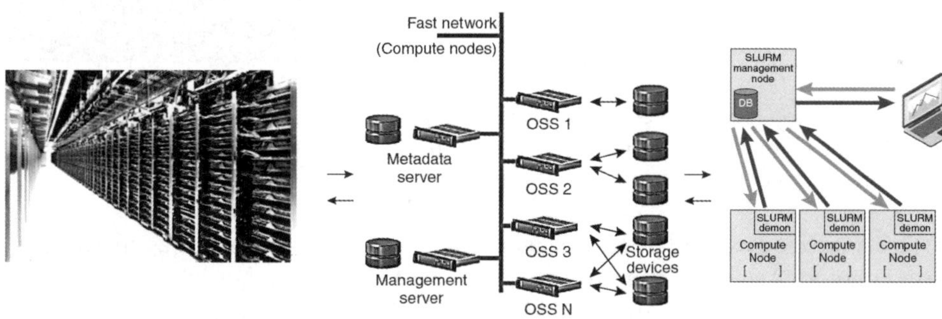

Figure S3.4.1. Data center (left = cluster) corresponding to the HPC network concepts (middle = compute nodes) and its scheduler (right = SLURM)

Here is an image of the cluster (Figure S3.4.1). The cluster itself is a collection of compute nodes. Each node has CPUs with a number of cores. The cores all share a specific amount of memory, and the nodes are all connected through a high-speed network.

- Node: A collection of cores with shared memory and storage (i.e., 128-512G memory with 16-128 cores)
- Core: Cluster computation unite (one processor core equals one thread of execution) and there is also a small set of login nodes.

You use the login nodes for moving files, setting up your work and running jobs on the compute nodes. The cluster is managed by a system (i.e., SLURM.) You ask the system to run jobs on the cluster on your behalf.

+ **Get an Account:** First, we will need to get an account from your institute. It is very easy to apply for a general account. Please go to the scientific computing and data analysis center and request the manager to 'Open Resources,' tell them you would like access and press submit. All research members will be able to get access, and the center will normally set it up for you within a couple of days. If you have access to the clusters, it should ask you for your password (i.e., Cisco AnyConnect Secure Mobility Client), then, perhaps you will need to have software installed to open a terminal (Figure S3.4.2):

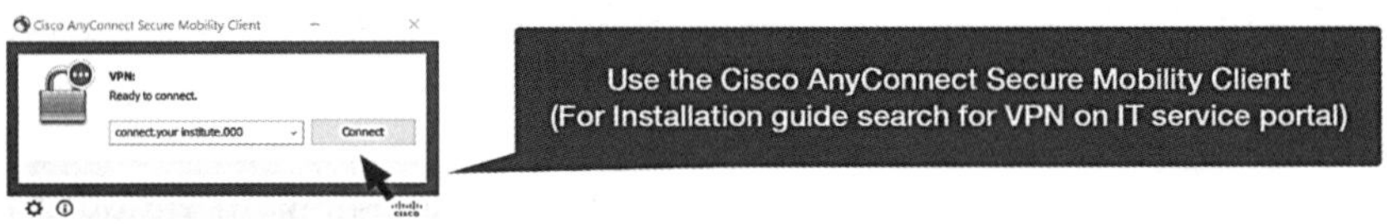

ssh -X user@server.domain.country

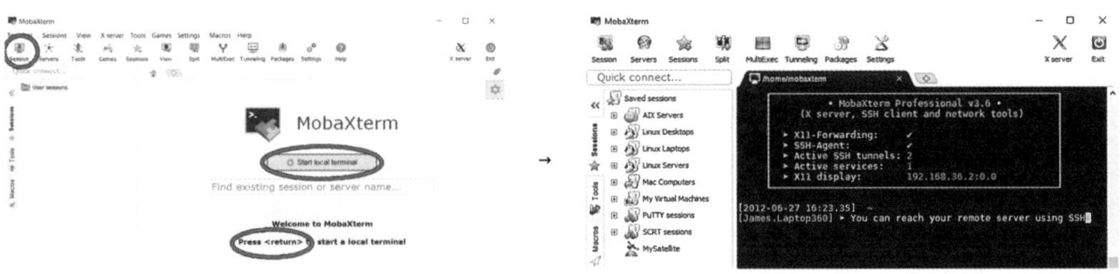

Figure S3.4.2. Schematic representation of the HPC process (VPN login = upper) corresponding to the software (MobaXterm) terminal (right) and its execution window examples (Source: https://mobaxterm.mobatek.net/)

There are numerous popular clients available, and you are encouraged to use your preferred choice. As an example, we recommend MobaXterm because it enables you to display graphics without requiring additional installation. However, feel free to explore alternative platforms that suit your computer's operating system. To download MobaXterm, visit the following page:

- http://mobaxterm.mobatek.net/download-home-edition.html.

After downloading, unpack the file in a desired location. To launch the program, simply click on it, or create a shortcut and place it on your desktop for easy access. Once the MobaXterm window appears, initiate a local terminal session. Click 'Start a local terminal.' If it does not appear, you can select 'Session,' then 'local terminal' in the list. Once the terminal appears, you can log in the same way:

```
[In]
ssh -X your - name@institute.00.00
```

This is necessary to access the clusters (even from outside of the institute). Now that we have set up HPC we can also log in from somewhere else without having to use the institute's virtual private network, then log in to the clusters: All your interactions with the clusters will be through the terminal. It may seem primitive at first, but there are some good reasons to learn and use the command line.

+ **Run Bash Jobs on the Cluster:** Now, there are two ways to run jobs. We can run it directly with the srun command. It is great for short jobs or for interactive use, but for long-running jobs it's better to use the 'sbatch' command. With sbatch you can create complicated jobs, and you do not have to wait for them to finish. Both srun and sbatch starts jobs, and the two share many options with each other. We will start by looking at running jobs with srun; then we look at how to use sbatch. Let us run a small interactive session with the Python package. Once running Python, you are on the right track already. After you have logged in, you start an interactive job (check the Slurm documentation above for details):

```
[In]
$ srun -p compute -t 0-4:00:00 -c 8 --mem=32G --x11 --pty bash
```

This will give you a new command line on a compute node, with 8 cores and 32 G memory (i.e., max 500 G), for 4 hours (0-4:00:00 = days-hours:minutes:seconds), allow applications to use graphics (--x11). Load the Jupyter module (you can load it before or after starting the job, it doesn't matter):

```
[In]
$ module load jupyter.py/3.7
$ jupyter-notebook
```

That will print out a bunch of text. To access the notebook, open this file in a browser: file:///home/ ~ /-open.html, Or copy and paste one of these URLs: http://127. ~

The instructions are used in a Slurm environment to launch a task that will run for 4 hours on a compute node with 8 cores and 32GB of memory, in accordance with the demands we made above. The Jupyter module is loaded once the task completes, and the 'jupyter-notebook' command is then used to launch a Jupyter notebook that can be viewed through a URL supplied in the command output. The URL is to the same compute node that you logged into. Copy that line and paste it into your browser, then this will open the Juyter session in your browser. For batch jobs, you put those commands in a text file, not on the command line. Then you use 'sbatch' to run that text file. On the command line open a text editor. 'nano' is a very simple, text-only editor (the '$' just represents typing into the command line. Do not actually type the $ itself):

```
[In]
$ nano myjob.slurm
```

These lines as below are written in the SBATCH script format and are used to configure and submit jobs to a high-performance computing (HPC) cluster using the Slurm workload manager.

- #!/bin/bash: This line specifies the script interpreter (Bash shell) to be used for executing the script.
- #SBATCH --partition=compute: This line sets the partition (queue) for the job to be submitted to the 'compute' partition.
- #SBATCH --job-name=py_job: This line assigns the name 'py_job' to the job.
- #SBATCH --output=py_%j.out: This line sets the name of the output file for the job, using the job ID as part of the name.
- #SBATCH --time=1-1:00:00: This line sets the maximum time limit for the job to run, which is 1 day and 1 hour.
- #SBATCH --mem=36G: This line sets the memory requirement for the job to 36 gigabytes.
- #SBATCH --ntasks=1: This line sets the number of tasks to be run in parallel for the job to 1.
- #SBATCH --cpus-per-task=1: This line sets the number of CPUs (cores) to be allocated per task to 1.

Then type the lines into the editor:

```
[In]
#!/bin/bash
#SBATCH --partition=compute
#SBATCH --job-name=py_job
#SBATCH --output=py_%j.out
#SBATCH --time=1-1:00:00
#SBATCH --mem=36G
#SBATCH --ntasks=1
#SBATCH --cpus-per-task=1
```

It includes several lines that specify various options and parameters for the job, such as the partition to run on, the name of the job, the output file, the time and memory limits, and the number of tasks and CPUs to use. After configuring the high-performance computing (HPC) cluster with the Slurm workload manager using the SBATCH script, we can load the Python module and execute your Python script within the same Bash script. Here is the script: load the python module, and run 'myprog.py' using python

```
[In]
module load python/3.7.3
python3 myprog.py
```

Now, we can save the SBATCH job commands by pressing Ctrl+o, and then exit the script by pressing Ctrl+x followed by Enter. Once you have completed these steps, exit the nano window by pressing Ctrl+x, which will bring you back to the MobaXterm window. Having returned to the MobaXterm window, we can now submit the job by running the 'sbatch' command: run this with 'sbatch'

```
[In]
$ sbatch myprog.slurm
```

Upon successful submission, the terminal (MobaXterm) will display a message containing the assigned batch job number, as shown below;

```
[Out]
Submitted batch job 00000000
```

There are several good reasons why we recommend to use batch jobs. You can submit a batch job, then log out without having to wait for it to finish. However, with srun, the job is cancelled if you lose the connection.

As previously mentioned, both 'srun' and 'sbatch' start jobs, and the two share many options. We can tell slurm how many nodes we want to use, how many tasks — that is, processes, or copies of our program — we want to run, and how many cores we want each task to have. The slurm will try to fulfill our request as best it can, while considering other pending jobs, how busy the cluster is, and so on.

[Table S3.4.1] Options – Shortcut Description

Option	Shortcut	Description
--ntasks	-n	Number of tasks or processes you wish to start. Each task uses at least one core.
--cpus-per-task	-c	Number of cores each task will use.
--nodes	-N	Minimum number of nodes. Each node needs to run at least one task.
--mem		Memory per node. Split among tasks. Shared between cores in a single task.
--mem-per-cpu		Memory per core. Total allocated memory per task is this value multiplied by the number of cores.
--time	-t	Amount of time you want to use or simulate.

In Table S3.4.1, the two main options you will use are '--ntasks' (short option form '-n') and '--cpus-per-task' ('-c'). ntasks tells slurm how many tasks (processes, or copies of your program) you want to run. The cpus-per-task sets the number of cores that each task will use. The cores for one task must all be on the same node (they need access to the same memory), so slurm will allocate enough nodes for you to get the tasks and cores per task that you ask for.

You specify the amount of memory as a number followed by a unit. Thus, to allocate 2 gigabytes per node, you would say '--mem=2G,' and to ask for 500 megabytes per core, you would say '--mem-per-cpu=500M.' You also specify the time with the --time (-t for shortcut) option. The full specification has the form 'days-hours:minutes:seconds.'

Now, try running the example above. Then copy your code to a text file and try doing the same with your own code.

[3.7]

Game Theory, Part 1

A Guide to Concepts, Famous Games and Payoff Matrices, Nash Equilibrium, Prisoner's Dilemma

Game theory is a means of mathematically understanding decision making. We are very interested in scenarios in which people appear to behave irrationally, and divergence between mathematical assumptions and real-world behavior is an object of the game theory approach.

It involves the study of mathematical models of strategic interactions among rational agents. It has applications in all fields of social science, as well as in logic, systems science, and computer science. The game in game theory is an abstract situation in which a player is faced with two or more choices or moves (Osborne & Rubinstein, 1994). These choices then have outcomes or payoffs given the abstract structure. The heart of game theory involves issues of utility. Utility is the amount of satisfaction or happiness a person achieves as a result of the outcomes of his or her choices in the game, and this utility is typically manifested in the player's preferences for certain outcomes over others (i.e., the player will pick the outcome that makes him or her truly happy).

Game theory is useful in analyzing decisions in an abstract, theoretical way. Even if the real world diverges from game theory, the analysis enables us to go searching for phenomena in the first place. Thus, this section has been written in a way that even beginners and nontechnical folks can follow along.

1. A Guide to Concepts

Here we discuss how game theory is used in the field of systems analysis. What Is Game Theory? You have surely come across the concept of game theory at some point but never truly dived into it. In general, it is an intriguing topic and an enlightening one in systems analysis (Myerson, 2013). Let us put out a formal definition of game theory first.

+ **Identification**: Game theory is the modeling of possible interactions between two or more rational agents or players. This discipline is the formal study of interactions that can occur among several individuals in a strategic situation. The term game is not a priori a playful term, John von Neumann having developed it for the modeling of conflict interactions. It has been used to model and simplify many problems, ranging from trade relations to armed conflicts.

Game theory decomposes games according to the number of players (usually two), then mainly into two main families: simultaneous games, in which players play simultaneously and therefore do not know when playing what the opponent is playing (e.g., rock-paper-scissors); and nonconcurrent games, in which everyone plays in turn (e.g., chess). Here we must stress the keyword rationality, as this acts as the foundation of game theory. But what exactly does rationality mean?

+ **Rationality:** Rationality is simply the understanding that every agent knows that all other agents are just as rational and have the same level of understanding and knowledge as he or she does. Also, rationality refers to the fact that agents always prefer the higher reward or payoff considering what other agents will do (Arthur, 1991). In short, every agent is selfish and tries to maximize the reward. Now that we know what rationality means, let us tackle some other keywords associated with game theory:

- Game: In a general sense, a game comprises a set of players, actions/strategies, and a final payoff (e.g., an auction, chess, politics).
- Players: Players are rational entities who participate in a game (e.g., bidders in an auction setting, players playing rock-paper-scissors, politicians participating in elections).
- Payoff: A payoff is the reward that all players get when a certain outcome is achieved. It can be either positive or negative. As we discussed before, each agent is selfish and wants to maximize his or her payoff.

+ **Payoff Matrix:** It is necessary to be able to code a payoff matrix. A game class will allow us to store the winnings of both players for a given situation. A game object takes as a parameter an array of pairs of integers corresponding to the scores of each outcome of the game as well as a table of the names of the corresponding actions. Many programming packages (e.g., NumPy) offer ideal array objects for storing and manipulating this. For a game with n choice options for each player, we create an $n \times n$ array of pairs of values (x, y), with x the win of player 1 and y the win of player 2.

First, we need to import libraries.

```
[In]
import numpy as np
import math
import pandas as pd
```

Then create a payoff matrix to simulate the game dynamics. A payoff matrix is a crucial part of game theory simulations because it shows the potential outcomes (payoffs) for each player in a game based on the strategies that both players have selected. We may understand the dynamics of the game and how players could act in various circumstances by evaluating the matrix.

```
[In]
class Game:
    def __init__(self, tab, actions):
        self.actions = actions
        m = np.array(tab, dtype=[('x,' object), ('y,' object)])
        self.size = int(math.sqrt(len(tab)))
        self.scores = m.reshape(self.size, self.size)
```

The 'Game' class is being created with a payoff matrix 'tab' and a list of 'actions'. The class has a few variables called 'actions', 'size' and 'scores'. The 'size' is being set to square root of the length of 'tab' and the 'scores' is being set to the reshaped version of 'tab' as an array with data type [(x object), (y object)]. The code is intended to print out the 'scores' matrix in a readable format using the 'prettyPrint' method by creating a new array filled with 'np.nan' values using 'self.actions' as the row and column labels. Then, it iterates over the elements in the 'scores' matrix and assigns each element to the corresponding position in the new array and print out.

```
[In]
    def prettyPrint(self):
        game = pd.DataFrame(np.nan, self.actions, self.actions, dtype=object)
        for i in range(self.size):
            for j in range(self.size):
                game.iat[i, j] = self.scores[i][j]

        print(game)
```

Now we can use a computerized system (i.e., Jupyter library) to represent these games. We start by building our own matrices as follows.

2. Famous Games and Their Payoff Matrices

We lay the groundwork for understanding the core ideas of games, such as strategies, payoffs. Here, by examining the payoff matrices of popular games like the Prisoner's Dilemma, Battle of the Sexes, and Chicken, we may learn more about how rational players could act in certain situations and how cooperation and competition might affect the game's results. In order to create and analyze increasingly complicated games and real-world scenarios, it is essential to comprehend these fundamental ideas.

+ **The Battle of the Sexes:** In game theory, the battle of the sexes is a two-player coordination game that also involves elements of conflict (Capel, 2000). Imagine that a man and a woman hope to meet this evening but have a choice between two events to attend: a prize fight or a ballet. The man would prefer to go to the prize fight. The woman would prefer the ballet. Both would prefer to go to the same event rather than different ones. If they cannot communicate, where should they go?

Code example of the battle of the sexes

```
[In]                                                  [Out: Payoff matrix]
gs=[(3,2),(1,1),(0,0),(2,3)]                                  Ballet   Fight
g = Game(gs,['Ballet','Fight'])              →       Ballet  [3, 2]   [1, 1]
g.prettyPrint()                                      Fight   [0, 0]   [2, 3]
```

The code creates a 'Game' object with the payoff matrix 'gs' and a list of actions ['Ballet', 'Fight']. Then, it calls the 'prettyPrint' method on the 'g' object to print out the 'scores' matrix in a readable format.

+ **The Chicken Game:** The principle of this game (also known as the hawk-dove game or snowdrift game) is that although the ideal outcome is for one player to yield (to avoid the worst outcome if neither yields), the individuals try to avoid it out of pride for not wanting to look like a chicken (Rapoport & Chammah, 1966). A formal version of the game of chicken has been the subject of serious research in game theory. Example versions of the payoff matrix for the game presented here are arbitrarily set numerical payoffs that conform theoretically to this situation. Here the benefit of winning is 2, the cost of losing is –2, and the cost of crashing is –100.

Code example of the chicken game

```
[In]                                                  [Out: Payoff matrix]
cg=[(-100,-100),(2,-2),(-2,2),(0,0)]                          Swerve    Straight
g = Game(cg,['Swerve','Straight'])           →   Swerve   [-100, -100]  [2, -2]
g.prettyPrint()                                  Straight    [-2, 2]     [0, 0]
```

The creates payoff matrix 'cg' and a list of actions ['Swerve', 'Straight']. Then, it calls the 'prettyPrint' method on the 'g' object to print out the 'scores' matrix in a readable format.

+ **Matching Pennies:** Matching pennies is the name of a simple game used in game theory. It is played between two players, Even and Odd. Each player has a penny and must secretly turn the penny to heads or tails (Mookherjee & Sopher, 1994). The players then reveal their choices simultaneously. If the pennies match (both heads or both tails), then Even keeps both pennies, and so wins one from Odd (+1 for Even, -1 for Odd). If the pennies do not match (one heads and one tails) Odd keeps both pennies, so receives one from Even (-1 for Even, +1 for Odd).

Code example of the matching pennies game

```
[In]                                                  [Out: Payoff matrix]
mp=[(1,-1),(-1,1),(-1,1),(1,-1)]                              Heads    Tails
g = Game(mp,['Heads','Tails'])               →       Heads  [1, -1]  [-1, 1]
g.prettyPrint()                                      Tails  [-1, 1]  [1, -1]
```

This code creates a new payoff matrix 'mp' and a list of actions ['Heads', 'Tails'].

Then, it calls the 'prettyPrint' method on the 'g' object to print out the 'scores' matrix in a readable format.

+ **Rock-Paper-Scissors:** Rock-paper-scissors is a hand game usually played between two people in which each player simultaneously forms one of three shapes with an outstretched hand (Sinervo & Lively, 1996). These shapes are a rock (a closed fist), paper (a flat hand), and a scissors (a fist with the index finger and middle finger extended, forming a V). Scissors is identical to the two-fingered V sign (also indicating victory or peace), except it is pointed horizontally instead of being held upright in the air.

Code example of rock-paper-scissors

```
[In]
rpc=[(0,0),(-1,1),(1,-1),(1,-
1),(0,0),(-1,1),(-1,1),(1,-1),(0,0)]
g =
Game(rpc,['Rock','Paper','Scissors'])
g.prettyPrint()
```

→

```
[Out: Payoff matrix]
              Rock    Paper   Scissors
Rock         [0, 0] [-1, 1] [1, -1]
Paper      [1, -1]  [0, 0] [-1, 1]
Scissors[-1, 1] [1, -1]  [0, 0]
```

This code creates a new payoff matrix 'rpc' and a list of actions ['Rock', 'Paper', 'Scissors']. Then, it calls the 'prettyPrint' method on the 'g' object to print out the 'scores' matrix in a readable format.

+ **Prisoner's Dilemma:** The prisoner's dilemma is a standard example of a game that shows why two completely rational individuals might not cooperate, even if it appears it is in their best interests to do so (Lacey, 2008). It was originally framed by a game with prison sentence rewards as follows:

Code example of the prisoner's dilemma

```
[In]
dp =[(3,3),(0,5),(5,0),(1,1)]
g = Game(dp,['Cooperates','Defects'])
g.prettyPrint()
```

→

```
[Out: Payoff matrix]
             Cooperates Defects
Cooperates    [3, 3]  [0, 5]
Defects       [5, 0]  [1, 1]
```

This code creates a new payoff matrix 'dp' and a list of actions ['Cooperates', 'Defects']. Then, it calls the 'prettyPrint' method on the 'g' object to print out the 'scores' matrix in a readable format. The above-mentioned payoff matrix is a useful tool in game theory that offers a framework for comprehending the possible outcomes of a game and the tactics that players could use. We can determine the Nash equilibrium, a notion that aids in understanding how rational players could behave in a game, by the examination of the payoff matrix.

We may forecast the outcomes of a game that are most likely to occur by knowing the Nash equilibrium, and we can devise plans to get the best possible result. It is crucial to keep researching the uses of the matrix and Nash equilibrium in order to create successful tactics for challenging games and real-world situations.

3. The Nash Equilibrium in Game Theory

The Nash equilibrium is the bedrock of the game theory approach and is named after its inventor, John Nash, an American mathematician. It is an action chosen by each player such that no player would want to change his or her action (Nash, 1950). Changing one's action from the Nash equilibrium means a player is not playing optimally.

To compute the 'no regret' situation corresponding to the Nash equilibrium, it is sufficient to note the answers of player 1 most adapted to the possible choices of player 2 (i.e., calculate the max(χ) in each column), then calculate the answers of player 2 most adapted to the possible choices of player 1 (i.e., calculate the max(y) in each line). If an issue has 2 max, then it is a Nash equilibrium. Depending on the game, there may of course be one, several, or none at all. Assuming the fundamental property previously discussed, we identify game theory using a simplistic 2∗2 matrix

- A game necessitates the involvement of multiple players (more than 2)
- Interaction amongst players is paramount, and there must exist a reward or payoff
- It is posited that players operate rationally, guided by their individual self-interest

	Player (B) silent	Player (B) betray
Player (A) silent	A:1 B:1	A:3 B:0
Player (A) betray	A:0 B:3	A:2 B:2

In the matrix presented, the dominant strategy for one player would be to 'betray' when the other player remains silent. This is because a dominant strategy implies a choice that offers better payoffs for a player than any other available strategies. Furthermore, an equilibrium point exists wherein neither party would gain from deviating from their chosen strategy. At this equilibrium, each player is fully aware of their opponent's strategy and remains consistent in their choices. This equilibrium is known as the Nash equilibrium. In the given matrix, the Nash equilibrium occurs when both players opt to remain silent. The significance of the Nash equilibrium lies in its underlying assumption: every game possesses at least one equilibrium point, and all players will benefit by identifying this point and shaping their strategies accordingly.

Using an np.Array makes things much easier because it is possible to have the max value vectors in *a* line or in *a* column. It is therefore sufficient to create matrices in each case and compute *a* or logical on them.

First, we need to import a few essential libraries (modules) to perform calculations.

```
[In]
import matplotlib.pyplot as plt
import numpy as np
import sympy as sy
```

Having imported the necessary libraries, we can now proceed to create a payoff matrix, which will help us in deciding between two options.

```
[In]
player_A = [0,2,4,1]
player_B = [0,0,2,5]
a = np.array([player_A,
player_B]).T
a
```

→

```
[Out: Payoff matrix]
Array
([[0, 0],
  [2, 0],
  [4, 2],
  [1, 5]])
```

The code produces two list 'player_A' and 'player_B', and then transposes them to create an array 'a'. The resulting array has the table as shown in the right side [Out]. This array is a payoff matrix for a game with two players, where the rows represent the actoins of player A and the columns represent the actions of player B. The element at position (i, j) in the matrix represents the payoff for player A when player A chooses action i and player B chooses action j. With the payoff matrix created, we can now analyze the results and make an informed decision based on the data. What we have is four rows and two columns, and this is how we present the information. Thus, for plotting we need to have just one command (except for the plotting function) that we are going to add the payoff matrix (Figure 3.7.1).

```
[In]
plt.plot(player_A, player_B)
```

→

[Out]

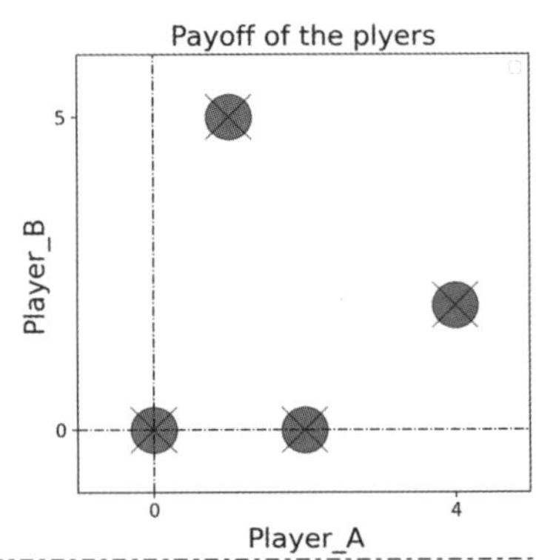

Figure 3.7.1. Simulation results of the payoff coordinates between two players

Now, when it comes to the Nash equilibrium, the solution has to be somewhere among these points. Nash's theorem says it is at the point that maximizes x times y. Hence, to maximize x times y, we need to express for y in terms of x, which can be found in the equation of the line joining the points. Here we take the slope of the line between the two points and use the x and y coordinates. One of them has slope $m = (4 - 2)/(2 - 4)$. Hence, if we use the point slope form to write the equation, we should have $y = -1(x - 4) + 2$. We are now interested in maximizing $x(-x + 6)$ or $-x^2 + 6x$.

```
[In]
def n(x):
    return -x**2 + 6*x
x = sy.Symbol('x')
```

We will find the solution by using the derivate and print these solutions out.

[In]
```
dn = sy.diff(n(x),x)
soln = sy.solve(dn)
```
→

[Out]
```
The derivative of the product is 6 - 2*x, this equals
zero at x =  [3], and this corresponds with y =  3
```

The code uses the 'sympy.diff' function to find the mathematical expression represented by the $n(x)$ with respect to the variable x and stored in the 'dn'. The 'sy.solve' is then used to find the values of x where 'dn' equals zero which the values of x are stored in the 'soln'. We calculated the slope of the line between the two points using the x and y coordinates, and then write the equation in point-slope form. Then, we found the solution by taking the derivative of the function and solving for x. With the derivative calculated and the solution found, we now have the optimal values for x and y that maximize the function. Finally, we add that point to our plot here (Figure 3.7.2, the solution is when x and y equal zero at 3).

[In]
```
plt.plot(player_A, player_B)
plt.plot(3,3,
label='solution_NE')
```
→

[Out]

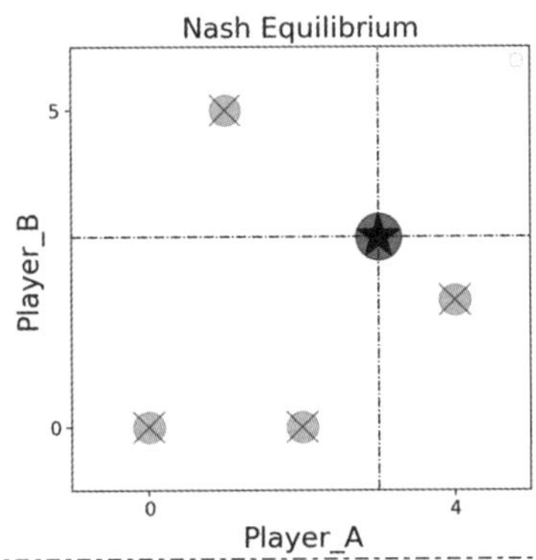

Figure 3.7.2. Simulation results of the solved Nash equilibrium (*)

Considering that all other agents are rational and choose the best action for themselves, the Nash equilibrium action for a given player is the best response. No player can increase his or her payoff by changing decisions within his or her action set. We can think of this as no regrets in the sense that once a decision is made, the player will have no regrets concerning decisions considering the consequences. To see Nash equilibrium in action, let us now tackle the most common problem of game theory, the prisoner's dilemma. This classic example illustrates the difficulty of acting together cooperatively for common or mutual benefit when agents are only concerned about their self-interest.

4. Prisoner's Dilemma (Empirical Details)

The prisoner's dilemma is a two-player game commonly studied in game theory (Kreps et al., 1982). The story goes like this: 'Two criminals have been arrested by the police. They are put in separate cells and have no way of communicating with each other.' Unfortunately for the police, there is not enough evidence to convict the criminals of a serious crime—only minor theft—but the officer comes up with a way to trick the two prisoners. Simultaneously, the police officer offers each prisoner a deal. Each prisoner is given the opportunity either to betray the other by testifying against him or to cooperate with the other prisoner by remaining silent. The offer is as follows:

- If A and B each betray the other, each of them serves 2 years in prison.
- If A betrays B but B remains silent, A will be set free, and B will serve 3 years in prison (and vice versa).
- If A and B both remain silent, both will serve only 1 year in prison (on a lesser charge).

In game theory we represent such games using a matrix. Instead of months in prison we use a positive payoff (gain) for each of the actions:

		Prisoner A	
		Cooperate	Defect
Prisoner B	Cooperate	(3, 3)	(0, 5)
	Defect	(5, 0)	(1, 1)

Now let us assume that the officer locks our two criminals back in prison and asks them the same question again the next day. If this is repeated for a given number of iterations, then we have what is called the iterated prisoner's dilemma. In this game, two prisoners, A and B, have been caught for the same crime and are being held in two different interrogation rooms. They have been given two choices. Thus, there are four outcomes in total, which can be conveniently represented as a game matrix.

		B	
		silent	confess
A	silent	A:-1 B:-1	A:-15 B:0
	confess	A:0 B:-15	A:-10 B:-10

In this representation, the payoffs are represented in the form of (A's payoff, B's payoff). In the rows we have A's actions, and in the columns, we have B's action. Take a good look at the payoffs. Why do you think the payoffs are negative? It is because based on their actions, they will receive a predetermined number of years of imprisonment (which is not at all desirable).

+ **The Possible Outcomes:** If both A and B remain silent, both get imprisonment for 1 year. If either one of them confesses, the confessor walks free, and the other prisoner gets 15 years' imprisonment. If both of them confess, both receive 10 years' imprisonment.

	B silent	B confess
A silent	A:-1 B:-1	A:-15 B:0
A confess	A:0 B:-15	A:-10 B:-10

The dilemma is that neither prisoner is aware of what the other prisoner is doing. What do you think is the Nash equilibrium action in this problem? It is very intuitive to think that the prisoners will collaborate with each other and stay silent, but we also know that prisoners have evident self-interest in minimizing the imprisonment they receive. And even if they remain silent, they still get 1 year's imprisonment each. Let us focus on what B is thinking. If we look at the game matrix, we can see that his thinking process makes perfect sense.

B(confess)

A	silent	confess
silent	A:-1 B:-1	A:-15 B:0
confess	A:0 B:-15	A:-10 B:-10

B(stay silent)

A	silent	confess
silent	A:-1 B:-1	A:-15 B:0
confess	A:0 B:-15	A:-10 B:-10

If B confesses, A's best choice is to confess. This will lead to the lesser punishment of 10 years rather than 15. If B stays silent, A is still better off confessing, as he will be a free man instead of facing 1 year of imprisonment if he also stays silent. This game matrix is in perfect congruence with what A is thinking. If B is also thinking the same, the game matrix would look like this for him:

A (stay silent)	B: silent	B: confess
silent	A:-1 B:-1	A:-15 B:0
confess	A:0 B:-15	A:-10 B:-10

A (confess)	B: silent	B: confess
silent	A:-1 B:-1	A:-15 B:0
confess	A:0 B:-15	A:-10 B:-10

Let us say that B goes through the same rational thought process as A. B also concludes that no matter what A chooses, he will always benefit from confessing. If we superimpose the rational thinking of both these prisoners, the result is something like this:

	B silent		B confess
A silent	A:-1 B:-1	→	A:-15 B:0
	↓		↓
A confess	A:0 B:-15	→	A:-10 B:-10

Looking at the results, we can see that the best strategy is (Confess, Confess). If either of them deviates from this action, he will be worse off than he would be after confessing. Hence, (Confess, Confess) is a Nash equilibrium strategy. Makes perfect sense, right?

The Nash equilibrium is a no regret solution for any game, but not necessarily the most optimal one.

+ **Classic Example with the Prisoner's Dilemma:** Assume that the two players are A and B. Their situation is that they were both arrested and taken to the police station, then separated into different interrogation rooms. The detectives interrogating our players do not have enough evidence; they need a confession. Here are the strategies and payoffs for our players:

If one player confesses while the other stays silent, the snitch goes free while the other player serves 10 years. If both players abide by the code and stay silent, each player gets a year in jail on bogus charges. If both players snitch on each other, they each get nine years (the court was kind enough to shave a year off for their cooperation). We can call staying silent the cooperate strategy and snitching the defect strategy.

Now, let us use Jupyter to create a payoff matrix, import the libraries:

```
[In]
import numpy as np
import pandas as pd
```

Create an array with Player_A's payoffs:

```
[In]
x = np.array([[-1, -10],[0, -9]])
```

Here, the x is set to an array with shape (2, 2) and elements [[−1, −10], [0, −9]] and the data is determined based on the input elements. The given code creates a 2D NumPy array named x hat represents Player_A's payoff matrix in a two-player game. The array's elements show the payoffs for Player_A when choosing different strategies and facing different opponent strategies. Label the matrix (rows and columns):

```
[In]
jpm=pd.DataFrame(x,
columns=['silent', 'confess'])
jpm.index = ['silent', 'confess']
jpm
```

[Out]

		B	
		silent	confess
A	silent	-1	-10
	confess	0	-9

Given the data x, the code sets the column labels to ['silent', 'confess'], and the index of the data is set to the ['silent', 'confess']. Players A and B are represented by the rows and columns. For example, if A plays silent while B plays confess, A's payoff is −10 (the number of years he goes to jail). If A plays confess and B plays silent, A's payoff is 0 (no jail time).

And let us create Player_B's payoff matrix. To achieve this, we need to swap the rows with the columns, because B's payoff matrix is the transpose (T) of A's.

This is very easy to do in code:

```
[In]
rpm=jpm.T
rpm
```

→

[Out]

		B	
		silent	confess
A	silent	-1	0
	confess	-10	-9

Here, the '.T' is a property of a data that returns the transpose of the data which is new data with the rows and columns of the original data swapped. It is a common operation that is often used to rearrange data for easier manipulation or visualization. Another way to create the payoff matrices is to use the nashpy package like so:

```
[In]
# A's payoffs (row player)
x = np.array([[-1, -10],[0, -9]])
# B's payoffs (column player)
y = x.T
import nashpy as nash
prisoners_dilemma = nash.Game(x,y)
prisoners_dilemma
```

→

```
[Out]
Bi matrix game with payoff matrices:
Row player:
[[ -1 -10]
 [  0  -9]]
Column player:
[[ -1   0]
 [-10  -9]]
```

Now we can create the payoff table. The payoff table is a bi-matrix that contains A's and B's payoff matrices written together using the 'quantecon' package. To install this package with 'conda' run one of the following:

```
[In]
conda install -c conda-forge quantecon
```

Then create a list containing both players' payoffs:

```
[In]
import quantecon as qe
pt = [[(-1,-1), (-10,0)], [(0,-10), (-9,-9)]]
g = qe.game_theory.NormalFormGame(pt)
print(g)
```

→

```
[Out]
2-player NormalFormGame with
payoff profile array:
[[[ -1,  -1],  [-10,   0]],
 [[  0, -10],  [ -9,  -9]]]
```

Here, the 'quantecon' is a collection of tools for quantitative modeling. It includes functions and classes for modeling dynamics (i.e., economic), solving dynamic optimization problems, and analyzing the data. The list of payoffs above were passed through a function called 'NormalFormGame' (it is in the output as well). The prisoner's dilemma represents a normal form game following three conditions: a set of players (A and B), a set of strategies for each player (silent, confess), each player has preferences over the set of strategy profiles.

Simply put, a strategy and a strategy profile are not the same. A strategy profile is a list containing one strategy for each player. Think of the game rock-paper-scissors. The strategy profiles are (rock, rock), (paper, rock), (scissors, rock), and so on. The strategy profiles for our prisoner's dilemma game are as follows: {(silent, silent), (silent, confess), (confess, silent), (confess, confess)}. Now that we understand the strategies and the payoffs, we can write a function of payoffs to further crystalize our understanding:

```
[In]
def payoff_function (x=str, y=str):
    if x == 'silent' and y == 'silent':
        print('Player_A {}'.format(-1),':','Player_B {}'.format(-1))
    elif x == 'silent' and y == 'confess':
        print('Player_A {}'.format(-10),':','Player_B {}'.format(0))
    elif x == 'confess' and y == 'confess':
        print('Player_A {}'.format(-9),':','Player_B {}'.format(-9), ':', "NE")
    else:
        print('Player_A {}'.format(0),':','Player_B {}'.format(-10))
```

The code defines a 'payoff_function' that takes two arguments x and y both of which are strings with four branches corresponding to a different combination of values for x and y. In the case of the first branch, for example, if x is equal to 'silent' and y is equal to 'silent', the function prints 'Player_A -1 : Player_B -1'. The other branches work as the same way and it only prints the payoffs for each combination of actions. We established earlier the possible strategy profile. If we wrote the function, this should be our output:

```
[In]                                          [Out]
payoff_function('confess', 'silent')    →     Player_A 0 : Player_B -10
[In]                                          [Out]
payoff_function('confess', 'confess')   →     Player_A -9 : Player_B -9 : NE
[In]                                          [Out]
payoff_function('silent', 'confess')    →     Player_A -10 : Player_B 0
```

*NE = Nash equilibrium

We might be able to get an understanding about game theory that complements the type of thinking one needs to develop to be successful.

Remarks: Game theory, in terms of systems analysis, basically helps us to make decisions (Hauert et al., 2002). This is not very difficult considering that rationality is the foundation of game theory. As a matter of fact, game theory has already started establishing its place in systems analysis—can you guess where? One such niche is around generative adversarial networks, called the coolest idea in the last 20 years in the field of systems analysis. How does game theory help us? Game theory's core implementation lies in games of imperfect information (Sasaki et al., 2012). As we have already discussed, player is a classic example and a proper benchmark problem for an application in imperfect information. Imperfect information is very important because real-world problems often fall under this category.

[3.8]

Game Theory, Part 2

Practical Aspects of Game Theory, Simulation of Game Theory Model

We just witnessed a compelling instance of game theory in action through the prism of the prisoner's dilemma. This concept, exemplified in the previous chapter, revealed that games can oscillate between competitive and cooperative dynamics.

For instance, the historical tension between the U.S. and Soviet Union during the Cold War, akin to a competitive political campaign where one nation's gain could signify the other's loss, mirrored a competitive game. However, insights from Robert Axelrod's tournaments in the context of the prisoner's dilemma highlighted a shift towards cooperative strategies (Axelrod, R. 1997). This cooperation is analogous to a basketball game, where players reap greater rewards through teamwork and mutual support. These examples from both nature and geopolitics underscore the intricate balance and strategic decision-making inherent in different game scenarios

1. Practical Aspects of Game Theory

To understand this, let us take the example of a problem called battle of the sexes. Consider that B and A are two friends who are fond of each other's company. They are well aware of each other's habit of going out for football games and dance parties, respectively. They decide to accompany each other. They can either discuss it with each other beforehand or just surprise each other. If they plan on surprising each other, they are unaware of each other's plans for the weekend.

This results in four different situations as depicted by the following game matrix:

	Football	Dance class
Football	A:5 B:10	A:0 B:0
Dance class	A:0 B:0	A:10 B:5

The game matrix clearly shows that both B and A get no payoff if they fail to coordinate with each other. This is an example of a simultaneous game in which both players make simultaneous moves and are unaware of the other player's actions in advance.

+ **Information**: In game theory, situations often arise in which players have different information. They might (not) know all of the other player's available strategies or potential payoffs. Players might (not) be aware of what kind of person they are dealing with or what their motivations are.

Thus, games can be broadly divided into three types depending on how much players know about the other agents: perfect, imperfect, and incomplete information.

- Perfect Information: In perfect information games, each participant is fully aware of all possible actions, their outcomes, and the choices made by others. Chess, with its open board and visible moves, is a classic example. Similarly, the Tit for Tat strategy (a reciprocal approach where a player initially cooperates and then mirrors the opponent's previous action in subsequent moves) in iterated games like the Prisoner's Dilemma also falls into this category, as players respond based on the known actions of their opponents. Despite their theoretical appeal, perfect information games are relatively rare in real-world scenarios, where uncertainties often prevail.

- Imperfect Information: These games, more common in real-world situations, feature agents who understand the intentions and potential payoffs of others but lack visibility on their current actions. For instance, a military commander might grasp the enemy's strategic objectives and the benefits they seek but remains unaware of their exact maneuvers. This scenario is less like Tit for Tat, where actions and responses are clear and more akin to a complex chess game with fog of war, where players must anticipate moves without full visibility.

- Incomplete Information: The most reflective of real-world complexities, incomplete information games encompass situations where players have limited insight into others' types, motivations, or rewards. Even if an agent observes others' actions, the underlying intentions or payoffs remain obscured. In this context, poker is an exemplary imperfect information game. Players cannot ascertain whether opponents hold strong or weak hands, mirroring the uncertainty of real-life decision-making. Unlike Tit for Tat, where actions and reactions are transparent, poker demands strategies that account for hidden information and unpredictability, challenging even advanced systems analysis and decision-making algorithms in the realm of imperfect information games.

+ **Calculating Payoffs (with Zero Sum Game):** Expanding on game theory, we explore zero-sum games, characterized by equally offsetting gains and losses between two players. This extends to game theory's application in systems analysis and real-world simulations.

In such games, players interact dynamically, each employing unique strategies informed by their behavior. The challenge is identifying the best strategy when players with different tactics interact, especially in perfect information scenarios like Tit for Tat, where each player's actions and payoffs are fully visible.

$$\begin{vmatrix} \alpha, & \beta \\ \gamma, & \delta \end{vmatrix}$$

Most conceptually straightforward games offer only two strategies (up and down, left and right) and four outcomes ($\alpha, \beta, \gamma, \delta$) between two players. However, a major challenge, even in a simple game, is to determine how to rank the payoff values. This practice approaches it simply by defining the dynamics of two strategies with four outcomes. Suppose that players must choose between n options or strategies.

To be specific, player 1 and player 2 each simultaneously reveal a penny. If both pennies show heads or both show tails, player 2 must pay player 1 $1; however, if one penny shows heads and the other shows tails, player 1 must pay player 2 $1.

The game is described as follows:

		Player 2	
		H	T
Player 1	H	$(-1, 1)$	$(1, -1)$
	T	$(1, -1)$	$(-1, 1)$

This matrix describes, in the ith row and jth column, the pair (a_{ij}, b_{ij}) of payoff values. It shows that if the outcome is (−1, 1), player 1, the row player, would have done better to choose the other strategy. However, if the outcome is (1, −1), it is player 2, the column player, who would have done better to switch.

Players 1 and 2 have diametrically opposed interests here. This is a very common strategic situation. For instance, in a game of soccer penalty kicks, if the striker and keeper are mismatched, then the striker must be happy, but if they are matched, then the keeper must be happy.

This same logic applies to many common situations of preference, and an efficient way to achieve these dynamics is to let the player decide to opt to play a strategy with a probability given as

$$x = (x_1, x_2, \dots, x_n),\; x_1 \geq 0 \; and \; x_1 + \cdots + x_n = 1,$$

where a probability distribution over two or more pure strategies that the players choose randomly among their options is denoted as the set of all such mixed-strategy probability distributions by

$$p_r(x_1) + p_r(x_2) + \cdots + p_r(x_n) = 1 = 100\%$$

$$p_r(x_i) \in [0,1]$$

For all events, if we follow the game (i.e., penny reveal or penalty kick), in every case the player should be happy half the time and sad half the time.

$$
\begin{array}{cccc}
 & & \multicolumn{2}{c}{Player\ 2} \\
 & & H(.5) & T(.5) \\
Player\ 1 & H(.5) & (-1,1) & (1,-1) \\
 & T(.5) & (1,-1) & (-1,1)
\end{array}
$$

The game matrix provided illustrates the preferences of two players, Player 1 and Player 2, in a given situation, such as a penny reveal or a penalty kick. Each player chooses their strategy with a probability of 0.5. It does not matter what they do, and they cannot change the outcome because they are just as happy to flip a coin or choose one direction between two.

The key is that the other player does not anticipate the decision, because if the striker knows that the keeper is playing heads, the striker does not want to play heads and should switch to tails. However, we cannot usually guess a person's coin-flipping strategy if we change the payoff a bit.

If the game expands to include arbitrarily mixed-strategy conditions $\{A > B > C\}$,

$$
\begin{array}{cccc}
 & & \multicolumn{2}{c}{Player\ 2} \\
 & & Left & Right \\
Player\ 1 & Up & (B,A) & (C,C) \\
 & Down & (C,C) & (A,B)
\end{array}
$$

The player's expected utility (which represents the motivations of players; a utility function for a given player assigns a number for every possible outcome of the game with the property that a higher number implies that the outcome is more preferred) is

$$
\sigma u = \frac{b-c}{a+b-2c}, \qquad a+b-2c > 0
$$

$$
\sigma u = \frac{b-c}{a+b-2c}, \qquad b-c > 0, \qquad a+b-2c > 0
$$

$$
\sigma u = \frac{b-c}{a+b-2c}, \qquad a+b-2c > b-c, \qquad a > c
$$

This function assigns a numerical value to each possible outcome of the game, with higher values indicating more preferred outcomes. The expected utility (σu) is defined by this equation under the conditions:

- b - c > 0: This condition implies that the benefit (b) should be greater than the cost (c) for the expected utility to be positive.
- a + b - 2c > 0: This condition ensures that the denominator of the equation does not equal zero, which would make the expected utility undefined.
- a + b - 2c > b - c: By rearranging this inequality, we get a > c, which means the value of a should be greater than the value of c.

The expected utility equation for a player, σu, is derived from the relative benefits and costs associated with different outcomes in a game.

The given conditions ensure that the expected utility remains positive and well-defined, reflecting the preferences of the player in a meaningful way; where σu represents the probability that player 1 plays. The expected utility of player 1is

$$EU_L = \sigma\mathrm{u}(\mathrm{a}) + (1 - \sigma\mathrm{u})(\mathrm{c}),\ EU_R = \sigma\mathrm{u}(\mathrm{c}) + (1 - \sigma\mathrm{u})(\mathrm{b})$$

The expected utility algorithms equal player 2's expected utility. Those two things are equal:

$$\begin{gathered} \sigma\mathrm{u}(\mathrm{a}) + (1 - \sigma\mathrm{u})(\mathrm{c}) = \sigma\mathrm{u}(\mathrm{c}) + (1 - \sigma\mathrm{u})(\mathrm{b}) \\ \mathrm{a}\sigma\mathrm{u} + \mathrm{c} - \mathrm{c}\sigma\mathrm{u} = \mathrm{c}\sigma\mathrm{u} + \mathrm{b} + \mathrm{b}\sigma\mathrm{u} \\ \mathrm{a}\sigma\mathrm{u} + \mathrm{b}\sigma\mathrm{u} - 2\mathrm{c}\sigma\mathrm{u} = \mathrm{b} + \mathrm{c} \\ \sigma\mathrm{u}(\mathrm{a} + \mathrm{b} - 2\mathrm{c}) = \mathrm{b} - \mathrm{c} \\ \sigma\mathrm{u} = (\mathrm{b} - \mathrm{c})/(\mathrm{a} + \mathrm{b} - 2\mathrm{c}) \end{gathered}$$

Even with a slightly different payoff under mixed-strategy conditions $\{A > B > C\}$, it is always positive between [0, 1], which can be confirmed by checking the above necessary conditions.

Based on this dynamic, if we calculate the probability that each outcome occurs given some percentage of the time, players 1 and 2 get the following:

$$\begin{array}{ccc} & Left(\frac{2}{3}) & Right(\frac{1}{3}) \\ Up(\frac{1}{3}) & (B = 1, A = 2) & (C = 0, C = 0) \\ Down(\frac{2}{3}) & (C = 0, C = 0) & (A = 2, B = 1) \end{array}$$

Because each outcome must be multiplied by the player's payoff

$$\begin{array}{ccc} & 2/3 & 1/3 \\ 1/3 & \begin{pmatrix} B = 1, A = 2 \\ \frac{1}{3} * \frac{2}{3} = \frac{2}{9} \end{pmatrix} & \begin{pmatrix} C = 0, C = 0 \\ \frac{1}{3} * \frac{1}{3} = \frac{1}{9} \end{pmatrix} \\ 2/3 & \begin{pmatrix} C = 0, C = 0 \\ \frac{2}{3} * \frac{2}{3} = \frac{4}{9} \end{pmatrix} & \begin{pmatrix} A = 2, B = 1 \\ \frac{2}{3} * \frac{1}{3} = \frac{2}{9} \end{pmatrix} \end{array}$$

and summed, player 1's payoff is

$$\tfrac{2}{3} = \tfrac{6}{9} = \tfrac{2}{9} + 0 + 0 + \tfrac{4}{9} = 1\left(\tfrac{2}{9}\right) + 0\left(\tfrac{1}{9}\right) + 0\left(\tfrac{4}{9}\right) + 2\left(\tfrac{2}{9}\right).$$

Player 1's earned mixed-strategy equilibrium is $\{2/3\}$. If player 2's payoff is

$$\tfrac{2}{3} = \tfrac{6}{9} = \tfrac{4}{9} + 0 + 0 + \tfrac{2}{9} = 2\left(\tfrac{2}{9}\right) + 0\left(\tfrac{1}{9}\right) + 0\left(\tfrac{4}{9}\right) + 1\left(\tfrac{2}{9}\right),$$

then that players earned mixed-strategy equilibrium is also {2/3}. Given the underlying assumption of the game's probability distribution $p_r(x_i) \in [0,1]$, all payoffs are always going to be positive between [0, 1] because all must sum to

$$\frac{2}{9}+\frac{1}{9}+\frac{4}{9}+\frac{2}{9}=1$$

Also, they satisfy the rules of probability distributions:

$$p_r(x_1)+p_r(x_2)+\cdots+p_r(x_n)=1=100\%$$

This relative two-by-two game allows us to set fundamental assumptions of evolutionary dynamics.

+ **Replicator dynamics (Practice with Public Good Games):** Now let us consider a population of players, each with a given payoff dynamic. The concept of replicator dynamics is used to express the evolutionary dynamics of an entity called a replicator (typical examples are the strategies of cooperation and defection in games) which can make more-or-less accurate copies of itself. From time to time, the players meet randomly and play a game according to a plan. If we suppose that each player is rational, we can consider several types of different payoffs for each:

$$x_i = \mathrm{Pr(i)}\pi(i)$$

Here each player has the payoff $\{\pi(i)\}$, which shows how well that type (i) is doing, and each type has a proportion {Pr(i)}. Then they choose a certain strategy they consider to be the best outcome from the entire population.

$$x'_i = \frac{\mathrm{Pr(i)}\pi(i)}{\sum_{j=1}^{N}\mathrm{Pr(j)}\pi(j)}$$

Pr(i) is the proportion of the different types, and $\pi(j)$ is the payoff for all of the types. Mathematically, this (replicator) dynamic has been expressed in the set of differential equations (where x'_i is the proportion of type i in the population) used to study dynamics in game theory. Here we consider a population of types that are succeeding at various levels. Some are doing well, and some are not.

The dynamics of the model suppose a series of change in distribution across types, such that there is a set of types {1, 2, …, N}, a payoff for each type (π_i), and a proportion of each as well (Pr_i). An individual player's strategy (x'_i) in each round is given a probability that is the ratio of this weighting to all possible strategies, where x'_i is the probability that an individual play will take a strategy times payoff $[\mathrm{Pr(i)}\pi(i)]$, divided by the sum of the weightings of all the different strategies $[\sum_{j=1}^{N}\mathrm{Pr(j)}\pi(j)]$. Thus, the probability that the individual player will act in a certain way in the next round is just that action's relative weighting. Specifically, let us propose that there are different probabilities Pr(i) of using different strategies $(x, y, and\ z)$: that is,

- strategy x has a probability of 40%
- strategy y also has a probability of 40%
- strategy z has a probability of 20%

This might lead us to guess that strategies x and y are better than z. However, we could also look at the payoffs $\pi(i)$ of the different strategies. For instance, it might be that when using

- strategy x players can obtain a payoff of 5
- strategy y there is a payoff of 4
- strategy z players take payoff of 6

This prompts us to consider which strategy we should use, and the answer depends on both the payoff and the probability. Proposed here is a description of how individuals might choose what to do or which strategies are best. After a certain move, some will appear to be doing better than others, and the ones that are doing the worst are likely to copy the ones that are doing better.

Based on cultural evolutionary assumptions for public goods game, we specify the respective frequencies of actions of cooperators (P_c), defectors (P_d), and loners (P_l). We assign a value to each player (Table 3.8.1), and then the players may contribute part (or all) of their value to the game (a common pool).

[Table 3.8.1] Strategies of Potential Players

Cooperators (P_c): Ready to join the group and contribute to its effort.
Defectors (P_d): Join the group but do not contribute.
Loners (P_l): Unwilling to join the public goods game.

In each round, a sample (S) of individuals is chosen randomly from the entire population N. S ($0 \leq S \leq N$) out of the individuals participate in the game, paying a cost (g). Each round requires at least two (cooperator and defector) participants, and others must be nonparticipants. The cooperators contribute a fixed amount of value $c > 0$ and share the outcomes multiplied by the interest rate r ($1 > r > N$) equally among all other $S - 1$ participants; defectors are in the round but do not contribute values. During the round, the payoffs for the strategies P_c, P_d, and P_l (denoted $x, y, and\ z$) are then determined, with a participation cost c and an interest rate rc, based on the relative frequencies of the strategies (n_c/N).

$$P_c = -c + rc\frac{n_c}{N}, \qquad P_d = rc\frac{n_c}{N}, \qquad 1 < r < N$$

Where n_c is the number of cooperators, the payoffs are defined as $P_c = -c + rc\frac{n_c}{N}$. And $P_d = rc\frac{n_c}{N}$. Here, the interest rate r must satisfy $1 < r < N$. Indeed, P_c is the payoff of the cooperators, P_d is the payoff of the defectors, and rc is the interest rate (r) multiplied by the fixed contribution cost (c) for the common good.

For the expected payoff values of the cooperators (P_C) and defectors (P_D), a defector in a group with $S-1$ coplayers ($S = 2, \dots, n$) obtains from the common good a payoff of $rcx/(1-z)$ on average because the nonparticipants (z) have a payoff of 0.

$$P_D = \left(rc\frac{x}{1-z} - g\right)(1 - z^{n-1})$$

Here z^{n-1} is the probability of finding no coplayer. In the abstract, we can write this as the derivative of z with respect to n for any power n, $nz^{(n-1)}$. This is known as the power rule and is symbolically represented as the exponent (i.e., $z^2 = 2z^1$, $z^3 = 3z^2, \dots, z^n = nz^{n-1}$). This is what it means for the derivative of z^n to be nz^{n-1} and for the population thus to be reduced to loners (nonparticipants). In addition, cooperators contribute effort c with a probability $1 - z^{n-1}$.

Hence,

$$P_D - P_C = c(1 - z^{n-1})$$

The average payoff ($\bar{P}$) in the population is then given in this way:

$$\bar{P} = (1 - z^{n-1})[(r-1)cx - (1-z)g]$$

The replicator equation, which describes the evolution of the strategies in the population, is given by:

$$x' = x(P_x - \bar{P}), \qquad y' = y(P_y - \bar{P}), \qquad z' = z(P_z - \bar{P})$$

The frequencies x_i of the strategies i can simply be represented by

$$x'_i = x_i(P_i - \bar{P})$$

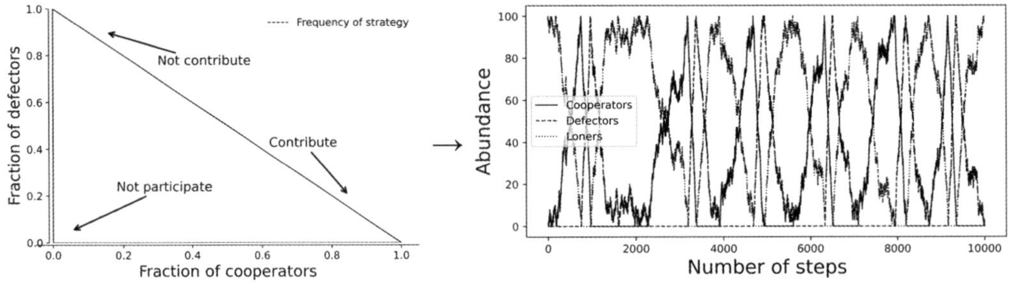

Figure 3.8.1. Illustration of the model (Public Goods Game) with replicator dynamics. Left side of the plot: lines indicate prototypes as a proportion of their implementation by cooperators (bottom right area), defectors (upper left area), and loners (bottom left area). The dotted line denotes their relative frequencies, where defectors dominate over cooperators, loners dominate over defectors, and cooperators dominate over loners. Right side of the plot: A well-mixed population with a finite number of strategies (cooperators, defectors, and loners) with incentive.

Here x_i denotes the frequency of strategy i, P_i is the payoff of strategy i, and $\bar{P}$ represents the average payoff in the population. Accordingly, a strategy will spread or dwindle depending on whether it does better or worse than average. This equation holds that populations can evolve, in the sense that the frequencies x_i of strategies change with time. Thus, we let the state $x(t)$ depend on time and denote this $x'_i(t)$, where x'_i changes dx_i/dt.

This practice is particularly interested in the growth rates of the relative frequencies of the strategies. In other words, the state of the population evolves according to the replicator equation, where the growth rate (x'_i/x_i) of a strategy's frequency corresponds to the difference between its payoff (P_i) and the average payoff $(\bar{P})$ in the population (Figure 3.8.1).

2. Simulation of Game Theory Model (Public Goods Game with Gillespie Algorithm)

The above procedures assume a well-mixed population with a finite number of strategies that are proportional to their relative abundances given that the fitness values are frequency-dependent, coexisting in steady or fluctuating frequencies with the evolutionnary games.

The mechanism is a combination of the rational process and the copying process; in order words, an individual rationally chooses from a nearby individual because it seems that it would lead to successful outcomes. Here we develop a mechanism that emphasizes a model's capacity to replicate empirical estimates of a real individual's behavior in the context of the public goods game.

To achieve this, we begin by importing the necessary libraries.

```
[In]
import matplotlib.pylab as plt
import numpy as np
from random import *
import math
```

Generate a specified number of populations, along with their corresponding sample groups and associated arrays.

```
[In]
M=100                     # number of population
N=5                       # number of sample group
pop=np.zeros(M)           # population array
```

In this case, 'pop' is a one-dimensional array with a length of 'M=100', where the value of 'N' does not influence the array's definition.

Begin by initializing the game parameters.

```
[In]
g=0.5                         # contribution cost
r=3                           # incentive
c=1                           # investment of cooperator
w=1                           # selection intensity
u=1e-10                       # mutation rate
```

For time domains

```
[In]
t=10000                       # number of steps
tt=500                        # number of realization
```

Now, we can define strategies with payoff dynamics (defector and loner)

```
[In]
def nCr(n,r):
    if (n>=r):
        f = math.factorial
        a=f(n) // f(r) // f(n-r)
    else:
        a=0
    return a
def PD(C,D,M):
    L=M-C-D
    a=0
    for S in range(2,N+1):
        for X in range(S):
          if (S-X<=M-L and X<=M-L-1):
            a+=nCr(N-1,S-1)*nCr(M-L-1,S-1)*nCr(L,N-S)*nCr(C,X)*nCr(D-1,S-X-
1)*nCr(S-1,X)*(r*c*X/(S-1)-g)/(nCr(M-1,S-1)*nCr(M,N-S)*nCr(M-L-1,X)*nCr(M-L-1,S-1-
X))
          else:
             break
    return a
```

The code calculates the probability of cooperation in a two-strategy iterated game with a population of size M where N players interact in each round. The function takes three arguments: C, D, and M. the C and D are the number of cooperators and defectors, respectively, in the population, and M is the population size. The function calculates the P_D for this population using the formula given in the iterated game. First, the function calculates the number of 'lonely' individuals in the population, which are individuals that do not interact with any other individuals in a given round. This is done using the L variable, which is calculated as 'M – C – D'.

The function then iterates over the number of strategies S that can be played in a given round from 2 to 'N=1' and the number of cooperators X that are present in that round from 0 to S. If the number of cooperators and the number of defectors is less than or equal to the number of lonely individuals in the population, the function calculates the P_D for this strategy. And number of cooperators using the given formula adds it to the running total. Otherwise, the function breaks out of the inner loop and moves on to the next value of S. The function returns the final value of P_D when all iterations are complete. And define strategies with payoff dynamics (cooperator and loner)

```
[In]
def PC(C,D,M):
    L=M-C-D
    a=0
    for S in range(2,N+1):
        for X in range(S):
          if (S-X<=M-L and X<=M-L-1):
            a+=nCr(N-1,S-1)*nCr(M-L-1,S-1)*nCr(L,N-S)*nCr(C-1,X)*nCr(D,S-X-1)*nCr(S-
1,X)*(r*c*X/(S-1)-g-c)/(nCr(M-1,S-1)*nCr(M,N-S)*nCr(M-L-1,X)*nCr(M-L-1,S-X-1))
          else:
              break
    return a
```

The P_C and P_D calculate some value given three inputs: C, D, and M which the functions use combinatorial operations 'nCr'. The 'nCr' is a function that calculates the number of ways to choose r items from a set of n items. The P_D and P_C are functions that calculate the probability of a cooperator player being replaced by a defector player and a defector player being replaced by a cooperator player, respectively. Both functions take in three arguments 'C=cooperators', 'D=defectors', and 'M=total players'. The functions also use the previously defined variables; N is the number of players in a sample group, r is an incentive, c is the investment of a cooperator, g is a contribution cost, u is a mutation rate, t is the number of steps, tt is the number of realizations, and ttt is the number of initial conditions. For the initial condition at each round

```
[In]
initial=[[0,0]]
for item in initial:
 x0=item[0]
 y0=item[1]
 ND=np.zeros([t])                #mean number of defector in each step
 NC=np.zeros([t])                #mean number of cooperator in each step
 NL=np.zeros([t])                #mean number of loner in each step
 for j in range(tt):             #realization
    C=x0                         #number of cooperator at zero time
    D=y0                         #number of defector at zero time
    L=M-C-D                      #number of loner at zero time
```

Here, the code initializes three arrays, 'ND', NC', and 'NL', to store the mean number of defectors, cooperators, and loners at each time step, respectively. It also starts a loop over tt where tt is the number of realizations. Within this loop, the variables C, D, and L are initialized to the initial number of cooperators, defectors, and loners, respectively. These variables are determined by the values of $x0$ and y0 which are the first and second elements of the 'item' list in the outer loop. This outer loop is iterating over the 'initial' list, which currently contains a single element [0, 0] which means that $x0$ is set to 0 and y0 is set to 0 in each iteration.

Calculate its dynamics using the Gillespie algorithm for the for loop.

```
[In]
for ii in range(t):  # steps
    pc = PC(C, D, M)
    pd = PD(C, D, M)
    pl = 0

    rates = [
        (L * u / M + L * D / (M**2 * (1 + math.exp(pl - pd)))), (1, -1, 0)),
        (L * u / M + L * C / (M**2 * (1 + math.exp(pl - pc)))), (-1, 0, 1)),
        (D * u / M + D * C / (M**2 * (1 + math.exp(pd - pc)))), (0, -1, 1)),
        (D * u / M + D * L / (M**2 * (1 + math.exp(pd - pl)))), (-1, 1, 0)),
        (C * u / M + C * D / (M**2 * (1 + math.exp(pc - pd)))), (0, 1, -1)),
        (C * u / M + C * L / (M**2 * (1 + math.exp(pc - pl)))), (1, 0, -1))
    ]

    total_rate = sum(rate[0] for rate in rates)
    rand_val = random()

    for rate, changes in rates:
        if rand_val < rate / total_rate:
            L += changes[0]
            D += changes[1]
            C += changes[2]
            Break

        rand_val -= rate / total_rate
```

We used a list of tuples to store the rates and the changes in L, D, and C. The for loop iterates through the rates and checks if the random value falls within the corresponding range. If it does, the corresponding changes are applied to L, D, and C, and the loop is terminated.

In particular, the Gillespie algorithm is a stochastic simulation method used to model the time evolution of a system, such as chemical reactions or game-theoretic scenarios. In the given code, the Gillespie algorithm is applied to this game simulation, where players can adopt different strategies (cooperator, defector, or loner) with probabilities based on payoffs and mutation rates. The algorithm runs for t steps, updating the number of players in each strategy based on the probabilities computed from the payoffs and mutation rates.

At each step, players can change their strategy by mutating (with probability u), or by imitating other players. The probability of a player changing its strategy to another strategy S is given by: P changing to S) = (u/M) + (number of players with strategy S/ M^2 ∗ (1 + exp(payoff of current strategy - payoff of strategy S))). The payoffs are given by the function 'PC' and 'PD', which compute the payoffs for cooperators and defectors, respectively. These payoffs depend on the current number of cooperators, defectors, and loners in the population (C, D, and L).

At each step, the code randomly selects a player to change its strategy, with probabilities proportional to the change rates 'al', 'at', etc. If the random number falls within the range for 'a1', for example, the player changes from being a loner to a defector. The total number of players and the number of players in each strategy is updated. The simulation runs for t steps, with tt realizations, and for each initial condition in the list 'initial'. The variables 'ND', 'ND', and 'NL' are arrays that store the mean number of players in each strategy at each step, across all realizations and initial conditions.

Finally, let us observe the results using a simple plot (at log scale).

```
[In]
for ii in range(t):
    ND[ii] += D
    NC[ii] += C
    NL[ii] += L

plt.plot(NC / M, ND / M, 'k')
```

The code defines a set of variables and functions related to a population of individuals that can be either 'cooperators', 'defectors', or 'loners.' The function calculates the probability of an individual changing from being a 'loner' to a 'cooperator', and the function calculates the probability of an individual changing from being a 'loner' to a 'defector.'

The 'for' loop iterates over a number of steps t and within each step it calculates the rate of change for each type of individual based on their current state and the probabilities calculated by C and D. It then randomly chooses an individual to change their state based on these rates of change.

The number of individuals of each type is updated accordingly. Then, the code plots the proportion of 'cooperators' and 'defectors' in the population over time.

[Out]

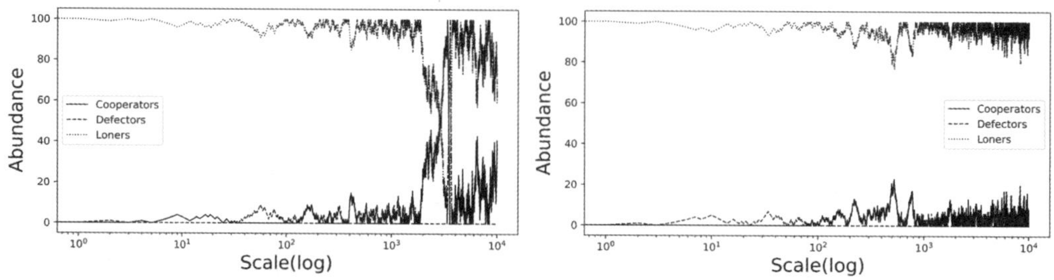

Figure 3.8.2. Replicator dynamics with parameters (participation cost). Simulation result with participation cost (g) and parameters: r = 3, c = 1, u = 1e-10, participation cost g = 2 (Left). Simulation result at g = 3 (Right). Both plots show the categorized oscillation of the strategies (cooperators = bottom line, defectors = bottom line, and loners = upper line).

In Figure 3.8.2, the simulation shows that the system (i.e., strategies) settles into different effects on the intermediate interest rate, with participation costs. As the interest rate increases (r), it prompts the population to undergo stable oscillations relative to a global attractor, where the players participate by contributing to the public good. However, if that contribution is too expensive, that is, if the participation cost is $g \geq (r-1)c + l$ for rewarding, or $g \geq (r-1)c$ for punishing, the players will not opt to participate. In this scenario, nonparticipation becomes the global attractor (see more detail in the Reference list; Park, 2022).

The simulation explores the Public Goods Game using the Discrete Gillespie Algorithm, examining cooperator, defector, and loner strategies. It demonstrates the impact of participation costs on population dynamics, showing stable oscillations at certain interest rates and nonparticipation when costs are high. This analysis offers insights into collective decision-making and helps identify factors promoting cooperation and participation, potentially guiding effective interventions or policies.

Remarks: We must be wondering what all this means in the context of strategic decision making. What do these different types of games and this information have to do with systems analysis? So far in the history of systems analysis, approaches have had very limited success when it comes to incomplete information games. Game theory is the key idea behind various analysis (Osborne, 2004). It uses relatively little computing power compared to deep-learning or reinforcement-learning methods. However, people often debate the carryover of research to real-world cases. Because real-world cases are often incomplete information games, most approaches (i.e., data science and deep-learning) struggle. Game theory approaches are gradually gaining momentum because of their generalizability to real-world cases. The best examples are work being undertaken in public safety, wildlife conservation, public health, and so on.

Supplemental Information 3.5: Systemic Risk Measurement

The world is being ruled by the quick development of technology and new models in various fields, while threats are interconnecting more than ever. As a result, using only the traditional approach (i.e., statistics and time sereise) to forecast the future is insufficient. Risk assessment has to take a more holistic approach (i.e., systemic) that goes beyond addressing risks independently since they are developing, merging, and spilling over into one another.

Notably, in various studies, systemic risk has been blamed for high profile disaster (i.e., for making a significant contribution to the financial crisis of 2008, Leduc, Poledna, & Thurner, 2017) and also for being a likely cause of cascading failures (i.e., socio-economic impact of COVID-19, World Health Organization, 'Coronavirus disease 2019').

+ **Define the Concept:** Systemic risk is a property of systems of interconnected components. This concept can be described as system instability, caused or exacerbated by idiosyncratic events, resulting in potential catastrophe.

The systemic risks including natural catastrophes, infectious illnesses, essential infrastructure failure, and climate change may not be viewed as particularly likely by organizations. However, if these risks do materialize, they may have detrimental short- and long-term effects on an organization, particularly if they set off other high-risk situations. Organizations should be equipped to measure these risks as a result. One commonly used measure of the systemic risk is the conditional value which measures the expected loss in the event of a crisis.

Here is one of the basic example;

$$\text{Systemic Risk (CoV)} = \alpha * \text{VaR} + (1 - \alpha) * \textit{expected loss}$$

Where, the VaR is the maximum expected loss is over a certain confidence interval, and the α is a factor that determines the importance of the expected loss regarding the VaR.

A distinguishing feature of such risks, sometimes called network risk, is that they emerge from the complex interactions among individual elements in a system or from their association with each other. The context-varying mechanical flux on a system's risk is actually very complex. The possibility to quantify systemic risk and capture its size need to be performd in the face of all these distortions and patterns of the influences.

+ **What to Measure:** By knowing the particular form in which an event could trigger instability or collapse an entire system regardless of individual capability at that point, it is possible to quantify with specificity the mechanisms underlying systemic risk using a computerized model. Here, we briefly highlight how the complexity of such inter-connected components. In particular, a failure proved to be the fundamental to under-standing the concept of the systemic risk.

- The term 'number of failures' describes the total number of nodes that failed throughout the simulation as a result of a certain probability of node failure. These failures might be viewed as the failure of certain individuals or institutions within a network.
- The term 'number of link failures,' on the other hand, describes the quantity of edges (or connections) that have been eliminated from the network throughout the simulation as a result of a certain probability of link failure. These mistakes might be seen as the breakdown or collapse of individuals or institutional connections.

We implement a mechanism focused on network-agent dynamics to reproduce empirical estimates of the number of failures.

First, let us import libraries

```
[In]
Import networkx as nx
Import numpy as np
```

and define number of individuals with their connections.

```
[In]
Num_indiv = 100
Num_connect = 10
```

We generate a specific network (scale-free structure) using the imported library

```
[In]
Scale_free_network = nx.barabasi_albert_graph(Num_indiv, Num_connect)
```

Intialize the number of individulas' failures and linked failures; all they are start with no failure as initial level

```
[In]
Num_indiv_failures = 0
Num_connect_failures = 0
```

We might think of hazards as nodes in a network that need to be evaluated. Each node (risk) will contain characteristics such as the failure that it will occur of its probability. We will create a link between two dangers to demonstrate their relationship if they influence or trigger one another.

Here, their probability of failure is originated at an individual with certain probability, define its simulation time periods.

```
[In]
Prob_indiv_failures = 0.1
Prob_connect_failures = 0.1
Simulation_time = 100
```

To observe the number of node and link failures over time

```
[In]
num_node_failures_over_time = []
num_link_failures_over_time = []
```

Then, let us simulate over the specified time period over the individuals and its connections. We will simulate node failures and increase the number of individuals as below;

```
[In]
for i in range(simulation_time):
    for node in list(scale_free_network.nodes()):
        if np.random.rand() < prob_node_failure:
            scale_free_network.remove_node(node)
            num_node_failures += 1
```

And simulate link failures and increase the number connections failure with random probability as below;

```
[In]
    for edge in list(scale_free_network.edges()):
        if np.random.rand() < prob_link_failure:
            scale_free_network.remove_edge(*edge)
            num_link_failures += 1
```

Now we calculate the number of failures and links simply as the defined equations above.

```
[In]
    num_node_failures_over_time.append(num_node_failures)
    num_link_failures_over_time.append(num_link_failures)
```

Finally, plot the number of node and link failures over time as below;

```
time = np.arange(0,
simulation_time)
plt.plot(time,
num_node_failures_over_time)
plt.plot(time,
num_link_failures_over_time)

plt.show()
```

→

[Out]

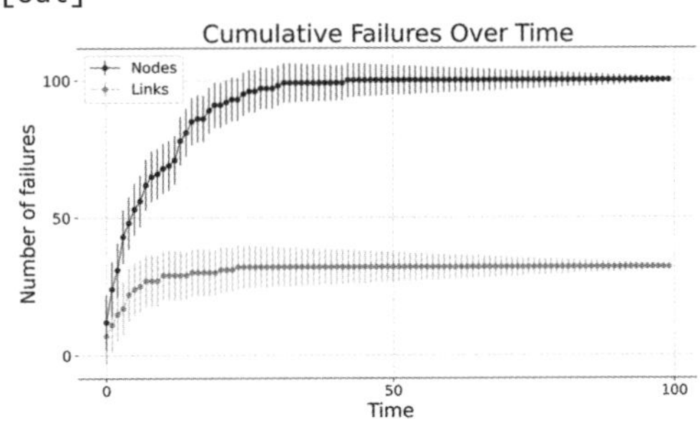

Figure S3.5.1. Schematic representation of the failure dynamics in systemic risk. We count the number failures and the links of failure, respectively, and initialize both of them at 0. Initialized parameters: number of nodes=100, number of connections per node=10, initial number of failures p_n=0.1, initial link of failure p_l=0.01, time steps=100, iteration=100. The error bars represent a simulated decreasing variance over time, signifying the stabilization of the system.

This simulation illustrates the cumulative number of node and link failures over time. The number of failures at each time step is added to the total of the previous steps, resulting in an ever-increasing trend (Figure S3.5.1). With respect to the fundamental characteristic of the model, we implement that an individual (node) can catch a failure if one of its neighbors is infected to measure how cascades of failure can propagate through the network. The elementary level of risk depends on the network units, which depend on the co-occurrence of the *i* and *j* of the nodes. This reflects that individuals are more biased when an individual is highly linked in its network. We add one more simulation to illustrate the independent number of node and link failures at each time step. Each time step is independent of others,

```
[In]
num_node_failures = 0
num_link_failures = 0
```

Then, repeat the same code procedures and observe the number of failures over time again as below;

[Out]

```
time = np.arange(0,
simulation_time)
plt.plot(time, num_node_failures)
plt.plot(time, num_link_failures)

plt.show()
```

→

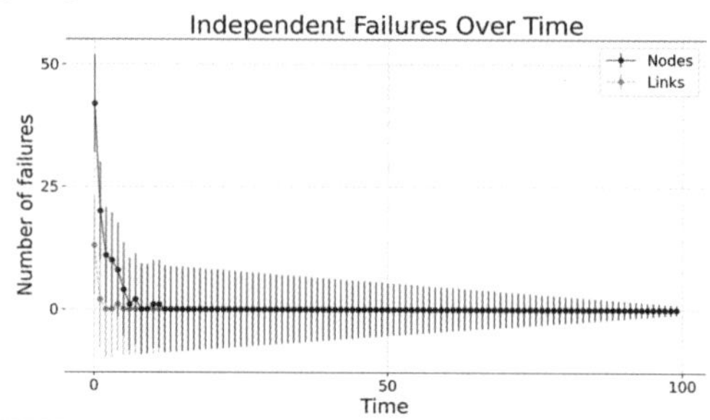

Figure S3.5.2. Schematic representation of the failure dynamics in systemic risk. We count the independent number failures over time, respectively, initialize both of them at 0.

The number of failures can fluctuate over time, showing a non-monotonic trend. The y-axis is more dramatic to emphasize the lower number of failures (Figure S3.5.2).

The establishment of a simple mechanism was proposed. At the same time, we would like to suggest that various strategic process and dynamic which acquired from the previous sections (i.e., Network, Agent-based Model, and Game theory) can be applied to explore the influence of the interconnected interaction on the elements; such as cultural evolution and social learning. Effective risk management requires that we take into account the links between hazards, the strength of those connections, the possibility for clustering, and the possible cumulative impact. Traditional assessment methods may not be sufficient, and their risk management strategies that only take into account one source of risk may not be appropriate in complex, concurrently networked marketplaces. Systemic methods that can do a multidimensional link analysis cover the gap left by conventional analytics techniques like the Graph networks. All of which will help us to characterize and understand the architecture of systems from the systemic risk perspective (see the Reading list 3.2 for more intuition).

[3.9]

Concluding Remarks

In preparation for writing this book, we circulated a questionnaire on the foundations, techniques, and applications of systems analysis and analysts even outside of the sciences. There were many responses, but they only served to confirm what everyone already knows: There is no agreement on what systems analysis is and how to deal with it. In fact, the answers showed that the conceptual basis of the discipline needs to be enriched and strengthened. We were thus determined to put our own stamp on the theory and practice of systems analysis at the earliest stage possible.

Early in our research career, systems analysis was viewed almost as rocket science, but at the International Institute for Applied Systems Analysis in Laxenburg, Austria, we taught that it should be the majority course to this approach in research. Systems analysis does not just involve classifying statistics or discovering properties that are common to various categories of study. It is not something used to discover the nature of social or environmental interaction. It is not just a branch of applied mathematics. It is also not a science, although science and technology are its cornerstones.

Systems analysis is in fact a problem-solving process in which many people take part: scientists of relevant disciplines, stakeholders, and decision makers. These are not just problems per se but problems along with all of the attendant factors and concepts they encompass. These factors include 'the knowledge and methods of modern science and technology, in combination with concepts of social goals and equities, elements of judgment and taste, and appropriate consideration of the larger contexts and uncertainties that inevitably attend such systems.' It can thus be said that systems analysis has both a quantitative and a qualitative side.

The central purpose of systems analysis is to help scientists, analysts, and even private decision makers resolve the problems they face in the short, medium, and long term. The problems we address here are large-scale, technical problems—problems affecting industry; problems of such immensity, complexity, and urgency that to neglect a single aspect could be very costly. Systems analysis is constantly transforming as it adapts to diverse problems and differing contexts. It is not static. However, we believe it is possible to explain systems analysis using a step-by-step framework.

First, systems analysis offers information and scientific knowledge on specific problems; if necessary, it gathers new evidence and develops new knowledge. Second, it determines what the goals of analysts are, both people and institutions. Third, it explores alternative ways of achieving those goals and designs or invents new options where appropriate. Fourth, it reconsiders the problems in relation to the knowledge accumulated. Fifth, it estimates the impacts of various possible scenarios of action, taking into account an uncertain future and the data structures that are required to implement the proposals. Sixth, it compares alternatives by making a detailed assessment of possible impacts and consequences. Seventh, it presents the results of this analysis in a framework that facilitates choice by the expert. Eighth, it provides follow-up assistance. Finally, it evaluates the results.

Systems analysis has been most helpful in addressing issues dominated by science and engineering, for industrial or military applications. It also works well for making budgetary decisions by pointing to the most cost-effective courses of action. It is more problematic where political, organizational, and social factors predominate and where goals may be obscure and authority diffuse and overlapping. As well as being used to craft good solutions, systems analysis can also be considered an art in terms of achieving an often-delicate balance that satisfies multiple stakeholders.

Although systems analysis, like any other scientific endeavor, has its limitations, and other means are available to assist the decision maker, it does have a number of virtues. It introduces a certain objectivity into the subjective process of data analysis and thus can help with the acceptance and implementation of decisions. It can take uncertainty explicitly into account, can determine interactions and side effects, and may reveal unexpected consequences of policies and actions. It can also provide insight into issues and suggest better alternatives. Systems analysis is inherently multi and interdisciplinary. Systems analysts, as specialists in research, are structured to reflect this way of thinking and researching.

Systems analysis is making an important—even essential—contribution to solving some of the most complex problems. We hope this book helps readers to see this vital role, to appreciate its intricacies and complexity, and to understand how to harness its power to meet the challenges that lie ahead.

Reading List 3.1: Applied Systems (Agent-Based Simulation) Research with Behavioral Bias

Purpose: The primary purpose of this study is to use a spatially explicit model of mobile agents in two-dimensional continuous space to understand the conditions that lead to the estimation of behavioral bias. The broader agenda of this model is to especially understand the conditions leading to the estimation of behavioral bias by including a fundamental modeling perspective with the cultural evolutionary process. **Methods:** By integrating behavioral algorithms with social dynamics, the model attempts to (i) capture emergent phenomena, (ii) provide a natural description of a pattern of behavior, and (iii) allow realistic adaptation to be understood. The behavioral pattern results from individual components being applied not only from an autonomous agent's internal trait (individual velocity/group velocity trade-off) and external trait (network characteristics) but also from its interconnected circumstance. **Results:** The range of different combinations of some initial bias values (scalar in the internal trait and the external trait) plays a part in the rapid propagation in the system or puts the system into even more jeopardy. However, when mutual relations between internal traits, which are the basis of external traits, are applied, widespread heterogeneity due to systemic bias can reduce the repertoire of displayed behaviors (Park, 2020a).

Background: To explore the rule of thumb from a broader perspective, computer simulation has been performed on the basis of a spatially explicit model of mobile agents in continuous space (Rangel et al., 2018) to determine the basic regulatory principles involved in the way they conceptualize their environment (Bonabeau, 2002). In other words, the implemented model shows the potential to infer how simple individual rules can lead to consistent group behavior and how slight changes in those mechanisms can have a dramatic impact on an individual's behavioral patterns (Reynolds, 1987). Although an analysis of simple implication is an apparent first step in providing proof of concept, this individual-based simulation has become significant enough to be tested in a broader range of applications for evolutionary dynamics for the following reasons. First, as the agents represent individuals who have participated from the bottom up, the actual state of their behavior tends to be more informative (agent-based modeling) than in other cases. Next, as the main point of this implementation is a description that deals with the state per time as the critical factor in its allocation of neighbors, the number of neighbors placed on the position is based on the moves scheduled for a given moment (context-varying cultural evolution; Oliva, 2016).

Methods (agent-based modeling): The broader agenda of this supplement is to show the mathematical process behind the fundamental modeling mechanisms used. This information is based on spatially explicit mobility, where the individuals can move around their environment. The rules and processes in the artificially modeled structure describe an individual's homogeneous drives and also how to apply the mechanism.

Model Detail: The agents are physically related to each other on some spatial representation allowing them to move anywhere in the space. The set of n-tuples of a real number, denoted $\mathbb{R}^n$, is called *n-spaces* $[x = (x_1, x_2, \dots, x_n) \in \mathbb{R}^n]$. A particular n-tuple in $\mathbb{R}^n$ is a point called the *coordinates*, *components*, or *elements* of x. This is one of the standard ways in which the agents can continue to move in the space. The agents then move within the boundaries of the plane, steering toward somewhere.

$$\vec{u} = \vec{a}_{avg} + \vec{b}_i, \qquad \vec{a}_{avg} = \|a_{avg}\| * \vec{d}_{avg}, \qquad \vec{b}_i = \|b_i\| * \vec{d}_i$$

Here $\vec{a}$ is the group's heading and $\vec{b}$ is each agent's (i) coherence toward the center of the group. The order ($\vec{u}$) is symmetric because all the agents are identical; thus, individuals are naturally heading together in a certain direction while at the same time maintaining a certain distance from each other as their inherited survival strategies (*1*).

The model, then, includes another operation with respect to the individual's current movement ($\vec{v}$) written simply by ($\vec{\omega} = \vec{u} + \vec{v}$). That new quantity $\vec{\omega}$ is the sum of $\vec{u} + \vec{v}$, where the vectors stay away from the origin. The way we define this is that each vector represents a certain movement, a step with a certain distance and direction in space. If we take a step along the first vector of the $\vec{u}$, and then a step in the direction and distance described by the second vector of the $\vec{v}$, the overall effect is just the same as if we had moved along the sum of those two vectors to begin with.

$$\vec{\omega} = \vec{u} = \begin{bmatrix}4\\1\end{bmatrix} + \vec{v} = \begin{bmatrix}1\\2\end{bmatrix} = \begin{bmatrix}4+1\\1+2\end{bmatrix} = \begin{bmatrix}5\\3\end{bmatrix}$$

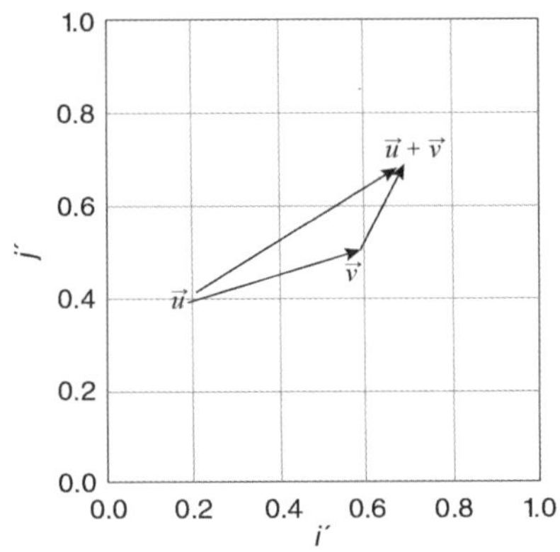

The first quantity here (green line) has the coordinate ($\vec{u} = \begin{bmatrix}4\\1\end{bmatrix}$), and the second quantity (blue line) has the coordinate ($\vec{v} = \begin{bmatrix}1\\2\end{bmatrix}$). When we take the sum of the two quantities, we can see a four-step path from the origin to the tip of the second quantity: move 4 to the right and 1 up, then move 1 to the right again and 2 up.

To rearrange these steps, first move 4 + 1 to the right, then move 1 + 2 up; the new quantity (black line) has coordinates 4 + 1 and 1 + 2 from the origin. Exhibited in this list-of-numbers conception is a matching up of their numerical terms and an adding of them both together ($\vec{\omega} = \begin{bmatrix}4+1\\1+2\end{bmatrix} = \begin{bmatrix}5\\3\end{bmatrix}$).

With this fundamental process, especially, we see that the second quantity $\vec{v}$ contains the individual's trait (agent-self interactions: agents can interact with themselves), based on its condition holds:

$$\vec{v} \rightarrow \vec{f} \rightarrow f(\vec{v}), \quad f(\vec{v}) = \begin{cases} \vec{v}\,(+)\; if\; f(\|v_i\|) < \|v_{avg}\| \\ \vec{v}\,(-)\; if\; f(\|v_i\|) > \|v_{avg}\| \end{cases}$$

where the function assumes that the attribute of the component is conditional on the value yield in the other direction (–) if the length of the magnitude $\|v_i\|$ is greater than the other length ($\|v_{avg}\|$). This trait is implemented according to quantity as follows:

$$\vec{v}_i = \|v_i\| * \vec{d}_i, \qquad \|v_i\| = \sqrt{v_{i_x}^2 + v_{i_y}^2}, \qquad \vec{d}_i = \frac{\left(v_{i_x}, v_{i_y}\right)}{\sqrt{v_{i_x}^2 + v_{i_y}^2}}$$

$$\vec{v}_{avg} = \frac{1}{N}\sum_{i=1}^{N} \vec{v}_i = \|v_{avg}\| * \vec{d}_{avg}$$

where $\vec{v}_i$ is the individual's velocity represented by the length of the individual's magnitude ($\|v_i\|$) with the direction of individual ($\vec{d}_i$). $\vec{v}_{avg}$ is the average velocity, which includes the entire population of N individuals' navigation. The result of $\vec{v}_i$ and $\vec{v}_{avg}$ produces a new quantity $\vec{v}_{inew}$, written simply as ($\vec{v} \rightarrow \vec{v}_{inew} = \vec{v}_i + \vec{v}_{avg}$).

Going back to the conditional (IF) assumption with this individual quantity, if object ($\vec{v}$) faces parameters ($\vec{u}$) with the states of their quantities ($\|v_i\|$ and $\|v_{avg}\|$), the condition set produces an opposite direction (±) depending on its norm as follows:

$$f(\|v_i\|) < \|v_{avg}\| \;\rightarrow\; \vec{\omega} = \vec{u} + \vec{v}$$

$$f(\|v_i\|) > \|v_{avg}\| \;\rightarrow\; \vec{\omega} = \vec{u} + (-\,\vec{v})$$

where $\vec{\omega}$ is a new position vector of the individual updated by the inherited trait $\vec{u}$ with the individual's current movement $\vec{v}$. This is a linear combination (or inverse transformation = 180° counterclockwise), something that takes in inputs and spits out an output for each one:

$$\underbrace{\begin{bmatrix} x_{in} \\ y_{in} \end{bmatrix}}_{input} \rightarrow f \rightarrow \underbrace{\begin{bmatrix} x_{out} \\ y_{out} \end{bmatrix}}_{output} = \underbrace{\begin{bmatrix} 0 & 1 \\ 1 & 0 \end{bmatrix}}_{matrix} \underbrace{\begin{bmatrix} x \\ y \end{bmatrix}}_{vector} \rightarrow f \rightarrow \underbrace{\begin{bmatrix} 0 & 1 \\ 1 & 0 \end{bmatrix}}_{matrix} \underbrace{\begin{bmatrix} x \\ y \end{bmatrix}}_{vector} \;or\; \underbrace{\begin{bmatrix} 0 & -1 \\ -1 & 0 \end{bmatrix}}_{matrix} \underbrace{\begin{bmatrix} x \\ y \end{bmatrix}}_{vector}$$

$$\begin{bmatrix} x_{out} \\ y_{out} \end{bmatrix} = \begin{bmatrix} 1\vec{x} \\ 1\vec{y} \end{bmatrix} \;or\; \begin{bmatrix} -1\vec{x} \\ -1\vec{y} \end{bmatrix}$$

Imagine that every possible input vector multiplied by the matrix moves over to its corresponding output vector multiplied by the matrix (or inverse) without becoming curved and that the origin must remain fixed in place; what the coordinates are is determined by where each basis vector lands.

For example:

$$\vec{\omega} = \vec{u} = \begin{bmatrix}4\\1\end{bmatrix} + \vec{v} = \begin{bmatrix}1\\2\end{bmatrix} = \begin{bmatrix}4+1\\1+2\end{bmatrix} = \begin{bmatrix}5\\3\end{bmatrix}$$

$$\vec{\omega} = \vec{u} = \begin{bmatrix}4\\1\end{bmatrix} + \vec{v} = \begin{bmatrix}(-)1\\(-)2\end{bmatrix} = \begin{bmatrix}4+(-)1\\1+(-)2\end{bmatrix} = \begin{bmatrix}3\\-1\end{bmatrix}$$

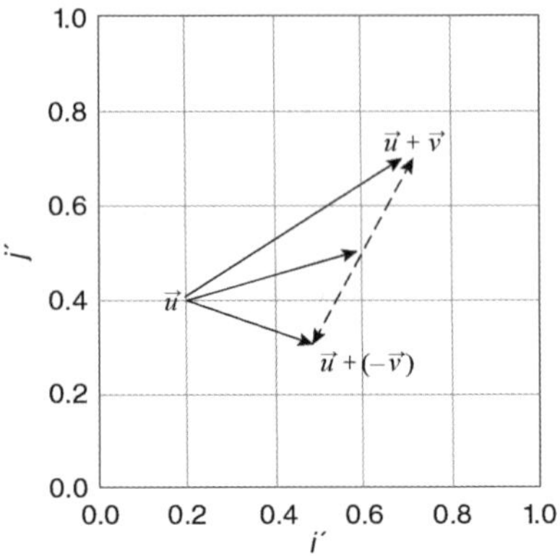

This refers to the fact that the model's basic pattern of group behavior depends on the value of $\vec{v}$ with a localized view of the initial conditions of the randomly initialized point; increasing the individual's quantity $\vec{v}$ underlying the group's initial condition $\vec{u}$ causes their portrait to diverge (dotted blue line) or converge (dotted red line). Such a fundamental operation allows us to reach every possible point (not as an arrow but actually as a single point) in the plane, considering every possible linear combination that we can obtain from the two-dimensional quantities.

Based on these functions, the element $\vec{v}$ then contains a more detailed algorithm of how the individual's new position was implemented in which a subset holds that (i) the zero vector belongs to $\vec{v}$, and (ii) vectors and any multiplication of scalars are also in $\vec{v}$.

Such a mechanism is dependent on quantity as follows.

First of all, the new position dynamic is augmented by the designated trade-off value (*2*) of the individual velocity/group heading [IGT: individual's velocity up ($\vec{v}_i$) resulting in a loss of group heading ($\vec{v}_{avg}$)] as follows:

$$\vec{v}_{iinew} = (1 - \|k\|) * \vec{v}_i + \|k\| * \vec{v}_{avg}, \qquad k \in [0,1]$$

where $\vec{v}_i$ is the velocity of each individual, $\vec{v}_{avg}$ is the average velocity about group heading, and k is a scalar that controls their trade-off. For example, the product of a $\vec{v}_i$ by a scalar k is a vector $\|k\|\vec{v}_i$ with magnitude $\|k\|$ times the magnitude of $\|v_i\|$ and with direction $\vec{d}_i$, the same as or opposite to that of $\vec{v}_i$, according to whether k is positive or negative (if $k = 0$ [null vector], then $\|k\|\vec{v}$ has zero magnitude and no specific direction). This means that every new position vector will be a combination multiplied by the scalar $[(c_1 = 1 - \|k\|), (c_2 = \|k\|)]$:

$$c_1\begin{bmatrix}4\\1\end{bmatrix} \pm c_2\begin{bmatrix}1\\2\end{bmatrix} = \begin{bmatrix}c_1 4\\c_1 1\end{bmatrix} \pm \begin{bmatrix}c_2 2\\c_2 2\end{bmatrix} = \begin{bmatrix}x_1\\x_2\end{bmatrix}$$

$$4 * c_1 \pm 1 * c_2 = x_1, \qquad 1 * c_1 \pm 2 * c_2 = x_2$$

and the mechanisms obtain:

$$f(\|v_i\|) < \|v_{avg}\| \quad \rightarrow$$

$$\vdots$$

$$f(\|v_i\|) > \|v_{avg}\| \quad \rightarrow$$

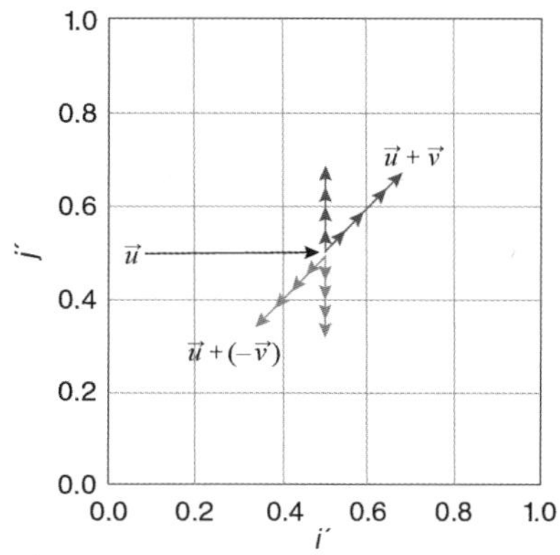

Here the different angles of the arrows are from $\vec{v}_{inew}$ and the spaces inside the arrows are from the multiplied scalar ($\|k\|,\ k \in [0,1]$). This gives us a global view so that we can conceptualize the list of quantities in a visual way and thereby simplify and clarify basic operational patterns.

The model then includes another characteristic for dealing with more or less distinct patterns of behavior. In social practice, just as individuals are more likely to change their decisions depending on the influences surrounding them (Marchal, 1975), so the individual's new quantity ($\vec{v}$) holds its network characteristics (Chen & Billings, 1992) as given:

$$\vec{v}_{iiinew} = \left((1 - \|k\|) * \vec{v}_i + \|k\| * \vec{v}_{avg}\right) + \vec{v}_s, \qquad \vec{v}_s = \|v_s\| * \vec{d}_s$$

where $\vec{v}_s$ is a vector with length $\|v_s\|$ and direction $\vec{d}_s$ as a function of the network density (*ND*). The network density arises from its social ties (Chen & Billings, 1992) based on the nodes (*N* = initial average exploration of the model) and calculated by its actual connection (*AC*) with the potential connection (*PC*) of the network.

$$\|v_s\| = \mathrm{ND} = AC/PC, \qquad AC = (2 * t)/N, \qquad PC = N(N-1)/2$$

Here the network density ($\|v_s\|$) describes the potential connections in a network that are actual connections (AC/PC). The potential connection ($PC = N(N-1)/2$) is a connection that could potentially exist between two individuals regardless of whether or not it actually does. *This* individual could know *that* individual; *this* object could connect to *that* object. Whether or not they do connect is irrelevant when we are talking about a potential connection.

By contrast, an actual connection ($AC = (2 * t)/N$) is one that actually exists. *This* individual does know *that* individual; *this* object is connected to *that* object. For example, in the living room of a house, actual connections may represent 100% of all potential relationships. In contrast, on a public bus, the number of actual connections is likely to be quite low relative to all potential relationships, because the number of people who actually know each other on the bus (actual connections) must be low (t =social ties).

Moreover, even on a public bus, any individual may connect to another even if it knows nothing about the other individuals that it actually connects to (if the other offers the highest payoff). In a house, too, anyone can bring a guest into their living room. This may make others modify the actual connection. These small linear contributions to their dynamics and this structural instability can be interpreted as network characteristics being influenced by the exploration rate (k' = scalar), which corresponds to a mutation term in genetics given as

$$\vec{v}_{iiinew} = \left((1-\|k\|) * \vec{v}_i + \|k\| * \vec{v}_{avg}\right) + \vec{v}_{ss}, \qquad \vec{v}_{ss} = \|v_{ss}\| * \vec{d}_{ss}$$
$$\|v_{ss}\| = [\|k'\|(1-\|v_s\|) - 2\|k'\|\|v_s\|], \qquad k' \in [0,1]$$

where k' controls how fast the transition function propagates in the network, and the new position vector considers its network density as another quantity $[(c_3 = \|k'\|)]$. For example, as previously:

$$c_1\begin{bmatrix}4\\1\end{bmatrix} \pm c_2\begin{bmatrix}1\\2\end{bmatrix} = \begin{bmatrix}c_1 4\\c_1 1\end{bmatrix} \pm \begin{bmatrix}c_2 2\\c_2 2\end{bmatrix} = \begin{bmatrix}x_1\\x_2\end{bmatrix}$$
$$4 * c_1 \pm 1 * c_2 = x_1, \qquad 1 * c_1 \pm 2 * c_2 = x_2$$

The new position is then

$$\begin{bmatrix}x_1\\x_2\end{bmatrix} \pm c_3\begin{bmatrix}2\\1\end{bmatrix} = \begin{bmatrix}x_1\\x_2\end{bmatrix} \pm \begin{bmatrix}c_3 2\\c_3 1\end{bmatrix} = \begin{bmatrix}x_{11}\\x_{22}\end{bmatrix}$$
$$x_1 \pm 2 * c_3 = x_{11}, \qquad x_2 \pm 1 * c_3 = x_{22}$$

In the presence of network density, the system settles down into a state with a more pronounced increasing (or decreasing) mutation rate in every update step, and it yields the following:

$f(\|v_i\|) < \|v_{avg}\| \;\rightarrow$

$\vdots$

$f(\|v_i\|) > \|v_{avg}\| \;\rightarrow$

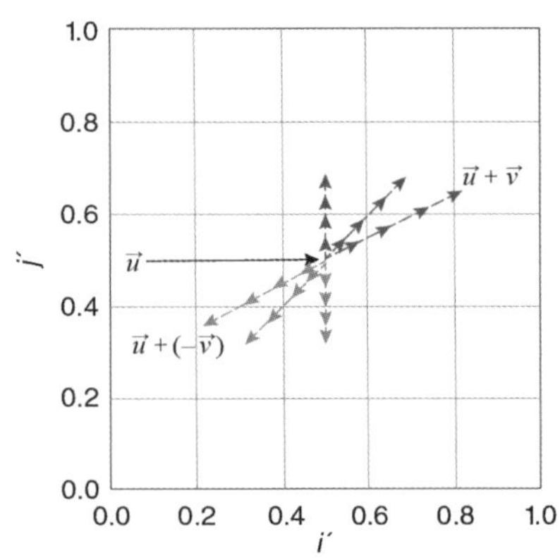

With these implementations, instead of the widespread extension underlying the combination, the model proposes to adopt an existing possible interconnected relationship between the network and its movement characteristics. Let us think about simple interdependency between the two components (the trade-off between individual velocity and group heading as an internal, network density multiplied by mutation as an external). If the individual's tendency is very remote from the group's purpose, its mutation in the system will not propagate to the individuals, and vice versa. For this application, the new position mechanism assumes that the social network characteristics ($\|v_{ss}\|$ = scalar) are a denominator applied by the index of difficulty (id = scalar) as a numerator.

$$\vec{v}_{iiiinew} = \left[(1 - \|k\|) * \vec{v}_i + \|k\| * \vec{v}_{avg}\right] * (\|v_{id}\|/\|v_{ss}\|)$$
$$\|v_{id}\| = \frac{2D}{W}, \qquad \|v_{ss}\| = [\|k'\|(1 - \|v_s\|) - 2\|k'\|\|v_s\|]$$

Here $\|v_{id}\|$ is the scalar as a function of the ratio between the two objects (2D = size of the trade-off $[k]$ between two objects about $\vec{v}_i$ and $\vec{v}_{avg}$) divided by the width of the object (W = arbitrary value corresponding to the individual's size or reputation; e.g., when $\|k\|$ is 0.1 applied to $(1 - \|k\|) * \vec{v}_i + \|k\| * \vec{v}_{avg}$, the 2D becomes large [i.e., 0.8]; in contrast, when $\|k\|$ is 0.4, the 2D becomes small [i.e., 0.2]).

This leads to a simple interpretation linking the vector as follows:

$$c_4 \begin{bmatrix} x_1 \\ x_2 \end{bmatrix} = \begin{bmatrix} c_4 x_1 \\ c_4 x_2 \end{bmatrix} = \begin{bmatrix} x_{111} \\ x_{222} \end{bmatrix} = \vec{v}_{iiiinew}, \qquad \begin{bmatrix} x_1 \\ x_2 \end{bmatrix} = \vec{v}_{iinew}, \qquad c_4 = \|v_{id}\|/\|v_{ss}\|$$

and the mechanisms yields

$$f(\|v_i\|) < \|v_{avg}\| \quad \rightarrow$$
$$\vdots$$
$$f(\|v_i\|) > \|v_{avg}\| \quad \rightarrow$$

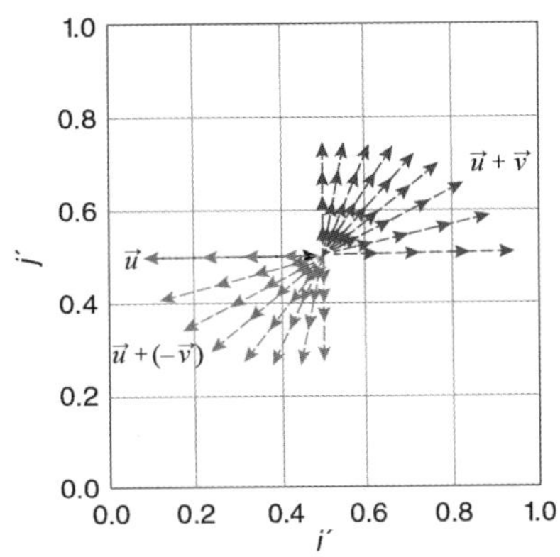

Notice that the combination of these arrows refers to this model's fundamental feature. The characteristics of every rule and process that are applied (or will be applied) in this model must be within this functional dynamic. This provides an excellent way of conceptualizing many lists of individuals in a visual way, which can clarify patterns in mechanisms. It also shows a global view of what certain operations do to describe how an individual is being manipulated in space using numbers that can be run through a computation. The model now considers an adoption probability that is given by an estimate of $\vec{v}_{inew}$ by the individuals. Indeed, although each individual may not know the exact value of the trait that has adopted the other's $\vec{v}_{inew}$, this model yields that they can estimate the value at every schedule of each generation via the comparison given.

$$p = [1 + e^{-\omega\Delta\pi}]^{-1}, \qquad \pi_r - \pi_f = \Delta\pi\big|_{\pi_r = role\ model}$$

Here p is the probability acceptance of the role model for imitation, π_f is a payoff (velocity) of the focal individual, π_r is a payoff of the role individual, e denotes the exponential, and ω is the intensity of the selection ($\omega < 1$ = weak selection, $\omega \to \infty$ = strong selection). The focal individual imitates the strategy of the nearby role individual, comparing its new position vector (large $\Delta\pi$ = velocity difference large, small $\Delta\pi$ = velocity difference small), and then the focal individual chooses to imitate the strategy of the role individual. Given that the individuals naturally navigate together and maintain a certain distance as their inherited survival strategies, the equation contains a detailed algorithm of how the individual's new payoff (velocity) was implemented.

[Table R3.1.1] Variables of Potential Behaviors

Match	neighbors' average heading
Coherence	toward the center of the neighbors
Movement	individual displacement

[Table R3.1.2] Model Parameters

Parameters for defaults	(Range of the value)	
Number of individuals	M	1000
Separation	D	1–10
Vision (exploration)	E	1–10
Velocity	V	1–10
Parameters for internal (individual movement) characteristics		
Individual velocity (from the velocity)	v_i	1–10
Group velocity (average individual velocity)	v_{avg}	1–10
Individual/group velocity trade-off (systemic scalar)	k	0.1–0.9
Index of difficulty (from the velocity trade-off)	id	0.1–0.9
Parameters for external (social network) characteristics		
Node (from the number of individuals)	n	1–10
Social ties (systemic scalar)	t	0.1–0.9
Mutation rate (systemic scalar)	k'	0.1–0.9
Selection intensity	ω	1–10

Results: The model provides a natural description of a pattern of behavior and allows us to understand a realistic adaptation incorporating behavioral algorithms with social dynamics. The mechanisms characterized in the agenda show that the model has the capacity to produce three types of factors.

First, the behavioral pattern is the result of the applied individual components; this comes not only from the initial conditions of the autonomous agents but also from the fact that they are interconnected. Second, the range of different combinations of internal and external states plays a part in the rapid propagation in the system (see Figure R3.1.1). Third, however, when the interconnected relation between the internal trait underlying its external trait is applied, the widespread heterogeneity of the mechanisms can abstract the repertoire of displayed behaviors.

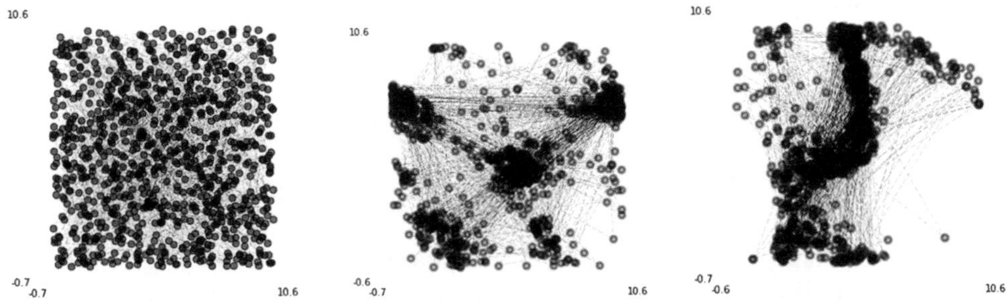

Figure R3.1.1. Behavioral dynamics underlying social characteristics. Following the simulation, the left plot shows a displacement that separates individuals with a relative position structure controlled by the initial setting. This implies that although the pattern of individual behavior depends on a localized view of the initial conditions, a slight change in individual characteristics (IGT: individual's velocity up $[\vec{v}_i]$ resulting in a different group heading $[\vec{v}_{avg}]$) underlying its social influence (ND: calculated from their social ties $[k = St]$ multiplied by mutation rate $[u]$) has a remarkably diverging (left) or converging (middle and right) effect on its displacement. The blue dots represent their position in an x, y coordinate plane, and the red lines denote links.

The agents' interactions in this abstract setting are primarily heterogeneous. As the topology of the interaction traits can lead to significant deviations from the predicted pattern of behavior, it may generate various effects that mimic the behavior of real individuals in social dynamics. At the points where these individuals interact, sensible decisions occur, in line with an empirical study showing that individuals learn how to keep relative velocity as a key factor for homogeneity (McLeod & Dienes, 1996).

Reading List 3.2: Applied Systems (Network-Agent Dynamic) Research with Systemic Risk

Purpose: We show how to construct a large variety of behaviors by applying a simple algorithm to networked agents, which could, conceivably, offer a straightforward way out of the complexity. **Method:** The model starts with the probability that systemic risk spreads. Even in a very random social structure, the propagation of risk is guaranteed by an arbitrary network property of a set of elements. Despite intensive systemic risk, the potential for the absence of failure could also be driven by a strong investment in protection through a heuristically evolved protection level. **Results:** It is very interesting to discover that many applications are still seeking the mechanisms through which networked individuals build many of these protection processes or mechanisms based on fitness due to evolutionary drift (Park, 2020b).

Background: Networks in any social system are dynamic entities (Vespignani, 2018), and in this sense the practical information of networks is continuously evolving (Pacheco et al., 2006). Thus, the primary purpose of this study is to examine mechanisms for the evolutionary origin of bias and make a heuristic assessment of protection against systemic risks in a contagious network. We established a modeling framework that can account for quantitative measurement in agented networks, which allows us to explore on a macro- and microscale how protecting potential affects risk potential. The mechanism tests what we can clearly state for different values of probability and how protection can be drawn by a set of entities against a cascading failure. To reach a better assessment of risk and how to reduce it, this model enables us not only to directly observe the spread of failure in agented network industries but also to understand how evolutionary heuristics protect against that spread (Helbing, 2013).

Methods (Model): We put forward the proposition regarding emerging protection and its potential far-reaching impacts on agented networks under the basic operating principle. We therefore suggest the following steps for modeling. First, (*network properties*) when constructing the basic data structure, the functionalities of the mechanism begins with a specific undirected relationship between agents. Next, (*primary risk influence*) when a parameter is used to evaluate the impact of risk for the networked agents, the influence of primary risk is estimated along the structure as a general failure property. Finally, (*protection against systemic risk*) a protection dynamic is embedded that emphasizes the role of payoff, failure, and strategy dynamics.

Network properties: Individuals in the model are considered vertices (fundamental elements drawn as nodes), and a set of two elements drawn as a line connecting two vertices (the lines are called edges) in the graph [usually controlled by (n, p)]. Networks have some additional properties in that edges can have a direction, which means that the relationship between two nodes only applies one way and not the other. In the present model, however, we use an undirected network, which has edges with no sense of direction (because if you have a network of individuals and edges are drawn between two individuals who have met before, directed edges might be unnecessary). Thus, the structure created (the network) is undirected and starts with the adding of edges between pairs of nodes one at a time randomly (Erdös-Renyi graph, 1959).

Notice that the graph is shown to us in any random way in which the computer generates the visuals. If we were to run it again, we would get a different picture, but no matter how we run it, the relationship between the vertices (nodes) and edges (line) creating its degrees would be the same.

$$A = \begin{bmatrix} a_{11} & a_{12} & \cdots & a_{1n} \\ a_{21} & a_{22} & \cdots & a_{2n} \\ \cdots & \cdots & \cdots & \cdots \\ a_{m1} & a_{m2} & \cdots & a_{mn} \end{bmatrix}$$

The resulting $m \times n$ matrix is obtained by $[A = G(n,p)]$. Any node (n) can be randomly exposed to the other nodes, which creates random connections. Given the collection of nodes influenced by the connection probability $[p \in (0,1)]$, the model asks what the distribution of the connection in the network is (probability of degree).

Primary risk influence: Next, to observe the process of propagation, the model uses an array (vector) as a probability of failure $[p \in (0,1)]$ with given initially influenced nodes $(1 \leq j \leq N)$ noted merely by (p_j). Each node can be in one of two states: not failed or failed. All nodes are initially without failure.

Extending the above output given the failure dynamics becomes the following:

$$A = \begin{bmatrix} a_{11} & a_{12} & \cdots & a_{1n} \\ a_{21} & a_{22} & \cdots & a_{2n} \\ \cdots & \cdots & \cdots & \cdots \\ a_{m1} & a_{m2} & \cdots & a_{mn} \end{bmatrix} \begin{bmatrix} \vec{a}_1 \\ \vec{a}_2 \\ \cdots \\ \vec{a}_n \end{bmatrix} = \begin{bmatrix} a_{11}\vec{a}_1 & + & a_{12}\vec{a}_2 & + & \cdots & a_{1n}\vec{a}_n \\ a_{21}\vec{a}_1 & + & a_{22}\vec{a}_2 & + & \cdots & a_{2n}\vec{a}_n \\ \cdots & & \cdots & & \cdots & \cdots \\ a_{m1}\vec{a}_1 & + & a_{m2}\vec{a}_2 & + & \cdots & a_{mn}\vec{a}_n \end{bmatrix} = B\vec{a} = \begin{bmatrix} B\vec{a}_1 \\ B\vec{a}_2 \\ \cdots \\ B\vec{a}_n \end{bmatrix}$$

i.e., States Matrix

[[1., 1., 1., 1., 0., 0., 1., 1., 1., 1.],
[0., 0., 0., 1., 1., 1., 1., 1., 1., 1.],
[0., 0., 0., 0., 0., 0., 0., 0., 0., 0.],
[0., 0., 0., 0., 1., 1., 1., 1., 0., 0.],
[0., 0., 0., 0., 1., 1., 1., 1., 1., 1.],
[0., 0., 0., 1., 1., 1., 1., 1., 1., 0.],
[0., 0., 0., 0., 0., 0., 0., 0., 0., 0.],
[0., 0., 1., 1., 1., 1., 1., 1., 1., 1.],
[0., 0., 0., 0., 0., 0., 0., 0., 0., 1.],
[0., 0., 0., 0., 0., 0., 1., 1., 1., 1.]]

Notice that the matrix is denoted by $B\vec{a}$ instead of A be representation because it no longer follows the adjacency matrix representation. $B\vec{a}$ is still going to be labeling our rows and columns via 1 and 0. The key difference is the possibility of showing the state of each node (m: 1 = failure, 0 = absence of failure) according to time steps (n).

Impose protection against the systemic risk: Along with the basic intuition mentioned above, protection dynamics is applied. First of all, we break the program into sub-dynamics (*payoff*, *failure*, and *strategy*). The result of each sub-dynamics is saved. These sub-dynamics are trivial problems that add complexity to the dynamics.

To implement each part, we use simple equations. These equations combine some previously computed variables and newly added or computed variables. In the following example, we use the values already stored in the table to compute new variables. This technique is often called *memorization*. For example:

a → stored in the table
b → stored in the table
a + b = c → look up a, b → compute c
d → stored in the table
a + d = e → look up a, d → compute e

Payoff dynamics: One agent is associated with each node and is characterized by its capital and strategy as below. In each time step, each agent receives one unit of payoff, which is added to its capital c, of which fractions f_m and f_p are spent on maintenance and protection, respectively, resulting in the updated capital $1 + (1 - f_m - f_p)c$. We use an element-wise computation by using arrays for the vectorization ($\vec{v}$) instead of using a loop.

$$f_p = \vec{v}_i = \begin{bmatrix} \vec{v}_{i1} \\ \vec{v}_{i2} \\ \cdots \\ \vec{v}_{in} \end{bmatrix}, f_m = \vec{v}_{ii} \begin{bmatrix} \vec{v}_{ii1} \\ \vec{v}_{ii2} \\ \cdots \\ \vec{v}_{iin} \end{bmatrix}, \vec{v}_{ii} + (-)\vec{v}_i = \begin{bmatrix} \vec{v}_{ii1} + (-\vec{v}_{i1}) \\ \vec{v}_{ii2} + (-\vec{v}_{i2}) \\ \cdots \\ \vec{v}_{iin} + (-\vec{v}_{in}) \end{bmatrix} = \begin{bmatrix} \vec{v}_{iii1} \\ \vec{v}_{iii2} \\ \cdots \\ \vec{v}_{iiin} \end{bmatrix},$$

$$c \begin{bmatrix} \vec{v}_{iii1} \\ \vec{v}_{iii2} \\ \cdots \\ \vec{v}_{iiin} \end{bmatrix} = \begin{bmatrix} c\vec{v}_{iii1} \\ c\vec{v}_{iii2} \\ \cdots \\ c\vec{v}_{iiin} \end{bmatrix} = \begin{bmatrix} \vec{v}_1 \\ \vec{v}_2 \\ \cdots \\ \vec{v}_n \end{bmatrix}$$

Because the previous initial random network property was

$$A \begin{bmatrix} a_{11} & a_{12} & \cdots & a_{1n} \\ a_{21} & a_{22} & \cdots & a_{2n} \\ \cdots & \cdots & \cdots & \cdots \\ a_{m1} & a_{m2} & \cdots & a_{mn} \end{bmatrix} + kA \begin{bmatrix} ka_{11} & ka_{12} & \cdots & ka_{1n} \\ ka_{21} & ka_{22} & \cdots & ka_{2n} \\ \cdots & \cdots & \cdots & \cdots \\ ka_{m1} & ka_{m2} & \cdots & ka_{mn} \end{bmatrix} = B, k \in [0,1]$$

the applied output given payoff dynamics becomes

$$B = \begin{bmatrix} b_{11} & b_{12} & \cdots & b_{1n} \\ b_{21} & b_{22} & \cdots & b_{2n} \\ \cdots & \cdots & \cdots & \cdots \\ b_{m1} & b_{m2} & \cdots & b_{mn} \end{bmatrix} \begin{bmatrix} \vec{v}_1 \\ \vec{v}_2 \\ \cdots \\ \vec{v}_n \end{bmatrix} = \begin{bmatrix} b_{11}\vec{v}_1 & + & b_{12}\vec{v}_2 & + & \cdots & b_{1n}\vec{v}_n \\ b_{21}\vec{v}_1 & + & b_{22}\vec{v}_2 & + & \cdots & b_{2n}\vec{v}_n \\ \cdots & & \cdots & & \cdots & \cdots \\ b_{m1}\vec{v}_1 & + & b_{m2}\vec{v}_2 & + & \cdots & b_{mn}\vec{v}_n \end{bmatrix} = B\vec{v} = \begin{bmatrix} B\vec{v}_1 \\ B\vec{v}_2 \\ \cdots \\ B\vec{v}_n \end{bmatrix}$$

where the vector ($\vec{v}$ = payoff_dynamics) components are equal to matrix B. Simply put, this product is equal to $B\vec{v}$.

Failure dynamics: According to the concise model description (operating principle), a failure potential can originate at each node with probability $p_n \in [0,1]$, and it also propagates along each link with probability $p_l \in [0,1]$ in each time step.

A failure potential turns into a failure with probability $1 - p_p$, depending on an agent's investment in protection; a possible choice is

$$p_p = p_{p,max}/(1 + c_{p,1/2}/(f_p c)$$

Here protection (p_p) is equal to the applied (saturation) function. $p_{p,max}$ is a designated protection maximum, $c_{p,1/2}$ denotes an allocated reference point, and $f_p c$ represents an evolutionary protection level multiplied by the updated capital.

$$p_p = \vec{u}_i = \begin{bmatrix} \vec{u}_{i1} \\ \vec{u}_{i2} \\ \cdots \\ \vec{w}_{in} \end{bmatrix}, \qquad f_p = \vec{v}_i \begin{bmatrix} \vec{v}_{i1} \\ \vec{v}_{i2} \\ \cdots \\ \vec{v}_{in} \end{bmatrix}, \qquad f_p c = c \begin{bmatrix} c\vec{v}_{i1} \\ c\vec{v}_{i2} \\ \cdots \\ c\vec{v}_{in} \end{bmatrix}$$

The applied output given failure dynamics becomes

$$B = \begin{bmatrix} b_{11} & b_{12} & \cdots & b_{1n} \\ b_{21} & b_{22} & \cdots & b_{2n} \\ \cdots & \cdots & \cdots & \cdots \\ b_{m1} & b_{m2} & \cdots & b_{mn} \end{bmatrix} \begin{bmatrix} \vec{u}_1 \\ \vec{u}_2 \\ \cdots \\ \vec{u}_n \end{bmatrix} = \begin{bmatrix} b_{11}\vec{u}_1 & + & b_{12}\vec{u}_2 & + & \cdots & b_{1n}\vec{u}_n \\ b_{21}\vec{u}_1 & + & b_{22}\vec{u}_2 & + & \cdots & b_{2n}\vec{u}_n \\ \cdots & & \cdots & & \cdots & \cdots \\ b_{m1}\vec{u}_1 & + & b_{m2}\vec{u}_2 & + & \cdots & b_{mn}\vec{u}_n \end{bmatrix} = B\vec{u} = \begin{bmatrix} B\vec{u}_1 \\ B\vec{u}_2 \\ \cdots \\ B\vec{u}_n \end{bmatrix}$$

where the vector ($\vec{u}$ = failure_dynamics) components are equal to matrix B. Simply put, this product is equal to $B\vec{u}$. In this section, a prewritten function (Erdös-Renyi) is used to make a short iterate 1 = D array for vectorization ($\vec{u}$) instead of using the adjacency matrix directly. This substitution makes the loop shorter. Failure lasts for one time step and causes the loss of an agent's capital.

Strategy dynamics: Each agent chooses its protection level according to the heuristics $f_p = f_{p0} + f_{p1}C$, truncated to the interval $(0,1 - f_m)$.

$$\vec{v} \to \vec{f} \to f(\vec{v}), \quad f(\vec{v}) = \begin{cases} 0 < f(\vec{v}) < 0.9, & f_m = 0.9 \\ 0 < f(\vec{v}) < 0.1, & f_m = 0.1' \end{cases} \qquad \vec{v}|_{f_p = f_{p0} + f_{p1}C}$$

For the initialization of the strategy values, two arrays are added for vectorization

$$(f_{p0} = \vec{w}_i), \qquad (f_{p1}C = C\vec{w}_i)$$

$$f_{p0} = \vec{w} = \begin{bmatrix} \vec{w}_1 \\ \vec{w}_2 \\ \cdots \\ \vec{w}_n \end{bmatrix}, \qquad f_{p1}C = C\vec{w} \begin{bmatrix} C\vec{w}_1 \\ C\vec{w}_2 \\ \cdots \\ C\vec{w}_n \end{bmatrix}, \quad C \in [0,1]$$

$\vec{w}_{i1}$ is a vectorization as the designated strategy of (f_{p0}), and $\vec{w}_{ii}$ is a vectorization as the designated strategy of (f_{p1}) multiplied by eigenvector centrality from the Erdös-Renyi graph (C), which is a measure of the centrality of the agent's node normalized to the interval $(0,1)$.

$$f_{p0} = \vec{w}_i = \begin{bmatrix} \vec{w}_{i1} \\ \vec{w}_{i2} \\ \cdots \\ \vec{w}_{in} \end{bmatrix}, \quad f_{p1}C = C\vec{w}_i \begin{bmatrix} C\vec{w}_{i1} \\ C\vec{w}_{i2} \\ \cdots \\ C\vec{w}_{in} \end{bmatrix}, \quad \vec{w}_i + C\vec{w}_i = \begin{bmatrix} \vec{w}_{i1} + C\vec{w}_{i1} \\ \vec{w}_{i2} + C\vec{w}_{i2} \\ \cdots \\ \vec{w}_{in} + C\vec{w}_{in} \end{bmatrix} = f_p = \begin{bmatrix} \vec{v}_{i1} \\ \vec{v}_{i2} \\ \cdots \\ \vec{v}_{in} \end{bmatrix}$$

Eigenvector centrality for node i is ($Ax = \lambda x$), where A is the matrix of the network with eigenvalue λ. The strategy values f_{p0} and f_{p1} evolve through social learning and strategy exploration as follows.

In each time step, each agent with probability $p_r \in [0,1]$ randomly chooses another agent as a role model and imitates that agent's strategy values with probability

$$p_i = [1 + e^{-\omega\Delta\pi}]^{-1}, \qquad \pi_r - \pi_f = \Delta\pi\big|_{\pi_r = role\ model}$$

where p_i is the probability of acceptance of the role model for imitation, π_f is a capital of the focal individual, π_r is a capital of the role individual, e denotes the exponential, and ω is the intensity of the selection

$$\omega < 1 = \text{weak selection}, \ \omega \to \infty = \text{strong selection}$$

The focal individual imitates the strategy of the nearby role individual, comparing its new capital (large $\Delta\pi$ = large difference in capital, small $\Delta\pi$ = small difference in capital), and then the focal individual chooses to imitate the strategy of the role individual.

Regarding this imitation part, a temporary matrix is used to avoid changing and using 1 matrix in the loop.

Finally, in each time step, each agent with probability $p_e \in [0,1]$ randomly chooses one of its two strategy values and alters it by a normally distributed increment with mean 0 and standard deviation as follows:

$$\sigma_e\ [f(x|\mu,\sigma^2) = \frac{1}{\sqrt{2\pi\sigma^2}} exp^{-\frac{(x-\mu)^2}{2\sigma^2}}\Bigg|\ x = individual\ capital,\ \mu \in R = mean(location),\ \sigma^2 > 0 = variance(squared\ scale)$$

The broader agenda of this supplement is to show the computerized process behind the fundamental modeling mechanisms used. The tables (Tables R3.2.1, R3.2.2, R3.2.3) in the artificially modeled structure account for the detail of the rules and procedures and also how to apply the mechanism.

[Table R3.2.1] Evolutionary Part (Four Parameters)

Imitation probability	p_r	$\in (0,1)$
Selection intensity	s	$\in (1,100)$
Exploration probability	p_e	$\in (0,1)$
Normally distributed increment	σ	$\in (0,1)$

[Table R3.2.2] Non-Evolutionary Part (Seven Parameters)

Number of individuals (nodes)	n	$\in (1,100)$
Connection probability	p	$\in (0,1)$
Maintenance	f_{m}	$\in (0,1)$
Propagation probability at each node	p_n	$\in (0,1)$
Propagation probability through each link	p_l	$\in (0,1)$
Protection maximum	$p_{p,max}$	$\in (0,1)$
Reference point	$c_{\mathrm{p},1/2}$	$\in (0,1)$

[Table R3.2.3] Time-Dependent Part (Four Parameters)

Time periods	t	1–10,000
Recovery rate	r_t	1–10
Recovery time delay	t_r	1–10
Realization	tt	1–100

Results: In line with this proposition, a protection dynamic is applied against the risk of failure. The model allows an agent to make a costly investment in protection. Note that we assume the systemic risk as a failure potential from the dynamics turns into a failure with probability $1 - p_{\mathrm{p}}$ depending on an agent's investment in protection; $p_{\mathrm{p}} = p_{\mathrm{p,max}}/(1 + c_{\mathrm{p},1/2}/(f_{\mathrm{p}}c))$ when a failure lasts for one time step.

In Figure R3.2.1, as a basic modeling framework, the probabilities of these plausible scenarios can be proved as follows.

First, scenario A's failure potential $1 - p_{\mathrm{p}}$ becomes 0.527 because protection p_{p} is 0.473 based on the possible choice p_{p} is $p_{\mathrm{p,max}}/(1 + c_{\mathrm{p},1/2}/(f_{\mathrm{p}}c)) = 1/(1 + 1/(0.9 * 1))$ with given parameter values ($p_{p,max} = 1$, $c_{\mathrm{p},1/2} = 1$, $p_r, p_e = 0.9, C = 1$).

Second, scenario B's failure potential $1 - p_{\mathrm{p}}$ becomes 0.953 because protection p_{p} is 0.047 based on the possible choice p_{p} is $p_{\mathrm{p,max}}/(1 + c_{\mathrm{p},1/2}/(f_{\mathrm{p}}c)) = 0.1/(1 + 1/(0.9 * 1))$ with given parameter values ($p_{p,max} = 0.1$, $c_{\mathrm{p},1/2} = 1$, $p_r, p_e = 0.9, C = 1$). As can be seen in the right-hand side of the upper plot, this scenario obtains the worst result, corresponding to all failure as time goes by.

Third, scenario C's failure potential $1 - p_{\mathrm{p}}$ becomes 0.95 because protection p_{p} is 0.05 based on the possible choice p_{p} is $p_{\mathrm{p,max}}/(1 + c_{\mathrm{p},1/2}/(f_{\mathrm{p}}c)) = 0.1/(1 + 0.1/(0.1 * 1))$ with given parameter values ($p_{p,max} = 0.1$, $c_{\mathrm{p},1/2} = 0.1$, $p_r, p_e = 0.1, C = 1$).

Finally, scenario D's failure potential $1 - p_{\mathrm{p}}$ becomes 0.5 because protection p_{p} is 0.5 based on the possible choice p_{p} is $p_{\mathrm{p,max}}/(1 + c_{\mathrm{p},1/2}/(f_{\mathrm{p}}c)) = 1/(1 + 0.1/(0.1 * 1))$ with given parameter values ($p_{p,max} = 1,\ c_{\mathrm{p},1/2} = 0.1, p_r, p_e = 0.1, C = 1$). As can be seen in the left-hand side of the bottom plot, this scenario obtains the best result, corresponding to the absence of failure.

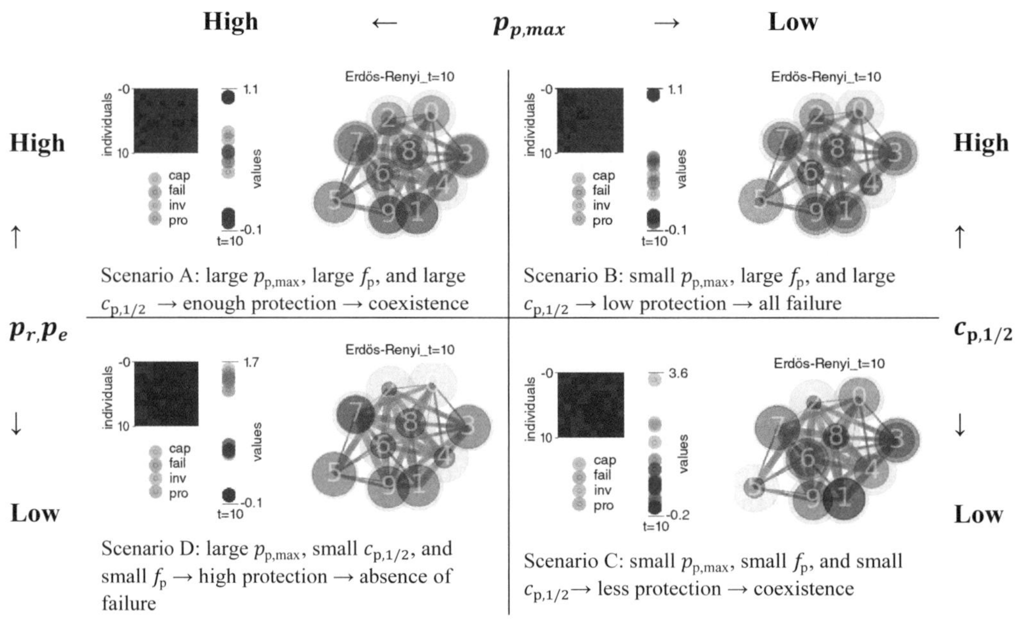

Figure R3.2.1: Protection dynamics against systemic risk. In each section, plots at left show the matrix: horizontal axis = time step from 1 to 10, vertical axis = individuals (10), failure state = failure (dark), failure state = absence of failure (light). The plots in the middle show each individual's parameter values at the time step t = 10: cap = capital (c), fail = failure, inv = investment (f_{p}), pro = protection potential (p_{p}). The plots at right represent individuals' dynamics within a random network: node number = random label of each node, line width = eigenvector centrality, node color = state (failure [dark] ←→ [light] absence of failure, green = protection potential, yellow = initial structure of the state without failure and protection). Initialized parameters of the simulations are nodes n = 10, connection p = 0.9, p = 1, f_{p0} = 0.4, f_{p1} = 0.5, f_{m} = 0.1, s = 1, μ = 0.0, σ = 0.1, p_n = 0.1, p_l = 0.3, t = 10.

Notice how it all came about and what the results mean to us. The model mechanism assumes that any individual that has a failure will devastate the whole system. Every one of them is in the situation of being what is called *systemically essential*, and all of them will systemically cascade. Let us specify what the simulation mechanism would say to reflect this point.

First, there is contagion. If one individual (= node) fails because of its interconnected relationship in a network (e.g., an industry), its failure will have an impact on other individuals in the industry. There is thus a contagion type of effect (reflected by links). Second, there is concentration. Even if only one player or a small entity with massive potential is more extensive than all the others, there will be significant concentration in the industry (reflected by the eigenvector centrality). Third, there is context, namely, what usually happens to an individual's function in an ordinarily operable environment; if the circumstance takes a turn for the worse, perhaps so too will all other individuals, as all of these individuals and institutions are in context (reflected by social learning and imitation). Any individual in the industry functions in the same way. It also takes the same kinds of risks as others take. Thus, if one goes down, they all go down.

We observe a significant correlation between capital and failure at both the micro- and macroscale, which is another critical fact. Note that if an individual without protection goes in the direction of eliminating risk through investment, every single individual in that context will fail.

References

Part 1. Applied Statistics Analysis

Abrams, D. M. & Strogatz, S. H. (2003). Modelling the dynamics of language death. *Nature, 424*(6951), 900-900.

Adleman, N. E., Menon, V., Blasey, C. M., White, C. D., Warsofsky, I. S., Glover, G. H., & Reiss, A. L. (2002). A developmental fMRI study of the stroop color-word task. *Neuroimage, 16*, 61-75.

Agresti, A. & Finlay, B. (2009). *Statistical methods for the social sciences* (No. 300.72 A3).

Agudo, A., Ahrens, W., Benhamou, E., Benhamou, S., Boffetta, P., Darby, S. C., et al. (2000). Lung cancer and cigarette smoking in women: A multicenter case-control study in Europe. *International journal of cancer*, *88*(5), 820-827.

Amenta, E. (2003). *Comparative historical analysis in the social sciences*. Cambridge University Press.

Anfara Jr. V. A., Brown, K. M., & Mangione, T. L. (2002). Qualitative analysis on stage: Making the research process more public. *Educational researcher*, *31*(7), 28-38.

Araújo, M. B., & New, M. (2007). Ensemble forecasting of species distributions. *Trends in ecology & evolution*, *22*(1), 42-47.

Asgari, S. & Baptista Nunes, J. M. B. (2011). Experimental and quasi-experimental research in information systems.

Ashby, W. R. (1947). Principles of the self-organizing dynamic system. *The Journal of general psychology*, *37*, 125-128.

Berkson, J. (1944). Application of the logistic function to bio-assay. *Journal of the American statistical association*, *39*(227), 357-365.

Billor, N., Hadi, A. S., & Velleman, P. F. (2000). BACON: Blocked adaptive computationnally efficient outlier nominators. *Computational statistics & data analysis*, *34*(3), 279-298.

Bland, J. M. & Altman, D. G. (1996). Statistics notes: measurement error. *Bmj*, *312*(7047), 1654.

Blaxter, L., Hughes, C., & Tight, M. (2010). *How to research*. McGraw-Hill Education.

Box, G. E. (1976). Science and statistics. *Journal of the American statistical association*, *71*(356), 791-799.

Bruce, P., Bruce, A., & Gedeck, P. (2020). *Practical statistics for data scientists: 50+ essential concepts using R and Python*. O'Reilly Media.

Buelow, J. M. (2006). Identifying a researchable problem. *Clinical nurse specialist*, *20*(4), 175-176.

Canning, P. M., Courage, M. L., & Frizzell, L. M. (2004). Prevalence of overweight and obesity in a provincial population of Canadian preschool children. *Cmaj*, *171*(3), 240-242.

Carmines, E. G. & Zeller, R. A. (1979). *Reliability and validity assessment*. Sage publications.

Cohen, J. (1988). Statistical power analysis for the behavioral sciences. Second Edition. Lawrence Erlbaum Associates.

Cohen, J. (1992). Statistical power analysis. *Current directions in psychological science*, *1*(3), 98-101.

Davis, W. S. & Yen, D. C. (Eds.) (2019). *The information system consultant's handbook: Systems analysis and design*. CRC press.

Feagin, J. R., Orum, A. M., & Sjoberg, G. (Eds.) (1991). *A case for the case study*. UNC Press Books.

Fischer, H. E., Boone, W. J., & Neumann, K. N. U. T. (2014). Quantitative research designs and approaches. In *Handbook of Research on Science Education, Volume II* (pp. 32-51). Routledge.

Fisher, R. A. (1992). Statistical methods for research workers. In *Breakthroughs in statistics* (pp. 66-70). Springer.

Fitts, P. M. & Seeger, C. M. (1953). SR compatibility: spatial characteristics of stimulus and response codes. *Journal of experimental psychology*, *46*, 199.

Franz, E. A., Zelaznik, H. N., & McCabe, G. (1991). Spatial topological constraints in a bimanual task. *Acta Psychologica, 77*, 137-151.

Galton, F. (1989). Kinship and correlation. *Statistical Science*, *4*(2), 81-86.

Gelo, O., Braakmann, D., & Benetka, G. (2008). Quantitative and qualitative research: Beyond the debate. *Integrative psychological and behavioral science*, *42*(3), 266-290.

Gibson, J. J. (1979). *The ecological approach to visual perception*. Houghton Mifflin.

Giele, J. Z. & Elder, G. H. (1998). Life course research: Development of a field. *Methods of life course research: Qualitative and quantitative approaches*, 5-27.

Gosset, W. S. (1908). Student. *The Application of the 'Law of Error' to the Work of the Brewery*, 3-6.

Grieve, S. M., Williams, L. M., Paul, R. H., Clark, C. R., & Gordon, E. (2007). Cognitive aging, executive function, and fractional anisotropy: A diffusion tensor MR imaging study. *American journal of neuroradiology*, *28*, 226-235.

Grove, S. K., Burns, N., & Gray, J. (2012). *The practice of nursing research: Appraisal, synthesis, and generation of evidence*. Elsevier Health Sciences.

Han, J. & Moraga, C. (1995, June). The influence of the sigmoid function parameters on the speed of backpropagation learning. In *International workshop on artificial neural networks* (pp. 195-201). Springer.

Hauke, J. & Kossowski, T. (2011). Comparison of values of Pearson's and Spearman's correlation coefficients on the same sets of data. *Quaestiones geographicae*, *30*(2), 87-93.

Heale, R. & Twycross, A. (2015). Validity and reliability in quantitative studies. *Evidence-based nursing*, *18*(3), 66-67.

Hedrick, T. E., Bickman, L., & Rog, D. J. (1993). *Applied research design: A practical guide*. Sage Publications.

Herrnstein, R. J. & Murray, C. (2010). *The bell curve: Intelligence and class structure in American life*. Simon and Schuster.

Johnson, B. (2001). Toward a new classification of nonexperimental quantitative research. *Educational Researcher*, *30*(2), 3-13.

Johnson, D. M. (1948). Applications of the standard-score IQ to social statistics. *The Journal of social psychology*, *27*(2), 217-227.

Kahneman, D. (2011). *Thinking, fast and slow*. Macmillan.

Kang, J. H. (2004). The role of self-esteem in initiating physical activity in consumption situations. *Journal of sport management*, *18*(3), 255-273.

Kao, L. S. & Green, C. E. (2008). Analysis of variance: Is there a difference in means and what does it mean? *Journal of surgical research*, *144*(1), 158-170.

Kiefer, M., Marzinzik, F., Weisbrod, M., Scherg, M., & Spitzer, M. (1998). The time course of brain activations during response inhibition: Evidence from event-related potentials in a go/no go task. *Neuroreport*, *9*, 765-770.

Kolstø, S. D. (2001). Scientific literacy for citizenship: Tools for dealing with the science dimension of controversial socioscientific issues. *Science education*, *85*(3), 291-310.

Kothari, C. R. (2004). *Research methodology: Methods and techniques*. New Age International.

Kruschke, J. K. (2013). Bayesian estimation supersedes the t test. *Journal of experimenttal psychology: General*, *142*(2), 573.

Kunn, D., Langer, J., Kohlberg, L., & Haan, N. (1977). The development of formal operations in logical and moral judgments. *Genetic Psychological monographs*, *98*, 97-188.

Labovitz, S. (1967). Some observations on measurement and statistics. *Social Forces*, *46*(2), 151-160.

Lindsey, J. K. (1996). *Parametric statistical inference*. Oxford University Press.

MacMahon, B. & Pugh, T. F. (1970). Epidemiology: Principles and methods. *Epidemiology: Principles and methods, 13*: 246-247.

Mahon, B. Z. & Caramazza, A. (2008). A critical look at the embodied cognition hypothesis and a new proposal for grounding conceptual content. *Journal of physiology-Paris*, *102*(1-3), 59-70.

Manochehr, N. N. (2006). The influence of learning styles on learners in e-learning environments: An empirical study. *Computers in higher education economics review*, *18*(1), 10-14.

Marshall, G., & Jonker, L. (1998). A COnCise, Synergy; Oct 2010; *ProQuest Nursing & Allied Health Source* pg. 22.

Menard, S. (2002). *Applied logistic regression analysis* (Vol. 106). Sage.

Mertler, C. A. & Reinhart, R. V. (2016). *Advanced and multivariate statistical methods: Practical application and interpretation*. Routledge.

Micceri, T. (1989). The unicorn, the normal curve, and other improbable creatures. *Psychological bulletin*, *105*(1), 156.

Moore, D. S. (1990). Uncertainty. *On the shoulders of giants: New approaches to numeracy*, 95-137.

Murphy, K. R., Myors, B., & Wolach, A. (2014). *Statistical power analysis: A simple and general model for traditional and modern hypothesis tests*. Routledge.

Mutz, D. C. & Pemantle, R. (2015). Standards for experimental research: Encouraging a better understanding of experimental methods. *Journal of experimental political science*, *2*(2), 192-215.

Newman, I., Benz, C. R., & Ridenour, C. S. (1998). *Qualitative-quantitative research methodology: Exploring the interactive continuum*. SIU Press.

Olkin, I., Gleser, L. J., & Derman, C. (1980). *Probability models and applications*. Macmillan.

Oudejans, R. R. D., Michaels, C. F., Bakker, F. C., & Dolne., M. A. (1996). The relevance of action in perceiving affordances: Perception of catchableness of fly balls. *Journal of experimental psychology: Human perception and performance, 22*(4), 879-891.

Park, C. (2018). Cognitive Function State Underlying Patterns of Movement Coordination during Behaviors by the Elderly: Applied to a System Approach. *International Journal of Human Movement Science*, *12*(1), 127-151.

Park, C. & Kim, S. (2014). Haptic perception accuracy depending on self-produced movement. *Journal of sports sciences*, *32*(10), 974-985.

Pearson, K. (1920). Notes on the history of correlation. *Biometrika*, *13*(1), 25-45.

Petersen, S. E., Fox, P. T., Posner, M. I., Mintun, M., & Raichle, M. E. (1988). Positron emission tomographic studies of the cortical anatomy of single-word processing. *Nature, 331*, 585-589.

Phillips, P. P. & Stawarski, C. A. (2008). *Data collection: Planning for and collecting all types of data* (Vol. 175). John Wiley & Sons.

Plackett, R. L. (1983). Karl Pearson and the chi-squared test. *International statistical review/Revue internationale de statistique*, 59-72.

Pontius, J., Richelle, J., & Wodak, S. J. (1996). Deviations from standard atomic volumes as a quality measure for protein crystal structures. *Journal of molecular biology*, *264*(1), 121-136.

Powell, R. R. (2006). Evaluation research: An overview. *Library trends*, *55*(1), 102-120.

Provost, F. & Fawcett, T. (2013). Data science and its relationship to big data and data-drven decision making. *Big data*, *1*(1), 51-59.

Quinlan, J. R. (2014). *C4. 5: Programs for machine learning*. Elsevier.

Raftery, A. E., Gilks, W. R., Richardson, S., & Spiegelhalter, D. (1995). Hypothesis testing and model. *Markov chain Monte Carlo in practice*, 165-187.

Randolph, J. (2009). A guide to writing the dissertation literature review. *Practical assessment, research, and evaluation*, *14*(1), 13.

Ranganathan, R. & Carlton, L. G. (2007). Perception-action coupling and anticipatory performance in baseball batting. *Journal of Motor Behavior, 39*(5), 369-380.

Rice, J. A. (2006). *Mathematical statistics and data analysis*. Cengage Learning.

Sawyer, S. F. (2009). Analysis of variance: the fundamental concepts. *Journal of manual & manipulative therapy*, *17*(2), 27E-38E.

Savin, N. E. (1984). Multiple hypothesis testing. *Handbook of econometrics*, *2*, 827-879.

Schenker, J. D. & Rumrill Jr, P. D. (2004). Causal-comparative research designs. *Journal of vocational rehabilitation*, *21*(3), 117-121.

Sekaran, U. & Bougie, R. (2019). *Research methods for business: A skill building approach*. John Wiley & Sons.

Silverman, D. (Ed.) (2020). *Qualitative research*. sage.

Singh, A. (2021). An introduction to experimental and exploratory research. SSRN electronic journal. https://Doi.Org/10.2139/Ssrn.3789360

Smith, J. M. (1968). *Mathematical ideas in biology*. CUP Archive.

Spearman, C. (1961). “General Intelligence” objectively determined and measured.

Stokes, P. D. (1995). Learned variability. *Animal learning & behavior*, *23*(2), 164-176.

Streiner, D. L. (1996). Maintaining standards: Differences between the standard deviation and standard error, and when to use each. *The Canadian journal of psychiatry*, *41*(8), 498-502.

Stroop, J. R. (1935). Studies of interference in serial verbal reactions. *Journal of experimental psychology: General, 121,* 15-23. Reprinted from *Journal of experimental psychology, 18*, 643-661).

Stuss, D. T, & Benson, D. F. (1986). *The frontal lobes*. Raven Press.

Sykes, A. O. (1993), *An introduction to regression analysis*, Coase-Sandor Working Paper series in Law and Economics, Coase-Sandor Institute for Law and Economics, University of Chicago Law School, viewed March 2018th.

Thyer, B. A. (2012). *Quasi-experimental research designs*. Oxford University Press.

Tranmer, M. & Elliot, M. (2008). Multiple linear regression. *The Cathie Marsh Centre for census and survey research (CCSR)*, *5*(5), 1-5.

Turvey, M. T. (1977). Preliminaries to a theory of action with reference to vision. In R. Shaw & J. Bransford (Eds.), *Perceiving, acting, and knowing: Toward an ecological psychology* (pp. 211-265). Lawrence Erlbaum.

Venable, J. (2006, February). The role of theory and theorising in design science research. In *Proceedings of the 1st International Conference on Design Science in Information Systems and Technology (DESRIST 2006)* (pp. 1-18).

Westfall, P. H. & Young, S. S. (1993). *Resampling-based multiple testing: Examples and methods for p-value adjustment* (Vol. 279). John Wiley & Sons.

Wilcox, R. R. (2009). *Basic statistics: Understanding conventional methods and modern insights*. Oxford University Press.

Wildt, A. R., Wildt, A. A., Ahtola, O., & Ahtola, O. T. (1978). *Analysis of covaryance* (Vol. 12). Sage.

Wilson, M. (2004). *Constructing measures: An item response modeling approach*. Routledge.

Zamel, V. (1982). Writing: The process of discovering meaning. *TESOL quarterly*, *16*(2), 195-209.

Zuur, A. F., Ieno, E. N., & Elphick, C. S. (2010). A protocol for data exploration to avoid common statistical problems. *Methods in ecology and evolution*, *1*(1), 3-14.

Part 2. Applied Time Series Analysis

Akaike, H. (1998). Autoregressive model fitting for control. In *Selected Papers of Hirotugu Akaike* (pp. 153-170). Springer.

Aschoff, J. (1983). Circadian control of body temperature. *Journal of thermal biology,* *8*(1), 143-147.

Bandt, C. & Shiha, F. (2007). Order patterns in time series. *Journal of time series analysis*, *28*(5), 646-665.

Bayes, T. (1763). LII. An essay towards solving a problem in the doctrine of chances. By the late Rev. Mr. Bayes, FRS communicated by Mr. Price, in a letter to John

Canton, AMFR S. *Philosophical transactions of the Royal Society of London*, (53), 370-418.

Beek, P. J., Peper, C. E., & Stegeman, D. F. (1995). Dynamical models of movement coordination. *Human movement science*, *14*(4-5), 573-608.

Berkes, F. (2007). Understanding uncertainty and reducing vulnerability: Lessons from resilience thinking. *Natural hazards*, *41*(2), 283-295.

Bingham, C., Godfrey, M., & Tukey, J. W. (1967). Modern techniques of power spectrum estimation. *IEEE transactions on audio and electroacoustics*, *15*(2), 56-66.

Bishop, J. C. & Bishop, J. (1989). *Natural agency: An essay on the causal theory of action*. Cambridge University Press.

Brown, R. (1827). Brownian motion. Unpublished experiment.

Burton, J. A. & Zinober, A. S. (1986). Continuous approximation of variable structure control. *International journal of systems science*, *17*(6), 875-885.

Cariani, P. (1993). To evolve an ear. Epistemological implications of gordon pask's electrochemical devices. *Systems research, 10*(3), 19-33.

Charpentier, A. (1891). Analyse experimentale de quelques elements de la sensation de poids. [Experimental analysis of some elements of weight sensations] *Archives de physiologie normales et pathologiques. 3*, 122-135.

Cox, D. R. & Miller, H. D. (2017). *The theory of stochastic processes*. Routledge.

Durbin, J. (1959). Efficient estimation of parameters in moving-average models. *Biometrika*, *46*(3/4), 306-316.

Ethier, S. N. & Kurtz, T. G. (2009). *Markov processes: Characterization and convergence* (Vol. 282). John Wiley & Sons.

Fan, J. & Yao, Q. (2008). *Nonlinear time series: Nonparametric and parametric methods*. Springer Science & Business Media.

Feindt, M. (2004). A neural Bayesian estimator for conditional probability densities. arXiv preprint physics/0402093.

Fine, A. (1982). Hidden variables, joint probability, and the Bell inequalities. *Physical review letters*, *48*(5), 291.

Fine, T. L. (2014). *Theories of probability: An examination of foundations*. Academic Press.

Fitts, P. M. (1954). The information capacity of the human motor system in controlling the amplitude of movement. *Journal of experimental psychology, 47*(6), 381.

Ford, B. J. (2008). Microscopical substantiation of intelligence in living cells. *Infocus Magazine,* (12), 4.

Frank, T. D. (2005). *Nonlinear Fokker-Planck equations: Fundamentals and applications*. Springer Science & Business Media.

Frank, T. D. (2011). Collective behavior of biophysical systems with thermodynamic feedback loops: A case study for a nonlinear Markov model, the Takatsuji system. *Modern physics letters,* B, *25*, 551-568.

Gelfand, A. E. & Smith, A. F. (1990). Sampling-based approaches to calculating marginal densities. *Journal of the American statistical association*, *85*(410), 398-409.

Gelman, A., Carlin, J. B., Stern, H. S., & Rubin, D. B. (1995). *Bayesian data analysis*. Chapman and Hall/CRC.

Gibson, J. J. (1950). The perception of visual surfaces. *The American journal of psychology,* 367-384.

Gillespie, D. T. (1996). Exact numerical simulation of the Ornstein-Uhlenbeck process and its integral. *Physical review E, 54*(2), 2084.

Grimmett, G. & Stirzaker, D. (2020). *Probability and random processes.* Oxford University Press.

Habbema, J. D. F., Van Oortmarssen, G. J., Lubbe, J. T. N., & Van der Maas, P. J. (1985). The MISCAN simulation program for the evaluation of screening for disease. *Computer methods and programs in biomedicine, 20*(1), 79-93.

Haken, H., Kelso, J. S., & Bunz, H. (1985). A theoretical model of phase transitions in human hand movements. *Biological cybernetics, 51*(5), 347-356.

Heckman, J. J. (1987). *The incidental parameters problem and the problem of initial conditions in estimating a discrete time-discrete data stochastic process and some Monte Carlo evidence* (pp. 114). University of Chicago Center for Mathematical studies in Business and Economics.

Hermanns, H. (2002). Markov Chains. In *Interactive Markov Chains* (pp. 35-55). Springer.

Hertwig, R. & Pleskac, T. J. (2010). Decisions from experience: Why small samples?. *Cognition, 115*(2), 225-237.

Howard, R. A. (1960). Dynamic programming and markov processes.

Howard, R. A. (1966). Information value theory. *IEEE transactions on systems science and cybernetics, 2*(1), 22-26.

Iberall, A. S. (2001). The physics, chemical physics, and biological physics of the origin of life on earth. *Ecological Psychology, 13,* 315-327.

Iberall, A. S. & Soodak, H. (1978). Physical basis for complex systems—some propositions relating levels of organizations. *Collective phenomena, 3,* 9-24.

Isenhower, R. W., Frank, T. D., Kay, B. A., & Carello, C. (2012). Capturing and quantifying the dynamics of valenced emotions and valenced events on the organism-environment System. *Nonlinear dynamics-psychology and life sciences, 16*(4), 397.

Jaynes, E. T. (1957). Information theory and statistical mechanics. *Physical review, 106*(4), 620.

Kandananond, K. (2012). A comparison of various forecasting methods for autocorrelated time series. *International journal of engineering business management, 4,* 4.

Klein, J. L. & Klein, D. (1997). *Statistical visions in time: A history of time series analysis, 1662-1938.* Cambridge University Press.

Kneale, W. (1951). Probability and induction. *Mind, 60*(239), 310-317.

Kugler, P. N. & Turvey, M. T. (1987). *Information, natural law and the self-assembly of rhythmic movement.* Erlbaum.

Lampard, D. G. (1954). Generalization of the Wiener-Khintchine theorem to nonstationary processes. *Journal of applied physics, 25*(6), 802-803.

Lorenz, E. (2000). The butterfly effect. *World scientific series on nonlinear science series A, 39,* 91-94.

Mahner, M. & Bunge, M. (1997). *Foundations of biophilosophy.* Springer Science & Business Media.

Manski, C. F. (2003). *Partial identification of probability distributions*. Springer Science & Business Media.

Masani, P. R. (2012). *Norbert Wiener 1894–1964* (Vol. 5). Birkhäuser.

Maury, E., Ramsey, K. M., & Bass, J. (2010). Circadian rhythms and metabolic syndrome from experimental genetics to human disease. *Circulation research, 106*(3), 447-462.

Medhi, J. (1994). Stochastic processes. New Age International.

Mörters, P., & Peres, Y. (2010). *Brownian motion* (Vol. 30). Cambridge University Press.

Myung, I. J. (2003). Tutorial on maximum likelihood estimation. *Journal of mathematical psychology*, *47*(1), 90-100.

Neumaier, A. & Schneider, T. (2001). Estimation of parameters and eigenmodes of multivariate autoregressive models. *ACM Transactions on Mathematical Software (TOMS)*, *27*(1), 27-57.

Norris, J. R. & Norris, J. R. (1998). *Markov chains* (No. 2). Cambridge University Press.

Packard, G. C., Birchard, G. F., & Boardman, T. J. (2011). Fitting statistical models in bivariate allometry. *Biological Reviews*, *86*(3), 549-563.

Palmer, T. N. (2009). Edward Norton Lorenz. 23 May 1917-16 April 2008.

Papoulis, A. & Saunders, H. (1989). Probability, random variables and stochastic processes.

Park, C. (2016). *Bi-manual Coordination Dynamics Embedding in Circadian Temperature Context* (Doctoral dissertation, Seoul National University press).

Park, C. (2019). Sensory dominance in extended haptics: Applied stochastic method. *International Journal of Applied Sports Sciences*, *31*(2).

Parzen, E. (1962). On estimation of a probability density function and mode. *The annals of mathematical statistics*, *33*(3), 1065-1076.

Rubin, D. B. (2005). Causal inference using potential outcomes: Design, modeling, decisions. *Journal of the American Statistical Association*, *100*(469), 322-331.

Schöner, G., Haken, H., & Kelso, J. A. S. (1986). A stochastic theory of phase transitions in human hand movement. *Biological cybernetics, 53*(4), 247-257.

Schrodinger, R., Schrödinger, E., & Dinger, E. S. (1992). *What is life?: With mind and matter and autobiographical sketches*. Cambridge university press.

Selya, A. S., Wakschlag, L. S., Dierker, L. C., Rose, J. S., Hedeker, D., & Mermelstein, R. J. (2013). Exploring alternate processes contributing to the association between maternal smoking and the smoking behavior among young adult offspring. *nicotine & tobacco research*, *15*(11), 1873-1882.

Shannon, C. E. (1948). A note on the concept of entropy. *Bell System Tech. J*, *27*(3), 379-423.

Shaw, R. & Kinsella-Shaw, J. (1988). Ecological mechanics: A physical geometry for intentional constraints. *Human movement science, 7*(2), 155-200.

Silverman, B. W. (1986). Monographs on statistics and applied probability. *Density estimation for statistics and data analysis*, (Vol. 26). CRC press.

Skyttner, L. (2005). *General systems theory: Problems, perspectives, practice*. World scientific.

Smith, B. C. (2011). *Fundamentals of Fourier transform infrared spectroscopy*. CRC press.

Soodak, H. & Iberall, A. (1978). Homeokinetics. *Science, 201,* 579-582.

Strong Jr, D. R., & Ray Jr, T. S. (1975). Host tree location behavior of a tropical vine (Monstera gigantea) by skototropism. *Science, 190*(4216), 804-806.

Taylor, R. (1990). Interpretation of the correlation coefficient: A basic review. *Journal of diagnostic medical sonography*, *6*(1), 35-39.

Tierney, L. (1996). Introduction to general state-space Markov chain theory. *Markov chain Monte Carlo in practice*, 59-74.

Twiss, B. C. (1992). *Forecasting for technologists and engineers: A practical guide for better decisions* (No. 15). IET.

Turvey, M. T. & Carello, C. (2012). On intelligence from first principles: Guidelines for inquiry into the hypothesis of physical intelligence (PI). *Ecological psychology, 24*(1), 3-32.

Uhlenbeck, G. E. & Ornstein, L. S. (1930). On the theory of the Brownian motion. *Physical review*, *36*(5), 823.

Uttley, A. M. (1959). The design of conditional probability computers. *Information and control*, *2*(1), 1-24.

Van Belle, G. (2011). *Statistical rules of thumb* (Vol. 699). John Wiley & Sons.

Van Kampen, N. G. (1981). The validity of nonlinear Langevin equations. *Journal of statistical physics*, *25*(3), 431-442.

Wickens, M. R. (1982). The efficient estimation of econometric models with rational expectations. *The review of economic studies*, *49*(1), 55-67.

Wing, A. M. & Kristofferson, A. B. (1973). Response delays and the timing of discrete motor responses. *Perception & Psychophysics*, *14*(1), 5-12.

Witt, A., Kurths, J., & Pikovsky, A. (1998). Testing stationarity in time series. *Physical Review E*, *58*(2), 1800-1810.

Yaffee, R. A. & McGee, M. (2000). *An introduction to time series analysis and forecasting: With applications of SAS® and SPSS®*. Elsevier.

Part 3. Applied Systems Analysis

Arthur, W. B. (1991). Designing economic agents that act like human agents: A behavioral approach to bounded rationality. *The American economic review*, *81*(2), 353-359.

Atkeson, A. (2020). *What will be the economic impact of COVID-19 in the US? Rough estimates of disease scenarios* (No. w26867). National Bureau of Economic Research.

Axelrod, R. (1997), Advancing the art of simulation in the social sciences, In *Simulating Social Phenomena*, Springer, pp. 21-40.

Barabási, A. L. & Bonabeau, E. (2003). Scale-free networks. *Scientific american*, *288*(5), 60-69.

Bernstein, D. S. (2018). *Scalar, vector, and matrix mathematics*. Princeton University Press.

Berry, M. W., Drmac, Z., & Jessup, E. R. (1999). Matrices, vector spaces, and information retrieval. *SIAM review*, *41*(2), 335-362.

Bollobás, B. (2013). *Modern graph theory* (Vol. 184). Springer Science & Business Media.

Bonabeau, E. (2002). Agent-based modeling: Methods and techniques for simulating human systems. *Proceedings of the national academy of sciences*, *99*(suppl. 3), 7280-7287.

Bonacich, P. (1972). Factoring and weighting approaches to status scores and clique identification. *Journal of mathematical sociology*. *2*: 113-120.

Bondy, J. A. & Murty, U. S. R. (1976). *Graph theory with applications* (Vol. 290). Macmillan.

Borsboom, D., Deserno, M. K., Rhemtulla, M., Epskamp, S., Fried, E. I., McNally, R. J., ... & Waldorp, L. J. (2021). Network analysis of multivariate data in psychological science. *Nature reviews methods primers*, *1*(1), 1-18.

Brazell, J. & Michael. B, (2005). "High performance computing". Texas State Technical College.

Broder, A., Kumar, R., Maghoul, F., Raghavan, P., Rajagopalan, S., Stata, R., ... & Wiener, J. (2011). Graph structure in the web. In *The structure and dynamics of networks* (pp. 183-194). Princeton University Press.

Capel, B. (2000). The battle of the sexes. *Mechanisms of development*, *92*(1), 89-103.

Chen, S. A. B. S. & Billings, S. A. (1992). Neural networks for nonlinear dynamic system modelling and identification. *International journal of control*, *56*(2), 319-346.

Clauser, C. E., McConville, J. T., & Young, J. W. (1969). *Weight, volume, and center of mass of segments of the human body*. Antioch Coll Yellow Springs OH.

Collette, M. R., Corey, I. R., & Johnson, J. R. (2005). *High performance tools and technologies* (No. UCRL-TR-209289). Lawrence Livermore National Lab.(LLNL), Livermore.

Conway, J. H. (2000). *On numbers and games*. CRC Press.

Courant, R. (2011). *Differential and integral calculus, Volume 2* (Vol. 2). John Wiley & Sons.

Crooks, A. T., & Heppenstall, A. J. (2011). Introduction to agent-based modelling. In *Agent-based models of geographical systems* (pp. 85-105). Dordrecht: Springer Netherlands.

DeAngelis, D. L. & Diaz, S. G. (2019). Decision-making in agent-based modeling: A current review and future prospectus. *Frontiers in ecology and evolution*, *6*, 237.

Diestel, J. & Faires, B. (1974). On vector measures. *Transactions of the American mathematical society*, *198*, 253-271.

Downs, R. M. & Stea, D. (Eds.) (2017). *Image and environment: Cognitive mapping and spatial behavior*. Transaction Publishers.

Erdős, P. & Rényi, A. (1959). On Cantor's series with convergent. *Eötvös de Budapest Sect. Math*, *2*, 93-109.

Euler, L. (1956). The seven bridges of Königsberg. *The world of mathematics*, *1*, 573-580.

Fox, R. L. & Kapoor, M. P. (1968). Rates of change of eigenvalues and eigenvectors. *AIAA journal*, *6*(12), 2426-2429.

Getchell, A. (2008). Agent-based modeling. *Physics*, *22*(6), 757-767.

Gilbert, N. & Troitzsch, K. (2005). *Simulation for the social scientist*. McGraw-Hill Education.

Gómez, S. (2019). Centrality in networks: Finding the most important nodes. In *business and consumer analytics: New ideas* (pp. 401-433). Springer, Cham.

Goodwin, R. M. (1967), "A Growth Cycle," In Feinstein, C.H. (Ed.), *Socialism, capitalism and economic growth*. Cambridge University Press.

Haken, H., Kelso, J. S., & Bunz, H. (1985). A theoretical model of phase transitions in human hand movements. *Biological cybernetics*, *51*(5), 347-356.

Hauert, C., De Monte, S., Hofbauer, J., & Sigmund, K. (2002). Replicator dynamics for optional public good games. *Journal of theoretical biology*, *218*(2), 187-194.

Heckmann, T., Schwanghart, W., & Phillips, J. D. (2015). Graph theory—Recent developments of its application in geomorphology. *Geomorphology*, *243*, 130-146.

Helbing, D. (2013). Globally networked risks and how to respond. *Nature*, *497*(7447), 51-59.

Hofbauer, J. & Sigmund, K. (1998). Dynamical systems and Lotka–Volterra Equations. *Evolutionary games and population dynamics*. Cambridge University Press.

Jackson, J. C., Rand, D., Lewis, K., Norton, M. I., & Gray, K. (2017). Agent-based modeling: A guide for social psychologists. *Social psychological and personality science*, *8*(4), 387-395.

Jenkins, B. C. (2020). A Python-based undergraduate course in computational macroeconomics. UC Irvine.

Johnson, S. (2002). *Emergence: The connected lives of ants, brains, cities, and software*. Simon and Schuster.

Keim, D., Andrienko, G., Fekete, J. D., Görg, C., Kohlhammer, J., & Melançon, G. (2008). Visual analytics: Definition, process, and challenges. In *Information visualization* (pp. 154-175). Springer.

Kitano, H. (2002). Computational systems biology. *Nature*, *420*(6912), 206-210.

Kramer, E. E. (1982). *The nature and growth of modern mathematics*. Princeton University Press.

Kreps, D. M., Milgrom, P., Roberts, J., & Wilson, R. (1982). Rational cooperation in the finitely repeated prisoners' dilemma. *Journal of economic theory*, *27*(2), 245-252.

Krishnaa, A. (2018). An example usage of graph theory in other scientific fields: On graph labeling, possibilities and role of mind/consciousness. *Graph theory: Advanced algorithms and applications*, 19.

Lacey, N. (2008). THE PRISONERS' DILEMMA. Cambridge UK, The Policy Press.

Leduc, M., Poledna, S., & Thurner, S. (2017). Systemic risk management in financial networks with credit default swaps.

Liben-Nowell, D. & Kleinberg, J. (2007). The link-prediction problem for social networks. *Journal of the American society for information science and technology*, *58*(7), 1019-1031.

Liu, S. B. (2009). Conceptual density functional theory and some recent developments. *Acta physico-chimica sinica*, *25*(3), 590-600.

Manning, C. & Schutze, H. (1999). *Foundations of statistical natural language processing*. MIT press.

Marchal, J. H. (1975). On the concept of a system. *Philosophy of science*, *42*(4), 448-468.

McGinn, C. (1989). Can we solve the mind-body problem? *Mind*, *98*(391), 349-366.

McLeod, P. & Dienes, Z. (1996). Do fielders know where to go to catch the ball or only how to get there? *Journal of experimental psychology: Human perception and performance*, *22*(3), 531.

Mookherjee, D. & Sopher, B. (1994). Learning behavior in an experimental matching pennies game. *Games and economic behavior*, *7*(1), 62-91.

Moretti, F. (2003). Graphs, maps, trees. *New left review*, *24*, 67.

Myerson, R. B. (2013). *Game theory*. Harvard University Press.

Nash, J. (1950). Equilibrium points in n-person games. *Proceedings of the national academy of sciences 36*(1), 48-49.

Newman, M. E. & Girvan, M. (2004). Finding and evaluating community structure in networks. *Physical review E*, *69*(2).

Nowak, M. A. & Sigmund, K. (2004). Evolutionary dynamics of biological games. *Science*, *303*(5659), 793-799.

Oliva, R. (2016). Structural dominance analysis of large and stochastic models. *System dynamics review*, *32*(1), 26-51.

Osborne, M. J. (2004). *An introduction to game theory* (Vol. 3, No. 3). Oxford University Press.

Osborne, M. J. & Rubinstein, A. (1994). *A course in game theory*. MIT Press.

O'Sullivan, D., Millington, J., Perry, G., & Wainwright, J. (2012). Agent-based models–because they're worth it?. In *Agent-based models of geographical systems* (pp. 109-123). Springer.

Pacheco, J. M., Traulsen, A., & Nowak, M. A. (2006). Coevolution of strategy and structure in complex networks with dynamical linking. *Physical review letters*, *97*(25).

Pal, C. V., Leon, F., Paprzycki, M., & Ganzha, M. (2020). A review of platforms for the development of agent systems. *arXiv:2007.08961*.

Park, C. (2020a). Evolutionary understanding of the conditions leading to estimation of behaviorral properties through system dynamics. *Complex Adaptive Systems Modeling*, *8*(1), 1-25.

Park, C. (2020b). Network and agent dynamics with evolving protection against systemic risk. *Complexity*, *2020*.

Park, C. (2022). Network Characteristic Control of Social Dilemmas in a Public Good Game: Numerical Simulation of Agent-Based Nonlinear Dynamics. *Processes*, *10*(7), 1-17.

Park, C. (2023). Cyclic relationship of mechanical likelihood: Coupling perception-action states in extended haptic accuracy. *Human Movement Science, 92*, 103140.

Råde, L., & Westergren, B. (1999). *Mathematics handbook for science and engineering* (Vol. 5), Springer.

Rangel, T. F., Edwards, N. R., Holden, P. B., Diniz-Filho, J. A. F., Gosling, W. D., Coelho, M. T. P., ... & Colwell, R. K. (2018). Modeling the ecology and evolution of biodiversity: Biogeographical cradles, museums, and graves. *Science*, *361*(6399).

Rapoport, A. & Chammah, A. M. (1966). The game of chicken. *American behavioral scientist*, *10*(3), 10-28.

Reynolds, C. W. (1987, August). Flocks, herds and schools: A distributed behavioral model. In *Proceedings of the 14th annual conference on Computer graphics and interactive techniques* (pp. 25-34).

Sasaki, T., Brännström, Å., Dieckmann, U., & Sigmund, K. (2012). The take-it-or-leave-it option allows small penalties to overcome social dilemmas. *Proceedings of the national academy of sciences*, *109*(4), 1165-1169.

Schelling, T. C. (1971). Dynamic Models of Segregation. *Journal of mathematical sociology*, *1*(2), 143-186.

Schöner, G., Haken, H., & Kelso, J. A. S. (1986). A stochastic theory of phase transitions in human hand movement. *Biological cybernetics*, *53*(4), 247-257.

Schwill, A. (1994). Fundamental ideas of computer science. *Bulletin-European association for theoretical computer science*, *53*, 274-274.

Scott, J. (1988). Social network analysis. *Sociology*, *22*(1), 109-127.

Silverman, E. K., Schmidt, H. H., Anastasiadou, E., Altucci, L., Angelini, M., Badimon, L., ... & Baumbach, J. (2020). Molecular networks in Network Medicine: Development and applications. *Wiley interdisciplinary reviews: Systems biology and medicine*, *12*(6), e1489.

Sinervo, B. & Lively, C. M. (1996). The rock–paper–scissors game and the evolution of alternative male strategies. *Nature*, *380*(6571), 240-243.

Skyttner, L. (2005). *General systems theory: Problems, perspectives, practice*. World Scientific.

Smith, E. R. & Conrey, F. R. (2007). Agent-based modeling: A new approach for theory building in social psychology. *Personality and social psychology review*, *11*(1), 87-104.

Steel, D. (2007). *Across the boundaries: Extrapolation in biology and social science*. Oxford University Press.

Strogatz, S. H. (2001). Exploring complex networks. *Nature*, *410*(6825), 268-276.

Tesfatsion, L. & Judd, K. L. (Eds.) (2006). *Handbook of computational economics: Agent-based computational economics*. Elsevier.

Tichy, N. M., Tushman, M. L., & Fombrun, C. (1979). Social network analysis for organizations. *Academy of management review*, *4*(4), 507-519.

Vespignani, A. (2018). Twenty years of network science.

Wahlquist, H. D. & Estabrook, F. B. (1975). Prolongation structures of nonlinear evolution equations. *Journal of mathematical physics*, *16*(1), 1-7.

Watts, D. J., & Strogatz, S. H. (1998). Collective dynamics of 'small-world' networks. *Nature*, *393*(6684), 440-442.

West, D. B. (2001). *Introduction to graph theory* (Vol. 2). Prentice Hall.

World Health Organization. Coronavirus disease 2019 (COVID-19): situation report. Weekly epidemiological and operational updates, *82*, 2020.

Appendix

Appendix 1

Statistical Tables

Cumulative Normal Distribution, Critical Values of the t Distribution, Critical Values of the F Distribution, Critical Values of the Chi-Squared Distribution, Critical Values of the Correlation Coefficient, Critical Values of Pearson's r for a Two-Tailed Test

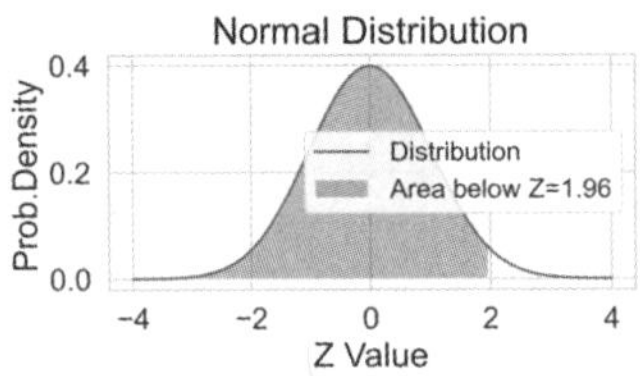

The **Z table** (also known as the standard normal table) provides the probability that a randomly selected value from a standard normal distribution (with a mean of 0 and a standard deviation of 1) will be less than or equal to a given Z value. Each cell in the table represents the cumulative probability up to the corresponding Z value (row) and significance level (column).

Cumulative Standardized Normal Distribution Table

	p=0.00	p=0.01	p=0.02	p=0.03	p=0.04	p=0.05	p=0.06	p=0.07	p=0.08	p=0.09
0.0	0.5000	0.5040	0.5080	0.5120	0.5160	0.5199	0.5239	0.5279	0.5319	0.5359
0.1	0.5398	0.5438	0.5478	0.5517	0.5557	0.5596	0.5636	0.5675	0.5714	0.5753
0.2	0.5793	0.5832	0.5871	0.5910	0.5948	0.5987	0.6026	0.6064	0.6103	0.6141
0.3	0.6179	0.6217	0.6255	0.6293	0.6331	0.6368	0.6406	0.6443	0.6480	0.6517
0.4	0.6554	0.6591	0.6628	0.6664	0.6700	0.6736	0.6772	0.6808	0.6844	0.6879
0.5	0.6915	0.6950	0.6985	0.7019	0.7054	0.7088	0.7123	0.7157	0.7190	0.7224
0.6	0.7257	0.7291	0.7324	0.7357	0.7389	0.7422	0.7454	0.7486	0.7517	0.7549
0.7	0.7580	0.7611	0.7642	0.7673	0.7704	0.7734	0.7764	0.7794	0.7823	0.7852
0.8	0.7881	0.7910	0.7939	0.7967	0.7995	0.8023	0.8051	0.8078	0.8106	0.8133
0.9	0.8159	0.8186	0.8212	0.8238	0.8264	0.8289	0.8315	0.8340	0.8365	0.8389
1.0	0.8413	0.8438	0.8461	0.8485	0.8508	0.8531	0.8554	0.8577	0.8599	0.8621
1.1	0.8643	0.8665	0.8686	0.8708	0.8729	0.8749	0.8770	0.8790	0.8810	0.8830
1.2	0.8849	0.8869	0.8888	0.8907	0.8925	0.8944	0.8962	0.8980	0.8997	0.9015
1.3	0.9032	0.9049	0.9066	0.9082	0.9099	0.9115	0.9131	0.9147	0.9162	0.9177
1.4	0.9192	0.9207	0.9222	0.9236	0.9251	0.9265	0.9279	0.9292	0.9306	0.9319
1.5	0.9332	0.9345	0.9357	0.9370	0.9382	0.9394	0.9406	0.9418	0.9429	0.9441
1.6	0.9452	0.9463	0.9474	0.9484	0.9495	0.9505	0.9515	0.9525	0.9535	0.9545
1.7	0.9554	0.9564	0.9573	0.9582	0.9591	0.9599	0.9608	0.9616	0.9625	0.9633
1.8	0.9641	0.9649	0.9656	0.9664	0.9671	0.9678	0.9686	0.9693	0.9699	0.9706
1.9	0.9713	0.9719	0.9726	0.9732	0.9738	0.9744	0.9750	0.9756	0.9761	0.9767
2.0	0.9772	0.9778	0.9783	0.9788	0.9793	0.9798	0.9803	0.9808	0.9812	0.9817
2.1	0.9821	0.9826	0.9830	0.9834	0.9838	0.9842	0.9846	0.9850	0.9854	0.9857
2.2	0.9861	0.9864	0.9868	0.9871	0.9875	0.9878	0.9881	0.9884	0.9887	0.9890
2.3	0.9893	0.9896	0.9898	0.9901	0.9904	0.9906	0.9909	0.9911	0.9913	0.9916
2.4	0.9918	0.9920	0.9922	0.9925	0.9927	0.9929	0.9931	0.9932	0.9934	0.9936
2.5	0.9938	0.9940	0.9941	0.9943	0.9945	0.9946	0.9948	0.9949	0.9951	0.9952
2.6	0.9953	0.9955	0.9956	0.9957	0.9959	0.9960	0.9961	0.9962	0.9963	0.9964
2.7	0.9965	0.9966	0.9967	0.9968	0.9969	0.9970	0.9971	0.9972	0.9973	0.9974
2.8	0.9974	0.9975	0.9976	0.9977	0.9977	0.9978	0.9979	0.9979	0.9980	0.9981
2.9	0.9981	0.9982	0.9982	0.9983	0.9984	0.9984	0.9985	0.9985	0.9986	0.9986
3.0	0.9987	0.9987	0.9987	0.9988	0.9988	0.9989	0.9989	0.9989	0.9990	0.9990
3.1	0.9990	0.9991	0.9991	0.9991	0.9992	0.9992	0.9992	0.9992	0.9993	0.9993
3.2	0.9993	0.9993	0.9994	0.9994	0.9994	0.9994	0.9994	0.9995	0.9995	0.9995
3.3	0.9995	0.9995	0.9995	0.9996	0.9996	0.9996	0.9996	0.9996	0.9996	0.9997
3.4	0.9997	0.9997	0.9997	0.9997	0.9997	0.9997	0.9997	0.9997	0.9997	0.9998
3.5	0.9998	0.9998	0.9998	0.9998	0.9998	0.9998	0.9998	0.9998	0.9998	0.9998
3.6	0.9998	0.9998	0.9999	0.9999	0.9999	0.9999	0.9999	0.9999	0.9999	0.9999

One-Tailed Critical Values of the *t* Distribution (Significance Level)

	5%	2.50%	1%	0.50%	0.10%
1	6.3138	12.7062	31.8205	63.6567	318.3088
2	2.92	4.3027	6.9646	9.9248	22.3271
3	2.3534	3.1824	4.5407	5.8409	10.2145
4	2.1318	2.7764	3.7469	4.6041	7.1732
5	2.015	2.5706	3.3649	4.0321	5.8934
6	1.9432	2.4469	3.1427	3.7074	5.2076
7	1.8946	2.3646	2.998	3.4995	4.7853
8	1.8595	2.306	2.8965	3.3554	4.5008
9	1.8331	2.2622	2.8214	3.2498	4.2968
10	1.8125	2.2281	2.7638	3.1693	4.1437
11	1.7959	2.201	2.7181	3.1058	4.0247
12	1.7823	2.1788	2.681	3.0545	3.9296
13	1.7709	2.1604	2.6503	3.0123	3.852
14	1.7613	2.1448	2.6245	2.9768	3.7874
15	1.7531	2.1314	2.6025	2.9467	3.7328
16	1.7459	2.1199	2.5835	2.9208	3.6862
17	1.7396	2.1098	2.5669	2.8982	3.6458
18	1.7341	2.1009	2.5524	2.8784	3.6105
19	1.7291	2.093	2.5395	2.8609	3.5794
20	1.7247	2.086	2.528	2.8453	3.5518
21	1.7207	2.0796	2.5176	2.8314	3.5272
22	1.7171	2.0739	2.5083	2.8188	3.505
23	1.7139	2.0687	2.4999	2.8073	3.485
24	1.7109	2.0639	2.4922	2.7969	3.4668
25	1.7081	2.0595	2.4851	2.7874	3.4502
26	1.7056	2.0555	2.4786	2.7787	3.435
27	1.7033	2.0518	2.4727	2.7707	3.421
28	1.7011	2.0484	2.4671	2.7633	3.4082
29	1.6991	2.0452	2.462	2.7564	3.3962
30	1.6973	2.0423	2.4573	2.75	3.3852
31	1.6955	2.0395	2.4528	2.744	3.3749
32	1.6939	2.0369	2.4487	2.7385	3.3653
33	1.6924	2.0345	2.4448	2.7333	3.3563
34	1.6909	2.0322	2.4411	2.7284	3.3479
35	1.6896	2.0301	2.4377	2.7238	3.34
36	1.6883	2.0281	2.4345	2.7195	3.3326
37	1.6871	2.0262	2.4314	2.7154	3.3256
38	1.686	2.0244	2.4286	2.7116	3.319
39	1.6849	2.0227	2.4258	2.7079	3.3128
40	1.6839	2.0211	2.4233	2.7045	3.3069
41	1.6829	2.0195	2.4208	2.7012	3.3013
42	1.682	2.0181	2.4185	2.6981	3.296
43	1.6811	2.0167	2.4163	2.6951	3.2909
44	1.6802	2.0154	2.4141	2.6923	3.2861
45	1.6794	2.0141	2.4121	2.6896	3.2815
46	1.6787	2.0129	2.4102	2.687	3.2771
47	1.6779	2.0117	2.4083	2.6846	3.2729
48	1.6772	2.0106	2.4066	2.6822	3.2689
49	1.6766	2.0096	2.4049	2.68	3.2651
50	1.6759	2.0086	2.4033	2.6778	3.2614
200	1.6525	1.9719	2.3451	2.6006	3.1315
1000	1.6464	1.9623	2.3301	2.5808	3.0984

Two-Tailed Critical Values of the *t* Distribution (Significance Level)

	10%	5%	2%	1%	0.20%
1	6.3138	12.7062	31.8205	63.6567	318.3088
2	2.92	4.3027	6.9646	9.9248	22.3271
3	2.3534	3.1824	4.5407	5.8409	10.2145
4	2.1318	2.7764	3.7469	4.6041	7.1732
5	2.015	2.5706	3.3649	4.0321	5.8934
6	1.9432	2.4469	3.1427	3.7074	5.2076
7	1.8946	2.3646	2.998	3.4995	4.7853
8	1.8595	2.306	2.8965	3.3554	4.5008
9	1.8331	2.2622	2.8214	3.2498	4.2968
10	1.8125	2.2281	2.7638	3.1693	4.1437
11	1.7959	2.201	2.7181	3.1058	4.0247
12	1.7823	2.1788	2.681	3.0545	3.9296
13	1.7709	2.1604	2.6503	3.0123	3.852
14	1.7613	2.1448	2.6245	2.9768	3.7874
15	1.7531	2.1314	2.6025	2.9467	3.7328
16	1.7459	2.1199	2.5835	2.9208	3.6862
17	1.7396	2.1098	2.5669	2.8982	3.6458
18	1.7341	2.1009	2.5524	2.8784	3.6105
19	1.7291	2.093	2.5395	2.8609	3.5794
20	1.7247	2.086	2.528	2.8453	3.5518
21	1.7207	2.0796	2.5176	2.8314	3.5272
22	1.7171	2.0739	2.5083	2.8188	3.505
23	1.7139	2.0687	2.4999	2.8073	3.485
24	1.7109	2.0639	2.4922	2.7969	3.4668
25	1.7081	2.0595	2.4851	2.7874	3.4502
26	1.7056	2.0555	2.4786	2.7787	3.435
27	1.7033	2.0518	2.4727	2.7707	3.421
28	1.7011	2.0484	2.4671	2.7633	3.4082
29	1.6991	2.0452	2.462	2.7564	3.3962
30	1.6973	2.0423	2.4573	2.75	3.3852
31	1.6955	2.0395	2.4528	2.744	3.3749
32	1.6939	2.0369	2.4487	2.7385	3.3653
33	1.6924	2.0345	2.4448	2.7333	3.3563
34	1.6909	2.0322	2.4411	2.7284	3.3479
35	1.6896	2.0301	2.4377	2.7238	3.34
36	1.6883	2.0281	2.4345	2.7195	3.3326
37	1.6871	2.0262	2.4314	2.7154	3.3256
38	1.686	2.0244	2.4286	2.7116	3.319
39	1.6849	2.0227	2.4258	2.7079	3.3128
40	1.6839	2.0211	2.4233	2.7045	3.3069
41	1.6829	2.0195	2.4208	2.7012	3.3013
42	1.682	2.0181	2.4185	2.6981	3.296
43	1.6811	2.0167	2.4163	2.6951	3.2909
44	1.6802	2.0154	2.4141	2.6923	3.2861
45	1.6794	2.0141	2.4121	2.6896	3.2815
46	1.6787	2.0129	2.4102	2.687	3.2771
47	1.6779	2.0117	2.4083	2.6846	3.2729
48	1.6772	2.0106	2.4066	2.6822	3.2689
49	1.6766	2.0096	2.4049	2.68	3.2651
50	1.6759	2.0086	2.4033	2.6778	3.2614
200	1.6525	1.9719	2.3451	2.6006	3.1315
1000	1.6464	1.9623	2.3301	2.5808	3.0984

Critical Values of the *F* Distribution ($\alpha = 0.05$)

	1	2	3	4	5	8	12	16	20	24	28	40	60	80	100	120	140	160	240
1	161.45	18.51	10.13	7.71	6.61	5.32	4.75	4.49	4.35	4.26	4.2	4.08	4	3.96	3.94	3.92	3.91	3.9	3.88
2	199.5	19	9.55	6.94	5.79	4.46	3.89	3.63	3.49	3.4	3.34	3.23	3.15	3.11	3.09	3.07	3.06	3.05	3.03
3	215.71	19.16	9.28	6.59	5.41	4.07	3.49	3.24	3.1	3.01	2.95	2.84	2.76	2.72	2.7	2.68	2.67	2.66	2.64
4	224.58	19.25	9.12	6.39	5.19	3.84	3.26	3.01	2.87	2.78	2.71	2.61	2.53	2.49	2.46	2.45	2.44	2.43	2.41
5	230.16	19.3	9.01	6.26	5.05	3.69	3.11	2.85	2.71	2.62	2.56	2.45	2.37	2.33	2.31	2.29	2.28	2.27	2.25
6	233.99	19.33	8.94	6.16	4.95	3.58	3	2.74	2.6	2.51	2.45	2.34	2.25	2.21	2.19	2.18	2.16	2.16	2.14
7	236.77	19.35	8.89	6.09	4.88	3.5	2.91	2.66	2.51	2.42	2.36	2.25	2.17	2.13	2.1	2.09	2.08	2.07	2.05
8	238.88	19.37	8.85	6.04	4.82	3.44	2.85	2.59	2.45	2.36	2.29	2.18	2.1	2.06	2.03	2.02	2.01	2	1.98
9	240.54	19.38	8.81	6	4.77	3.39	2.8	2.54	2.39	2.3	2.24	2.12	2.04	2	1.97	1.96	1.95	1.94	1.92
10	241.88	19.4	8.79	5.96	4.74	3.35	2.75	2.49	2.35	2.25	2.19	2.08	1.99	1.95	1.93	1.91	1.9	1.89	1.87
12	243.91	19.41	8.74	5.91	4.68	3.28	2.69	2.42	2.28	2.18	2.12	2	1.92	1.88	1.85	1.83	1.82	1.81	1.79
15	245.95	19.43	8.7	5.86	4.62	3.22	2.62	2.35	2.2	2.11	2.04	1.92	1.84	1.79	1.77	1.75	1.74	1.73	1.71
20	248.01	19.45	8.66	5.8	4.56	3.15	2.54	2.28	2.12	2.03	1.96	1.84	1.75	1.7	1.68	1.66	1.65	1.64	1.61
24	249.05	19.45	8.64	5.77	4.53	3.12	2.51	2.24	2.08	1.98	1.91	1.79	1.7	1.65	1.63	1.61	1.6	1.59	1.56
30	250.1	19.46	8.62	5.75	4.5	3.08	2.47	2.19	2.04	1.94	1.87	1.74	1.65	1.6	1.57	1.55	1.54	1.53	1.51
40	251.14	19.47	8.59	5.72	4.46	3.04	2.43	2.15	1.99	1.89	1.82	1.69	1.59	1.54	1.52	1.5	1.48	1.47	1.44
60	252.2	19.48	8.57	5.69	4.43	3.01	2.38	2.11	1.95	1.84	1.77	1.64	1.53	1.48	1.45	1.43	1.41	1.4	1.37
120	253.25	19.49	8.55	5.66	4.4	2.97	2.34	2.06	1.9	1.79	1.71	1.58	1.47	1.41	1.38	1.35	1.33	1.32	1.29
240	253.78	19.49	8.54	5.64	4.38	2.95	2.32	2.03	1.87	1.76	1.68	1.54	1.43	1.37	1.33	1.31	1.29	1.27	1.24
500	254.06	19.49	8.53	5.64	4.37	2.94	2.31	2.02	1.86	1.75	1.67	1.53	1.41	1.35	1.31	1.28	1.26	1.24	1.2
1000	254.19	19.49	8.53	5.63	4.37	2.93	2.3	2.02	1.85	1.74	1.66	1.52	1.4	1.34	1.3	1.27	1.25	1.23	1.19

Explanation of each column and row: Criticla Values of the *F* distribution ($\alpha = 0.05$)

- Columns: df_d values (degrees of freedom of denominator)
- Rows: df_n values (degrees of freedom of numerator)
- *F* critical values: The value in each cell is the critical value of *F* at the intersection of the corresponding dfn and dfd values. The critical value of *F* is used to test whether the means of two populations are significantly different when analyzing variance. It compares the variance between two samples to the variance within each sample.
- $\alpha = 0.05$: The significance level of 0.05 indicates that there is a 5% chance of rejecting the null hypothesis when it is actually true. This level of significance is commonly used in statistical hypothesis testing.

Critical Values of the *F* Distribution (α = 0.01)

	1	2	3	4	5	8	12	16	20	24	28	40	60	80	100	120	140	160	240
1	4052.18	98.5	34.12	21.2	16.26	11.26	9.33	8.53	8.1	7.82	7.64	7.31	7.08	6.96	6.9	6.85	6.82	6.8	6.74
2	4999.5	99	30.82	18	13.27	8.65	6.93	6.23	5.85	5.61	5.45	5.18	4.98	4.88	4.82	4.79	4.76	4.74	4.69
3	5403.35	99.17	29.46	16.69	12.06	7.59	5.95	5.29	4.94	4.72	4.57	4.31	4.13	4.04	3.98	3.95	3.92	3.91	3.86
4	5624.58	99.25	28.71	15.98	11.39	7.01	5.41	4.77	4.43	4.22	4.07	3.83	3.65	3.56	3.51	3.48	3.46	3.44	3.4
5	5763.65	99.3	28.24	15.52	10.97	6.63	5.06	4.44	4.1	3.9	3.75	3.51	3.34	3.26	3.21	3.17	3.15	3.13	3.09
6	5858.99	99.33	27.91	15.21	10.67	6.37	4.82	4.2	3.87	3.67	3.53	3.29	3.12	3.04	2.99	2.96	2.93	2.92	2.88
7	5928.36	99.36	27.67	14.98	10.46	6.18	4.64	4.03	3.7	3.5	3.36	3.12	2.95	2.87	2.82	2.79	2.77	2.75	2.71
8	5981.07	99.37	27.49	14.8	10.29	6.03	4.5	3.89	3.56	3.36	3.23	2.99	2.82	2.74	2.69	2.66	2.64	2.62	2.59
9	6022.47	99.39	27.35	14.66	10.16	5.91	4.39	3.78	3.46	3.26	3.12	2.89	2.72	2.64	2.59	2.56	2.54	2.52	2.48
10	6055.85	99.4	27.23	14.55	10.05	5.81	4.3	3.69	3.37	3.17	3.03	2.8	2.63	2.55	2.5	2.47	2.45	2.43	2.4
12	6106.32	99.42	27.05	14.37	9.89	5.67	4.16	3.55	3.23	3.03	2.9	2.66	2.5	2.42	2.37	2.34	2.31	2.3	2.26
15	6157.28	99.43	26.87	14.2	9.72	5.52	4.01	3.41	3.09	2.89	2.75	2.52	2.35	2.27	2.22	2.19	2.17	2.15	2.11
20	6208.73	99.45	26.69	14.02	9.55	5.36	3.86	3.26	2.94	2.74	2.6	2.37	2.2	2.12	2.07	2.03	2.01	1.99	1.96
24	6234.63	99.46	26.6	13.93	9.47	5.28	3.78	3.18	2.86	2.66	2.52	2.29	2.12	2.03	1.98	1.95	1.93	1.91	1.87
30	6260.65	99.47	26.5	13.84	9.38	5.2	3.7	3.1	2.78	2.58	2.44	2.2	2.03	1.94	1.89	1.86	1.84	1.82	1.78
40	6286.78	99.47	26.41	13.75	9.29	5.12	3.62	3.02	2.69	2.49	2.35	2.11	1.94	1.85	1.8	1.76	1.74	1.72	1.68
60	6313.03	99.48	26.32	13.65	9.2	5.03	3.54	2.93	2.61	2.4	2.26	2.02	1.84	1.75	1.69	1.66	1.63	1.61	1.57
120	6339.39	99.49	26.22	13.56	9.11	4.95	3.45	2.84	2.52	2.31	2.17	1.92	1.73	1.63	1.57	1.53	1.5	1.48	1.43
240	6352.61	99.49	26.17	13.51	9.07	4.9	3.41	2.8	2.47	2.26	2.12	1.86	1.67	1.57	1.5	1.46	1.43	1.41	1.35
500	6359.5	99.5	26.15	13.49	9.04	4.88	3.38	2.78	2.44	2.24	2.09	1.83	1.63	1.53	1.47	1.42	1.39	1.36	1.3
1000	6362.68	99.5	26.14	13.47	9.03	4.87	3.37	2.76	2.43	2.22	2.08	1.82	1.62	1.51	1.45	1.4	1.37	1.34	1.28

Explanation of each column and row: Criticla Values of the *F* distribution (α = 0.01)

- Columns: df_dvalues (degrees of freedom of denominator)
- Rows: df_nvalues (degrees of freedom of numerator)
- *F* critical values: The value in each cell is the critical value of *F* at the intersection of the corresponding dfn and dfd values. The critical value of *F* is used to test whether the means of two populations are significantly different when analyzing variance. It compares the variance between two samples to the variance within each sample.
- α = 0.01: The significance level of 0.01 indicates that there is a 1% chance of rejecting the null hypothesis when it is actually true. This level of significance is a more stringent threshold than 0.05 and is often used when a high degree of confidence is required in the statistical analysis.

Critical Values of the Chi-Squared Distribution
Significance Level

	5.0%	1.0%	0.1%
1	3.84	6.63	10.83
2	5.99	9.21	13.82
3	7.81	11.34	16.27
4	9.49	13.28	18.47
5	11.07	15.09	20.52
6	12.59	16.81	22.46
7	14.07	18.48	24.32
8	15.51	20.09	26.12
9	16.92	21.67	27.88
10	18.31	23.21	29.59
11	19.68	24.72	31.26
12	21.03	26.22	32.91
13	22.36	27.69	34.53
14	23.68	29.14	36.12
15	25	30.58	37.7
16	26.3	32	39.25
17	27.59	33.41	40.79
18	28.87	34.81	42.31
19	30.14	36.19	43.82
20	31.41	37.57	45.31
25	37.65	44.31	52.62
30	43.77	50.89	59.7
40	55.76	63.69	73.4
50	67.5	76.15	86.66
60	79.08	88.38	99.61
80	101.88	112.33	124.84
100	124.34	135.81	149.45
120	146.57	158.95	173.62
150	179.58	193.21	209.26
200	233.99	249.45	267.54
500	553.13	576.49	603.45
1000	1074.68	1106.97	1143.92

Explanation of each column and row: Critical Values of the Chi-Squared Distribution

- Columns: α values (Significance level)
- Rows: df (degrees of freedom)
- α values: The significance level is the probability of making a Type I error (rejecting the null hypothesis when it is actually true). The three different significance levels provided in the table are 5%, 1%, and 0.1%, which represent different thresholds for the probability of making a Type I error. A lower significance level requires stronger evidence to reject the null hypothesis.

Critical Values of the Correlation Coefficient
Significance Level

	10%	5%	25%	1%
2				
3	6.314	12.706	2.414	63.657
4	2.92	4.303	1.604	9.925
5	2.353	3.182	1.423	5.841
6	2.132	2.776	1.344	4.604
7	2.015	2.571	1.301	4.032
8	1.943	2.447	1.273	3.707
9	1.895	2.365	1.254	3.499
10	1.86	2.306	1.24	3.355
11	1.833	2.262	1.23	3.25
12	1.812	2.228	1.221	3.169
13	1.796	2.201	1.214	3.106
14	1.782	2.179	1.209	3.055
15	1.771	2.16	1.204	3.012
16	1.761	2.145	1.2	2.977
17	1.753	2.131	1.197	2.947
18	1.746	2.12	1.194	2.921
19	1.74	2.11	1.191	2.898
20	1.734	2.101	1.189	2.878
21	1.729	2.093	1.187	2.861
22	1.725	2.086	1.185	2.845
23	1.721	2.08	1.183	2.831
24	1.717	2.074	1.182	2.819
25	1.714	2.069	1.18	2.807
26	1.711	2.064	1.179	2.797
27	1.708	2.06	1.178	2.787
28	1.706	2.056	1.177	2.779
29	1.703	2.052	1.176	2.771
30	1.701	2.048	1.175	2.763

Explanation of each column and row: Critical Values of the Correlation Coefficient

- Columns: Significance level (α)
- Rows: Sample size (n)
- n: The sample size is the number of observations in a sample.
- α: The significance level is the probability of making a Type I error (rejecting the null hypothesis when it is actually true). It is the threshold used to determine whether a result is statistically significant. The four values of α used in this table are 0.1, 0.05, 0.025, and 0.01, which correspond to 90%, 95%, 97.5%, and 99% confidence levels, respectively.

Critical Values of the Correlation Coefficient
Significance Level

	10%	5%	2%	1%
1				
2				
3	0.988	0.997	1	1
4	0.9	0.95	0.98	0.99
5	0.805	0.878	0.934	0.959
6	0.729	0.811	0.882	0.917
7	0.669	0.754	0.833	0.875
8	0.621	0.707	0.789	0.834
9	0.582	0.666	0.75	0.798
10	0.549	0.632	0.715	0.765
11	0.521	0.602	0.685	0.735
12	0.497	0.576	0.658	0.708
13	0.476	0.553	0.634	0.684
14	0.458	0.532	0.612	0.661
15	0.441	0.514	0.592	0.641
16	0.426	0.497	0.574	0.623
17	0.412	0.482	0.558	0.606
18	0.4	0.468	0.543	0.59
19	0.389	0.456	0.529	0.575
20	0.378	0.444	0.516	0.561
21	0.369	0.433	0.503	0.549
22	0.36	0.423	0.492	0.537
23	0.352	0.413	0.482	0.526
24	0.344	0.404	0.472	0.515
25	0.337	0.396	0.462	0.505
26	0.33	0.388	0.453	0.496
27	0.323	0.381	0.445	0.487
28	0.317	0.374	0.437	0.479
29	0.311	0.367	0.43	0.471
30	0.306	0.361	0.423	0.463

The table we provided gives the critical values of Pearson's r for a two-tailed test, which is used to test the hypothesis that the population correlation coefficient is equal to a specific value (usually zero). This table also includes the degrees of freedom, which is calculated as $(n-2)$ for the sample size n.

Note: the above table (Critical Values of the Correlation Coefficient) is used to test the significance of the correlation coefficient, while this table (Critical Values for a two-tailed test) is used to test the hypothesis that the population correlation coefficient is equal to zero.

Appendix 2

Glossary of Terms

Preface

Probability is a branch of mathematics that enables you to compute what will happen when you run an experiment based on assumptions about how the world works.
Statisticians try to reason about what the world is like based on the outcome of an investigation. Statistics is a method of understanding from observation.
Data science involves taking large data sets and trying to make them useful or informative, especially for understanding the world or making informed decisions.
Machine learning is about not programming every detail but instead involves using the data to make decisions or choices in a program.
System is a collection of individual elements; these elements exhibit nontrivial interactions as having coherent behavior (and they are not always positively synergetic) such that individuals must interact not just within themselves but also with all external players.
System dynamics are interconnected elements that function together to create structure or adapt to changes in the environment.

Part 1. Applied Statistics Analysis

Descriptive statistics are used to organize or summarize a particular set of measurements. In other words, a descriptive statistic describes information that is compiled to describe a population at a given point in time.
Inferential statistics use data gathered from a sample to make inferences about the larger population from which the sample was drawn.
Statistical bias refers to measurement or sampling errors that are systematic in nature and produced by the measurement or sampling process.
Selection biases are statistical biases in which each member of a population—consciously or unconsciously—is chosen selectively, which leads to a misleading conclusion.
Constant denoted as C, represents an attribute or condition consistent across all subjects within a research.
Variables denoted by V, signify attributes that manifest variably among subjects.
Independent variable are variables that exert influence on the dependent variables.
Dependent variable is a variable whose values are contingent upon another.
Scales of measurement describe how variables and values are defined and categorized. Each scale of measurement has specific properties that in turn determine its appropriateness for use in accurate statistical analyses.
Data are values that convey information, other basic units of meaning, or simply sequences of symbols that may be further interpreted.

Numerical data are expressed with digits while categorical data are collections of information that is divided into groups.
Distribution is a particular value or range of values for a variable. Generalized functions are used to formulate solutions to partial differential equations.
Central tendency is a summary statistic that represents the typical value of a data set. These measures indicate the center of a data set, where most values fall.
Mode is the value that appears most frequently in a data set.
Median is the value that lies in the middle of a list of ordered numbers.
Mean is the arithmetic average (the most commonly used measure of central tendency), or the sum of all values divided by the total number of values.
Variability refers to how spread-out a group of values is. The terms variability, spread, and dispersion are synonyms and refer to how spread-out data are.
Range is simply the highest value minus the lowest value.
Variance is the average squared difference of the scores from the mean.
Standard deviation is simply the square root of the variance.
Standard score (more commonly referred to as a Z score or a t score) is a handy statistic because it (a) allows us to calculate the probability of a score occurring within a normal distribution and (b) enables us to compare two scores from different distributions.
***Z* score** is a statistical measure that tells you how far a data point is from the rest of a data set.
***t* score** is a standard Z score shifted and scaled to have a mean of 50 and a standard deviation of 10.
Coefficient of variation is the standard deviation relative to the mean of the data set which is obtained by dividing the standard deviation by the mean of the sample.
Distribution is the range of data/population/sample and how data are spread out over that range.
Normal distribution, sometimes called the bell curve, is a distribution that resembles a bell. The bell curve is symmetrical: half of the data will fall to the left of the mean, and half will fall to the right.
Central limit theorem posits that the distribution of sample means approximates a normal distribution as the sample size increases, regardless of the population's distribution; the average of the sample means and standard deviations will equal the population mean and standard deviation.
Outlier is an observation that lies an abnormal distance from other values in a random sample from a population.
Confidence interval is a range of values that is likely to include a population value with a certain degree of confidence; it is often expressed as a percent whereby a population mean lies between an upper and lower interval.
Hypothesis is an assumption, an idea that is proposed for the sake of argument so that it can be tested to see whether it might be right.
Hypothesis testing makes a hypothesis and examines whether it is true using statistics. Because it is challenging to prove that a particular hypothesis is true, we usually prove that a reverse hypothesis (null hypothesis) is false.
Null hypothesis (or statistical hypothesis) is a hypothesis of no difference or no relationship.
Alternative hypothesis is a hypothesis of a difference or relationship.

Type I error occurs when a researcher rejects the null hypothesis but that hypothesis is true. The probability of type I error is called the significance level.
Type II error occurs when a researcher fails to reject the null hypothesis but that hypothesis is false. In practice, type II error (β) error is more serious than type I error.
Research is an organized method of finding answers to questions. It is systematic because it is a process broken up into clear steps that lead to conclusions.
Independent variable is what the scientist changes or what changes on its own.
Dependent variable is what is being studied or measured.
Validity is an indication of how sound your research is. It applies to both the design and the methods of your analysis.
Reliability is the overall consistency of a measure. It produces similar results under consistent conditions.
***t* test** is a statistical model (method) that compares two means (between groups, repeated measures mean). It tells you how significant differences between groups are.
One-sample *t* test is used to compare a sample mean with a known population mean or some other meaningful, fixed value.
Independent-samples *t* test is used to compare two means from independent groups.
Paired-samples *t* test is used to compare two means that are repeated measures for the same participants—scores might be repeated across different measures or across time.
***t* score** is a ratio of the difference between two groups and the difference within the groups. The larger the score, the greater the difference between the groups. The smaller the score, the greater the similarity between the groups.
***p* value** is the probability that you will obtain a test result given an actual distribution where p indicates a certain confidence level. You often hear people say, 'I am 90% confident I will get that job' or something like this involving a confidence level (p).
Analysis of variance (ANOVA) is the generalization (expansion) of the t test and a statistical method that compares three or more means. It is a test of a hypothesis that is appropriate to compare means of a continuous variable in two or more independent comparison groups.
***F* statistic** is calculated by dividing the difference between the group means (variance) by the difference between subjects within groups (variance). If there are no treatment differences (i.e., if there is no actual effect), we expect F to be 1. If there are treatment differences, we expect F to be greater than 1.
Degrees of Freedom is the maximum number of logically independent values, which are values that have the freedom to vary, in the data sample. The value is calculated by subtracting one from the number of items within the data sample.
Two-way ANOVA compares the differences in means between groups that have been split on two independent variables (called factors). It is to determine whether there is an interaction between two independent variables on the dependent variable.
Experimental research is research that includes an experiment. It contains one or more experimental treatments (experimental variables), a control condition, and dependent variables.
Experimental design is a plan for the experiment. It specifies the identification and manipulation of the experimental variable(s) and other independent variable(s) and the data-collection procedures.

Quasi-experimental research is research that resembles experimental research but is not true experimental research. Although the independent variable is manipulated, participants are not randomly assigned to conditions or orders of conditions.
Nonexperimental quantitative research is research that includes the collection and analysis of quantitative data but does not manipulate an independent variable.
Qualitative research is used to investigate complex phenomena in depth. Priori assumptions and conclusions (i.e., the directional research hypothesis in quantitative research) are avoided; there is no manipulation or intervention, and it is mostly conducted in natural settings.
Case study is a detailed investigation of a case, person, organization, specific event, and so on. It is commonly used in historical organizational studies and observational studies.
Correlation is the degree to which two variables go together. Statistics can show the relationship between two continuous variables; the relationship between two variables is referred to as their correlation.
Correlation coefficient is a number that summarizes the direction and degree (closeness) of the linear relationship between two variables.
Covariance is a measure of the joint variability of two random variables. The sign of the covariance can reflect whether the two variables change in the same direction (positive) or in different directions (negative).
Linear regression is for predicting the value of one dependent variable from the value of one independent variable; it is the most widely used type of regression.
Multiple linear regression is used to estimate the relationship between two or more independent variables and one dependent variable.
Logistic regression is a generalized linear regression in the sense that it does not output the weighted sum of the inputs but instead passes through a sigmoid function that can map any real value between 0 and 1.
Chi-square statistic (χ^2) is used to investigate whether distributions of categorical variables differ from one another. Chi-square is a single number that tells you the relationship between two categorical variables.
Critical value is a cut-off value that constitutes the boundary of the rejection region(s). In other words, critical values divide the scale of your test statistic into the rejection region and non-rejection region.

Part 2. Applied Time Series Analysis

Time series is a sequence of observations recorded at regular time intervals. It is a set of data points ordered in time, where time is usually the independent variable.
Time series analysis involves understanding various aspects of the inherent nature of a series so that you are better informed to create meaningful and accurate forecasts.
Probability is how likely something is to happen. Whenever we are unsure about the outcome of an event, we can talk about the probabilities of certain outcomes—how likely they are.
Variable is a symbolic name associated with a value and whose associated value may be changed. It may represent a number, a vector, a matrix, a function, the argument of a function, a set, or an element of a set.

System variables are predefined variables that control the number of random samples used to represent uncertain quantities.
Parameter is an arbitrary numerical or other measurable factor whose value character-rizes a member of a system. It can be any value, and it is used as a referent to determine other variables and to describe the conditions of a system.
Deterministic system is a conceptual doctrine of determinism applied to a system to understand everything that has occurred and will occur in the system based on the physical outcomes of causality; every action (or cause) produces a reaction or effect, and every reaction in turn becomes the cause of subsequent reactions.
Frequency is the number of occurrences of a repeating event per unit time.
Relative frequency is the ratio of the number of outcomes in which a specified event occurs to the total number of trials.
Histogram is a graph that shows the frequency of numerical data using rectangles. The height of a rectangle represents the distribution frequency of a variable.
Probability distribution is a statistical function that links or lists all possible outcomes a random variable can take, in any random process, with its corresponding probability of occurrence.
Random variable is a variable whose value is unknown or a function that assigns values to each of an experiment's outcomes; its value depends on the outcomes of a random phenomenon.
Probability density is the relationship between observations and their probability. And a probability density function is a probability distribution of a continuous random variable, or density of a continuous random variable.
Empirical probability is the ratio of the number of outcomes in which the event occurs to the total number of trials, not in a theoretical sample space but in an actual experiment. In a more general sense, the empirical probability estimates probabilities from experien-ce and observation.
Symmetry is a sense of harmony, proportion, and balance. It refers to an object that is invariant under some transformations, including translation, reflection, rotation, or scaling.
Symmetrical distribution occurs when the values of variables appear at regular frequencies and often the mean, median, and mode all occur at the same point.
Skewed distribution is neither symmetric nor normal because the data values trail off more sharply on one side than on the other.
Modality of a distribution is determined by the number of peaks it contains (i.e., how many maxima are there?). Most distributions have only one peak, but it is possible that you will encounter distributions with two or more peaks.
Kurtosis is a measure of the combined weight of a distribution's tails relative to the center of the distribution. It is the degree to which the tails of a distribution are heavy or light compared to a normal distribution.
Log is the inverse function to exponentiation. It is the exponent or power to which a base must be raised to yield a given number. A logarithm can be defined with respect to a base (b) where the base b logarithm of x is equal to y because x equals b to the power of y ($\log(x) = y$ because of $x = b^y$.
Log transformation is a data transformation method that replaces each variable x with a $\log(x)$.

Expected value is the n-th moment of a distribution (or set of data) about a number and is the expected value of the n-th power of the deviations about that number.
Exponential function is an essential function that is often used to model the time elapsed between events. To create an exponential function, we let the independent variable (X) be the exponent.
Boltzmann distribution is a distribution or measure that gives the probability that a system will be in a certain state as a function of that state's energy and the temperature of the system.
Entropy is a measure of disorder, or unpredictability, in a system. It refers to lack of knowledge and a degree of uncertainty. High entropy means low information gain, and low entropy means high information gain.
Uniform distribution, also known as a rectangular distribution, has a constant probability for the interval $[a, b]$. The uniform distribution is mainly used for generating random numbers.
Beta distribution is used in Bayesian inference, as it appears as the a priori distribution of other distributions. It is defined for values between 0 and 1.
Cauchy distribution, also known as the Lorentz distribution, is a family of continuous probability distributions that resemble the normal distribution family of curves.
Frequentist probability explains situations in which we know that an event can be simulated or triggered again (e.g., drawing a card).
Bayesian probability explains situations that are difficult to simulate and trigger, yet because of its uncertain nature, we treat it as a frequentist probability.
Marginal probability is a subset of variables. For example, if we know p_{xy}, and we want to know p_x, we sum over the ys of p_{xy}. For a continuous distribution, we integrate over the y of p_{xy}.
Conditional probability is a probability of an event happening given another event happening. $p_y|x = p_{yx}/p_x$, where p_x must be greater than 0.
Deterministic model: Deterministic models, by definition, follow a structured and predictable path. When a specific set of initial conditions is provided, these models invariably produce a singular trajectory that the system will follow.
Stochastic model: stochastic models represent an element of randomness, uncertainty and variability often encountered in real-world systems. Even with the same initial conditions, stochastic models can yield varying outcomes upon each run.
Stochastic process is a collection of random variables indexed by t and a collection of deterministic functions of time indexed by outcome s.
Trajectory is a sequence that describes the evolution of a state variable or vector (i.e., that describes a state variable or state vector as a function of time).
Joint probability is a measure of two events occurring together and at the same point in time. Simply it is the probability of event Y occurring at the same time that event X occurs.
Bayes' theorem is one of the most prominent implementations of conditional probability. It describes the probability of an event, based on prior knowledge of conditions that might be related to the event.
Joint probability distribution is the joint probability of two or more random variables. The joint probability of events A and B is calculated as the probability of event A given

event B multiplied by the probability of event B. This can be stated formally as P(A and B) = P(A given B) × P(B).

Stationarity means that the statistical properties of the process do not change over time. Most methods are designed to work on a stationary time series.

Maximum likelihood estimation is a method used for determining values for the parameters of a model. It is a relatively simple method of constructing an estimator for an unknown parameter.

Markov property is the concept that at a given time step and knowing the current state, we will not obtain any additional information about the future by gathering information about the past.

Markov process is a random process employing the Markov property; the future behavior of a Markov process depends only on the current state and not on past states.

Markov chain is a Markov process with discrete time and a discrete state space. It is a discrete sequence of states, each drawn from a discrete state space (finite or not), and it follows the Markov property.

Pattern is a particular way in which something is done, is organized, or happens. Most time series data usually have at least one of three patterns: trend, seasonality, and/or cycles.

Trend describes the general behavior of a time series. There can be an up-to-down trend or a down-to-up trend. A stationary or horizontal trend defines a time series with neither positive nor negative long-term patterns.

Seasonality is any kind of fluctuation (change) in a time series that is caused by calendar-related events. These events can be the time of year (like winter or summer) or the time of day or week.

Cyclical pattern in a time series is a change that is not related to seasonal factors. These are rises and falls with nonfixed magnitudes that can last for more than a calendar year.

Random walk is any sequence of discrete steps (always of the same length) moving in random directions. Random walks are used to describe a discrete-time process.

Monte Carlo simulation is a numerical solution method involving a probabilistic element. The simulation is based on a repetitive drawing of random variables.

Moving average model is a common approach for modeling univariate time series. It specifies that the output variable depends linearly on the current and various past values of a stochastic (imperfectly predictable) term.

Autoregressive model is used for forecasting when there is a correlation between the values in a time series and the values that precede and succeed them. It predicts future behavior based on past behavior.

Autoregressive moving average is a model that uses time series data to either better understand the data set or to predict future trends. It can be broken down into two main parts: the autoregressive stage takes advantage of the connection between a predefined number of lagged observations and the current one; the moving average stage takes advantage of the relationship between the residual error and the observations.

Autocorrelation is a linear relationship between states X_1 and X_2 occupied at two different times t_1 and t_2. It refers to the degree of correlation of the same variables between two successive time intervals and measures how the lagged version of the value of a variable is related to the original version of it in a time series.

Power spectrum indicates the power of each frequency component of a source time domain. It shows at which frequencies variations are strong and at which frequencies variations are weak.
Fourier transform is a way of splitting something up into a bunch of sine waves. In mathematical terms, the Fourier transform is a technique that transforms a signal into its constituent components and frequencies.
Master equations are the continuous-time versions of recurrence relations for Markov chains. They represent a phenomenological set of first-order differential equations describing the time evolution of (usually) the probability of a system occupying each one of a discrete set of states with regard to a continuous-time variable t.
Wiener process is a continuous-time stochastic process of one-dimensional Brownian motion. It characterizes erratic motion (i.e., diffusion) and the resulting motion, which can be viewed as a scaled random walk on any finite time interval, is almost surely continuous.
Ornstein-Uhlenbeck process is a continuous-time stochastic process defined as the solution of a special kind of stochastic differential equation called the Langevin equation.
Drift-diffusion analysis is a model of sequential sampling with diffusion (Brownian) signals, where the decision maker accumulates evidence until the process hits a stopping boundary and then stops and chooses the alternative that corresponds to that boundary.
Parametric data analysis refers to the use of methods when the data being studied comes from a sample or population of data that is normally distributed. It assumes that the variance is homogeneous and the type of data is supposed to be interval or ratio (sometimes referred to as continuous variables/data).

Part 3. Applied Systems Analysis

Systems analysis is fundamental to almost all aspects of life. It is a method for identifying and solving problems that looks at each component in the overall system for the purpose of achieving specific goals. This approach can inform the decision-making process or enhance efficient use of time and resources in life. It includes defining process steps toward an end result.
Scalar is a number. Examples of scalars are temperature, distance, speed, and mass—all quantities that have a magnitude but no direction, other than perhaps positive or negative.
Vector is a list of numbers. It is as a magnitude and a direction (e.g., a quantity-like velocity: the fighter jet's velocity is 250 mph north-by-northwest). In this way, a vector is a directed arrow pointing from the origin to the end point given by the list of numbers.
Matrix is a table of numbers, with columns and rows, rather than a list (a vector is a single column of a data set). They are usually denoted with a capital letter, like this: A.
Eigenvalues are a special set of scalars associated with a matrix equation that are sometimes also known as characteristic roots or characteristic values.
Eigenvectors are vectors that do not change their orientation when multiplied by the transition matrix but just scale by a factor of the corresponding eigenvalues.
Function is one of the basic concepts in mathematics. It defines a relationship between a set of inputs and a set of possible outputs where each input is related to one output.

Differential calculus is a fundamental division of calculus. It is a subfield of calculus that deals with infinitesimal change in some varying quantity. Equations that define the relationship between variables and their derivatives are called differential equations.
Derivative of a function represents an instantaneous rate of change in the value of a dependent variable with respect to the change in value of an independent variable. Entities that change are called variables, and the rate of change in one variable with respect to another is called the derivative.
Integral of a function can be interpreted geometrically as the area under the curve of the mathematical function $f(x)$ plotted as a function of x. It gives you a mathematical way of drawing an infinite number of blocks and getting a precise analytical expression for the area.
Graph is represented by a set of points, known as vertices or nodes, connected by a set of lines known as edges.
Graph theory is ultimately the study of relationships. Given a set of nodes and connections, it can abstract anything from layouts to computer data. Graph theory provides a helpful tool for quantifying and simplifying the many moving parts of dynamic systems.
Network is a collection of nodes connected by links (i.e., a graph).
Nodes are the parts of graphs that store data. Nodes contain any amount of data.
Edges are the connections between nodes, but edges can store data too. Edges include data about the strength of a connection.
Adjacency list represents the edge list of a graph, as the nodes normally have a number that corresponds to the index in an array.
Degree is the number of links incident on a node.
Clustering indicates how frequently links of a node are linked with each other. The more links a node has, the more clustering there is.
Regular network, each vertex has the same number of neighbors (i.e., every vertex has the same *degree* or valence).
Random network, there is a population of *n* nodes by randomly picking one of the nodes and linking them randomly with another node.
Small-world network is a type of graph in which most nodes are not neighbors of one another, but the neighbors of any given node are likely neighbors themselves and most nodes can be reached from every other node by a small number of hops or steps.
Scale-free networks are a type of network characterized by the presence of large hubs. A scale-free network is one with a power-law degree distribution, at least asymptotically.
Network analysis is used to identify the most important nodes within a graph; in other words, how central a node is within its parent graph.
Degree centrality assigns a score based simply on the number of links held by each node—the higher the degree of a node, the more important it is in the graph.
Closeness centrality (or closeness) of a node is a measure of its centrality in a network that is calculated as the normalized average of all of its geodesic distances.
Betweenness centrality shows which nodes are bridges between nodes in a network; it is calculated by identifying all shortest paths (geodesic distances) between all pairs of nodes and then by counting how many times each node falls on one of these paths.
Module is basically a bunch of related code saved in a file with the extension (i.e., ~.py). For example, we can define a function to welcome new students to a particular course.

Package is basically a directory of a collection of modules. Packages allow the hierarchical structure of the module namespace.
Library is an umbrella term referring to a reusable chunk of code. Usually, a library contains a collection of related modules and packages.
Agent-based modeling is a computational simulation in which artificial entities interact over time within customized environments.
Agents are programmed to represent entities (i.e., individuals, humans, objects, cells etc.) that behave in precisely specified ways. They are adaptive in that they respond to their environment through learning and evolution and are autonomous in that they control their own goals, states, and behaviors.
Environment (in agent-based modeling) is a grid space in which the agents can move around, one cell at a time, in a neighboring cell (according to Moore's definition of a neighborhood).
Cellular automata is a mathematical model that contains a grid of cells, usually (but not always) two dimensions.
High-performance computing (HPC) represents the tools and systems used to implement and create high-performance computing systems. It employs computer clusters to solve advanced computation problems and integrates multidisciplinary fields, combining system software, programming languages, algorithms, and computational techniques.
Node (in HPC) is a collection of cores with shared memory and storage (i.e., 128~512G memory with 16~128 cores).
Core (in HPC) is capable of running processes, and has its own processing logic and floating point. It is a unit of cluster computation (one processor core equals one thread of execution) and there is also a small set of login nodes.
Game theory is a means of mathematically understanding decision making. It is the study of mathematical models of strategic interactions among rational agents applied in many fields of social science, as well as systems science.
Game in game theory is an abstract situation in which a player is faced with two or more choices or moves. These choices then have outcomes or payoffs.
Rationality is simply the understanding that every agent knows that all other agents are just as rational and have the same level of understanding and knowledge as he or she does. It refers to the fact that agents always prefer the higher reward or payoff considering what other agents will do.
Players are rational entities who participate in a game (e.g., bidders in an auction setting, players playing rock-paper-scissors, politicians participating in elections).
Payoff is the reward that all players get when a certain outcome is achieved. It can be either positive or negative.
Nash equilibrium is an action chosen by each player such that no player would want to change his or her action. Changing one's action from the Nash equilibrium means a player is not playing optimally.
Prisoner's dilemma is a two-player game commonly studied in game theory. It is also a thought experiment (in strategic situation) that challenges two completely rational agents to a dilemma: cooperate with police and disclose information, or do not cooperate and remain silent.

Perfect information in games is when every agent has knowledge of all of the possible actions other agents can take, what actions they are performing, and what reward they will get.
Imperfect information in games is when agents are aware of the nature and motives of other agents and what payoff they will get in all possible outcomes but they do not know what actions they are performing.
Incomplete information in games is when agents do not have information about the type of other agents. Even if a given agent is able to see the actions taken by the other agents, he or she does not know about the motivation of the other agents or the reward the agents will get for performing that action.
Strategies are about how players act according to their behavioral phenotypes. The goal is to decide what is best in terms of interactions between two (or more) players with different views of strategy that translate into payoffs.
Utility in game theory represents the motivations of players; a utility function for a given player assigns a number for every possible outcome of the game where a higher number implies that the outcome is more preferred.

Appendix 3

Data File Instructions

The datasets used in 'Practical Foundation of Applied Systems Analysis' originate from lecture materials that the author utilized at Seoul National University, the University of Connecticut, and the International Institute for Applied Systems Analysis. Each dataset contains numerous data points.

To gain access to these datasets, you can directly visit the book's dedicated GitHub repository:

- Visit https://github.com/pcw8531/PFASA

You will see a list of files and directories. Navigate through them to find the desired data. To download individual files, click on the file name, then click on the button, and save the page that opens. To download the entire repository, click on the green 'Code' button on the main page of the repository, then click 'Download ZIP'.

Note: Your access can only be used to complete a single registration.

The 'Practical Foundation of Applied Systems Analysis' provides all the necessary data sets, so upon logging in, you will gain easy access to all data for your analysis platform, such as Jupyter notebook (or Python). All data will also be automatically updated whenever there are updates in accordance with the guide in the book.

The data lists follow:

Advertisign.csv
AirPassengers.csv
dataset_stress.txt
dataset_velocity.txt
facebook_combined.txt
HousePrice.csv
housePrice.txt
HW2phasedataA.txt
HW2phasedataB.txt
HW3phasedataA.txt
HW3phasedataB.txt
iris.txt
outestdata.txt
outestdata.txt.docx

phasedataS1.txt.docx
phasedataS2.txt.docx
phasedataS3.txt.docx
phasedataS4.txt.docx
phasedataS5.txt.docx
pokemon.csv
sales-of-shampoo-over-a-three-ye.csv
timeseries_ac_data.txt
timeseries_PA_periodtwo.txt
weight-height.csv
winemag-data_first150k.csv

If you have a problem with any aspect of this download, please let us know, and we will do our best to address the problem (pcw8531@sun.ac.kr). We appreciate any contributions, questions, and/or corrections. Also, please contact us if you feel your use of our code examples falls outside fair use or the permissions given above.

Appendix 4

Codebook Instructions

This book is here to guide you with your systems analysis work. To achieve this, you may use the code presented in this book through your computers and platforms. You do not need to contact us for permission unless you are reproducing a significant portion of the code. Indeed, feel free to explore the code corresponding to your work.

To gain access datasets for some of these codes, you can directly request the book's dedicated GitHub repository:

- Visit https://github.com/pcw8531/PFASA

All of the discussed codebooks are included in the book. A list of the code practice examples follows:

Part 1. Applied Statistics Analysis

1.1. Data mining.ipynb
1.2. Histogram function with Frequency.ipynb
1.2. Central tendency; Mode, Mean, Median.ipynb
1.2. Range, Interquartile range, Variance. Ipynb
1.2. Standard Deviation, *Z* score.ipynb
1.3. Measures of Central tendency.ipynb
1.3. Measures of Dispersion.ipynb
1.3. Measures of Standard error.ipynb
1.3. Outlier detection.ipynb
1.4. Sampling for inference.ipynb
1.5. *t* distribution.ipynb
1.5. *t* test.ipynb
1.6. *F* distribution.ipynb
1.6. ANOVA test.ipynb
1.7. Covariance.ipynb
1.7. Linear relationship between two variables.ipynb
1.7. Pearson's correlation.ipynb
1.7. Spearman's correlation.ipynb
1.8. Simple linear regression.ipynb
1.8. Multiple regression.ipynb
1.8. Logistic regression.ipynb
1.9. Chi-square test.ipynb

Part 2. Applied Time Series Analysis

2.1. Continuous-time and continuous-state programing example.ipynb
2.1. Graphical representation of the X-exact solution. ipynb
2.2. Frequency histogram. ipynb
2.2. Relative frequency. ipynb
2.2. Histograms created using different bin_size.ipynb
2.3. Discrete event values (flipping a coin).ipynb
2.3. Discrete variable example.ipynb
2.3. Continuous variable example.ipynb
2.3. Probability distribution.ipynb
2.3. Probability density. ipynb
2.3. Probability density with error bands. Ipynb
2.3. Data-estimated probability density with standard deviation. ipynb
2.3. Various discrete and continuous probability distributions.ipynb
2.4. Lognormal distribution. ipynb
2.4. Relationship between two variables. ipynb
2.4. Expectation values (cluster methods). Ipynb
2.5. Exponential decay. ipynb
2.5. Boltzmann distribution of velocities. ipynb
2.5. Information gain.ipynb
2.5. Information gain and probability.ipynb
2.5. Entropy production. Ipynb
2.5. Maximum likelihood estimation.Ipynb
2.5. Maximum likelihood with regression.Ipynb
S2.1. Gaussian distribution.ipynb
S2.1. Binomial distribution.ipynb
S2.1. Negative Binomial distribution.ipynb
S2.1. Hypergeometric distribution.ipynb
S2.1. Lognormal distribution.ipynb
S2.1. Uniform distribution.ipynb
S2.1. Beta distribution.ipynb
S2.1. Student's t-distribution.ipynb
S2.1. Chi-Square distribution.ipynb
S2.1. F-distribution.ipynb
S2.1. Cauchy distribution.ipynb
S2.2. Deterministic and Stochastic Models.ipynb
2.6. Three cards are shuffled into a sequence of states. ipynb
2.6. Monte Carlo trajectory. ipynb
2.6. Student attendance with Markov chain. ipynb
2.6. Time-dependent moments and expectation values. ipynb
2.7. Stationary process and practice.ipynb
2.9. Markov chain model. ipynb
2.9. Evolution of Markov chain. ipynb
2.10. Moving Average(0) trajectory. ipynb

2.10. Moving Average(1) trajectory. ipynb
2.10. Autoregressive model.ipynb
2.11. Autoregressive Moving Average model. ipynb
2.11. Autoregressive Moving Average model pracitce.ipynb
S2.3. Autocorrelation simulation example. ipynb
S2.4. Power spectrum simulation example. Ipynb
S2.4. Fourier Transoms. ipynb
2.12. Master equation simulation. ipynb
2.13. Wiener process simulation. ipynb
2.14. Ornstein-Uhlenbeck process. ipynb

Part 3. Applied Systems Analysis

3.2. Vectors addition.ipynb
3.2. Scalar multiplication.ipynb
3.2. Matrix multiplication.ipynb
3.2. Matrix transposition.ipynb
3.2. Determinants.ipynb
3.2. Inverse matrices.ipynb
S3.1. Summation and Product.ipynb
S3.1. Factorial.ipynb
S3.1. Conditional.ipynb
S3.1. Cartesian Multiplication.ipynb
S3.2. Define the differential equation.ipynb
S3.2. Define the function $f(x)$ and its derivative $f'(x)$.ipynb
S3.2. Define the integral function.ipynb
S3.2. Differential equation (predator-prey relationships).ipynb
S3.2. Derivative (SEIR model).ipynb
3.3. Regular network.ipynb
3.3. Random network.ipynb
3.3. Small-world network.ipynb
3.3. Scale-free network.ipynb
3.3. Degree centrality.ipynb
3.3. Closeness centrality.ipynb
3.3. Betweenness centrality.ipynb
3.4. Networkx simulation.ipynb
3.4. Symmetry and asymmetry.ipynb
3.4. Network connectivity.ipynb
3.4. Network analysis example (social media).ipynb
S3.3. Modules, packages, and libraries.ipynb
3.5. Agent-Based Model basic.ipynb
3.6. Schelling's segregation model.ipynb
3.6. Monetary policy model.ipynb
3.6. Distributed behavioral model.ipynb
3.6. Cellular automata model.ipynb

S3.4. High performance computing.ipynb
3.7. Payoff matrix with battle of the sexes.ipynb
3.7. Payoff matrix with chicken game.ipynb
3.7. Payoff matrix with matching pennies.ipynb
3.7. Payoff matrix with rock-paper-scissors.ipynb
3.7. Payoff matrix with prisoner's dilemma.ipynb
3.7. Nash equilibrium.ipynb
3.7. Prisoner's dilemma example.ipynb
3.8. Public goods game.ipynb
S3.5. Systemic risk measurement.ipynb

If you have a problem with any aspect of this code list, please write pcw8531@sun.ac.kr, and we will do our best to address the problem. We appreciate any contributions, questions, and/or corrections. Also, please contact us if you feel your use of our code examples falls outside fair use or the permissions given above.

Appendix 5

Tables and Figures

We present facts using tables and figures. Tables give numerical or textual data in rows and columns, whereas figures typically provide non-textual information. After the part where they are stated, they are incorporated in the key concepts of the material.

Tables and figures were utilized to present and illustrate statistical data in Part. Tables and figures were utilized to display and illustrate time series data in Part 2. Tables and figures were utilized to present and display systems analysis data in Part 3.

Tables

Part 2. Applied Time Series Analysis

Part 3. Applied Systems Analysis

Figures

Part 1. Applied Statistics Analysis

Part 2. Applied Time Series Analysis

Part 3. Applied Systems Analysis

Appendix 6

Index

Chulwook Park, a Research Professor at Seoul National University in South Korea, earned his Ph.D. in 2016, focusing on elementary coordination dynamics in ecological contexts. His academic journey included a significant period from 2014 to 2015 at the Center for the Ecological Study of Perception and Action, University of Connecticut (CESPA, USA). In 2017, he joined the International Institute for Applied Systems Analysis (IIASA, Austria) as a postdoctoral research scholar, a position he held until 2019. Since 2022, Dr. Park has been a part of the Okinawa Institute of Science and Technology (OIST, Japan), contributing to the Complexity Science and Evolution Unit. He has made notable contributions to applied systems analysis, publishing various influential papers and employing dynamic systems-based approaches in his research.